Mathematics of Optimization: Smooth and Nonsmooth Case

Mathematics of Optimization: Smooth and Nonsmooth Case

G. Giorgi

Faculty of Economics
University of Pavia
Pavia, Italy

A. Guerraggio

Faculty of Economics
Insubria University
Varese, Italy

J. Thierfelder

Institute of Mathematics
Technical University Ilmenau
Ilmenau, Germany

2004

ELSEVIER

Amsterdam – Boston – Heidelberg – London – New York – Oxford
Paris – San Diego – San Francisco – Singapore – Sydney – Tokyo

ELSEVIER B.V.
Sara Burgerhartstraat 25
P.O. Box 211, 1000 AE
Amsterdam, The Netherlands

ELSEVIER Inc.
525 B Street
Suite 1900, San Diego
CA 92101-4495, USA

ELSEVIER Ltd
The Boulevard
Langford Lane, Kidlington,
Oxford OX5 1GB, UK

ELSEVIER Ltd
84 Theobalds Road
London WC1X 8RR
UK

First edition 2004

Library of Congress Cataloging in Publication Data
A catalog record is available from the Library of Congress.

British Library Cataloguing in Publication Data
A catalogue record is available from the British Library.

ISBN: 0-444-50550-4

♾ The paper used in this publication meets the requirements of ANSI/NISO Z39.48-1992 (Permanence of Paper).
Printed in The Netherlands.

Whatever happens in the world, within can be observed the meaning of maximum or minimum; hence there is no doubt that all phenomena in nature can be explained via the maximum and minimum method...

L. Euler, Methodus inveniendi lineas curvas maximi minimive propietate gaudentes, sive solutio problematis isoperimetrici latissimo sensu accepti, Lausannae et Genevae, 1744.

PREFACE

The present book is primarily dedicated to the study of the optimality conditions for a *nonlinear programming problem*, also known as a *mathematical programming problem*. Indeed one of the main subjects of mathematical optimization (a relatively modern branch of applied mathematics which has grown at an exponential rate, both from a theoretical point of view and on the side of applications) is the study of a class of extremum problems where the objective function f is to be optimized under some restrictions, usually in the form of equalities and/or inequalities.

As an example for a nonlinear programming problem in R^n, we mean an extremum problem of the following type:

$$\operatorname*{Min}_{x \in S} \ f(x) \qquad (P)$$

where $S = \{x \in X \mid g_i(x) \leqq 0, \ i = 1, ..., m; \ h_k(x) = 0, \ k = 1, ..., p\}$, $X \subset R^n$, $f, g_i, h_k : X \to I\!R$.

If all the functions involved in (P) are linear, we speak of *linear programming problems*. A more general version of (P) is one where the objective function f is a *vector-valued* function. This latter case (especially important from an economic point of view) will be treated in the last chapter of the book.

Problems similar to (P) arise in several contexts: the building and interpreting of economic models; the study of various technological processes; the development of optimal choices in finance; operations research; management science; production processes; transportation models; statistical decisions, etc. Therefore it is of the utmost importance to study the existence of solutions for (P) and to study the effective methods (i.e. numerical algorithms) for finding solutions of the problem. This second aspect of the study of (P) will not be treated here; we shall be concerned only with the study of the optimality conditions for (P), i.e. of the necessary and sufficient conditions such that a point $x^* \in S$ is a solution (local or global) for (P).

We have decided to conduct our analysis in the Euclidean space $I\!R^n$ (there is only one exception in Section 13 of Chapter III, where problems defined in infinite dimensional spaces are briefly mentioned). We have renounced to a more general mathematical setting, because of two main reasons:

a) Mathematical programming in $I\!R^n$ is a relatively self-contained field within the larger class of optimization problems (which includes static and dynamic optimization problems, calculus of variations, optimal control problems, etc.). The reader can quickly become familiar with the main topics; only some previous knowledge in linear algebra, mathematical analysis and convex analysis is necessary in order to understand the subject.

b) Many optimization problems in practice are defined on $I\!R^n$ or can be approximated by a problem defined on R^n; moreover, the numerical solution of optimization problems in general spaces by means of computers usually requires a former embedding of the problem in a finite dimensional space.

Besides the classical optimality conditions for a problem (P), where all the functions involved are assumed differentiable or even continuously differentiable (i.e. "smooth"), special attention has been given to the study of optimality conditions when the functions of (P) are "nonsmooth". The title itself of this book obviously points out this aspect. Moreover, we have dedicated an entire chapter (the last one) to the study of so-called *vector optimization problems*, which are rarely considered in books treating mathematical programming problems. Indeed, in several practical problems we have to optimize simultaneously, on the basis of certain criteria, several objective functions (e.g. costs of production, production times, degrees of mechanization and automatization, etc.). Such problems are obviously more complex than problem (P) and, accordingly, their mathematical handling is more complex.

In preparing this book a special effort has been made to obtain a self-contained treatment of the subjects; so we hope that this may be a suitable book as a textbook and a reference (every chapter ends with an extensive bibliography that should be useful to the reader for further investigations).

We do not claim to have produced an "advanced" book on mathematical programming; this book is addressed to those researchers, graduate students and post-graduates wishing to begin the study of nonlinear programming problems, in view of further deepenings. We think and hope that this book may also be useful to theoretical economists, engineers and applied researchers involved in such problems.

The book is divided into six chapters:

Chapter I briefly deals with the notion of nonlinear programming problems and with basic notations, conventions, definitions and results pertaining to set theory, mathematical analysis and linear algebra.

Chapter II deals with convex sets, convex functions and generalized convex functions; indeed in the study of optimality conditions for (P) and in the study of the so-called "dual problems" of (P), a fundamental role is played by the convexity or also by the generalized convexity of the functions involved. This explains why we have dedicated an entire chapter to the following topics: topological properties of convex sets, separation of convex sets, theorems of the alternative, properties of convex functions and generalized convex functions.

Chapter III covers the classical optimality conditions for a nonlinear programming problem in the smooth case, i.e. when all the functions involved are differentiable or continuously differentiable. The Fritz John and Karush-Kuhn-Tucker optimality conditions are developed, both for equality and/or inequality constrained problems. Both first-order and second-order optimality conditions are developed; some fundamental material on (first-order) constraint qualifications is presented. Sufficient optimality conditions are obtained by use of various generalizations of convex functions. The last section of the chapter deals with so-called "saddle point problems".

Chapter IV deals with the study of optimality conditions for (P) in the nonsmooth case, i.e. when the functions involved in (P) are not differentiable in the classical sense. Thus, different kinds of generalized differentiability are introduced and compared. After some general remarks about classical differentiability notions, we present some generalized directional derivatives and generalized subdifferential mappings and discuss these notions in connection with their application in mathematical optimization. In this context, the notion of "local cone approximation" of sets, introduced in Chapter II, is of great importance. Using this notion, we develop the so-called K-directional derivatives and K-subdifferentials and give an abstract approach for the development of general necessary optimality conditions for the nonsmooth case.

In Chapter V we give a survey about the most important duality concepts in mathematical optimization. For different kinds of optimization problems we discuss properties of their dual problems and develop different duality assertions. Starting with the well-known duality in linear optimization, we describe the Wolfe duality for convex optimization problems, the Lagrange duality in connection with saddle-point assertions and the perturbation concept, using the notion of conjugate functions. In this manner we shall see that the concepts will be generalized step by step.

Chapter VI is concerned with vector nonlinear programming problems. The presentation of the main optimality notions (weak efficient points, efficient points, proper efficient points) is followed by providing theorems giving conditions about the existence of said points. A classical problem of vector optimization is then discussed by means of the Fritz John and Kuhn-Tucker conditions, both for the smooth and nonsmooth (Lipschitzian) case. Links with a scalar optimization problem are emphasized (especially by means of some generalized convex functions, described in Chapter II) through various scalarization techniques and in a vector version of the Wolfe dual problem presented in Chapter IV.

The idea to write the present book emerged several years ago, during a trip of the two Italian authors to the Department of Mathematics of the Technical University of Ilmenau (Germany), in those days a relatively small university department among the woods of Thuringia ("Das grüne Thüringen), where Prof. Karl-Heinz Elster had founded the now renowned review "Optimization" and created an important center for studies in mathematical programming.

We recognize our intellectual debt toward many people from whom we have borrowed many ideas and results; indeed we owe much to the authors cited in the bibliographical list at the end of each chapter. We want, however, to pay a particular respect to the memory of Prof. Karl-Heinz Elster.

We want to express our gratitude to the Italian Ministry for University, Scientific and Technological Research for the financial support provided.

Finally we want to thank Drs. A. Sevenster of Elsevier BV (Amsterdam) for his patience and precious collaboration and express our most deep and sincere thanks to Mrs. Anita Klooster for her diligent and hard work in typing our almost unreadable manuscript.

The book has been planned and discussed throughout its various phases of developments by all three authors. However, Chapters I, II and III have been written by G. Giorgi; Chapters IV and V by J. Thierfelder; Chapter VI by A. Guerraggio.

The authors.

CONTENTS

CHAPTER I. INTRODUCTION

1.1. Optimization Problems

In the analysis of numerous mathematical problems (arising in Economics, Statistics, Operations Research, Engineering, Physics, etc.) situations often occur where a decison maker must make a decision in order to rule a system in an optimal way (on the basis of specified criteria), a model containing several alternative possibilities and which is, at least partially, under the control of the decision maker.

Such a problem can be considered as an *optimization problem*; "optimization" is a catch-all term for maximization or minimization and lies at the heart of several applied sciences, but perhaps the more natural application of optimization theory is in the field of economic sciences. Indeed optimization (or better, optimization subject to constraints) has been considered by many authors as defining the essential nature of Economics. We quote only the famous classical statement of Robbins (1932, p. 16): "Economics is the science which studies human behaviour as a relationship between ends and scarce means which have alternative uses".

The first mathematicians dealing with optimization problems were Fermat, Euler, the Bernoullis, Lagrange and others in connection with the development of Calculus in the 17th and 18th centuries. However, the first results in the modern theory of mathematical optimization were presented by the Russian Nobel prize winner L.V. Kantorovich and the American mathematician G.B. Dantzig, in 1939 and 1947 respectively. In 1951 the American mathematicians H.W. Kuhn and A.W. Tucker published important theoretical results in the theory of mathematical optimization, as extensions of the classical methods of Euler-Lagrange for the solution of optimization problems with equality constraints and in presence of differentiability assumptions.

Prior to the work of Kuhn and Tucker and in the same direction, we mention the contributions of Fritz John in 1948 and the unpublished Master's thesis of W. Karush in 1939.

Since then, studies in mathematical optimization developed rapidly, from a both theoretical and practical point of view, stimulated by the necessity of dealing with various organizational and planning problems used in Economics, Engineering and Natural Science. Such studies often made use of several tools of some modern mathematical fields as, e.g.. functional analysis, topology, linear algebra, convex analysis. Conversely, the development of optimization theory has considerably forced the development of these same fields. So we can affirm that Mathematics has received many impulses from the development of the studies on extremum problems.

As we mentioned in the Preface, almost all the problems we handle in the following chapters will be described by real-valued functions defined on $\mathbb{R}^n$. The optimization problems in which we shall be concerned may often be formalized as follows:

a) The behaviour of a system (in the most general meaning) depends on some variables, some of them beyond the control of the decision maker (these are namely the "data" of the problem) and the other ones under his control (these latter are the true variables of the problem, variables usually described by a vector $x \in \mathbb{R}^n$).

b) The various alternative possibilities for the decision maker are described by a set $S \subset \mathbb{R}^n$: so one has to choose, in an optimal way, a vector $x^0 \in S$ or more than one vector, in case of a problem with several solutions.

c) The "advantage" of this choice is shown by the value assumed by a function f, usually defined on a set containing S.

d) Let us consider for the moment the case of $f : D \subset \mathbb{R}^n \to \mathbb{R}$, i.e. the case of *scalar functions* (only in the last chapter of this book shall we face the *vector optimization problem*, where $f : D \subset \mathbb{R}^n \to \mathbb{R}^m$, $m > 1$). Then the choice of vector $x^0 \in S$ is considered an optimal one and we say that x^0 solves our optimization problem when it is

$$f(x^0) \leqq f(x) , \qquad \text{for each } x \in S , \tag{1}$$

in case of a *minimization problem*;

$$f(x^0) \geqq f(x) , \qquad \text{for each } x \in S , \tag{2}$$

in case of a *maximization problem*.
When the previous inequalities hold strictly, for each $x \in S$, $x \neq x^0$, we say that x^0 is a *strict* solution of the related optimization problem. More precisely: we have an *optimization problem* or a *mathematical programming problem* when we are given a set $S \subset I\!R^n$ and a real-valued scalar function $f : I\!R^n \to I\!R$ (some authors also consider extended real-valued scalar functions $f : I\!R^n \to I\!R \cup \{\pm\infty\}$), defined on a set D containing S, and when we are looking for some element $x^0 \in S$ such that (1) holds. In this case we shall write the problem as

$$\operatorname*{Min}_{x \in S} f(x) . \tag{3}$$

If we are looking for some element $x^0 \in S$ such that (2) holds, we shall write the problem as

$$\operatorname*{Max}_{x \in S} f(x) . \tag{4}$$

It is better to distinguish between minimization and maximization problems and the corresponding optimal values:

$$f(x^0) = \min_{x \in S} f(x) = \min \{f(x) \mid x \in S\} ;$$

$$f(x^0) = \max_{x \in S} f(x) = \max \{f(x) \mid x \in S\} .$$

Obviously it is always

$$\min_{x \in S} f(x) = - \max_{x \in S} \{-f(x)\} .$$

Therefore we can study, without loss of generality, only one type of optimization problems. In the present book we shall study minimization problems i.e. problems expressed by relation (3). The set S is called the *set of feasible solutions* or simply the *feasible set* or *constraint set*. A point $x \in S$ is a *feasible point* or also a *decision vector* or a *program vector*. The function f is named *objective function*. A point x^0 such that (1) holds is called an *optimal solution* or *optimal point* for problem (3) (i.e. a global, minimum point of f on S) or simply a *solution* (perhaps not unique); $f(x^0)$ is the corresponding *optimal value* of the problem. If $S = \emptyset$ the optimization problem is *not feasible* or *not consistent*; if $S \neq \emptyset$ the problem is *feasible* or *consistent*, even if in this case the problem may admit no solution.

Many authors term (3) an *unconstrained* or *free* minimization problem, when S coincides with the domain D of f, D *open* set, or also when S is an open subset of D. More generally, a free minimization problem consists in the search of minimizers which are interior points of S. Otherwise we have a *constrained* minimization problem. In the case where S (not open) is not given by explicit functional constraints, it may be viewed as a true *set constraint*. We shall treat separately the cases where S is given also by specific functional constraints; these functional constraints may be expressed by a system of equalities and/or by a system of inequalities.

The following terminology is well known. Consider a point $x^0 \in S$; then f is said to have a *local minimum* at x^0 if

$$f(x^0) \leqq f(x) , \qquad \forall x \in S \cap N(x^0) ,$$

where $N(x^0)$ is some neighbourhood of x^0. If

$$f(x^0) < f(x) , \qquad \forall x \in S \cap N(x^0) , \qquad x \neq x^0 ,$$

then we have a *strict* local minimum.

Of course each global minimum is a local minimum, but not conversely.

If the feasible set $S \subset \mathbb{R}^n$ is not empty, is bounded and closed (i.e. compact) and the objective function f is continuous on S, the well-known *Weierstrass theorem* assures us that problems (3) and (4) admit solution. If we are concerned only with problem (3) or only with problem (4), the above-mentioned assumptions of the Weierstrass theorem are weakened by means of the semicontinuity notions.

Definition 1.1. A function $f : D \subset \mathbb{R}^n \to \mathbb{R}$ is called *lower semicontinuous at* $x^0 \in D$, if for each $\varepsilon > 0$ there exists a neighbourhood $N(x^0)$ of x^0 such that

$$f(x) > f(x^0) - \varepsilon \;, \qquad \forall x \in D \cap N(x^0) \;.$$

If it is

$$f(x) < f(x^0) + \varepsilon \;, \qquad \forall x \in D \cap N(x^0)$$

f is called *upper semicontinuous at* $x^0 \in D$. The function f is said to be lower (upper) semicontinuous on D (with respect to D) if it is lower (upper) semicontinuous (with respect to D) at each point $x^0 \in D$ or, equivalently if one of the following conditions holds:

a) The set $L(f, \alpha) = \{x \mid x \in D, \ f(x) \leqq \alpha\}$ $(U(f, \alpha) = \{x \mid x \in D, \ f(x) \geqq \alpha\})$ is closed relative to D for each $\alpha \in \mathbb{R}$.

b) The set $SU(f, \alpha) = \{x \mid x \in D, \ f(x) > \alpha\}$ $(SL(f, \alpha) = \{x \mid x \in D, \ f(x) < \alpha\})$ is open relative to D for each $\alpha \in \mathbb{R}$.

We recall that, given Γ and A two sets such that $\Gamma \subset A \subset \mathbb{R}^n$, Γ is said to be *open* (*closed*) *relative to* A if $\Gamma = A \cap \Omega$, where Ω is some open (closed) set in $\mathbb{R}^n$. Then we have the following generalized Weierstrass theorem: if $S \subset \mathbb{R}^n$ is nonempty and compact and f is lower semicontinuous on S, then it admits a global minimum over S; if f is upper semicontinuous on S, then it admits a global maximum over S.

If we also take into consideration cases where f does not admit a minimum or a maximum over S, we shall write, instead of (3) or (4), respectively

$$\operatorname*{Inf}_{x \in S} f(x) \;; \qquad \operatorname*{Sup}_{x \in S} f(x) \;.$$

In this case, when $S = \emptyset$, then the following convention is used:

$$\inf_{x \in S} f(x) = +\infty \;; \qquad \sup_{x \in S} f(x) = -\infty \;.$$

We shall be mainly concerned with those constrained minimization problems where the feasible set S has a specified structure, i.e. with the so-called *nonlinear programming problems* (NLP). These problems can be formalized as follows:

$$\underset{x \in S}{\text{Min}} \ f(x) \ , \tag{5}$$

$$S = \{x \mid x \in X, \ g_i(x) \leqq 0, \ i = 1, ..., m, \ h_j(x) = 0, \ j = 1, ..., r\} \ ,$$

where all functions g_i, h_j are defined on a set $D \subset I\!R^n$ and $X \subset I\!R^n$ is any set contained in D. In vector notation:

$$S = \{x \mid x \in X, \ g(x) \leqq 0, \ h(x) = 0\},$$

where $g = [g_1, ..., g_m]$, $h = [h_1, ..., h_r]$.

The functions g, h are the *constraints* or *constraint functions* whereas X, when it is not open, may be considered a nonfunctional constraint, i.e. a *set constraint*. The set X might typically include lower and upper bounds on the variables (in this case X is often called a "box") or might represent a specially structured abstract constraint. If the optimal point $x^0 \in S$ is not interior to X, some complications may arise in the study of optimality conditions for problem (5) (see Chapter III).

Each of the constraints $g_i(x) \leqq 0$ is called an *inequality constraint* and each of the constraints $h_j(x) = 0$ is called an *equality constraint*. NLP problems containing only equality constraints and where X is open are also called "classical constrained optimization problems", as they were first considered by J.L. Lagrange more than two centuries ago.

In the following we shall assume that all functions involved in (NLP) are continuous, at least at a feasible point x^0. Thus, if we want to check whether a feasible point x^0 is a local solution or not, only those constraints for which $g_i(x^0) = 0$ holds true are relevant, among the constraints $g_i(x) \leqq 0$. In case of $g_i(x^0) < 0$, because of the continuity, there exists a suitable neighbourhood of x^0 which belongs to $\{x \mid g_i(x) < 0\}$, i.e. which does not contradict this constraint. So, given a point $x^0 \in S$, the set

$$I(x^0) = \{i \mid g_i(x^0) = 0\}$$

is called the set of the *active* or *effective* or *binding* constraints for problem (5) at x^0, whereas the set

$$J(x^0) = \{i \mid g_i(x^0) < 0\}$$

is the set of *non active* constraints at $x^0 \in S$. If in (5) all functions are differentiable, we speak of *smooth* optimization problems, otherwise they are *nonsmooth* optimization problems. (More precisely: a function is called "smooth of order k" if it is differentiable and the partial derivatives of order k are continuous.) We also note that the term "nonlinear programming" is a little confusing; indeed here "nonlinear" means "not necessarily linear" and therefore formulation (5) also includes the case where all the functions involved are linear, i.e. the *linear programming problem.*

The scheme used to describe a nonlinear programming problem is, in spite of its appearance, quite general, as in it we may include variants and generalizations, described, e.g., as follows:

1) If some constraint is in the form $g_i(x) \leqq b_i$, $b_i \neq 0$, it is sufficient to put $\hat{g}_i(x) = g_i(x) - b_i$ to obtain the standard formulation (5).

2) If some constraint is in the form $g_i(x) \geqq 0$, it is sufficient to put $\hat{g}_i(x) = -g_i(x)$ to obtain the standard formulation (5).

3) Also the *non negativeness conditions* imposed on vector x, i.e. $x_1 \geqq 0$, $x_2 \geqq 0, ..., x_n \geqq 0$, may be included in formulation (5) by setting $g_{m+1}(x) = -x_1 \leqq 0, ..., g_{m+n}(x) = -x_n \leqq 0$.

Moreover, we note that sometimes it is useful to substitute an equality constraint $h_j(x) = 0$ with two opposite inequality constraints $h_j(x) \leqq 0$, $h_j(x) \geqq 0$. Similarly, it is sometimes convenient to substitute an inequality constraint $g_i(x) \leqq 0$ with an equality constraint of the type

$$g_i(x) + z_i = 0 \ ,$$

where $z_i \geqq 0$ is a so-called *slack variable* (see also Chapter III).

We give only the following simple example of an NLP problem.

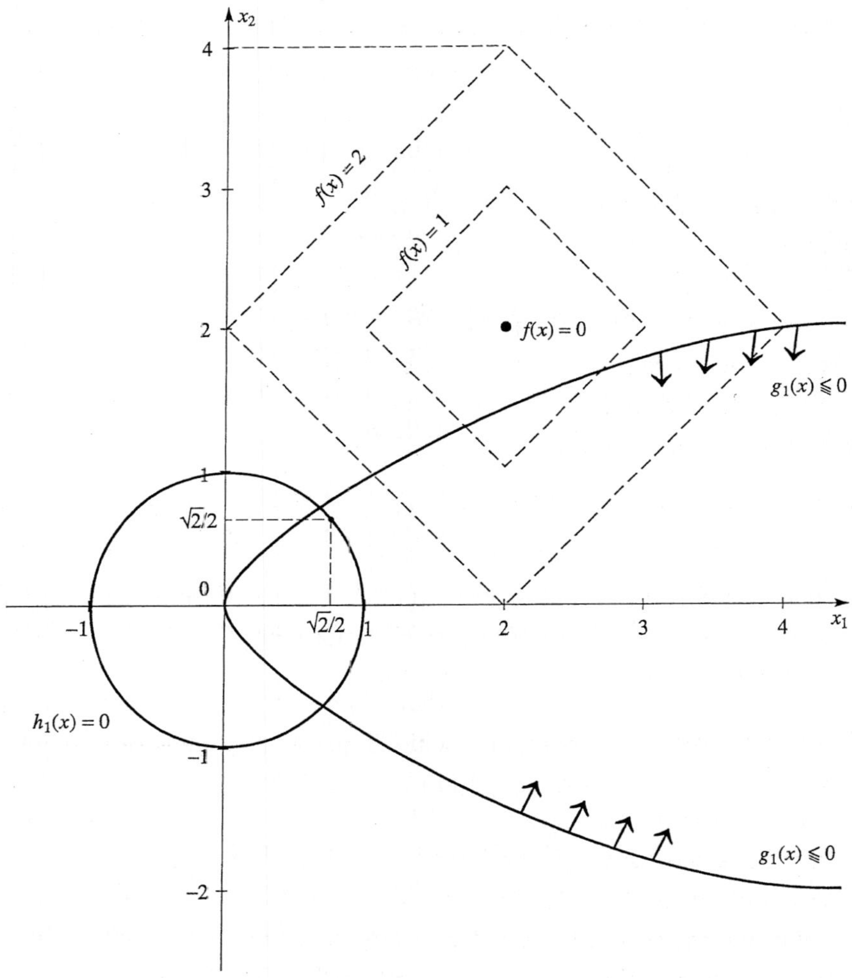

x_2
4
3
2
1
$\sqrt{2}/2$
0
−1
−2
$f(x) = 2$
$f(x) = 1$
$f(x) = 0$
$g_1(x) \leqslant 0$
$g_1(x) \leqslant 0$
$h_1(x) = 0$
−1
$\sqrt{2}/2$
1
2
3
4
x_1

Figure 1.

Example 1.1.

$$\operatorname*{Min}_{x \in S} f(x) = |x_1 - 2| + |x_2 - 2|$$

$$S = \{x \mid x \in I\!R^2,\ g_1(x) = (x_2)^2 - x_1 \leqq 0\,;\ h_1(x) = (x_1)^2 + (x_2)^2 - 1 = 0\}\,.$$

The dashed lines in Figure 1 represent two level sets of the objective function, i.e. points at which the objective function f has constant values. The feasible set is the arc of the circumference lying within the parabola. It is easy to see that the solution of this problem is the point $x^0 = (\sqrt{2}/2\,;\ \sqrt{2}/2)$. If the equality constraint is removed, the solution is seen to be at $x^0 = (2, \sqrt{2})$. If both constraints are removed, i.e. if we have a free or unconstrained minimization problem, the solution is $x^0 = (2, 2)$.

Several classes of NLP problems have been dealt with in the past years; we briefly mention some of them.

a) The case where in (5) all functions are linear (i.e. we have a *linear programming problem*) has been treated extensively. There is enormous literature on this subject, both from a theoretical and a computational point of view. We shall not be particularly concerned with this kind of mathematical programming problem.

b) The case where f and every g_i are convex functions and all h_j are linear functions, has received particular attention. In this case we have a *convex programming problem*. When f, g_i, h_j are generalized convex functions, we have a *generalized convex programming problem*. These assumptions are usually taken in order to make the computation of the solutions easier and to assure that any local solution is also global. Many mathematical programming models, arising from practical problems, exhibit convex or generalized convex functions and numerous computational algorithms are "well behaved" under convexity or generalized convexity assumptions.

c) When in (5) $X \subset I\!R^n$ is open and there are no inequality constraints, we have a "classical optimization problem". These problems are indeed the oldest mathematical programming problems since their analysis goes back to the eighteenth century (see Chapter III).

d) When in (5) f is a quadratic form and the constraints are linear, we have a *quadratic programming problem*. Also for this case, particular results and algorithms have been obtained.

e) When f, g and h are separable functions (i.e. of the form $f(x) = \sum_{k=1}^{n} f_k(x_k)$; $g_i(x) = \sum_{k=1}^{n} g_{i,k}(x_k)$; $h_j(x) = \sum_{k=1}^{n} h_{j,k}(x_k)$), we have a *separable nonlinear programming problem*. Also for this case, important for some applications, special algorithms have been developed.

f) When in (5) the objective function is of the type

$$f(x) = \frac{u(x)}{v(x)}$$

we have a *fractional* of *hyperbolic programming problem*. This is another important case of nonlinear programming problems, arising from practical problems.

If f is a *vector-valued* function, it is again possible to define optimization problems, but in this case the concept of optimal point has to be redefined; indeed in this case we have several definitions of an optimal point. When f is a vector-valued function, we speak of *vector optimization*. Vector optimization is treated in Chapter VI.

We cannot let pass unnoticed the limits of our formulation of an optimization problem; in effect we shall focus our analysis only on nonlinear programming problems and on vector programming problems, mostly defined on $\mathbb{R}^n$, whereas some other important mathematical programming problems are excluded from our analysis. These include the case of the *stochastic programming problems*, i.e. those optimization problems where random variables are to be taken into account (we shall treat only *deterministic* problems), and *integer programming problems*, i.e. those optimization problems where some or all components of $x \in S$ must be integer numbers, a condition typical of several problems of Management Science or of

Operations Research. Finally, it is convenient to distinguish between *static* optimization problems and *dynamic* optimization problems. The first concern situations at a certain moment and therefore no variation of time is taken into account; in the second type of problems the time evolution of the considered quantities assumes a fundamental importance. In this class of problems we may include the ones typical of *Calculus of Variations, Control Theory, Dynamic Programming*. We shall not even be concerned with these types of optimization problems which form independent (and very wide) subjects of studies. The interested reader may consult the classical works listed in the bibliographical references of the present chapter.

1.2. Basic Mathematical Preliminaries and Notations

We review in this section, for the reader's convenience, some basic concepts from set theory, linear algebra and real functions theory. We list also some notations and conventions used throughout the book (some of them were already introduced in the previous section). For more details, the reader is referred to the books cited at the end of this chapter.

A) *SET THEORY AND LOGICAL NOTATIONS*

- The *empty set* is denoted by $\emptyset$.
- If X and Y are two sets and X is contained in Y, we shall write $X \subset Y$ (or $Y \supset X$); if X is *properly* contained in Y, we shall write $X \subset Y$, $X \neq Y$.
- The *difference* of two sets X and Y is denoted by $X \backslash Y$: $X \backslash Y = \{x \mid x \in X,\ x \notin Y\}$.
- If $Y \subset X$, then $X \backslash Y$ is called the *complement* of Y (relative to X) and is also denoted by Y^c_X. When X is the whole space we use also the notation Y^c or $\sim Y$. Do not confuse with the algebraic difference of two sets: $X - Y$.
- The *algebraic sum* (*difference*) of two sets X, Y is denoted by $X \pm Y$ and it is $X \pm Y = \{z \mid z = x \pm y;\, x \in X,\ y \in Y\}$.

- In the same manner it is $\lambda X = \{z | z = \lambda x,\ x \in X\}$, for an arbitrary real number λ, and it is $\sum_{i=1}^{p} X_i = \{z \mid z = x^1 + ... + x^p,\ x^i \in X_i,\ i = 1, ..., p\}$, for the given sets $X_1, ..., X_p$.
- The *Cartesian product* of n sets $S_1, S_2, ..., S_p$ is denoted by $S_1 \times S_2 \times ... \times S_p$ or also by $\prod_{i=1}^{p} S_i$; the p-times repeated Cartesian product of a set S by itself is denoted by S^p.
- $I\!R$ denotes the set of *real numbers*; Q denotes the set of *rational numbers*; $Z\!\!Z$ denotes the set of *integer numbers*; $I\!N$ denotes the set of *natural numbers*; $I\!N_+$ is the set of *positive* natural numbers.
- The *extended real number system*, i.e. $I\!R \cup \{\pm\infty\}$, is also denoted by $\bar{I\!R}$.
- The *closed real interval* is denoted by $[a, b] = \{x \in I\!R \,|\, a \leqq x \leqq b\}$, with $a, b \in I\!R$; the *open real interval* is denoted by $(a, b) = \{x \in I\!R \mid a < x < b\}$, with $a, b \in I\!R$. The significance of $(a, b]$, $[a, b)$ is obvious.
- $I\!R^n$ denotes the n-dimensional real euclidean space. $I\!R^n_+$ and $I\!R^n_-$ denote the set of elements of $I\!R^n$ having respectively nonnegative, nonpositive components. $\overset{o}{I\!R}{}^n_+$ and $\overset{o}{I\!R}{}^n_-$ are, respectively, the positive and the negative orthant of $I\!R^n$.
- The origin of $I\!R^n$ is simply denoted by 0 and from the context it will be clear if it is a scalar or a vector.
- Given a point $x^0 \in I\!R^n$, an (open) *neighbourhood* of x^0 or (open) *ball of centre* x^0 is denoted by $N(x^0)$ or $B(x^0)$ or $V(x^0)$, etc. If we need to specify the *radius* $\delta > 0$ of the ball we shall write $N_\delta(x^0)$, $B_\delta(x^0)$, $V_\delta(x^0)$ or also $N(x^0, \delta)$, $B(x^0, \delta)$, etc.
- $\mathrm{card}(S)$ is the *cardinality* of S.
- $\mathrm{int}(S)$ denotes the *interior* of S, i.e. the set of all its interior points.
- $\mathrm{relint}(S)$ denotes the *relative interior* of S (see Section 2.1 of Chapter II).
- $\mathrm{bd}(S)$ denotes the *boundary* of S.
- $\mathrm{cl}(S)$ or $\bar{S}$ denotes the *closure* of S.

- ext(S) denotes the *exterior* of S, i.e. the set of all its exterior points.
- If A and B are two logical propositions, then:
  $A \Rightarrow B$ denotes the implication (if A holds then B holds; A implies B). The reverse implication is denoted by $A \Leftarrow B$.
  $A \Leftrightarrow B$ denotes the logical equivalence (A holds if and only if B holds; sometimes "if and only if" is shortened by "iff").

B) *LINEAR ALGEBRA*

- A vector $x \in \mathbb{R}^n$ is an array of scalars $x_1, x_2, ..., x_n$ which are the *components* or *elements* of the vector; usually we use lower case latin letters to denote vectors and we write $x = [x_1, x_2, ..., x_n]$ or $x = (x_1, x_2, ..., x_n)$. Note that $x_1, x_2, ..., x_n$ are the elements of $x \in \mathbb{R}^n$, whereas we shall write $x^1, x^2, ..., x^n, x^0, x^*$, etc. to denote different vectors.
- The *zero vector* is simply denoted by 0; the *unit vector* is denoted by e ($e = [1, 1, ..., 1]$); e^i is the i-th *unit vector* or *coordinate vector* and consists of all zeroes, except for a 1 at the i-th position.
- All vectors will be used interchangeably as *row* and *column* vectors and the *inner product* of two vectors x and y will be denoted by $x \cdot y$ or simply xy:

  $$x \cdot y = \sum_{i=1}^{n} x_i \, y_i \ .$$

- The (euclidean) *norm* of a vector $x \in \mathbb{R}^n$ is denoted by $\|x\| = (x \cdot x)^{\frac{1}{2}}$.
- If we need to specify transposition of a vector x, we shall write x^T.
- A *sequence* of vectors of $\mathbb{R}^n$ is denoted by $\{x^k\}$, $k = 1, 2, ...$ and if a sequence of vectors converges to the limit vector x^0 we shall write $\lim_{k \to \infty} x^k = x^0$ or, more briefly, $x^k \to x^0$.
- A (real) matrix A of *order* or *dimension* m, n (or $m \cdot n$) is an array of $m \cdot n$ real numbers; matrices are usually denoted by capital latin letters. A matrix A is also denoted by $A = [a_{ij}]$ $i = 1, ..., m$; $j = 1, ..., n$. Its j-th column is denoted by A^j whereas its i-th row

is denoted by A_i. A matrix is *square* if $m = n$; in this case n is the order (or dimension) of the square matrix.

- If A is an m by n matrix, its *transpose* is denoted by A^T. We recall that a square matrix A is *symmetric* if $A = A^T$ and is *skew symmetric* if $A = -A^T$.
- I_n is the *identity* matrix of order n; if there is no need to specify its dimension, we shall write I. The zero (or null) matrix will be denoted simply by 0.
- If A is a square matrix, $\det(A)$ or $|A|$ denotes its *determinant*.
- Let A and B be two m by n matrices; we shall write:

$A \geqq B$, if $a_{ij} \geqq b_{ij}$, $\forall i, j$.
$A \geq B$, if $A \geqq B$, but $A \neq B$
(i.e. there exists at least an element $a_{ij} > b_{ij}$).
$A > B$, if $a_{ij} > b_{ij}$, $\forall i, j$.

If in the above inequalities B is the m by n zero matrix, we say that

A is *nonnegative* if $A \geqq 0$.
A is *semipositive* if $A \geq 0$.
A is *positive* if $A > 0$.

The same convention and terminology apply also when comparing two vectors of the same dimension. It is apparent the significance of $A \leqq B$, $A \leq B$, $A < B$. Thus the following (already introduced) subsets of $\mathbb{R}^n$

$$\mathbb{R}^n_+ = \{x \mid x \in \mathbb{R}^n,\ x \geqq 0\}$$

$$\overset{o}{\mathbb{R}}{}^n_+ = \{x \mid x \in \mathbb{R}^n,\ x > 0\}$$

are, respectively, the nonnegative and the positive orthant of $\mathbb{R}^n$. The set

$$\mathbb{R}^n_+ \backslash \{0\} = \{x \mid x \in \mathbb{R}^n,\ x \geq 0\}$$

is the semipositive orthant of $\mathbb{R}^n$.

- The *inverse* of a square matrix A (when it exists) is denoted by A^{-1}; the *rank* of a matrix, of order m, n, is denoted by $\operatorname{rank}(A)$.
- We recall that a *quadratic form* $Q(x) = x^T Ax$ (A symmetric) is *positive (negative) definite* if $Q(x) > 0$ (< 0), $\forall x \in \mathbb{R}^n$, $x \neq 0$. It is *positive (negative) semidefinite* if $Q(x) \geqq 0$ ($\leqq 0$), $\forall x \in \mathbb{R}^n$, $x \neq 0$. It is *indefinite* if $\exists \bar{x}, \bar{\bar{x}}$ such that $Q(\bar{x}) > 0$, $Q(\bar{\bar{x}}) < 0$.

C) *FUNCTIONS*

- $f : X \to Y$, with $X \subset \mathbb{R}^n$, $Y \subset \mathbb{R}^m$, denotes a function defined on X and with *range* Y. The set X is also called its *domain* and sometimes it is also denoted by $\operatorname{dom}(f)$. If $m > 1$ we also write $f = [f_1, f_2, .., f_m]$ and speak then of *vector functions* or *vector-valued functions*.
- $\operatorname{epi} f = \{(x,y) \in \mathbb{R}^{n+1} \mid x \in X,\ y \geqq f(x)\}$ is the *epigraph* of $f : X \to \mathbb{R}$ and the set $\operatorname{hypo} f = \{(x,y) \in \mathbb{R}^{n+1} \mid x \in X,\ y \leqq f(x)\}$ is the *hypograph* of f.
  The *strict epigraph* and *strict hypograph* of f are respectively the sets

$$\operatorname{epi}^0 f = \{(x,y) \in \mathbb{R}^{n+1} \mid x \in X,\ y > f(x)\} \ ;$$

$$\operatorname{hypo}^0 f = \{(x,y) \in \mathbb{R}^{n+1} \mid x \in X,\ y < f(x)\} \ .$$

- The *α-lower level set* of $f : X \to \mathbb{R}$ is denoted by $L(f, \alpha)$ and it is defined as $L(f, \alpha) = \{x \in X \mid f(x) \leqq \alpha\}$, with $\alpha \in \mathbb{R}$.
  The *α-upper level set* of $f : X \to \mathbb{R}$ is denoted by $U(f, \alpha)$ and is defined as $U(f, \alpha) = \{x \in X \mid f(x) \geqq \alpha\}$.
  The *α-level set* of $f : X \to \mathbb{R}$ or *α-isocontour set of f* is $Y(f, \alpha) = L(f, \alpha) \cap U(f, \alpha)$, i.e. $Y(f, \alpha) = \{x \in X \mid f(x) = \alpha\}$. $SL(f, \alpha) = \{x \in X \mid f(x) < \alpha\}$ is the *α-strict lower level set* of f and $SU(f, \alpha) = \{x \in X \mid f(x) > \alpha\}$ is the *α-strict upper level set* of f. We have

$$L(f, -\infty) = SL(f, -\infty) = U(f, +\infty) = SU(f, +\infty) = \emptyset \ ;$$

$$U(f, -\infty) = SU(f, -\infty) = L(f, +\infty) = SL(f, +\infty) = X \ .$$

- Sometimes we shall use the "small o" of Landau: let $f : I\!\!R^n \to I\!\!R$ and $g : I\!\!R^n \to I\!\!R$ be defined in a neighbourhood $N(x^0)$ of x^0, with $g(x) \neq 0$ in $N(x^0)$. Then we may write $f = o(g)$ for $x \to x^0$, when

  $$\lim_{x \to x^0} \frac{f(x)}{g(x)} = 0 \ .$$

  So, $f = o(1)$, for $x \to x^0$, is equivalent to $\lim_{x \to x^0} f(x) = 0$, $f = o(\|x\|)$ and $f = o(\|x\|^2)$, for $x \to x^0$, mean respectively that $\lim_{x \to x^0} \frac{f(x)}{\|x\|} = 0$ and $\lim_{x \to x^0} \frac{f(x)}{\|x\|^2} = 0$.
- Let $f : I\!\!R^n \to I\!\!R$; then the *non deleted limit superior* of f at $x^0 \in I\!\!R^n$, denoted by $\limsup_{x \to x^0} f(x)$ (also $\overline{\lim}_{x \to x^0} f(x)$) is defined by

  $$\Lambda = \limsup_{x \to x^0} f(x) = \inf_{N(x^0)} \sup_{x \in N(x^0)} f(x) \, .$$

  The value $\Lambda \in \bar{I\!\!R}$ may be equivalently characterized as the maximum value of the so-called *limit class* of Ω, where $\omega \in \Omega$ when there exists a sequence $\{x^k\} \subset I\!\!R^n$, $x^k \to x^0$ such that $f(x^k) \to \omega$, i.e. when, for each $\varepsilon > 0$ and each $N(x^0)$, there exists $\bar{x} \in N(x^0)$ such that $|f(\bar{x}) - \omega| < \varepsilon$. It is easy to verify that the limit superior Λ is characterized by the proposition $\forall \, \varepsilon > 0$, $\exists \, N(x^0)$ such that $f(x) < \Lambda + \varepsilon$, $\forall \, x \in N(x^0)$.
  The qualifying "non deleted" refers to the fact we have used non deleted neighbourhoods $N(x^0)$ of x^0. (We note however that in the literature the *deleted* limit superior of f at x^0 is also often used; this one has the same definition, except that the point x^0 is deleted from $N(x^0)$.)
  Similarly, the *non deleted limit inferior* of f at x^0, denoted by $\liminf_{x \to x^0} f(x)$ (also $\underline{\lim}_{x \to x^0} f(x)$), is the extended real number

  $$\liminf_{x \to x^0} f(x) = \sup_{N(x^0)} \inf_{x \in N(x^0)} f(x) \ .$$

  We have

  $$\liminf_{x \to x^0} (-f(x)) = - \limsup_{x \to x^0} f(x) \ ;$$

$$\limsup_{x \to x^0} (-f(x)) = -\liminf_{x \to x^0} f(x) \; ;$$

$$\liminf_{x \to x^0} f(x) = \limsup_{x \to x^0} f(x) \Leftrightarrow \lim_{x \to x^0} f(x) \text{ exists.}$$

A useful result related to these quantities is given by the following:

Theorem. Let f and g be bounded in a neighbourhood of x^0; then

i) $\limsup_{x \to x^0} (f(x) + g(x)) \leqq \limsup_{x \to x^0} f(x) + \limsup_{x \to x^0} g(x)$;

ii) $\liminf_{x \to x^0} (f(x) + g(x)) \geqq \liminf_{x \to x^0} f(x) + \liminf_{x \to x^0} g(x)$;

iii) if $\lim_{x \to x^0} g(x)$ exists, then

$$\limsup_{x \to x^0} (f(x) + g(x)) = \limsup_{x \to x^0} f(x) + \lim_{x \to x^0} g(x)$$

(similarly for the $\liminf$ operation).

It is not difficult to see that $f : \mathbb{R}^n \to \bar{\mathbb{R}}$ is upper semicontinuous at $x^0 \in \mathbb{R}^n$ if and only if $\limsup_{x \to x^0} f(x) = f(x^0)$ (but if we would adopt the deleted definition we have to write $\limsup_{x \to x^0} f(x) \leqq f(x^0)$).

Similarly f is lower semicontinuous at $x^0 \in \mathbb{R}^n$ if and only if $\liminf_{x \to x^0} f(x) = f(x^0)$ ($\liminf_{x \to x^0} f(x) \geqq f(x^0)$ if we adopt the deleted definition). Another notion of generalized "mixed" limits has been introduced in optimization theory by Rockafeller. If $g : \mathbb{R}^n \times \mathbb{R}^m \to \bar{\mathbb{R}}$, with $x^0 \in \mathbb{R}^n$, $y^0 \in \mathbb{R}^m$, then

$$\limsup_{x \to x^0} \inf_{y \to y^0} g(x, y) = \sup_{U(y^0)} \lim_{V(x^0)} \sup_{x \in V(x^0)} \inf_{y \in U(y^0)} g(x, y) \; ;$$

$$\liminf_{x \to x^0} \sup_{y \to y^0} g(x, y) = \inf_{U(y^0)} \sup_{V(x^0)} \inf_{x \in V(x^0)} \sup_{y \in U(y^0)} g(x, y) \; .$$

- Let $f : X \to \mathbb{R}$, with $X \subseteq \mathbb{R}^n$; then f is said to satisfy a *Lipschitz condition on* X if there exists a number $L > 0$ such that

$$|f(x^1) - f(x^2)| \leqq L \, \|x^1 - x^2\| \; , \qquad \forall \, x^1, x^2 \in X \; .$$

We shall say that f is *Lipschitzian around* x^0 (or *near* x^0 or *locally Lipschitzian at* x^0) if there exists a neighbourhood of x^0, $N(x^0)$, and a number $L > 0$ such that

$$|f(x^1) - f(x^2)| \leqq L\,\|x^1 - x^2\|\ , \qquad \forall\, x^1, x^2 \in N(x^0)\ .$$

• The terms "non decreasing" and "increasing" will be considered as equivalent; similarly for the terms "non increasing" and "decreasing". We shall write "strictly increasing" and "strictly decreasing" for the strict relations.

- Let $f : D \subset \mathbb{R}^n \to \mathbb{R}$ and $x^0 \in \operatorname{int}(D)$. We say that f is *differentiable* at x^0 if there exists a vector $y^0 \in \mathbb{R}^n$ such that, for each $x \in D$, then

  $$f(x) - f(x^0) = (x - x^0) \cdot y^0 + o(\|x - x^0\|)$$

  for $x \to x^0$.
  Then, the vector y^0 (which is unique) is formed by the n *partial derivatives* of f evaluated at x^0

  $$\frac{\partial f(x^0)}{\partial x_1}, \ldots, \frac{\partial f(x^0)}{\partial x_n}$$

  (sometimes the following notation is also used: $f_{x_1}(x^0), \ldots, f_{x_n}(x^0)$). The *gradient vector* or *gradient* of f at x^0 consists of all n partial derivatives evaluated at x^0 and is denoted by $\nabla f(x^0)$.

- We say that $f : \mathbb{R}^n \to \mathbb{R}$ is *continuously differentiable at* x^0 or that it is of *class* C^1 *at* x^0 (and we shall write $f \in C^1(x^0)$) when $\nabla f(x)$ is continuous at x^0. The following implications hold:

$$\begin{array}{ccccc} f \in C^1(x^0) & \Rightarrow & f \text{ is differentiable at } x^0 & \Rightarrow & \nabla f(x^0) \text{ exists} \\ & & \Downarrow & & \\ & & f \text{ is continuous at } x^0 & & \end{array}$$

A vector function $f : \mathbb{R}^n \to \mathbb{R}^m$ is said to be differentiable at
... to have partial

derivatives if all its components admit partial derivatives. The m by n matrix of the partial derivatives is called the *Jacobian matrix* of f and is denoted by $\nabla f(x)$ of $J f(x)$. Thus in this case

$$J f(x^0) = \nabla f(x^0) = \begin{bmatrix} \frac{\partial f_1(x^0)}{\partial x_1} & \cdots\cdots & \frac{\partial f_1(x^0)}{\partial x_n} \\ \vdots & & \vdots \\ \frac{\partial f_m(x^0)}{\partial x_1} & \cdots\cdots & \frac{\partial f_m(x^0)}{\partial x_n} \end{bmatrix} =$$

$$= \left[\frac{\partial f_i(x^0)}{\partial x_j}\right], \qquad i = 1, ..., m; \quad j = 1, ..., n .$$

- We say that $f : D \subseteq \mathbb{R}^n \to \mathbb{R}$ is *twice differentiable* at $x^0 \in \operatorname{int}(D)$ if, for each $x \in D$ it is

$$f(x) - f(x^0) = (x - x^0) \nabla f(x^0) + \tfrac{1}{2} \cdot$$
$$\cdot (x - x^0)^T H f(x^0)(x - x^0) + o(\|x - x^0\|)^2 , \qquad \text{for } x \to x^0 .$$

The n by n matrix $H f(x^0)$ is called the *Hessian matrix* of f at x^0; it is unique and its elements are the *second-order partial derivatives* of f, evaluated at x^0:

$$H f(x^0) = \left[\frac{\partial^2 f(x^0)}{\partial x_i \, \partial x_j}\right]; \qquad i, j = 1, ..., n .$$

- We say that f is *twice-continuously differentiable* at x^0 or that f is of *class* $\mathcal{C}^2$ at x^0 when f admits second-order partial derivatives, continuous at x^0.

- The following implications hold:

  a) ∇f is differentiable at $x^0 \Leftrightarrow f$ is twice differentiable at x^0.

  b) ∇f is of class $\mathcal{C}^1$ at $x^0 \Leftrightarrow f$ is of class $\mathcal{C}^2$ at x^0.

  c) $f \in \mathcal{C}^2(x^0) \Rightarrow H f(x^0)$ is symmetric.

- If $f : \mathbb{R}^n \times \mathbb{R}^m \to \mathbb{R}$ is differentiable at (x^0, y^0), $x^0 \in \mathbb{R}^n$, $y^0 \in \mathbb{R}^m$, then we define

$$\nabla_x f(x^0, y^0) = \left[\frac{\partial f(x^0, y^0)}{\partial x_1}, ..., \frac{\partial f(x^0, y^0)}{\partial x_n}\right]$$

and

$$\nabla_y f(x^0, y^0) = \left[\frac{\partial f(x^0, y^0)}{\partial y_1}, \ldots, \frac{\partial f(x^0, y^0)}{\partial y_m}\right].$$

So

$$\nabla f(x^0, y^0) = [\nabla_x f(x^0, y^0), \nabla_y f(x^0, y^0)] .$$

Similar notations are used to describe Jacobian matrices and Hessian matrices for analogous cases.

- We recall the *implicit function theorem*, useful, e.g., to draw the necessary optimality conditions for the classical constrained optimization problem.

Implicit Function Theorem. Let $X \subset \mathbb{R}^n$, $Y \subset \mathbb{R}^m$ be open sets and $f : X \times Y \to \mathbb{R}^n$; let $(x^0, y^0) \in X \times Y$.
Suppose the following conditions hold:

a) $f(x^0, y^0) = 0$;

b) f has continuous partial derivatives in a neighbourhood of (x^0, y^0);

c) $\nabla_y f(x^0, y^0)$ is non singular, i.e. $|\nabla_y f(x^0, y^0)| \neq 0$.

Then there exists a unique m-vector function $y = g(x)$, defined on an open set $\Omega \subset \mathbb{R}^n$ containing x^0, with the following properties:

i) $y^0 = g(x^0)$;

ii) $f[x, g(x)] = 0$, $\forall x \in \Omega$;

iii) $g \in C^1(\Omega)$;

iv) $\nabla_y f(x, y)$ is non singular for all (x, y) in an open ball $B(x^0, y^0)$.

v) It is $\nabla g(x) = -[\nabla_y f(x)]^{-1} \cdot \nabla_x f(x)$, $\forall x \in \Omega$.

References to Chapter I

M. AOKI (1976), Optimal Control and System Theory in Dynamic Economic Analysis, North Holland, Amsterdam.

T.M. APOSTOL (1974), Mathematical Analysis, Addison-Wesley Publishing Co., Reading, Mass.

M. ATHANS and P.L. FALB (1966), Optimal Control, McGraw-Hill, New York.

R.G. BARTLE (1976), The Elements of Real Analysis, J. Wiley & Sons, New York.

R. BELLMAN (1957), Dynamic Programming, Princeton Univ. Press, Princeton.

R. BELLMAN (1967), Introduction to the Mathematical Theory of Control Processes. Academic Press, New York.

R. BELLMAN and S. DREYFUS (1962), Applied Dynamic Programming, Princeton Univ. Press, Princeton.

C. BERGE (1963), Topological Spaces, The MacMillan Company, New York.

G.C. BLISS (1946), Lectures on the Calculus of Variations, Univ. of Chicago Press, Chicago.

E. BODEWIG (1956), Matrix Calculus, North Holland, Amsterdam.

C. CARATHEODORY (1967), Calculus of Variations and Partial Differential Euqations, Holden Day, San Francisco.

B.D. CRAVEN (1981), Functions of Several Variables, Chapman and Hall, London.

W.H. FLEMING (1965), Functions of Several Variables, McGraw-Hill, New York.

L.M. GRAVES (1956), The Theory of Functions of Real Variables, McGraw-Hill, New York.

G. HADLEY (1961), Linear Algebra, Addison-Wesley, Reading, Mass.

G. HADLEY (1964), Nonlinear and Dynamic Programming, Addison-Wesley, Reading, Mass.

M.R. HESTENES (1966), Calculus of Variations and Optimal Control Theory, J. Wiley, New York.

M.I. KAMIEN and N.L. SCHWARTZ (1981), Dynamic Optimization, North Holland, Amsterdam.

E.B. LEE and L. MARKUS (1966), Foundations of Optimal Control Theory, J. Wiley, New York.

G. LEITMANN (1981), The Calculus of Variations and Optimal Control, Plenum Press, New York.

B. NOBLE (1969), Applied Linear Algebra, Prentice Hall, Englewood Cliffs, N.J.

L.S. PONTRYAGIN, V.G. BOLTYANSKII, R.V. GAMKRELIDZE and E.F. MISCHENKO (1962), The Mathematical Theory of Optimal Processes, Wiley-Interscience, New York.

L. ROBBINS (1932), An Essay on the Nature and Significance of Economic Science MacMillan, London.

W. RUDIN (1964), Principles of Mathematical Analysis, McGraw-Hill, New York.

G.F. SIMMONS (1963), Introduction to Topology and Modern Analysis, McGraw-Hill, New York.

H. TOLLE (1975), Optimization Methods, Springer-Verlag, Berlin.

CHAPTER II. CONVEX SETS. CONVEX AND GENERALIZED CONVEX FUNCTIONS

2.1. Convex Sets and Their Main Properties

Let $x^1, x^2 \in \mathbb{R}^n$ and $\lambda \in \mathbb{R}$; we define the

i) *closed line segment* $[x^1, x^2] = \{x \mid x = \lambda x^1 + (1-\lambda) x^2,\ 0 \leqq \lambda \leqq 1\}$;

ii) *open line segment* $(x^1, x^2) = \{x \mid x = \lambda x^1 + (1-\lambda) x^2.\ 0 < \lambda < 1\}$.

The set $X \subset \mathbb{R}^n$ is a *convex set* if it contains the closed line segment joining every two pcints $x^1, x^2 \in X$, i.e. if

$$x^1, x^2 \in X,\ 0 \leqq \lambda \leqq 1 \Rightarrow \lambda x^1 + (1-\lambda) x^2 \in X .$$

Example 2.1.1. Examples of convex sets are given below.

i) Every *hyperplane*, i.e. every set defined as

$$H = \{x \mid x \in \mathbb{R}^n,\ ax = \alpha\} ,$$

where $a \neq 0$ is a given vector of $\mathbb{R}^n$ and $\alpha \in \mathbb{R}$ is a given real number.

ii) Every *affine set* or *linear manifold*, i.e. the sets of the form

$$V = \{x \mid x \in \mathbb{R}^n,\ Ax = b\} ,$$

where $b \in \mathbb{R}^m$ is a given vector and A is a given (real) matrix of order m, n.

(*closed* halfspace) or defined as

$$H^{<} = \{x \mid x \in \mathbb{R}^n,\ ax < \alpha\}$$

(*open* halfspace).

iv) Every *polyhedron*, i.e. every set of the form

$$T = \{x \mid x \in \mathbb{R}^n,\ Ax \leqq b\}\ ,$$

where $b \in \mathbb{R}^m$ and A is a matrix of order m, n. A polyhedron is therefore given by the intersection of a finite number of closed half-spaces.

v) Every *ball*, i.e. the sets defined as

$$\overline{B(a)} = \{x \mid x \in \mathbb{R}^n,\ \|x - a\| \leqq r\}\ ,$$

where $a \in \mathbb{R}^n$, $r > 0$ (*closed ball*); or defined as

$$B(a) = \{x \mid x \in \mathbb{R}^n,\ \|x - a\| < r\}$$

(*open ball*).

vi) Every *ellipsoid*, i.e. the sets defined as

$$E = \{x \mid x \in \mathbb{R}^n,\ x^T C\, x \leqq \gamma\}\ ,$$

where C is a symmetric definite positive matrix of order n and $\gamma > 0$.

vii) The empty set $\emptyset$; any singleton.

<u>Theorem 2.1.1</u>. The set $X \subset \mathbb{R}^n$ is convex if and only if

$$k \in \mathbb{N},\ x^1, x^2, ..., x^k \in X,\ \lambda_k \geqq 0\ \forall k\ ,$$

$$\sum_{i=1}^{k} \lambda_i = 1 \Rightarrow \sum_{i=1}^{k} \lambda_i\, x^i \in X \tag{1}$$

i.e. if and only if every *convex combination* of the points $x^1, x^2, ..., x^k \in X$ is contained in X.

Proof. The sufficiency is trivial: take $k = 2$, then X is convex. The necessity is proved by induction. For $k = 1$ relation (1) holds trivially; for $k = 2$ relation (1) holds as consequence of the definition of convex set. Assume that (1) holds for $k = m$; we will show that it holds for $k = m+1$. Let $\bar{x} = \lambda_1 x^1 + \lambda_2 x^2 + ... + \lambda_m x^m + \lambda_{m+1} x^{m+1}$. If $\lambda_{m+1} = 0$, then $\bar{x} \in X$ by assumption; if $\lambda_{m+1} = 1$, then $(\lambda_1 x^1 + \lambda_2 x^2 + ... + x^{m+1}) = x^{m+1} \in X$; if $0 < \lambda_{m+1} < 1$ we can write

$$\bar{x} = (1 - \lambda_{m+1})\, z + \lambda_{m+1} x^{m+1} ,$$

where

$$z = \frac{\lambda_1}{1 - \lambda_{m+1}} x^1 + ... + \frac{\lambda_m}{1 - \lambda_{m+1}} x^m .$$

Since

$$\frac{\lambda_1}{1 - \lambda_{m+1}} \geqq 0, ..., \frac{\lambda_m}{1 - \lambda_{m+1}} \geqq 0 ; \quad \sum_{i=1}^{m} \frac{\lambda_i}{1 - \lambda_{m+1}} = 1 ,$$

it follows, by the induction hypothesis, that $z \in X$; and by the convexity of X we have $\bar{x} \in X$. □

Theorem 2.1.2. The set $X \subset \mathbb{R}^n$ is convex if and only if the set $T_{x,y} = \{t \mid t \in \mathbb{R},\ x + ty \in X\}$ is convex for every $x \in X$ and $y \in \mathbb{R}^n$.

Proof. Let X be convex and $t_1, t_2 \in T_{x,y}$; let $\lambda \in [0,1]$. We shall prove that $\lambda t_1 + (1-\lambda)\, t_2 \in T_{x,y}$. As $t_1, t_2 \in T_{x,y}$ it results that $\bar{x} = x + t_1 y \in X$, $\bar{y} = x + t_2 y \in X$ and as X is convex it results $\lambda \bar{x} + (1 - \lambda)\, \bar{y} \in X$. Hence $\lambda \bar{x} + (1 - \lambda)\, \bar{y} = x + [\lambda t_1 + (1 - \lambda)\, t_2]\, y \in X$ from which it results $\lambda t_1 + (1 - \lambda)\, t_2 \in T_{x,y}$. Conversely, let $\bar{x}$ and $\hat{x} = \bar{x} + \bar{t} y$ be two points of X; clearly $t = 0$ belongs to $T_{x,y}$ as well as $t = \bar{t}$. Being $T_{x,y}$ convex, it contains the closed line segment $[0, \bar{t}]$, i.e. it is $\lambda \bar{x} + (1 - \lambda)(\bar{x} + \bar{t} y) \in X$, $\forall \lambda \in [0,1]$, i.e. $\lambda \bar{x} + (1 - \lambda)\, \hat{x} \in X$, $\forall \lambda \in [0,1]$, i.e. X is convex. □

Theorem 2.1.3. The intersection of an arbitrary family of convex sets is also a convex set.

Proof. Let x^1 and x^2 be points contained in the intersection. Then they are also contained in every member of the family and so is $\bar{x} = \lambda x^1 + (1-\lambda)\, x^2$, $\lambda \in [0,1]$.Hence $\bar{x}$ is contained in the intersection. □

Note that the above property does not hold true with respect to the union, i.e. the union of a family of convex sets is not, in general, a convex set.

If X, Y are two convex sets, then

$$\alpha X = \{z \mid z = \alpha x, \; x \in X\}$$

$$X + Y = \{z \mid z = x + y, \; x \in X, \; y \in Y\}$$

are convex sets. Then we denote that the set $X - Y = X + (-1)\, Y$ is a convex set and that if $X_1, X_2, ..., X_k$ are convex sets and $\alpha_1, \alpha_2, ..., \alpha_k \in \mathbb{R}$, then $\sum_{i=1}^{k} \alpha_i X_i$ is a convex set. If X is a convex set, then for every subset X_i of X, $i = 1, 2, ..., k$, and every scalar $\lambda_i \geqq 0$, $i = 1, 2, ..., k$, with $\sum_{i=1}^{k} \lambda_i = 1$, we have $\sum_{i=1}^{k} \lambda_i X_i \subset X$.

We call *convex hull* of an arbitrary set $A \subset \mathbb{R}^n$, the intersection of all convex sets containing A. The convex hull of A is denoted by $\operatorname{conv}(A)$ or $[A]$ and is actually the smallest convex set in $\mathbb{R}^n$ containing A. The following properties hold.

i) $X = \operatorname{conv}(X)$ if and only if X is convex.

ii) If X is bounded, $\operatorname{conv}(X)$ is bounded; if X is compact, $\operatorname{conv}(X)$ is compact (for the proof see, e.g., Nikaido (1968), Rockafeller (1970)). Note that the convex hull of a closed set of $\mathbb{R}^n$ is not necessarily closed, as is shown by the following example:

$$X = \{x \in \mathbb{R}^2 \mid x_1 \geqq 0, \; x_1 \cdot x_2 \geqq 1 \cup x_1 \leqq 0, \; x_1 \cdot x_2 \leqq -1\} \; .$$

Moreover, it is always $\operatorname{conv}(\bar{X}) \subset \operatorname{cl}(\operatorname{conv}(X))$ and $\operatorname{cl}(\operatorname{conv}(X)) = \operatorname{cl}(\operatorname{conv}(\bar{X}))$, but it is $\operatorname{conv}(\bar{X}) = \operatorname{cl}(\operatorname{conv}(X))$ if X is bounded. In fact, in this case the sets $\bar{X}$ and $\operatorname{conv}(\bar{X})$ are compact and we have $\operatorname{conv}(\bar{X}) = \operatorname{cl}(\operatorname{conv}(\bar{X})) = \operatorname{cl}(\operatorname{conv}(X))$.

iii) $\operatorname{conv}(X)$ is equal to the set of all convex combinations of any given number of points of X. In effect, let Λ denote this latter set, i.e.

$$\Lambda = \Big\{x \mid x = \sum_{i=1}^{k} \lambda_i \, a^i, \ \lambda_i \in I\!R, \ a^i \in X,$$
$$\lambda_i \geqq 0 \ \forall i, \ \sum_{i=1}^{k} \lambda_i = 1, \ k \geqq 1\Big\} .$$

If $x^1, x^2 \in \Lambda$, then

$$x^1 = \sum_{i=1}^{k} \lambda_i \, a^i, \ \lambda_i \in I\!R, \ a^i \in X, \ \lambda_i \geqq 0, \ \sum_{i=1}^{k} \lambda_i = 1 \ ,$$

$$x^2 = \sum_{i=1}^{m} \mu_i \, b^i, \mu_i \in I\!R, \ b^i \in X, \ \mu_i \geqq 0, \ \sum_{i=1}^{m} \mu_i = 1 \ .$$

Hence for $0 \leqq \lambda \leqq 1$ we have

$$\lambda x^1 + (1-\lambda)\, x^2 = \sum_{i=1}^{k} \lambda \, \lambda_i \, a^i + \sum_{i=1}^{m} (1-\lambda)\, \mu_i \, b^i$$

and

$$\lambda \, \lambda_i \geqq 0, \ (1-\lambda)\, \mu_i \geqq 0, \ \sum_{i=1}^{k} \lambda \, \lambda_i + \sum_{i=1}^{m} (1-\lambda)\, \mu_i = 1 \ .$$

Thus $\lambda x^1 + (1-\lambda)\, x^2 \in \Lambda$ and Λ is convex. It is also clear that $X \subset \Lambda$. Since Λ is convex, then $\operatorname{conv}(X) \subset \Lambda$. We also have by Theorem 2.3.1 that the convex set $\operatorname{conv}(X)$ containing X must also contain all convex combinations of points of X. Hence $\Lambda \subset \operatorname{conv}(X)$; therefore $\Lambda = \operatorname{conv}(X)$. □

Property iii) can be made clearer by the following property:

iv) Every point of $\text{conv}(X)$ can be written as a convex combination of at most $n+1$ points of X, i.e.

$$\text{conv}(X) = \Big\{ x \mid x \in \mathbb{R}^n, \; x = \sum_{i=1}^{n+1} \alpha_i x^i ;$$

$$x^i \in X, \; \alpha_i \geqq 0, \; i = 1, 2, ..., n+1; \; \sum_{i=1}^{n+1} \alpha_i = 1 \Big\} .$$

This property is a direct consequence of iii) and of the following:

Theorem 2.1.4 (Charatheodory's theorem). If $\bar{x}$ is a convex combination of points of the convex set $X \subset \mathbb{R}^n$, then $\bar{x}$ is a convex combination of $n+1$ or fewer points of X.

Proof. Let $\bar{x} = \sum_{i=1}^m \lambda_i x^i$, $x^i \in X$, $\lambda_i \in \mathbb{R}$, $\lambda_i \geqq 0$, $\sum_{i=1}^m \lambda_i = 1$. If $m \leqq n+1$, then nothing need be proved. We shall show now that if $m > n+1$ then $\bar{x}$ can be written as a convex combination of $m-1$ points of X. This would establish the theorem then, for we could repeatedly apply the result until $\bar{x}$ is a convex combination of $n+1$ points of X. If any λ_i in the above expression is zero, then $\bar{x}$ is a convex combination of $m-1$ or fewer points of X. So let $\lambda_i > 0 \; \forall i = 1, 2, ..., m$. Since $m > n+1$, there exist $r_1, ..., r_{m-1} \in \mathbb{R}$, not all zero, such that

$$r_1(x^1 - x^m) + ... + r_{m-1}(x^{m-1} - x^m) = 0 .$$

Define $r_m = -(r_1 - ... + r_{m-1})$; then

$$\sum_{i=1}^m r_i = 0 ; \quad \sum_{i=1}^m r_i x^i = 0 .$$

Define

$$q_i = \lambda_i - \alpha r_i , \qquad \text{for } i = 1, ..., m ,$$

where α is some positive number chosen so that $q_i \geqq 0$ for all i, and at least one q_i, say q_k, is equal to zero. In particular we choose α such that

$$\frac{1}{\alpha} = \max_i \left\{ \frac{r_i}{\lambda_i} \right\} = \frac{r_k}{\lambda_k} .$$

Then $q_i \geqq 0$, $i = 1, 2, ..., m$; $q_k = 0$;

$$\sum_{\substack{i=1 \\ i \neq k}}^{m} q_i = \sum_{i=1}^{m} q_i = \sum_{i=1}^{m} \lambda_i - \alpha \sum_{i=1}^{m} r_i = \sum_{i=1}^{m} \lambda_i = 1$$

and

$$\bar{x} = \sum_{i=1}^{m} \lambda_i x^i = \sum_{i=1}^{m} q_i x^i + \alpha \sum_{i=1}^{m} r_i x^i = \sum_{\substack{i=1 \\ i \neq k}}^{m} q_i x^i .$$

Hence $\bar{x}$ is a convex combination of $m-1$ points in X. □

Another property of the convex hull is the following one:

v) If $x^1, x^2, ..., x^m \in I\!R^n$ $(m > n+1)$, denoted by X the set of vectors $\{x^1, x^2, ..., x^m\}$ and by X_i the set $\{x^j\}$, with $j \neq i$, then we have

$$\operatorname{conv}(X) = \bigcup_{i=1}^{m} \operatorname{conv}(X_i) .$$

In effect if $x \in \operatorname{conv}(X)$, on the ground of property iv) on the convex hull, then x can be written as a convex combination of at most $m-1$ points x^i.

Consider now a subset A of $I\!R^n$ with the following property: for every two points $x^1, x^2 \in A$ and for every value of the real number α, also $\alpha x^1 + (1-\alpha) x^2 \in A$. A set satisfying this property is called an *affine set.* Examples of affine sets are single points, lines and hyperplanes of $I\!R^n$. Given a set $X \subset I\!R^n$, the intersection of all affine sets containing X is called the *affine hull* of X and denoted by $\operatorname{aff}(X)$. It turns out that $\operatorname{aff}(X)$ is the minimal affine set containing X. We may also say that X is an affine set if X contains all its *affine combinations,* i.e the linear combinations $\sum_{i=1}^{k} \lambda_i x^i$, where $\sum_{i=1}^{k} \lambda_i = 1$, k positive integer. Similarly, $\operatorname{aff}(X)$ may also be defined as the collection of all affine combinations of vectors of X, i.e.

$$\operatorname{aff}(X) = \Big\{ x \mid x = \sum_{i=1}^{k} \lambda_i x^i,\ x^i \in X,\ \lambda_i \in I\!R,\ \sum_{i=1}^{k} \lambda_i = 1 \Big\} .$$

It happens that X is an affine set if and only if $X = \text{aff}(X)$. We will now introduce the concept of *dimension* of a subset X of $\mathbb{R}^n$. First we would remark that an affine set X is merely the result of the translation of a linear subspace. This is the result of the following proposition.

Theorem 2.1.5. A subset X of $\mathbb{R}^n$ is affine if and only if $X - x^0$ is a linear subspace of $\mathbb{R}^n$, for every $x^0 \in X$.

Proof. First suppose that X is affine and that $x^0 \in X$. We want to show that $X - x^0$ is a subspace of $\mathbb{R}^n$. To do this let $y, z \in X - x^0$ and $\alpha, \beta \in \mathbb{R}$. Since $y, z \in X - x^0$, then $y + x^0, z + x^0 \in X$. But then $x^0 \in X$, $\alpha + \beta + (1 - \alpha - \beta) = 1$ and X affine all imply that

$$\alpha(y + x^0) + \beta(z + x^0) + (1 - \alpha - \beta)\, x^0 \in X$$

$$\alpha y + \beta z + x^0 \in X$$

$$\alpha y + \beta z \in X - x^0 \ .$$

Hence $X - x^0$ is a linear subspace of $\mathbb{R}^n$. Conversely, suppose $y, z \in X$ and $\alpha, \beta \in \mathbb{R}$ with $\alpha + \beta = 1$. Then $y - x^0, z - x^0 \in X - x^0$. Since $X - x^0$ is a linear subspace and $\alpha + \beta = 1$, it follows that

$$\alpha y + \beta z - x^0 = \alpha y + \beta z - (\alpha + \beta)\, x^0 =$$
$$= \alpha(y - x^0) + \beta(z - x^0) \in X - x^0 \ ,$$

i.e. $\alpha y + \beta z \in X$; hence X is affine. □

Thus an affine set is just given by the translation of a linear subspace. It is an easy exercise to show that this subspace is unique.
We define *dimension* of an affine set X as the dimension of the subspace L parallel to X, i.e. of the subspace of which it is a translation: $\dim(X) = \dim(L)$. The dimension of *any* set X is defined to be the dimension of its affine hull, $\text{aff}(X)$. We say that the dimension of the empty set is -1.

The *relative interior* of $X \subset \mathbb{R}^n$, denoted by $\text{relint}(X)$ or $\text{ri}(X)$, is defined as the interior of X, viewed as a subset of its affine hull (it is therefore the interior of X with respect to the topology of $\mathrm{aff}(X)$). Hence it is $\mathrm{relint}(X) = \{x \mid \exists N(x) : N(x) \cap \mathrm{aff}(X) \subset X\}$. It is $\mathrm{relint}(X) = \mathrm{int}(X)$ if and only if $\mathrm{aff}(X) = \mathbb{R}^n$. If the vectors $x^1, x^2, ..., x^k \in \mathbb{R}^n$, $k \leqq n$, are linearly independent, then

a) $$\mathrm{relint}\left\{x \mid x = \sum_{i=1}^{k} \lambda_i x^i,\ \lambda_i \geqq 0,\ i = 1, 2, ..., k\right\} =$$

$$= \left\{x \mid x = \sum_{i=1}^{k} \lambda_i x^i,\ \lambda_i > 0,\ i = 1, 2, ..., k\right\}.$$

b) $$\mathrm{relint}\left\{x \mid x = \sum_{i=1}^{k} \lambda_i x^i,\ \lambda_i \geqq 0,\ i = 1, 2, ..., k,\ \sum_{i=1}^{k} \lambda_i \leqq 1\right\} =$$

$$= \left\{x \mid x = \sum_{i=1}^{k} \lambda_i x^i,\ \lambda_i > 0,\ i = 1, 2, ..., k,\ \sum_{i=1}^{k} \lambda_i < 1\right\}.$$

c) More generally, for every set $\Lambda \subset \mathbb{R}^k$, we have

$$\mathrm{relint}\left\{\sum_{i=1}^{k} \lambda_i x^i,\ \lambda \in \Lambda\right\} = \left\{\sum_{i=1}^{k} \lambda_i x^i,\ \lambda \in \mathrm{int}(\Lambda)\right\}.$$

The previous results follow from the fact that the transformation $T(\lambda) = \sum_{i=1}^{k} \lambda_i x^i$ is an homeomorphism between $\mathbb{R}^k$ and the k-dimensional subspace $T(\mathbb{R}^k) \subset \mathbb{R}^n$; therefore $\mathrm{relint}(T(\Lambda)) = T(\mathrm{int}(\Lambda))$ for each $\Lambda \subset \mathbb{R}^k$.

Theorem 2.1.6. Every convex set $X \subset \mathbb{R}^n$, with $\mathrm{int}(X) = \emptyset$, is contained in a hyperplane.

Proof. Let $X \subset \mathbb{R}^n$ be convex with $\mathrm{int}(X) = \emptyset$. We may suppose that $0 \in X$ without loss of generality (in the negative we perform the proof on the set $X - x^0$, with $x^0 \in X$). Let d ($d \leqq n$) be the maximum number of linearly independent vectors of X and let $x^1, x^2, ..., x^d \in X$ be vectors with the said property. We have $d < n$, as if $d = n$, denoting $A =$

$\{0, x^1, ..., x^n\}$, we have $\text{conv}(A) \subset X$, therefore $\text{int}(\text{conv}(A)) \subset \text{int}(X)$, with

$$\text{int}(\text{conv}(A)) = \Big\{\sum_{i=1}^{n} \lambda_i x^i,\ \lambda_i > 0,\ i = 1, 2, ..., n;\ \sum_{i=1}^{n} \lambda_i < 1\Big\} \neq \emptyset\ ,$$

in contradiction with the assumptions. Therefore there exists a vector $a \in I\!R^n$, $a \neq 0$, such that $a\, x^i = 0$, $i = 1, 2, ..., d$.
We prove that for every $x \in X$ it is $a\, x = 0$; as the vectors $x, x^1, x^2, ..., x^d$ are linearly dependent, we have $\lambda x + \sum_{i=1}^{d} \lambda_i x^i = 0$, with $\lambda \neq 0$ as $x^1, x^2, ..., x^d$ are linearly independent. Therefore it results $0 = \lambda\, a\, x + \sum_{i=1}^{d} \lambda_i\, a\, x^i = \lambda\, a\, x$; so it is $a\, x = 0$. □

Theorem 2.1.7.

i) The closure $\bar{X}$ and the relative interior $\text{relint}(X)$ of a nonempty convex set $X \subset I\!R^n$ are nonempty convex sets.

ii) If X is a nonempty convex set in $I\!R^n$ and if $x^1 \in \text{relint}(X)$ and $x^2 \in \bar{X}$, then $\lambda x^1 + (1 - \lambda)\, x^2 \in \text{relint}(X)$, $\forall\, \lambda \in (0, 1)$.

Proof.

i) Note first that the convexity of $\bar{X}$ is trivial if X is empty, If $X \neq \emptyset$, let $x, y \in \bar{X}$; then there exist two sequences $\{x^k\}$, $\{y^k\}$ in X with the respective limits x and y. For any $\alpha \geqq 0$, $\beta \geqq 0$, $\alpha + \beta = 1$, the continuity of linear operations implies the convergence of $\alpha x^k + \beta y^k$ to $\alpha x + \beta y$. Moreover, by the convexity of X, $\alpha x^k + \beta y^k$ belongs to X in the course of convergence. This proves that $\alpha x + \beta y \in \bar{X}$. The fact that $\text{relint}(X)$ is convex follows from ii); let us prove that it is $\text{relint}(X) \neq \emptyset$ if $X \neq \emptyset$. We may suppose that $0 \in X$ without loss of generality; let d be the maximum number of linearly independent vectors of X and consider the linearly independent vectors $x^1, x^2, ..., x^d \in X$. Denote by A the set $\{0, x^1, ..., x^d\}$; from Theorem 2.1.1 we have $\text{conv}(A) \subset X$ and therefore, as $\text{aff}(\text{conv}(A)) = \text{aff}(X)$, we have $\text{relint}(\text{conv}(A)) \subset \text{relint}(X)$ and, thanks to the properties of the relative interior, $\text{relint}(\text{conv}(A)) \neq \emptyset$.

ii) Without loss of generality, we may suppose $\mathrm{int}(X) \neq \emptyset$; in the opposite case we can perform the proof on $\mathrm{aff}(X)$, where it is $\mathrm{int}(X) = \mathrm{relint}(X) \neq \emptyset$. Indeed closures and relative interiors are preserved under translations and more generally under any one-to-one affine trasformation of $I\!R^n$ onto itself. This often enables us to reduce a question about general convex sets to the case where the convex set is of full dimension, i.e. has the whole space as its affine hull. As $x^1 \in \mathrm{int}(X)$ there exists a neighbourhood $B(x^1, \delta)$ about x^1 of radius $\delta > 0$ such that $B(x^1, \delta) \subset X$. As $x^2 \in \bar{X}$, there exists a sequence $\{z^k\} \subset X$, with $z^k \to x^2$ and therefore, denoting $y = \lambda x^1 + (1 - \lambda) x^2$, $y^k = \lambda x^1 + (1 - \lambda) x^k$, with $\lambda \in (0, 1)$, we have $y^k \to y$. Therefore there exists a natural number k such that $\|y - y^k\| < \lambda\delta$ and so $y \in B(y^k, \lambda\delta) = \lambda B(x^1, \delta) + (1 - \lambda)\{x^k\} \subset X$. Hence y is within the neighbourhood $B(y^k, \lambda\delta)$ which is contained in X; therefore $y \in \mathrm{int}(X)$. □

Corollary 2.1.1. Let $X \in I\!R^n$ be a nonempty convex set; let $x^1 \in \mathrm{int}(X)$ and $x^2 \in \bar{X}$; then $y = \lambda x^1 + (1 - \lambda) x^2 \in \mathrm{int}(X)$, $\forall \lambda \in (0, 1)$.

Corollary 2.1.1 is also called "accessibility lemma" (see Peressini, Sullivan and Uhl (1988)).

Corollary 2.1.2. If $X \subset I\!R^n$ is convex, then

i) $\mathrm{relint}(X) = \mathrm{relint}(\bar{X})$;

ii) $\bar{X} = \mathrm{cl}(\mathrm{relint}(X))$.

Proof.

i) By definition it is (since $\mathrm{aff}(X) = \mathrm{aff}(\bar{X})$)

$$\mathrm{relint}(X) \subset \mathrm{relint}(\bar{X}) .$$

For the converse inclusion, let $x \in \mathrm{relint}(\bar{X})$. Taking a point $y \in \mathrm{relint}(X) \neq \emptyset$, we can find also a point $z \in \bar{X}$ "opposite" to x, such that x is between z and y, i.e. we have

$$x = \lambda z + (1 - \lambda) y$$

with $\lambda \in (0,1)$. Since $z \in \bar{X}$ and $y \in \text{relint}(X)$, by Theorem 2.1.7 it is $x \in \text{relint}(X)$.

ii) By definition it is

$$\text{cl}(\text{relint}(X)) \subset \bar{X} .$$

For the converse inclusion, let $x \in \bar{X}$. Taking a point $y \in \text{relint}(X) \neq \emptyset$, it is

$$\lambda x + (1 - \lambda) y \in \text{relint}(X)$$

for all $\lambda \in (0,1)$. Hence $x \in \text{cl}(\text{relint}(X))$. □

Corollary 2.1.3. If $X \subset \mathbb{R}^n$ is convex, then

i) $\text{int}(X) = \text{int}(\bar{X})$;

ii) $\text{int}(X) \neq \emptyset \Rightarrow \bar{X} = \text{cl}(\text{int}(X))$.

Proof. Since $\text{aff}(\bar{X}) = \text{aff}(X)$, we have the equivalence

$$\text{int}(X) = \emptyset \Leftrightarrow \text{int}(\bar{X}) = \emptyset .$$

If $\text{int}(X) \neq \emptyset$, then $\text{int}(X) = \text{relint}(X)$ and the assertions are consequences of Corollary 2.1.2. □

2.2. Separation Theorems

Separation theorems play an extremely important role in mathematical programming. Given any hyperplane H, we can associate to H the two (closed) halfspaces

$$H^{\leqq} = \{x \mid ax \leqq \alpha, \ \alpha \in \mathbb{R}\} ,$$

$$H^{\geqq} = \{x \mid ax \geqq \alpha, \ \alpha \in \mathbb{R}\} .$$

Obviously it is $H^{\leqq} \cup H^{\geqq} = \mathbb{R}^n$ and one can consider open halfspaces $H^{<}$ and $H^{>}$ simply by changing the weak inequalities into strict inequalities. Let S be a subset of $\mathbb{R}^n$; a hyperplane $H = \{x \mid ax = \alpha, \ a \neq 0, \ \alpha \in \mathbb{R}\}$ is called a *supporting hyperplane* of S if $S \subset H^{\leqq}$ (or $S \subset H^{\geqq}$) and $\bar{S} \cap H \neq \emptyset$. H is called a *proper* supporting hyperplane of S if in addition to the said properties it is $S \cap H \neq S$. From the above definition of supporting hyperplane, we must impose two requirements in order for the hyperplane H to support a set S. The first of these is that the set S should be completely contained in one of the halfspaces associated to H. The second requirement is that the closure of S and the hyperplane H should have at least one point in common. If in this case $x^0 \in \bar{S} \cap H$ then we say that H supports the set S at x^0. It is obvious that a hyperplane H may support a set S at several distinct points.

The following result proves the intuitively obvious fact that $\inf_{x \in S} \{ax\}$ is indeed α if the hyperplane $H = \{x \mid ax = \alpha\}$ supports S and S is contained in $H^{\geqq}$.

<u>Theorem 2.2.1</u>. Let $H = \{x \mid ax = \alpha\}$ be a supporting hyperplane of S such that $S \subset H^{\geqq}$. Then $\inf_{x \in S} \{ax\} = \alpha$.

<u>Proof</u>. Since $S \subset H^{\geqq}$ then $ax \geqq \alpha$ for each $x \in S$, i.e. $\inf_{x \in S} \{ax\} \geqq \alpha$. If the equality does not hold, there must exist an $\varepsilon > 0$ with $ax \geqq \alpha + \varepsilon$ for each $x \in S$. But by definition of a supporting hyperplane $\bar{S} \cap H \neq \emptyset$, i.e. there exists $y \in \bar{S}$, with $ay = \alpha$. But $y \in \bar{S}$ implies that there exists a sequence $\{x^k\}$ in S which converges to y and hence $\{ay\} = \lim_{k \to +\infty} \{ax^k\} \geqq \alpha + \varepsilon$, a contradiction. □

It is quite obvious that equivalently we can say that if $H = \{x | ax = \alpha\}$ is a supporting hyperplane for S such that $S \subset H^{\leqq}$, then $\sup_{s \in S} \{ax\} = \alpha$.

The next theorem is a projection theorem on closed convex sets; we need first the following:

Definition 2.2.1. Let X be a subset of $\mathbb{R}^n$ and $y \in \mathbb{R}^n$; $x^0 \in X$ is the *best approximation of* y *on* X or *projection of* y *on* X if

$$\|y - x^0\| = \min_{x \in X} \|y - x\| \ .$$

If $y \in X$, then the best approximation is unique and is $x^0 = y$; it must be noted that the best approximation of y on X not always exists (e.g. if X is an open set) and when it exists it may not be unique (e.g. if $X = \{x \in \mathbb{R}^2 \mid \|x\| \geqq 1\}$ and y is the origin). However, under closedness and convexity assumptions the following assertion holds.

Theorem 2.2.2. Let X be a nonempty closed convex set of $\mathbb{R}^n$ and y a point of $\mathbb{R}^n$, with $y \notin X$. Then there exists a unique point $x^0 \in X$ such that

$$\|y - x^0\| = \min_{x \in X} \|y - x\| \ .$$

Moreover, the point x^0 verifies the following variational inequality

$$(y - x^0)(x - x^0) \leqq 0 \ , \qquad \forall x \in X$$

and is the unique point of X to verify the same.

Proof. Let us prove first the following equivalence:

i) $x^0 \in X$ and $\|y - x^0\| = \min_{x \in X} \|y - x\|$

$\Leftrightarrow$

ii) $x^0 \in X$ and $\forall x \in X$ it is $(y - x^0)(x - x^0) \leqq 0$.

i) $\Rightarrow$ ii): Let us fix $x \in X$ and $\lambda \in (0, 1)$; as X is convex $[(1-\lambda)\, x^0 + \lambda x] \in X$.

From i) we obtain

$$\frac{1}{\lambda} \left(\|y - \{(1 - \lambda)\, x^0 + \lambda x\}\|^2 - \|y - x^0\|^2\right) =$$

$$= \frac{1}{\lambda} \left(\|(y - x^0) - \lambda(x - x^0)\|^2 - \|y - x^0\|^2\right) \geqq 0 \ .$$

Simplifying this expression we obtain

$$\lambda \|x - x^0\|^2 - 2(y - x^0)(x - x^0) \geqq 0 .$$

Letting $\lambda \to 0^+$, the result follows.
ii) $\Rightarrow$ i): Let us fix $x \in X$ and apply ii):

$$0 \geqq (y - x^0)(x - x^0) = (y - x^0)\{(y - x^0) - (y - x)\} =$$
$$= \|y - x^0\|^2 - (y - x^0)(y - x) .$$

By means of the Cauchy-Schwartz inequality we have

$$\|y - x^0\|^2 \leqq (y - x^0)(y - x) < \|y - x^0\| \cdot \|y - x\| .$$

Therefore it is $\|y - x^0\| \leqq \|y - x\|$ (it is also true if $\|y - x^0\| = 0$), i.e. we obtain i).
Let us now prove the existence of a minimizing point x^0, i.e. of a point x^0, verifying i), Choose a closed ball $\overline{B(y, \alpha + 1)}$, where $\alpha = \inf_{x \in X} \|y - x\|$. As $X \cap \overline{B(y, \alpha + 1)}$ is compact, the continuous function $\|y - x\|$ has its minimum at a point x^0, according to the Weierstrass theorem. It is also evident that $\|y - x^0\| = \alpha$. It remains to prove the uniqueness of the point x^0 verifying i). Let us suppose the existence of two points x^1 and x^2 which verify i); we have just shown that also ii) will be verified by the said points. Apply then relation ii) to x^1, choosing $x = x^2$ and subsequently apply the same relation to x^2, choosing $x = x^1$. It results

$$(y - x^1)(x^2 - x^1) \leqq 0$$
$$(y - x^2)(x^1 - x^2) \leqq 0 .$$

If we add these inequalities we obtain

$$(x^1 - x^2)(x^1 - x^2) = \|x^1 - x^2\| \leqq 0$$

which implies $x^1 = x^2$. □

The following result shows that if we have a closed convex set and a point outside the set, then we can find a hyperplane such that the convex set lies in one of the halfspaces generated by the hyperplane and the point lies in the interior of the other halfspace.

Theorem 2.2.3. Let X be a nonempty closed convex set in $I\!R^n$ and $y \notin X$; then there exists a nonzero vector a in $I\!R^n$ such that $\inf_{x \in X} \{ax\} > ay$.

Proof. By Theorem 2.2.2, there exists a unique projection $x^0 \in X$ with $(y - x^0)(x - x^0) \leqq 0$, $\forall x \in X$. Note that $y \neq x^0$ and so

$$0 < \|y - x^0\|^2 = (y - x^0)(y - x^0) = y(y - x^0) - x^0(y - x^0) \ .$$

From the first inequality we get

$$x(y - x^0) \leqq x^0(y - x^0)$$

and, taking the second inequality into account,

$$x(y - x^0) \leqq x^0(y - x^0) < y(y - x^0) \ , \qquad \forall x \in X \ .$$

Letting $x^0 - y = a$ it follows that

$$ax \geqq ax^0 > ay \ , \qquad \forall x \in X$$

and the desired result follows. □

The following theorem is a consequence of the last result and may be called "theorem of the supporting hyperplane".

Theorem 2.2.4.

i) If y is a point of the boundary of the nonempty convex set $X \subset I\!R^n$, then there exists a nonzero vector $a \in I\!R^n$ such that $\inf_{x \in X} \{ax\} = ay$, i.e. there exists a hyperplane $H = \{x \mid a(x - y)\}$ which supports X at y.

ii) If the set $X \subset I\!R^n$ admits a supporting hyperplane at each point y of its boundary, then $\operatorname{int}(X)$ is a convex set.

Proof.

i) As $y \in \mathrm{bd}(X)$, there exists a sequence $\{x^k\}$ outside $\bar{X}$ such that $x^k \to y$. From the previous theorem we have that there exists a sequence

$$\{a^k\} \subset S = \{a \mid \|a\| = 1\}$$

such that, for each k and for each $x \in X$, $a^k x \geqq a^k x^k$. As S is compact, there will exist a converging subsequence $a^{k'} \to a$ and therefore, for each $x \in X$, we have $ax = \lim_{k' \to \infty} a^{k'} x \geqq \lim_{k' \to \infty} a^{k'} x^{k'} = ay$.
Note that this result is equivalent to the fact that in each boundary point of a convex set there is a supporting hyperplane passing through it.

ii) Let $x^1, x^2 \in \mathrm{int}(X)$; let us suppose by contradiction that the set

$$\Lambda = \{\lambda \mid 0 < \lambda < 1,\ (1-\lambda)\, x^1 + \lambda x^2 \notin \mathrm{int}(X)\}$$

is nonempty. Denote $\lambda_0 = \inf \Lambda$; we have $0 < \lambda_0 < 1$ and $x^0 = (1-\lambda_0)\, x^1 + \lambda_0 x^2 \in \mathrm{bd}(X)$, as $x^0 \in \mathrm{int}(X)$ or $x^0 \in \mathrm{ext}(X)$ contradict the definition of λ_0.
Let $H = \{x \mid ax = ax^0\}$ be a supporting hyperplane at x^0 $(a \neq 0)$; therefore $x^1 \in H$, i.e. $ax^1 = ax^0$, as if we suppose, e.g., $ax^1 > ax^0$, it results $ax^2 < ax^0$, in contradiction to the definition of H. Therefore x^1 must be a boundary point of X, in contradiction to the assumption. Therefore $\mathrm{int}(X)$ is convex. □

As a consequence of the above theorem we deduce that an open set $X \subset \mathbb{R}^n$ or also a closed set $X \subset \mathbb{R}^n$, with $\mathrm{int}(X) \neq \emptyset$, is convex if and only if it admits a supporting hyperplane at each point $x \in \mathrm{bd}(X)$. If X is open the assertion is immediate. If X is closed, with $\mathrm{int}(X) \neq \emptyset$, under the assumption of Theorem 2.2.4, we have already shown that $\mathrm{int}(X)$ is a convex set; since it is $\overline{\mathrm{int}(X)} \subset \bar{X} = X$, we now have to prove that it is $X = \bar{X} \subset \overline{\mathrm{int}(X)}$, which implies that $X = \bar{X} = \overline{\mathrm{int}(X)}$ is convex by Theorem 2.1.7. Suppose by contradiction the existence of $x^1 \in X$,

$x^1 \notin \overline{\text{int}(X)}$; choose a point $x^2 \in \text{int}(X)$ ($\text{int}\, X \neq \emptyset$ by assumption) and let

$$\lambda_0 = \inf \{\lambda \mid 0 \leqq \lambda \leqq 1,\ \lambda x^1 + (1 - \lambda)\, x^2 \notin \text{int}(X)\} .$$

Clearly it is $0 < \lambda_0 \leqq 1$. Moreover, $\lambda_0 = 1$ would imply $x^1 \in \overline{\text{int}(X)}$ which is a contradiction; let $x^0 = \lambda_0 x^1 + (1 - \lambda_0)\, x^2$; it is

$$x^0 \in \text{bd}(\text{int}(X)) \subset \text{bd}(X) .$$

So we can find a supporting hyperplane $H = \{x \mid ax = ax^0\}$, with $a \neq 0$. Since $x^2 \in \text{int}(X)$ it is $ax^2 > ax^0$ and since $x^1 \in X$ it is $ax^1 \geqq ax^0$. Now

$$x^1 = \frac{1}{1 - \lambda_0}\, x^0 - \frac{\lambda_0}{1 - \lambda_0}\, x^2$$

and so we have

$$a \left[\frac{1}{1 - \lambda_0}\, x^0 - \frac{\lambda_0}{1 - \lambda_0}\, x^2 \right] \geqq ax^0 ,$$

i.e. $ax^0 \geqq ax^2$, a contradiction.

We can sharpen the above assertion avoiding the assumption $\text{int}(X) \neq \emptyset$. Indeed, from Theorem 2.2.3 we can deduce that a set $X \subset \mathbb{R}^n$ is closed and convex if and only if it coincides with the intersection of all its (closed) supporting halfspaces; then every boundary point of X lies on a supporting hyperplane.

The study of optimality conditions makes often use of the concept of *separation* between two sets. Let S_1 and S_2 be arbitrary sets of $\mathbb{R}^n$; a hyperplane H is said to *separate* S_1 and S_2 if $S_1 \subset H^{\geqq}$ and $S_2 \subset H^{\leqq}$ (or else $S_1 \subset H^{\leqq}$ and $S_2 \subset H^{\geqq}$). The hyperplane H is then called a *separation hyperplane* and sets S_1 and S_2 are said to be *separable* if at least a separation hyperplane exists. H is called a *proper separation hyperplane* if in addition it is $S_1 \cup S_2 \not\subset H$. H is said to be a *strict separation hyperplane* of S_1 and S_2 if it is a separation hyperplane and in addition, at least one of the following relations hold:

$$S_1 \cap H = \emptyset ; \qquad S_2 \cap H = \emptyset .$$

H is said to be a *strong separation hyperplane* for S_1 and S_2 if there exists a ball B of radius $\varepsilon > 0$ such that the sets

$$S_1 + B \; ; \qquad S_2 + B$$

are separable.

It appears that strong separability implies strict separability and that strict separability implies proper separability, but not conversely. If S_1 and S_2 are open, then separability implies strict separability. Moreover, the empty intersection of S_1 and S_2 is neither necessary nor sufficient for separability and even for proper separability. The empty intersection of S_1 and S_2 is necessary but not sufficient for strict separability and, much more, for strong separability.

Figure 1 shows different types of separation of two sets in $\mathbb{R}^2$.

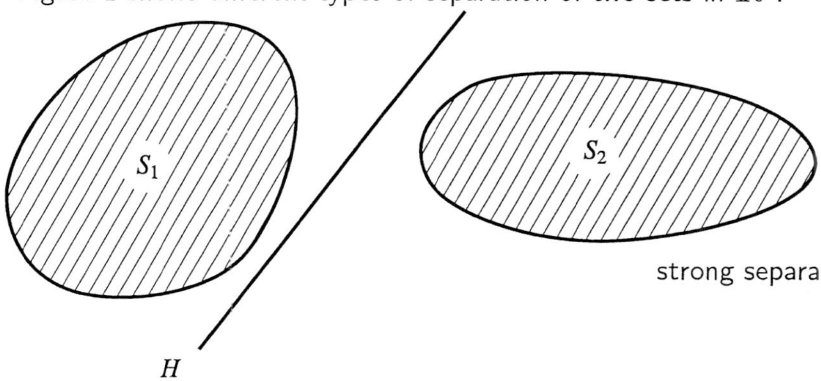

strong separation

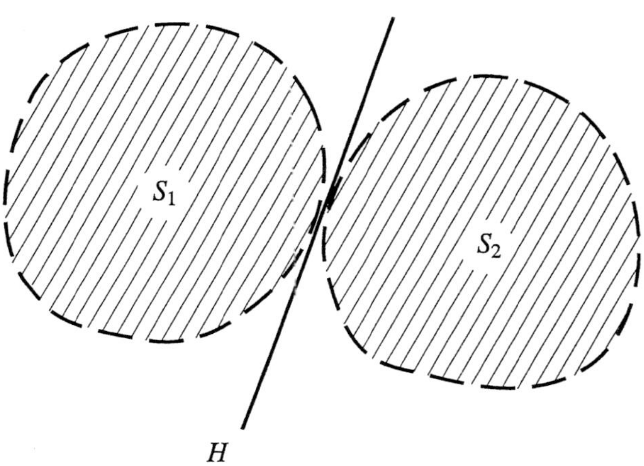

strict separation

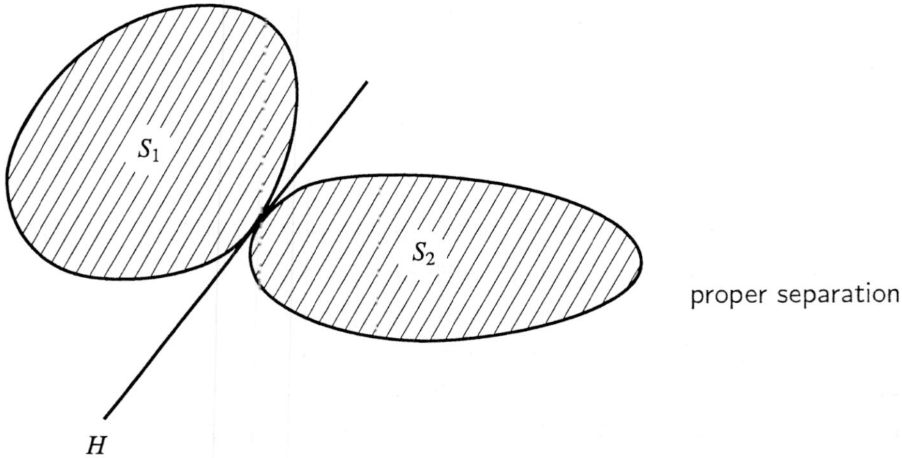

Figure 1.

From the above definitions it is quite immediate to prove that the same definitions are equivalent to the following ones:

- Separability between S_1 and S_2: there exists a hyperplane $H = \{x|ax = \alpha\}$ such that

  $$\sup_{s \in S_1} \{ax\} \leqq \inf_{x \in S_2} \{ax\} .$$

- Proper separability between S_1 and S_2: there exists a hyperplane $H = \{x \mid ax = \alpha\}$ such that

  $$\sup_{x \in S_1} \{ax\} \leqq \inf_{x \in S_2} \{ax\}$$

  and

  $$\inf_{x \in S_1} \{ax\} < \sup_{x \in S_2} \{ax\} .$$

- Strict separability between S_1 and S_2: there exists a hyperplane $H = \{x \mid ax = \alpha\}$ such that

  $ax < \alpha$

– Strong separability between S_1 and S_2: there exists a hyperplane $H = \{x \mid ax = \alpha\}$ such that

$$\sup_{x \in S_1} \{ax\} < \inf_{x \in S_2} \{ax\} .$$

The notion of separability is of utmost importance in the case of convex sets; indeed we have the following fundamental results.

Theorem 2.2.5 (Theorem of separation). If X, Y are nonempty convex sets of $\mathbb{R}^n$ and $(X \cap Y) = \emptyset$, then there exists a hyperplane that separates X and Y, i.e. there exists a vector $a \in \mathbb{R}^n$, $\|a\| = 1$, such that for every $x \in X$ and for every $y \in Y$, $ax \geqq ay$, i.e.

$$\inf_{x \in X} \{ax\} \geqq \sup_{y \in Y} \{ay\} .$$

Proof. By assumption the set $X - Y$ is convex and the point $x^0 = 0 \notin (X - Y)$. We have two cases:

a) $0 \in \overline{X - y}$; then 0 is a boundary point of $X - Y$ and we can use Theorem 2.2.4.

b) $0 \notin \overline{X - Y}$; then we can use Theorem 2.2.3.

In both cases, the result is that there exists a vector $a \neq 0$ such that $a(x - y) \geqq a \cdot 0 = 0$, $\forall\, x \in X$, $\forall\, y \in Y$. □

The same result holds with the assumption $0 \notin \mathrm{int}(X - Y)$ or also with the assumption $0 \notin \mathrm{relint}(X - Y)$ instead of $(X \cap Y) = \emptyset$.

Theorem 2.2.6 (Theorem of strong separation). If X, Y are nonempty closed convex sets in $\mathbb{R}^n$, with $X \cap Y = \emptyset$ and at least one of them is bounded (i.e. is a convex compact set), then there exists a hyperplane that strongly separates X and Y, i.e. there exists a vector $a \in \mathbb{R}^n$, $\|a\| = 1$, such that

$$\inf_{x \in X} \{ax\} > \sup_{y \in Y} \{ay\} .$$

Proof. By assumption the set $X - Y$ is convex; moreover it is closed: indeed consider the sequence $\{x^k - y^k\} \to z$, with $\{x^k\} \subset X$, $\{y^k\} \subset Y$; we prove that $z = x^0 - y^0$, with $x^0 \in X$, $y^0 \in Y$. If X is compact, there exists a convergent subsequence $x^{k'} \to x^0 \in X$; then also the subsequence $\{y^{k'}\}$ is convergent and it is $y^{k'} = x^{k'} - (x^{k'} - y^{k'}) \to y^0 \in Y$ as Y is closed. Moreover, it is $0 \notin X - Y$. Apply then Theorem 2.2.3: there exists a nonzero vector $a \in I\!R^n$ such that $\inf_{z \in X-Y} \{az\} > a \cdot 0 = 0$. This implies that $\inf_{x \in X} \{ax\} > \sup_{y \in Y} \{ay\}$. □

Corollary 2.2.1. If the nonempty convex set $X \subset I\!R^n$ does not intersect the nonnegative orthant $I\!R^n_+ = \{x \mid x \in I\!R^n,\ x \geqq 0\}$, then there exists a semipositive vector $a \geq 0$ (i.e. $a \geqq 0$, $a \neq 0$) such that $ax \leqq 0$, $\forall x \in X$, i.e. there exists a hyperplane separating X and $I\!R^n_+$.

Proof. Apply Theorem 2.2.5 to the sets X and $Y = I\!R^n_+$. □

The following theorem gives necessary and sufficient conditions for proper separation between two sets:

Theorem 2.2.7. The nonempty convex sets of $I\!R^n$, X and Y, are properly separable if and only if $\mathrm{relint}(X) \cap \mathrm{relint}(Y) = \emptyset$.

Proof. See Rockafeller (1970), Theorem 11.3. □

We note that the notion of separability holds also in nonfinite dimensional spaces; consider, e.g., the following version of the *Hahn-Banach separation theorem*: let X be a Banach space with disjoint nonempty convex subsets A and B; if A is open and X^* is the space of all linear continuous functionals on X (dual space), then there is an $x^* \in X^*$, $x^* \neq 0$, such that

$$\sup_{a \in A} \{x^*(a)\} \leqq \inf_{b \in B} \{x^*(b)\} .$$

See, e.g., Schaefer (1966).

From the above theorems we can deduce several important results about the intersection of convex sets: Theorems 2.2.8 and 2.2.9 are due to Berge, whereas Theorem 2.2.10 is due to Helly (see, e.g., Berge and Ghouila-Houri (1965)).

Theorem 2.2.8. Let $X_1, X_2, ..., X_m$ $(m \geqq 2)$ be nonempty closed convex sets of $I\!R^n$, with $X = \bigcup_{i=1}^m X_i$ a convex set. If the intersection of any $m-1$ of the above sets is nonempty, then $\bigcap_{i=1}^m X_i$ is nonempty.

Proof. We can, without loss of generality, suppose that the sets $X_1, X_2, ...,$ X_m are compact; otherwise it suffices to take the points $a^1, a^2, ..., a^m$ with $a^i \in \bigcap_{i \neq j} X_j$ and to write

$A = \{a^1, a^2, ..., a^m\}$ and $X_i' = X_i \cap \text{conv}(A)\ , \qquad i = 1, 2, ..., m\ .$

We shall therefore prove the theorem for compact convex sets X_i, by induction over m.

1) Let $m = 2$; let X_1 and X_2 be convex compact sets, with $X_1 \neq \emptyset$, $X_2 \neq \emptyset$ and $X_1 \cup X_2$ convex. If $X_1 \cap X_2 = \emptyset$, there would exist a hyperplane that strongly separates them. There would then be points of $X_1 \cup X_2$ on both sides of the hyperplane. Hence there would be points on this hyperplane, since $X_1 \cup X_2$ is convex. This cannot be since the hyperplane must not intersect either X_1 or X_2. Therefore the property is true for $m = 2$.

2) Let us suppose that the theorem is true for $m = p$ $(p \geqq 2)$ and let us prove that it is true also for $m = p + 1$.
Let $X_1, X_2, ..., X_{p+1}$ be convex and compact and let $\bigcup_{i=1}^{p+1} X_i$ be convex. Let every intersection of p of the above sets be nonempty. Put $X = \bigcap_{i=1}^p X_i$; by assumption $X \neq \emptyset$ and $X_{p+1} \neq \emptyset$: if the two sets are disjoint there exists a hyperplane H that strongly separates them. Apply the induction hypothesis to the sets $X_i' = X_i \cap H$, $i = 1, 2, ..., p$, convex, closed, with convex union given by

$$\bigcup_{i=1}^p X_i' = H \cap \Big(\bigcup_{i=1}^p X_i\Big) = H \cap \Big(\bigcup_{i=1}^{p+1} X_i\Big)\ ,$$

being $H \cap X_{p+1} = \emptyset$.
By assumption, the intersection of any $p-1$ of the sets $X_1, X_2, ..., X_p$

contains X and intersects X_{p+1} and consequently intersects H. It follows that any $p-1$ of the sets $X'_1, X'_2, ..., X'_p$ has a nonempty intersection and hence from the induction hypothesis their intersection is nonempty. Thus $\bigcap_{i=1}^{p} X'_i = X \cap H \neq \emptyset$, which is absurd. □

Theorem 2.2.9. Let X be a nonempty convex set and $X_1, X_2, ..., X_m$ $(m \geqq 1)$ be nonempty closed convex sets in $\mathbb{R}^n$. If $A_i = X \cap (\bigcap_{j \neq i} X_j) \neq \emptyset$, $i = 1, 2, ..., m$, but $X \cap (\bigcap_{i=1}^{m} X_i) = \emptyset$, then $X \not\subset \bigcup_{i=1}^{m} X_i$.

Proof. The result is trivial for $m = 1$, as the intersection of any $m-1$ of these convex sets is identical with the intersection of an empty family of sets, namely with $\mathbb{R}^n$. If, then

$$X \cap \mathbb{R}^n = X \neq \emptyset$$

and if $X \cap X_1 = \emptyset$, clearly we cannot have $X \subset X_1$.
In the case where $m \geqq 2$ we have $\operatorname{conv}(A) \subset X$. The sets $X'_i = \operatorname{conv}(A) \cap X_i$, $i = 1, 2, ..., m$ are closed and convex; every $m-1$ of them have a non-empty intersection, namely

$$a^i \in \bigcap_{j \neq i} X'_j = \operatorname{conv}(A) \cap \left(\bigcap_{j \neq i} X_j\right),$$

but their intersection is empty, otherwise $\operatorname{conv}(A)$ and hence X intersect $\bigcap_{i=1}^{m} X_i$. Their union $\bigcup_{i=1}^{m} X'_i = \operatorname{conv}(A) \cap (\bigcup_{i=1}^{m} X_i)$ is not convex (see Theorem 2.2.8) and therefore $\operatorname{conv}(A) \not\subset \bigcup_{i=1}^{m} X_i$, as otherwise $\bigcup_{i=1}^{m} X'_i = \operatorname{conv}(A)$ and hence would be convex. Therefore there exists $x^0 \in \operatorname{conv}(A) \subset X$, such that $x^0 \notin X_i$, $i = 1, 2, ..., m$. □

The following theorem of Helly is a consequence of the above results.

Theorem 2.2.10 (Helly's theorem). Let $X_1, X_2, ..., X_m$, with $m > n+1$, be convex sets of $\mathbb{R}^n$. If the intersection of any $(n+1)$ of these sets is nonempty, the intersection $\bigcap_{i=1}^{m} X_i$ is nonempty.

Proof. Let us suppose that the intersection of any p $(n+1 \leqq p < m)$ of the sets $X_1, X_2, ..., X_m$ is nonempty; we shall show that the intersection of any $p+1$ of these, for example $X_1, X_2, ..., X_{p+1}$, is nonempty. Let $a^1, a^2, ..., a^{p+1}$ be vectors such that

$$a^j \in \bigcap_{\substack{k \neq j \\ 1 \leqq k \leqq p+1}} X_k \ , \qquad j = 1, 2, ..., p+1 \ ,$$

and let

$$A_i = \{a^j\} \ , \qquad j \neq i \ , \quad i = 1, 2 ..., p+1 \ .$$

The closed convex sets $\mathrm{conv}(A_i) \subset X_i$ have a convex union, as it is $p+1 > n+1$ (see property v) of the convex hull of a set; we have $\bigcup_{i=1}^{p+1} \mathrm{conv}(A_i) = \mathrm{conv}\,\{a^1, a^2, ..., a^{p+1}\}$) and every p sets of these sets have a nonempty intersection; indeed $a^j \in \bigcap_{i \neq j} \mathrm{conv}(A_i)$; therefore from Theorem 2.2.8 we have $\bigcap_{i=1}^{p+1} \mathrm{conv}(A_i) \neq \emptyset$. But $\mathrm{conv}(A_i) \subset X_i$, $i = 1, 2, ..., p+1$, and hence $\emptyset \neq \bigcap_{i=1}^{p+1} \mathrm{conv}(A_i) \subset \bigcap_{i=1}^{p+1} X_i$. □

We note that, as a consequence of Helly's theorem, it results that the assumptions of Theorem 2.2.9 are consistent only for $m \leqq n$ if $\bigcap_{i=1}^{m} X_i \neq \emptyset$.

2.3. Some Particular Convex Sets. Convex Cones

The notion of extreme point of a convex set plays an important role in mathematical programming, especially in linear programming. Let X be a convex set in $I\!R^n$; each point $x \in X$ for which there do not exist two distinct points $x^1, x^2 \in X$ different from x such that $x \in (x^1, x^2)$ is called an *extreme point* of X, i.e. x cannot be represented as a strict convex combination of two distinct points in X. In other words, if $x = \lambda x^1 + (1-\lambda)\, x^2$; $\lambda \in (0,1)$; $x^1, x^2 \in X \Rightarrow x = x^1 = x^2$.

Clearly the extreme points of a convex set X are those points of X that do not lie on the interior of any line segment connecting any other pair of points of X.

Any convex set $X \subset I\!R^n$ may have no extreme points (for example, the hyperplane $H = \{x \,|\, x \in I\!R^n, \ ax = \alpha\}$, $a \in I\!R^n$, $a \neq 0$, and any open ball $B(x^0, \varepsilon)$ has no extreme points; more generally, every open set has no extreme points), a finite number of extreme points (for example, the set $A = \{x \mid x \in I\!R^n,\ x \geqq 0,\ ex = 1\}$, where $e = [1, 1, ..., 1]$) or an infinite number of extreme points (for example, any closed ball $B(x^0, \varepsilon)$ has an infinite number of extreme points given by $\{x \mid x \in I\!R^n,\ \|x - x^0\| = \varepsilon\}$). An important result concerning extreme points is the following:

Theorem of Krein-Milman. A closed bounded convex set $S \subset I\!R^n$ is the convex hull of its extreme points.
A set in $I\!R^n$ which is given by the intersection of a finite number of closed halfspaces in $I\!R^n$ is called a (*convex*) *polyhedron* or *polyhedral set*. The term "convex" is in fact superfluous as it follows from convexity of the halfspaces that polyhedra are (closed) convex sets.

A polyhedron can be identified by means of its algebraic representation, which, on the ground of the definition, is given by the set $X = \{x \mid x \in I\!R^n,\ Ax \leqq b\}$, where A is a real (m, n) matrix and b is a vector of $I\!R^n$. If a polyhedron is bounded, it is often called a *polytope*.

Beware! Some authors call polyhedron what we have called polytope and vice versa (see, e.g., Mangasarian (1969), Martos (1975)). The definition we have accepted seems more the standard in books and papers specifically concerned with convex analysis and linear programming (see, e.g., Stoer-Witzgall (1970), Rockafeller (1970), Bazaraa-Jarvis and Sherali (1990), Gale (1960)).

As a consequence of the Krein-Milman theorem we have that a nonempty polytope is given by the convex hull of its extreme points, i.e. if $X \subset I\!R^n$ is a polytope and $\hat{x}^1, \hat{x}^2, ..., \hat{x}^p$ are its extreme points (the number of extreme points of a polytope is always finite and nonzero), then

$$X = \left\{x \mid x \in I\!R^n,\ x = \sum_{i=1}^{p} \beta_i \hat{x}^i\, ;\ \sum_{i=1}^{p} \beta_i = 1,\ \beta_i \geqq 0,\ i = 1, 2, ..., p\right\}.$$

Let $x^0, x^1, ..., x^m$ be $m + 1$ distinct vectors of $I\!R^n$, with $m \leqq n$. If the vectors $x^1 - x^0, ..., x^m - x^0$ are linearly independent, then the convex hull of $x^0, x^1, ..., x^m$ is called an m-*simplex* in $I\!R^n$ with extreme points (called also vertices) $x^0, x^1, ..., x^m$:

$$S = \left\{ x \mid x \in \mathbb{R}^n, \ x = \sum_{i=0}^{m} \lambda_i x^i, \ \lambda_i \geqq 0, \ i = 0, 1, ..., m, \ \sum_{i=0}^{m} \lambda_i = 1 \right\} .$$

The numbers $\lambda_0, ..., \lambda_m$ are called the *barycentric coordinates* of X; note that, as in $\mathbb{R}^n$ the maximum number of linearly independent vectors is n, then there could be no simplex in $\mathbb{R}^n$ with more than $n+1$ vertices.

A 0-simplex is a point; a 1-simplex is a closed line segment; a 2-simplex is a triangle; a 3-simplex is a tetrahedron.

We now briefly discuss the notion of *cone* and especially *convex cone*; indeed convex cones are of utmost importance in several questions of optimization theory.

Definition 2.3.1. A nonempty set $K \subset \mathbb{R}^n$ is called a *cone with vertex at* $\bar{x} \in \mathbb{R}^n$ if $\bar{x} + \alpha(x - \bar{x}) \in K$, $\forall x \in K$, $\forall \alpha > 0$. In the special case where $\bar{x} = 0$, the cone with vertex at zero is called simply *cone* (i.e. it is $\alpha x \in K$, $\forall x \in K$, $\forall \alpha > 0$). Such a set is a union of halflines emanating from the origin.

From now on, unless otherwise specified, the vertex of the cones considered is the origin. From the previous definition it appears that the vertex of a cone may or may not belong to the cone (however, it belongs to the closure of the cone). Many authors, however, do include the vertex in the cone by letting $\alpha \geqq 0$ in the definition. Note that our definition implies that the interior of a cone is again a cone. Anyhow, we shall specify the cases where the origin is required to belong to the cone (when it is not clear from the context). Again it may be worth nothing that a cone K may or may not be convex and that a cone K may be open, closed or neither open nor closed.

If in addition the cone K is convex, then it is called a *convex cone*. A cone K is said to be *pointed* if, whenever $x \neq 0$ is in this cone, then $-x$ is not in the cone, i.e., in case K contains the origin, if $K \cap (-K) = \{0\}$. The following result is often used to characterize convex cones and is a rather immediate consequence of the above definition.

Theorem 2.3.1. $K \subset \mathbb{R}^n$ is a convex cone if and only if:

a) $\alpha x \in K$, $\forall x \in K$, $\forall \alpha > 0$;

b) $x^1 + x^2 \in K$, $\forall x^1, x^2 \in K$.

Proof. Suppose that K is a convex cone, then $x^1, x^2 \in K$ implies that $\lambda x^1 + (1-\lambda)\, x^2 \in K$, $\forall \lambda \in (0,1)$. Letting $\lambda = \frac{1}{2}$, we get $\frac{1}{2}\, x^1 + \frac{1}{2}\, x^2 \in K$ and hence $x^1 + x^2 \in K$. Conversely, assume a) and b); if $x^1, x^2 \in K$, then from a) we get $\lambda x^1 \in K$ and $(1-\lambda)\, x^2 \in K$ for each $\lambda \in (0,1)$. From b) it follows that also $\lambda x^1 + (1-\lambda)\, x^2 \in K$ and hence K is a convex cone.□
Some examples of convex cones are:

i) Hyperplanes through a point x^0,i.e. $H = \{x \mid a(x - x^0)\} = 0\}$, where $a \in \mathbb{R}^n$, $a \neq 0$. Here the cone has vertex at x^0.

ii) Closed halfspaces, for example $H^{\geqq} = \{x \mid a(x - x^0) \geqq 0\}$. Here again the cone has vertex at x^0.

iii) The set $C = \{x \mid Ax \geqq 0\}$, where A is a real m, n matrix.

It results that if K_1, K_2 are convex cones, then $K_1 \cap K_2$ and $K_1 + K_2$ are again convex cones; if K is a cone, $\operatorname{conv}(K)$ is a convex cone. If K_1 and K_2 are convex cones, then $K_1 + K_2 = \operatorname{conv}(K_1 \cup K_2)$. We note that, given a set $S \subset \mathbb{R}^n$, we can associate to the same a cone, called a *cone generated* (or *spanned*) *by* S or *conical hull of* S or *projection cone of* S, and defined as:

$$K(S) \text{ or } \operatorname{cone}(S) = \{x \mid x = \lambda y,\ \lambda > 0,\ y \in S\}\ .$$

It is easy to see that $\operatorname{cone}(S)$ is given by the intersection of all cones containing S.

The *convex conical* (or *positive*) *hull* of S or *convex cone generated* (or *spanned*) by S is defined as:

$$C(S) = \left\{x \mid x = \sum_{i=1}^{k} \lambda_i x^i,\ k \in \mathbb{N},\ x^i \in S,\ \lambda_i > 0,\ i = 1, ..., k\right\}.$$

Evidently it is $\operatorname{cone}(S) \subset C(S)$ and $\{\{0\} \cup C(S)\} = \{\sum \lambda_i x^i \mid x^i \in S,\ \lambda_i \geqq 0\}$.

It is easy to see that $C(S)$ is given by the intersection of all convex cones containing S (note the parallelism with the definition of convex hull of a set).

Theorem 2.3.2.

i) Given $S \subset I\!R^n$, it results

$$C(S) = \text{cone}(\text{conv}(S)) = \text{conv}(\text{cone}(S)) \ .$$

Therefore, if S is a convex set, it is

$$C(S) = \text{cone}(S) \ .$$

If S is a cone, it is

$$C(S) = \text{conv}(S) \ .$$

ii) If $S \subset I\!R^n$ is a convex compact set, with $0 \notin S$, then $\{0\} \cup \text{cone}(S)$ is closed.

Proof.

i) By definition it is $\text{conv}(S) \subset C(S)$ and $\text{cone}(\text{conv}(S)) \subset C(S)$, as $C(S)$ is a cone. Analogously, we have $\text{cone}(S) \subset C(S)$ and hence $\text{conv}(\text{cone}(S)) \subset C(S)$, as $C(S)$ is a convex set. For the reverse inclusions, let $x \in C(S)$; then $x = \sum_{i=1}^{k} \lambda_i x^i$, with $\lambda_i > 0$ and $x^i \in S$. We set $\mu = \sum \lambda_i > 0$ and $\mu_i = \lambda_i \,/\, \sum \lambda_j = (\lambda_i \,/\, \mu) > 0$. Of course we have $\sum \mu_i = 1$ and we get:

$$x = \mu(\sum \mu_i x^i) \in \text{cone}(\text{conv}(S)) \ ;$$

$$x = \sum \mu_i(\mu x^i) = \text{conv}(\text{cone}(S)) \ .$$

The other assertions of i) are straightforward.

ii) Let $\{y^k\} \subset \text{cone}(S)$ be a sequence with $y^k \to y^0$; we have to show that $y^0 \in \text{cone}(S)$. By definition there exist sequences $\{\lambda_k\} \subset I\!R_+$ and $\{x^k\} \subset S$ with $y^k = \lambda_k x^k$, $\forall k$. Because of the compactness of S and $0 \notin S$, it is $\|x^k\| \geqq \alpha > 0$, $\forall k$. Therefore the sequence $\{\lambda_k\}$ is bounded. Otherwise, in case of $\lambda_k \to +\infty$, we would get $\|y^k\| = \lambda_k \|x^k\| \geqq \lambda_k \cdot \alpha \to +\infty$ in contradiction to the convergence of $\{y^k\}$. Again because of the compactness of S and the boundedness of $\{\lambda_k\}$ we can assume (without loss of generality) that both sequences are convergent, i.e. $x^k \to x^0 \in S$, $\lambda_k \to \lambda_0 \geqq 0$. Now, of course it is

$$y^k = \lambda_k x^k \to \lambda_0 x^0 \in \{0\} \cup \text{cone}(S) . \qquad \square$$

Summing up: a set $X \subset I\!R^n$ is

(1) a *linear subspace*,
(2) an *affine set*,
(3) a *convex set*,
(4) a *convex cone*,

if for any $x^1, x^2 \in X$, also $\lambda_1 x^1 + \lambda_2 x^2 \in X$, for every

(1) $\lambda_1, \lambda_2 \in I\!R$;
(2) $\lambda_1, \lambda_2 \in I\!R$, $\lambda_1 + \lambda_2 = 1$;
(3) $\lambda_1, \lambda_2 \in I\!R_+$, $\lambda_1 + \lambda_2 = 1$;
(4) $\lambda_1, \lambda_2 \in \overset{o}{I\!R}_+$.

Moreover $X \subset I\!R^n$ is respectively (1), (2), (3), (4) if for any $x^1, ..., x^m \in X$ also $\sum_{i=1}^m \lambda_i x^i \in X$ for every

(1) $\lambda_i \in I\!R$;
(2) $\lambda_i \in I\!R$, $\sum_i \lambda_i = 1$;
(3) $\lambda_i \in I\!R_+$, $\sum_i \lambda_i = 1$;
(4) $\lambda_i \in \overset{o}{I\!R}_+$.

The *linear hull of* X (1'), denoted by $\text{span}(X)$, the *affine hull of* X (2'), denoted by $\text{aff}(X)$, the *convex hull of* X (3'), denoted by $\text{conv}(X)$, the *convex conical hull of* X (4'), denoted by $C(S)$, are respectively:

– the smallest linear subspace containing X (1');
– the smallest affine set containing X (2');
– the smallest convex set containing X (3');
– the smallest convex cone containing X (4').

Moreover,

$$\operatorname{span}(X) = \left\{ x \mid x = \sum_i \lambda_i x^i, \ \lambda_i \in I\!R, \ x^i \in X \right\} ;$$

$$\operatorname{aff}(X) = \left\{ x \mid x = \sum_i \lambda_i x^i, \ \lambda_i \in I\!R, \ \sum_i \lambda_i = 1, \ x^i \in X \right\} ;$$

$$\operatorname{conv}(X) = \left\{ x \mid x = \sum_i \lambda_i x^i, \ \lambda_i \geqq 0, \ \sum_i \lambda_i = 1, \ x^i \in X \right\} ;$$

$$C(X) = \left\{ x \mid x = \sum_i \lambda_i x^i, \ \lambda_i > 0, \ x^i \in X \right\} .$$

Now let S be an arbitrary set in $I\!R^n$; the (negative) *polar cone* of S, denoted by S^* is the set $S^* = \{y \mid y \in I\!R^n, \ yx \leqq 0, \ \forall x \in S\}$. If S is empty we will interpret S^* as the whole space $I\!R^n$.
It is immediate to note that the polar cone of a set contains the origin. Some of the following results are a direct consequence of the definition of polar cone; see Bazaraa and Shetty (1976), Ben-Israel (1969), Fenchel (1953) for the less obvious proofs. Let S, S_1 and S_2 be nonempty sets in $I\!R^n$, then it holds:

i) S^* is a closed convex cone with vertex at the origin; therefore this cone is called the *polar cone* (sometimes also the *dual cone*) of S.

ii) $S^* = (\bar{S})^* = (\operatorname{conv}(S))^* = (\operatorname{cl}(\operatorname{conv}(S)))^* = (\operatorname{cone}(S))^* =$
$= (\operatorname{cl}(\operatorname{cone}(S)))^* = (C(S))^* = (\operatorname{cl}(C(S)))^*$.

iii) $S_1 \subset S_2$ implies $S_2^* \subset S_1^*$.

iv) $S \subset S^{**}$, where $S^{**} = (S^*)^*$.

v) $S^* = S^{***}$, where $S^{***} = (S^{**})^*$.

vi) $S_1^* \cup S_2^* \subset (S_1 \cap S_2)^*$.

vii) $S_1^* \cap S_2^* = (S_1 \cup S_2)^*$.

viii) $S_1^* \cap S_2^* \subset (S_1 + S_2)^*$; $(S_1 + S_2)^* \subset S_1^* \cap S_2^*$ if $0 \in \bar{S}_1 \cap \bar{S}_2$.

Therefore if $0 \in \bar{S}_1 \cap \bar{S}_2$ it is $(S_1 + S_2)^* = S_1^* \cap S_2^*$. Indeed, if $0 \in \bar{S}_1 \cap \bar{S}_2$ then $C(\bar{S}_1 \cup \bar{S}_2) = C(\bar{S}_1 + \bar{S}_2)$. Now

$$(C(\bar{S}_1 \cup \bar{S}_2))^* = (C(\bar{S}_1 + \bar{S}_2))^*$$

and from

$$(C(\bar{S}_1 \cup \bar{S}_2))^* = (\bar{S}_1 \cup \bar{S}_2)^* = (\overline{S_1 \cup S_2})^* = (S_1 \cup S_2)^*$$

and from

$$(C(\bar{S}_1 + \bar{S}_2))^* = (\bar{S}_1 + \bar{S}_2)^* = (\bar{S}_1 + \bar{S}_2)^* = (S_1 + S_2)^*$$

we get

$$(S_1 + S_2)^* = (S_1 \cup S_2)^* = S_1^* \cap S_2^* .$$

ix) $S_1^* + S_2^* = \mathrm{cone}(S_1^* \cup S_2^*) = \mathrm{conv}(S_1^* \cup S_2^*) \subset (S_1 \cap S_2)^*$.

E.g. in $\mathbb{R}^2$ the polar of a convex cone K is seen to consist of all vectors making a non acute angle with all vectors of the cone K (see Figure 2).

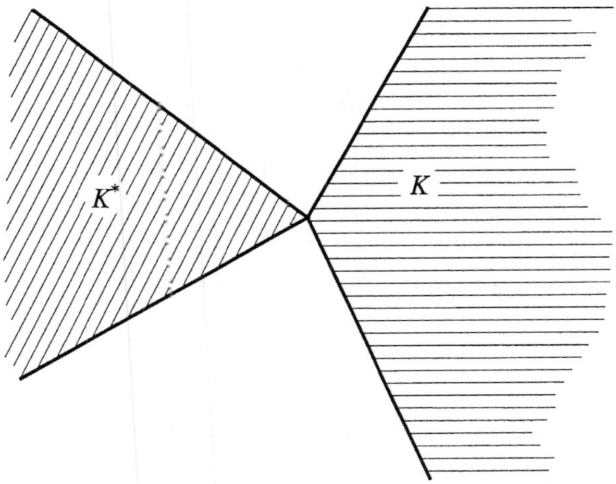

Figure 2.

We will now focus further attention on convex cones and will prove an important result for closed convex cones. This result is also known as the *polarity property* or *duality property* for closed convex cones.

Theorem 2.3.3. Let K be a nonempty convex cone in $\mathbb{R}^n$; then $K^{**} = \bar{K}$.

Proof. Let $x \in \bar{K}$; then $ax \leqq 0 \ \forall a \in K^*$ (note that $K^* = (\bar{K})^*$) and hence $x \in K^{**}$. To prove that $K^{**} \subset \bar{K}$, let $x \in K^{**}$ and suppose that $x \notin \bar{K}$. By Theorem 2.2.3 there exists a nonzero vector a such that $ay \leqq \alpha$, $\forall y \in \bar{K}$ and $ax > \alpha$ for some α. But since $y = 0 \in \bar{K}$, then $\alpha \geqq 0$ and so $ax > 0$. We will show that this is impossible by showing that $a \in K^*$ (note that $x \in K^{**}$ by hypothesis).
Suppose by contradiction that $a \notin K^*$; then there exists a vector $y \in K$ with $ay > 0$. But then $a \cdot \lambda y$ can be made arbitrarily large by choosing λ sufficiently large, which violates the fact that $a \cdot \lambda y \leqq \alpha$ for each $\lambda > 0$. This completes the proof. □

Corollary 2.3.1. The set $K \subset \mathbb{R}^n$, $K \neq \emptyset$, is a closed convex cone if and only if $K = K^{**}$ (*polarity property*).

As a consequence of the previous theorems and properties we get the following other useful results:

a) If $K \subset \mathbb{R}^n$, $K \neq \emptyset$, is any set, then K^{**} is the closed convex conical hull of K (K^{**} is also called *bipolar cone* of A).

b) If K_1 and K_2 are nonempty *closed* convex cones in $\mathbb{R}^n$ we have the following *modularity properties*:

$$K_1^* \cap K_2^* = (K_1 + K_2)^*$$

$$(K_1 \cap K_2)^* = \overline{K_1^* + K_2^*}\ .$$

The first one is a direct consequence of the previous property vi). To prove the second one, note that it results

$$(K_1 \cap K_2)^* = (K_1^{**} \cap K_2^{**})^* = (K_1^* \cup K_2^*)^{**} =$$

$$= \mathrm{cl}(\mathrm{conv}(K_1^* \cup K_2^*)) = \overline{K_1^* + K_2^*} \ .$$

c) If K_1, K_2 are closed convex cones with $(\mathrm{int}(K_1) \cap K_2) \neq \emptyset$, then

$$K_1^* + K_2^* = (K_1 \cap K_2)^* \ .$$

Indeed, let $a \in (K_1 \cap K_2)^*$ and

$$X = \{(x, \alpha) \in \mathbb{R}^n \times \mathbb{R} : | \, x \in K_1, \ \alpha \leqq ax\} \ ;$$

$$Y = \{(y, \beta) \in \mathbb{R}^n \times \mathbb{R} \, | \, y \in K_2, \ \beta \geqq 0\} \ .$$

Then $\mathrm{int}(X) \cap Y = \emptyset$ and by the separation theorem we can therefore find a vector $(u, \xi) \in \mathbb{R}^n \times \mathbb{R}$, $(u, \xi) \neq (0, 0)$ such that

$$ux + \alpha\xi \leqq uy + \beta\xi \ , \qquad \forall \, (x, \alpha) \in X \ , \ \ \forall \, (y, \beta) \in Y \ .$$

Obviously, it is $\xi \geqq 0$; assuming $\xi = 0$ we get $ux \leqq uy$, $\forall \, x \in K_1$, $\forall \, y \in K_2$, which is impossible since $(\mathrm{int}\, K_1) \cap K_2 \neq \emptyset$ and $u \neq 0$. Without loss of generality we can thus set $\xi = 1$ and get (set $\alpha = ax$, $\beta = 0$) $ux + ax \leqq uy$, $\forall \, x \in K_1$, $\forall \, y \in K_2$, i.e. $u + a \in K_1^*$, $-u \in K_2^*$ and therefore $a = (u + a) + (-u) \in K_1^* + K_2^*$.

d) More generally, if K_i, $i = 1, ..., m$, are nonempty closed convex cones,

$$\Big(\bigcap_{i=1}^m K_i\Big)^* = \overline{\sum_{i=1}^m K_i^*}$$

and if

$$\bigcap_{i \neq j} \mathrm{int}(K_i) \cap K_j \neq \emptyset$$

then

$$\Big(\bigcap_{i=1}^m K_i\Big)^* = \sum_{i=1}^m K_i^* \ .$$

We shall now discuss a special class of convex cones, namely the (convex) *polyhedral cones* or *finite cones*. We have seen that the intersection of a finite number of closed halfspaces is called a (convex) polyhedron. Now a cone which is a polyhedron is called a *polyhedral* or *finite cone.* A polyhedral cone is thus given by the intersection of a finite number of halfspaces whose hyperplanes pass through the origin.

In other words, the solution set of a homogeneous system of inequalities represents a polyhedal cone $C = \{x \mid Ax \leqq 0\}$, where A is an m,n real matrix. On the other hand, we have the following converse result.

Theorem 2.3.4. Every polyhedral cone is the solution set of a homogeneous system of inequalities.

Proof. Assume that $C = \{x \mid Ax \leqq b\}$ is a polyhedral cone; since every polyhedral cone contains the origin we must have $b \geqq 0$. Homogeneity implies $Ax \leqq \alpha b$ for every $x \in C$ and $\alpha > 0$. Hence $Ax \leqq 0$ for all $x \in C$. □

From this theorem it follows that every polyhedral cone is convex and closed (therefore it contains the origin); moreover, it can be proved the following fundamental result of Weyl (see Gale (1951)): every convex cone, generated by a *finite* number of points, is a polyhedral cone and vice versa, i.e. the polyhedral cone C can be described by the set

$$C = \left\{x \mid x = \sum_{i=1}^{k} \lambda_i x^i,\ \lambda_i \geqq 0,\ i = 1, 2, ..., k,\ k \in \mathbb{N}\right\} .$$

Therefore the polyhedral cone C is the convex cone generated by a finite set $X = \{x^1, x^2, ..., x^k\}$. For instance $\mathbb{R}^n_+$ is the convex polyhedral cone generated by $(e^1, e^2, ..., e^n)$, where e^i is the i-th unit vector of $\mathbb{R}^n$. Stated differently, a polyhedral convex cone is given by the set

$$C = \{x \mid x = By,\ y \geqq 0\} ,$$

where B is a given real matrix. Polyhedral cones have been extensively treated, e.g. by Gale (1951, 1960); for the proof of the theorem of Weyl, see also Stoer and Witzgall (1970). See the same authors also for the following interesting result: the solution set of a linear inequality system $Ax \leqq b$ (i.e. a polyhedron) is given by the sum of a polytope and a polyhedral cone (this result appears originally in Goldman (1956)).

The following results hold with respect to polyhedral cones (recall Corollary 2.3.1):

i) If K is a polyhedral cone, then $K = K^{**}$.

ii) The set $K = \{x \mid Ax = 0,\ x \geqq 0\}$ where A is a given real matrix, is a polyhedral cone.

iii) If K_1 and K_2 are polyhedral cones, then:

$K_1 + K_2$ is a polyhedral cone;
K_1^* and K_2^* are polyhedral cones;
$K_1 \cap K_2$ is a polyhedral cone;
$(K_1 + K_2)^* = K_1^* \cap K_2^*$;
$(K_1 \cap K_2)^* = K_1^* - K_2^*$ and hence
$K_1^* + K_2^* = \overline{K_1^* + K_2^*}$.

We have mentioned the notion of cone and convex cone generated by an arbitrary set $S \subset \mathbb{R}^n$ ("conical hull" and "convex conical hull" of that set). If we make a suitable translation of the set S we obtain the following two cones.

I) Let $S \subset \mathbb{R}^n$ and $x^0 \in \mathbb{R}^n$; the *(projection) cone of* S *at* x^0, denoted $K(S, x^0)$, is the cone generated by $S - x^0$, i.e. the set

$$\operatorname{cone}(S - x^0) = \{z \mid z = \lambda(x - x^0),\ x \in S,\ \lambda > 0\}\ .$$

$K(S, x^0)$ is therefore given by the intersection of all cones containing the set $S - x^0$; moreover, it is clear that, changing the point x^0, we can define different cones of this type. From the above definition it is also clear that if x^0 is an interior point of S, then $K(S, x^0)$ is the whole space $\mathbb{R}^n$. This follows from the fact that $S - x^0$ will have 0 as an interior point and any cone with vertex zero containing $S - x^0$ is $\mathbb{R}^n$.

II) Let $S \subset I\!R^n$ and $x^0 \in I\!R^n$; the *convex cone of* S *at* x^0, denoted by $C(S, x^0)$ is the convex cone generated by $S - x^0$, i.e. the set we have denoted also by $C(S - x^0)$. In other words $C(S, x^0)$ is given by the intersection of all convex cones containing the set $S - x^0$.
From Theorem 2.3.2 we have that if S is convex, then $C(S, x^0) = K(S, x^0)$.

A vector $y \in I\!R^n$ is said to be *normal* to a convex set S at a point $x^0 \in S$, when $\forall x \in S$ it is $y \cdot (x - x^0) \leqq 0$. The set of normal vectors to the convex set S at a point x^0 is said to be a *normal cone* to S at $x^0 \in S$ and is denoted by $N(S, x^0)$. Therefore $N(S, x^0) = \{y \in I\!R^n \mid y \cdot (x - x^0) \leqq 0,\ \forall x \in S\}$. The reader can verify easily that this cone is always convex and closed; moreover, from the definition it is $N(S, x^0) = (S - x^0)^* = (K(S, x^0))^* = (C(S, x^0))^*$. Other cones associated with a set will be discussed in Chapter III; these cones are particularly important in obtaining general optimality conditions for smooth and nonsmooth optimization problems.

2.4. Theorems of the Alternative for Linear Systems

In this section we shall be concerned with a series of theorems related to the occurrence of one of two mutually exclusive events: the existence of solutions of two systems of linear relations. These theorems, known as *theorems of the alternative* (for linear systems), are numerous, but we shall prove that all known theorems of this kind can be deduced from a unique generator: the so-called theorem (or lemma) of Farkas-Minkowski.

The general form of a theorem of the alternative (in the present case, for linear systems) is the following one: Given two linear systems (of inequalities and/or equalities) S and S^* (S^* may be called "the dual of S"), it results that S admits solution if and only if S^* is impossible. In other words: either S admits solution or S^* admits solution, but never both systems admit or do not admit solution.

If we denote by $\exists S$ ($\exists S^*$) the existence of solutions of S (of S^*) and by $\nexists S$ ($\nexists S^*$) the impossibility of S (of S^*), a typical theorem of the alternative can thus be stated as follows:

$\exists S \Leftrightarrow \nexists S^*$

or, equivalently,

$\nexists S \Leftrightarrow \exists S^*$.

It follows that a typical proof of such a theorem is:

$\exists S \Rightarrow \nexists S^*$ and $\nexists S \Rightarrow \exists S^*$;

or, equivalently,

$\exists S^* \Rightarrow \nexists S$ and $\nexists S^* \Rightarrow \exists S$.

<u>Theorem 2.4.1</u> (Theorem of Farkas-Minkowski). For each given matrix A of order (m, n) and each given vector $b \in \mathbb{R}^m$, either the system

$$S_1 : Ax = b \; , \qquad x \geqq 0$$

has a solution $x \in \mathbb{R}^n$, or the system

$$S_1^* : yA \geqq 0 \; , \qquad yb < 0$$

has a solution $y \in \mathbb{R}^m$, but never both.

<u>Proof</u>. First of all recall that the convex polyhedral cone $Z = \{z \mid z \in \mathbb{R}^m, \; z = Ax, \; x \geqq 0\}$ is a closed set and then observe that S_1 and S_1^* cannot admit both solution, as in this case it would be $0 \leqq yAx = yb < 0$, which is absurd. It remains therefore to prove that if S_1 does not admit solution, i.e. if $b \notin Z$ (therefore $b \in \text{ext}(Z)$, since Z is closed), then S_1^* admits solution. But this follows at once from Theorem 2.2.3: there exists a vector $y \neq 0$ such that $yb < yz$, $\forall z \in Z$, i.e., equivalently, $yb < yAx$, $\forall x \geqq 0$. From this relation it results $yb < 0$ (for $x = 0$) and also $yA \geqq 0$, as if $(yA)_i < 0$ for some i, we could choose x_i sufficiently large, in order to contradict the inequality obtained. □

System S_1^* may be equivalently written as

$$S_1^* : yA \leqq 0 \; , \qquad yb > 0 \; .$$

Moreover, S_1 and S_1^* may obviously be rewritten in the form

$S_1 : xA^T = b\ , \quad x \geqq 0\ ; \qquad S_1^* : A^T y \geqq 0\ , \quad by < 0\ .$

Another equivalent formulation of Theorem 2.4.1 is the following one:

Theorem 2.4.1 bis. A necessary and sufficient condition for system S_1 to admit solution is:

$yA \geqq 0 \Rightarrow yb \geqq 0\ .$

Note moreover that Theorem 2.4.1 is trivial for $b = 0$, as in this case S_1 admits the solution $x = 0$ and of course S_1^* is impossible. A geometric interpretation in $\mathbb{R}^2$ of the theorem is given in the following Figure 3:

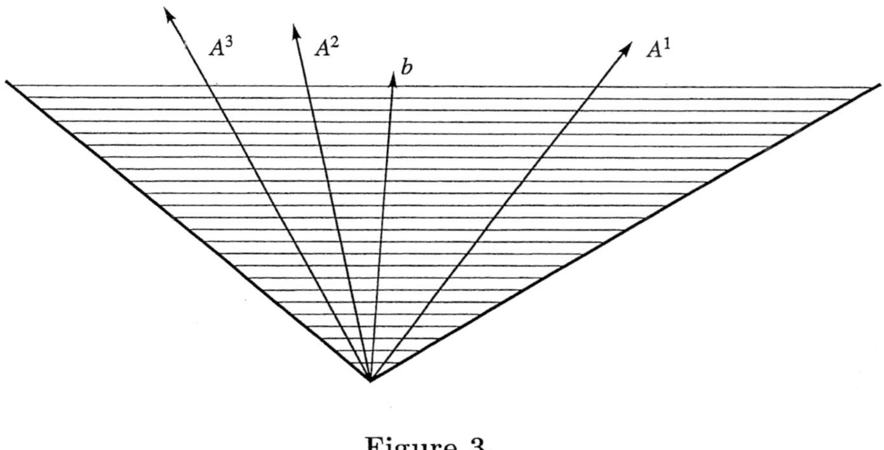

Figure 3.

The shaded region represents the set of vectors $y \in \mathbb{R}^2$ such that $yA \geqq 0$, i.e. the vectors making a non obtuse angle with every column vector of A. If all these vectors form a non obtuse angle with a vector b, i.e. $yb \geqq 0$, then b must belong to the convex polyhedral cone generated by A ($b = Ax$, $x \geqq 0$), i.e. b must belong to the nonnegative linear combinations of the columns of A: $b = x_1A^1 + x_2A^2 + ... + x_nA^n$, $x_1 \geqq 0, ..., x_n \geqq 0$.

The following result is nothing but a formal modification of the previous theorem and will be useful in deducing a first series of theorems of the alternative.

Theorem 2.4.2. Let be given the positive integers m_1, m_2, m_3, n_1, n_2; the matrices A_{ij} of order (m_i, n_j) and the vectors $b^i \in \mathbb{R}^{m_i}$, with $i = 1, 2, 3$; $j = 1, 2$. Then either the system

$$\begin{cases} A_{11}x^1 + A_{12}x^2 & \leqq & b^1 \\ A_{21}x^1 + A_{22}x^2 & = & b^2 \\ A_{31}x^1 + A_{32}x^2 & \geqq & b^3 \\ x^1 \in \mathbb{R}^{n_1}, \quad x^2 \in \mathbb{R}^{n_2}, \quad x^1 \geqq 0 \end{cases} \tag{1}$$

admits solution (x^1, x^2), either the system

$$\begin{cases} y^1 A_{11} + y^2 A_{21} + y^3 A_{31} & \geqq & 0 \\ y^1 A_{12} + y^2 A_{22} + y^3 A_{32} & = & 0 \\ y^1 b^1 + y^2 b^2 + y^3 b^3 & < & 0 \\ y^1 \in \mathbb{R}^{m_1}, \quad y^2 \in \mathbb{R}^{m_2}, \quad y^3 \in \mathbb{R}^{m_3}, \quad y^1 \geqq 0, \quad y^3 \leqq 0 \end{cases} \tag{2}$$

admits solution (y^1, y^2, y^3), but never both.

Proof. The result is easily obtained from the theorem of Farkas-Minkowski; set $x^2 = v^1 - v^2$, with $v^1 \geqq 0$, $v^2 \geqq 0$ and transform the inequalities into equalities by means of the "slack" vectors $s^1 \geqq 0$, $s^2 \geqq 0$. Then system (1) may be rewritten in the form

$$\begin{bmatrix} A_{11} & A_{12} & -A_{12} & I & 0 \\ A_{21} & A_{22} & -A_{22} & 0 & 0 \\ A_{31} & A_{32} & -A_{32} & 0 & -I \end{bmatrix} \cdot \begin{bmatrix} x^1 \\ v^1 \\ v^2 \\ s^1 \\ s^2 \end{bmatrix} = \begin{bmatrix} b^1 \\ b^2 \\ b^3 \end{bmatrix} \tag{3}$$

with $[x^1, v^1, v^2, s^1, s^2]^T \geqq 0$.
Then apply to (3) the theorem of Farkas-Minkowski. □

Theorem 2.4.2 is useful in obtaining a first series of theorems of the alternative for linear systems, here presented in the synthetic form:

$S_k : \ldots; \quad S_k^* : \ldots$

by which we mean that system S_k admits solution if and only if S_k^* is impossible. In all the following formulations, matrices are denoted by capital letters and vectors by small letters. Matrices and vectors are all real and chosen so that the conformability rules of the products are respected:

1) $S_2 : Ax = b$; $S_2^* : yA = 0,\ yb \neq 0$.
(Put $A = A_{22}$, $x = x^2$, $b = b^2$, $y = y^2$; then apply Theorem 2.4.2, making void in (1) and (2) all other quantities. All other theorems here listed are obtained in a similar way.) Note that this result gives necessary and sufficient conditions for the existence of a solution of a linear nonhomogeneous system. Indeed this theorem of the alternative may be equivalently reformulated as: the linear system $Ax = b$ admits solution if and only if $by = 0$ for each vector y such that $yA = 0$.

2) $S_3 : Ax \leqq b$; $S_3^* : yA = 0,\ y \geqq 0,\ yb < 0$.

3) $S_4 : Ax \leqq b,\ x \geqq 0$; $S_4^* : yA \geqq 0,\ y \geqq 0,\ yb < 0$.

4) (Theorem of the alternative of Ky Fan.)
$S_5 : Ax \leqq b^1,\ Bx = b^2$; $S_5^* : y^1A + y^2B = 0,\ y^1 \geqq 0,\ y^1b^1 + y^2b^2 < 0$.

5) $S_6 : Ax \leqq b^1,\ Bx = b^2,\ x \geqq 0$; $S_6^* : y^1A + y^2B \geqq 0,\ y^1 \geqq 0,\ y^1b^1 + y^2b^2 < 0$.

6) $S_7 : Ax + Bz = b,\ x \geqq 0$; $S_7^* : yA \geqq 0,\ yB = 0,\ yb < 0$.

7) $S_8 : Ax + Bz \leqq b,\ x \geqq 0$; $S_8^* : yA \geqq 0,\ yB = 0,\ yb < 0,\ y \geqq 0$.

Another series of theorems of the alternative for linear systems, some of them as classical as the one of Farkas-Minkowski, may be obtained, directly of indirectly, from the following general result, which in turn is deduced from other formal modifications of the theorem of Farkas-Minkowski.

Theorem 2.4.3. Let be given

a) the partitions

$$M_i,\ i = 1, 2, ..., p;\quad \bigcup_{i=1}^{p} M_i = \{1, 2, ..., m\}\ ;$$

$$N_j,\ j=1,2,...,q\,;\quad \bigcup_{j=1}^{q} N_j=\{1,2,...,n\}\,;$$

b) the matrices

$$A_{ij}=[a_{kr}]\,,\qquad k\in M_i\,,\quad r\in N_j\,;$$

c) the vectors

$$b^i=[b_k],\ k\in M_i\,;\quad x^j=[x_r],\ r\in N_j\,;\quad y^i=[y_k],\ k\in M_i\ .$$

Then the system

$$\begin{cases} A_{11}x^1+A_{12}x^2+A_{13}x^3+\sum\limits_{j=4}^{q}A_{1j}x^j=b^1\\ A_{21}x^1+A_{22}x^2+A_{23}x^3+\sum\limits_{j=4}^{q}A_{2j}x^j\leqq b^2\\ A_{31}x^1+A_{32}x^2+A_{33}x^3+\sum\limits_{j=4}^{q}A_{3j}x^j<b^3\\ A_{i1}x^1+A_{i2}x^2+A_{i3}x^3+\sum\limits_{j=4}^{q}A_{ij}x^j\leq b^i,\ i=4,...,p\\ x^1\ \text{arbitrary},\ x^2\geqq 0,\ x^3>0,\ x^j\geq 0,\ j=4,...,q \end{cases} \tag{3}$$

admits solution $(x^1,x^2,x^3,x^j,(j=4,...,q))$ if and only if the system

$$\begin{cases} y^1A_{11}+\sum\limits_{i=2}^{p}y^1A_{i1}=0\\ E^j=y^1A_{1j}+\sum\limits_{i=2}^{p}y^iA_{ij}\geqq 0,\ j=2,...,q\\ yb=y^1b^1+\sum\limits_{i=2}^{p}y^ib^i\leqq 0\\ y^1\ \text{arbitrary},\ y^i\geqq 0,\ i=2,...,p \end{cases} \tag{4}$$

does not admit solution y such that $yb<0$ or $y^3\geq 0$ or

$y^4>0$... or ...$y^p>0$

or $E^3 \geq 0$ or

$E^4 > 0$... or ...$E^q > 0$.

The proof of this theorem will be performed by means of the following three theorems of the alternative described in 8), 9), 10), all obtained from the first series of results.

8) $S_9 : Ax + Bv + Cz = 0,\ v \geqq 0,\ z > 0;\quad S_9^* : yA = 0,\ yB \geqq 0,$ $yC \geq 0$.

Proof. If S_9^* is impossible it will be impossible also the system

$\tilde{S}_9^* : yA = 0,\ yB \geqq 0,\ yC \geqq 0,\ yCu > 0$,

where $u = [1, 1, ..., 1]^T$, and conversely. But if $\tilde{S}_9^*$ is impossible, by means of the result described in 6), it will be possible the system

$\tilde{S}_9 : Ax + Bv + C\bar{z} = -Cu,\ v \geqq 0,\ \bar{z} \geqq 0$.

Hence it will be possible also S_9 as soon as we put $z = (\bar{z} + u) > 0$. It is then easy to obtain the converse result: if S_9^* is impossible, then S_9 admits solution. □

9) $S_{10} : A_1x^1 + A_2x^2 + A_3x^3 + \sum_{j=1}^{q} A_jx^j = 0;\ x^2 \geqq 0;\ x^3 > 0;\ x^j \geq 0,$ $j = 4, ..., q$.

(Here A_j denotes the j-th matrix, *not* the j-th row of A.)

$S_{10}^* : yA_1 = 0;\ yA_j \geqq 0,\ j = 2, ..., q,\ yA_3 \geq 0$ or $yA_j > 0$ for at least one j, $4 \leqq j \leqq q$.

Proof. Note first that $x^j \geq 0$ is equivalent to $\{x^j \geqq 0,\ u^jx^j - w_j = 0,\ w_j > 0\}$, where $w_j \in I\!R$ and $u^j = [1, 1, ..., 1]$ has, for each $j = 4, ..., q$, the same dimension of x^j. Then system S_{10} may be rewritten in the equivalent form

$$\begin{bmatrix} A_1 \\ 0 \\ 0 \\ \vdots \\ 0 \end{bmatrix} x^1 + \begin{bmatrix} A_2 & A_4 & A_5 & \dots & A_q \\ 0 & u^4 & 0 & \dots & 0 \\ 0 & 0 & u^5 & \dots & 0 \\ \dots & \dots & \dots & \dots & \dots \\ 0 & 0 & 0 & \dots & u^q \end{bmatrix} \cdot \begin{bmatrix} x^2 \\ x^4 \\ x^5 \\ \vdots \\ x^q \end{bmatrix} +$$

$$+ \begin{bmatrix} A_3 & 0 \\ 0 & -1 \end{bmatrix} \begin{bmatrix} x^3 \\ w \end{bmatrix} = 0 ;$$

$$(x^2, x^4, x^5, ..., x^q)^T \geqq 0; \quad (x^3, w)^T > 0 .$$

If the said system is impossible, then by means of the result described in 8), we can affirm the existence of vectors y and $v = [v_j]$, $j = 4, ..., q$, such that:

a) $yA_1 = 0$;

b) $yA_2 \geqq 0$;

c) $yA_j + u^j v_j \geqq 0, \ j = 4, ..., q$;

d) $[yA_3 ; -v] \geq 0$.

From d) we have $yA_3 \geqq 0$ and $v \leqq 0$ and from c) it results $yA_j \geqq -u^j v_j \geqq 0$. Moreover: if $v = 0$, then $yA_3 \geq 0$ but if there exists a j_0 such that $v_{j_0} < 0$, then $yA_{j_0} \geqq -u^{j_0} v_{j_0} > 0$. Again by means of the result in 8) we can assert that if S_{10} admits solution, then S_{10}^* is impossible. □

10) $S_{11} : A_1 x^1 + A_2 x^2 + A_3 x^3 + \sum_{j=4}^{q} A_j x^j = b; \ x^2 \geqq 0; \ x^3 > 0; \ x^j \geq 0,$
$j = 4, ..., q$.

(Again A_k is the k-th matrix.)

$S_{11}^* : yA_1 = 0; \ yA_j \geqq 0, \ j = 2, ..., q; \ yb \leqq 0$ and $yb < 0$ or $yA_3 \geq 0$ or $yA_j > 0$ for at least one j, $4 \leqq j \leqq q$.

Proof. S_{11} may be rewritten in the equivalent form:

$$A_1x^1 + A_2x^2 + [A_3\,|\,-b] \cdot \begin{bmatrix} x^3 \\ 1 \end{bmatrix} + \sum_{j=4}^{q} A_jx^j = 0\,;$$

$$x^2 \geqq 0;\ [x^3\,|\,1]^T > 0;\ x^j \geq 0,\ j = 4, \ldots, q\,.$$

The present result is obtained at once by applying to this form the previous theorem of the alternative. □

Proof of Theorem 2.4.3. Let us introduce in the inequalities appearing in system (3) the following "slack" vectors: $w^2 \geqq 0$, $w^3 > 0$, $w^i \geq 0$, $i = 4, \ldots, p$. Then (3) may be rewritten in the form:

$$\begin{bmatrix} A_{11} \\ A_{21} \\ A_{31} \\ \vdots \\ A_{p1} \end{bmatrix} x^1 + \begin{bmatrix} A_{12} & 0 \\ A_{22} & I \\ A_{32} & 0 \\ \vdots & \vdots \\ A_{p2} & 0 \end{bmatrix} \cdot \begin{bmatrix} x^2 \\ w^2 \end{bmatrix} + \begin{bmatrix} A_{13} & 0 \\ A_{23} & 0 \\ A_{33} & I \\ \vdots & \vdots \\ A_{p3} & 0 \end{bmatrix} \cdot \begin{bmatrix} x^3 \\ w^3 \end{bmatrix} +$$

$$+ \sum_{j=4}^{q} \begin{bmatrix} A_{1j} \\ A_{2j} \\ A_{3j} \\ \vdots \\ A_{pj} \end{bmatrix} x^j + \sum_{i=4}^{p} \begin{bmatrix} 0 \\ 0 \\ \vdots \\ I \\ \vdots \\ 0 \end{bmatrix} w^i = b\,,$$

with x^1 arbitrary, $[x^2\,|\,w^2]^T \geqq 0$, $[x^3\,|\,w^3]^T > 0$, $x^j \geq 0$, $j = 4, \ldots, q$; $w^j \geq 0$, $j = 4, \ldots, p$.
Then apply the result described in 10) in order to obtain that (3) is possible if and only if (4) is impossible. □

From Theorem 2.4.3 we obtain a second numerous series of theorems of the alternative; here we list only the most quoted in the literature.

11) (Theorem of the alternative of Gordan.)
$S_{12} : Ax = 0,\ x \geq 0;\quad S_{12}^* : yA > 0\,.$

12) (Theorem of the alternative of Stiemke.)
$S_{13} : Ax = 0,\ x > 0\,;\quad S^*_{13} : yA \geq 0\,.$

13) (Theorem of the alternative of Ville.)
$S_{14} : Ax \leqq 0,\ x \geq 0\,;\quad S^*_{14} : yA > 0,\ y \geqq 0\,.$

14) $S_{15} : Ax \leqq 0,\ x > 0\,;\quad S^*_{15} : yA \geq 0,\ y \geqq 0\,.$
(This theorem is due to Gale (1960).)

15) $S_{16} : Ax < 0,\ x > 0\,;\quad S^*_{16} : yA \geqq 0,\ y \geq 0$ or $yA \geq 0,\ y \geqq 0\,.$

16) $S_{17} : Ax < b\,;\quad S^*_{17} : yA = 0,\ y[I \mid -b] \geq 0\,.$

17) $S_{18} : Ax \leqq b,\ x > 0\,;\quad S^*_{18} : y[A \mid -b] \geq 0,\ y \geqq 0\,.$

18) $S_{19} : Ax = b,\ x > 0\,;\quad S^*_{19} : y[A \mid -b] \geq 0\,.$

19) (Theorem of the alternative of Motzkin or transposition theorem of Motzkin.)
$S_{20} : Ax = 0\,;\quad Bx \geqq 0\,;\quad Dx > 0\,;$
$S^*_{20} : yA + vB + wD = 0\,;\quad v \geqq 0\,;\quad w \geq 0\,.$

20) (Theorem of the alternative of Tucker.)
$S_{21} : Ax = 0\,;\quad Bx \geqq 0\,;\quad Cx \geq 0\,;$
$S^*_{21} : yA + vB + wC = 0\,;\quad v \geqq 0\,;\quad w > 0\,.$

21) (Theorem of the alternative of Slater.)
$S_{22} : Ax = 0\,;\quad Bx \geqq 0\,;\quad Cx \geq 0\,;\quad Dx > 0\,;$
$S^*_{22} : \{yA + vB + w^1C + w^2D = 0\,;\quad v \geqq 0\,;\quad w^1 \geqq 0\,;\quad w^2 \geq 0\}$
or $\{yA + vB + w^1C + w^2D = 0\,;\quad v \geqq 0\,;\quad w^1 > 0\,;\quad w^2 \geqq 0\}\,.$

22) (First theorem of the alternative of Mangasarian.)
$S_{23} : \{Ax \geq 0\,;\quad Bx \geqq 0\,;\quad Cx \geqq 0\,;\quad Dx = 0\}$
or $\{Ax \geqq 0\,;\quad Bx > 0\,;\quad Cx \geqq 0\,;\quad Dx = 0\}\,;$
$S^*_{23} : y^1A + y^2B + y^3C + y^4D = 0\,;\quad y^1 > 0\,;\quad y^2 \geq 0\,;\quad y^3 \geqq 0\,.$

23) (Second theorem of the alternative of Mangasarian.)
$S_{24} : Ax \leq b\,;$
$S^*_{24} : \{yA = 0\,;\quad y \geqq 0\,;\quad by = -1\}$
or $\{yA = 0\,;\quad y > 0\,;\quad yb \leqq 0\}\,.$

24) (First theorem of the alternative of Fenchel.)
$S_{25} : Ax + Bz = 0\,; \quad x \geqq 0\,; \quad z \geq 0\,;$
$S^*_{25} : yA \geqq 0\,; \quad yB > 0\,.$

25) (Second theorem of the alternative of Fenchel.)
$S_{26} : Ax + Bz = 0\,; \quad x \geqq 0\,; \quad z > 0\,;$
$S^*_{26} : yA \geqq 0\,; \quad yB \geq 0\,.$

26) (Theorem of the alternative of Duffin or nonhomogeneous theorem of the alternative of Farkas-Minkowski.)
$S_{27} : Ax \leqq b\,; \quad cx > \gamma\,;$
$S^*_{27} : \{yA = 0\,; \quad yb < 0\,; \quad y \geqq 0\}$
or $\{yA = c\,; \quad yb \leqq \gamma\,; \quad y \geqq 0\}\,.$

We report also another theorem, due to A.W. Tucker (1956), which is easily obtained from the previous results.

Theorem 2.4.4. For any given matrix A of order (m, n), the systems

$\{Ax \geqq 0\}\,; \quad \{yA = 0\,; \quad y \geqq 0\}$

admit, respectively, solutions x^0, y^0 such that $Ax^0 + y^0 > 0$.

Proof. In order to prove that the system

$$\begin{cases} A^T y = 0 \\ -Ax \leqq 0 \\ -Ax - Iy < 0 \\ y \geqq 0 \end{cases}$$

admits solution, it is necessary and sufficient to show, on the ground of Theorem 2.4.3, that the dual system

$$\begin{cases} z^2 A + z^3 A = 0 \\ z^1 A^T - z^3 \geqq 0 \\ z^1 \in \mathbb{R}^n\,; \quad z^2 \geqq 0\,; \quad z^3 \geq 0 \end{cases}$$

is impossible. This is easily proved by means of Theorem 2.4.3. Conversely, suppose that the dual system admits solution: we obtain

$0 = (z^2 + z^3)\, Az^1 \geqq (z^2 + z^3)\, z^3 \geqq z^3 \cdot z^3 > 0\,,$

absurd. □

As Theorem 2.4.4 is in turn a basic result which generates many theorems of the alternative (see Tucker (1956)), among which the theorem of Farkas-Minkowski, we have so obtained a kind of "loop".

2.5. Convex Functions

Let $f : X \subset I\!R^n \to I\!R$, where X is a convex set; the function f is said to be *convex* on X if:

$$f(\lambda x^1 + (1-\lambda)\, x^2) \leqq \lambda f(x^1) + (1-\lambda)\, f(x^2)\,, \quad \forall x^1, x^2 \in X\,,\ \forall \lambda \in [0,1]\,. \tag{1}$$

The function $f : X \to I\!R$ is called *concave* on X if and only if $-f$ is convex on X.

Convex (and concave) functions of one real variable have the familiar geometric interpretation encountered in almost all the textbooks on elementary mathematical analysis. See Figures 4 and 5 below.

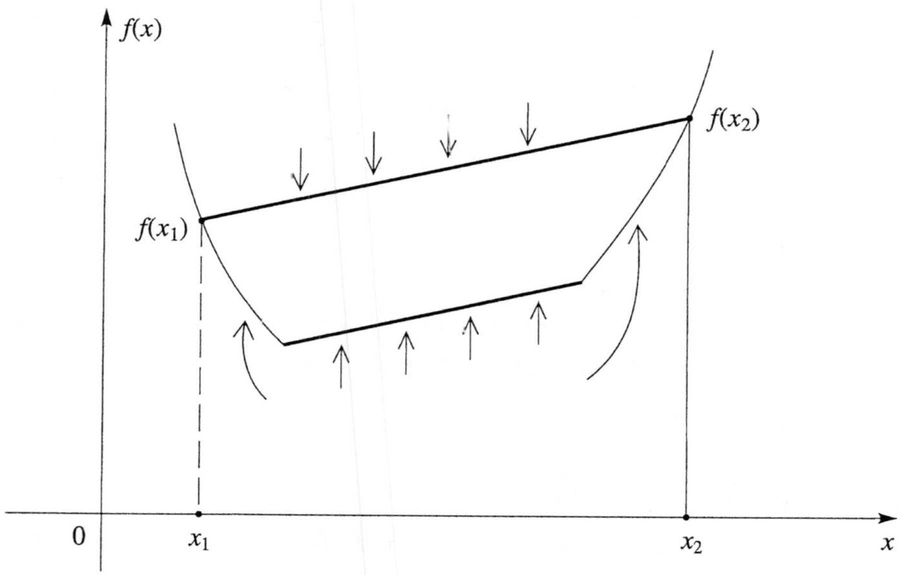

Figure 4. Example of convex function of one variable.

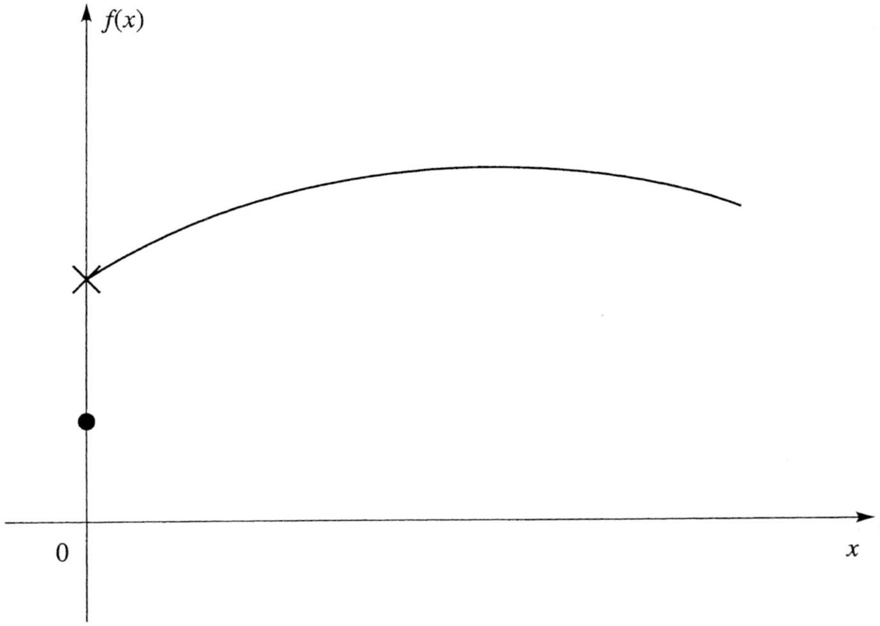

Figure 5. Example of concave function of one variable.

We recall here the basic characterizations and properties for convex functions of one real variable. If φ is a function defined on the interval $(a, b) \subset \mathbb{R}$, i.e. $\varphi : (a, b) \to \mathbb{R}$, then the following conditions are equivalent:

i) φ is convex on (a, b).

ii) $\varphi(t_2) \leqq \dfrac{t_3 - t_2}{t_3 - t_1} \varphi(t_1) + \dfrac{t_2 - t_1}{t_3 - t_1} \varphi(t_3)$,
$\forall\, t_1, t_2, t_3$, with $a < t_1 < t_2 < t_3 < b$.

iii) $\dfrac{\varphi(t_2) - \varphi(t_1)}{t_2 - t_1} \leqq \dfrac{\varphi(t_3) - \varphi(t_1)}{t_3 - t_1}$,
$\forall\, t_1, t_2, t_3$, with $a < t_1 < t_2 < t_3 < b$.

iv) $\dfrac{\varphi(t_3) - \varphi(t_1)}{t_3 - t_1} \leqq \dfrac{\varphi(t_3) - \varphi(t_2)}{t_3 - t_2}$,
$\forall\, t_1, t_2, t_3$, with $a < t_1 < t_2 < t_3 < b$.

v) $\begin{vmatrix} 1 & t_1 & \varphi(t_1) \\ 1 & t_2 & \varphi(t_2) \\ 1 & t_3 & \varphi(t_3) \end{vmatrix} \geqq 0,$
$\forall t_1, t_2, t_3$, with $a < t_1 < t_2 < t_3 < b$.

vi) $\varphi(t) \leqq \dfrac{1}{2h} \displaystyle\int_{t-h}^{t+h} \varphi(u)\, du,$
$\forall t, h$, with $a \leqq t - h < t + h \leqq b$.

If φ is differentiable on (a, b), then φ is convex on (a, b) if and only if:

vii) φ' is nondecreasing on (a, b).

If φ admits second-order derivative on (a, b), then φ is convex on (a, b) if and only if:

viii) $\varphi''(t) \geqq 0, \quad \forall t \in (a, b)$.

Moreover if φ is convex on $[a, b]$, then:

a) φ is continuous on (a, b);

b) $\varphi'(t)$ exists on (a, b), except for a countable set of points.

c) $\varphi''(t)$ exists on (a, b), except on a set of points of Lebesgue measure equal to zero.

For the proof of these properties the reader may consult Fenchel (1953), Valentine (1964), Roberts and Varberg (1973), Rockafellar (1970).

Now we give several equivalent definitions of a real-valued convex function of several real variables, i.e. defined on a convex set $X \subset \mathbb{R}^n$.

Theorem 2.5.1. If $X \subset \mathbb{R}^n$ is a convex set and if $f : X \to \mathbb{R}$, then the following conditions are equivalent:

i) $f(\lambda x^1 + (1-\lambda)\, x^2) \leqq \lambda f(x^1) + (1-\lambda)\, f(x^2)$,
$\forall x^1, x^2 \in X; \quad \forall \lambda \in [0, 1]$.

ii) $f(\lambda x^1 + (1-\lambda)\,x^2) \geqq \lambda f(x^1) + (1-\lambda)\,f(x^2)$,
$\forall\, x^1, x^2 \in X$, $\forall\, \lambda > 1$ such that $\lambda x^1 + (1-\lambda)\,x^2 \in X$.

iii) $f(\lambda x^1 + (1-\lambda)\,x^2) \geqq \lambda f(x^1) + (1-\lambda)\,f(x^2)$,
$\forall\, x^1, x^2 \in X$, $\forall\, \lambda < 0$ such that $\lambda x^1 + (1-\lambda)\,x^2 \in X$.

iv) For any $x^i \in X$, $i = 1, ..., m$, and for any $t_i \in [0,1]$, $i = 1, ..., m$, $\sum t_i = 1$, the following inequality holds (*Jensen's inequality*):

$$f\Big(\sum_{i=1}^{m} t_i x^i\Big) \leqq \sum_{i=1}^{m} t_i\, f(x^i)\ .$$

v) The function $\varphi_{x,y}(t) = f(x+ty)$ is convex, $\forall\, x \in X$, $\forall\, y \in I\!R^n$, on the interval $T_{x,y} = \{t \mid t \in I\!R,\ x+ty \in X\}$.

vi) The function $\psi_{x^1,x^2}(\lambda) = f(\lambda x^1 + (1-\lambda)\,x^2)$ is convex, $\forall\, x^1, x^2 \in X$, on the interval $[0,1]$.

vii) The *epigraph* (or *epigraphical set*) of f, defined as

$$\operatorname{epi} f = \{(x,\alpha) \mid (x,\alpha) \in X \times I\!R,\ f(x) \leqq \alpha\}$$

is a convex set in $I\!R^{n+1}$.

Proof. The equivalence among i), ii) and iii) is quite evident. Also it is clear that if iv) holds, then i) holds. Let us suppose that i) holds and let us prove by induction the validity of iv).
For $m = 1$ and $m = 2$ the property holds trivially true; let us suppose the property true for $r-1$ and let $t = t_1 + ... + t_{r-1} \geqq 0$. If $t = 0$, obviously the property holds true for r. If $t \neq 0$, denoting $t'_i = t_i/t$, we obtain $t'_i \geqq 0$, $\sum_{i=1}^{r-1} t'_i = 1$ and hence, by the introduction hypothesis,

$$f\Big(\sum_{i=1}^{r} t_i x^i\Big) = f\Big(t \sum_{i=1}^{r-1} t'_i x^i + t_r x^r\Big) \leqq t\, f\Big(\sum_{i=1}^{r-1} t'_i x^i\Big) + t_r\, f(x^r) \ \leqq$$

$$\leqq\ t \sum_{i=1}^{r-1} t'_i\, f(x^i) + t_r\, f(x^r) = \sum_{i=1}^{r} t_i\, f(x^i)\ .$$

i) $\Rightarrow$ v). Let $t_1, t_2 \in T_{x,y}$ and $\lambda \in [0,1]$; on the grounds of Theorem 2.1.2 $T_{x,y}$ is convex; therefore $\lambda t_1 + (1-\lambda)\, t_2 \in T_{x,y}$. Taking the convexity of f into account, we have

$$\varphi_{x,y}[\lambda t_1 + (1-\lambda)\, t_2] = f[x + (\lambda t_1 + (1-\lambda)\, t_2)\, y] =$$

$$= f[\lambda(x + t_1 y) + (1-\lambda)(x + t_2 y)] \leqq \lambda f(x + t_1 y) +$$

$$+ (1-\lambda)\, f(x + t_2 y) = \lambda \varphi_{x,y}(t_1) + (1-\lambda)\, \varphi_{x,y}(t_2)\ .$$

v) $\Rightarrow$ vi). Let $x^1, x^2 \in X$ and $\lambda \in [0,1]$. We have

$$\psi_{x^1,x^2}(\lambda) = f[\lambda x^1 - (1-\lambda)\, x^2] = f[x^2 + \lambda(x^1 - x^2)] =$$

$$= \varphi_{x^2, x^1 - x^2}(\lambda)\ ,$$

from which the convexity of the function ψ_{x^1,x^2} results, as the function φ_{x^2,x^1-x^2} is convex.

vi) $\Rightarrow$ vii). Let (x^1, α_1) and (x^2, α_2) belong to $\operatorname{epi} f$. We have then $f(x^1) \leqq \alpha_1$, $f(x^2) \leq \alpha_2$. Then

$$f(\lambda x^1 + (1-\lambda)\, x^2) = \psi_{x^1,x^2}(\lambda) \leqq \lambda \psi_{x^1,x^2}(1) + (1-\lambda)\, \psi_{x^1,x^2}(0) =$$

$$= \lambda f(x^1) + (1-\lambda)\, f(x^2) \leqq \lambda \alpha_1 + (1-\lambda)\, \alpha_2\ ,$$

owing to the convexity of ψ_{x^1,x^2}. Therefore $\lambda(x^1, \alpha_1) + (1-\lambda)(x^2, \alpha_2) \in \operatorname{epi} f$.

vii) $\Rightarrow$ i). Let $x^1, x^2 \in X$; then

$$(x^1, f(x^1)) \in \operatorname{epi} f\ , \quad (x^2, f(x^2)) \in \operatorname{epi} f$$

for any $x^1, x^2 \in X$. By the convexity of $\operatorname{epi} f$ we have

$$[\lambda x^1 + (1-\lambda)\, x^2,\ \lambda f(x^1) + (1-\lambda)\, f(x^2)] \in \operatorname{epi} f$$

for $0 \leqq \lambda \leqq 1$, or

$$f[\lambda x^1 + (1-\lambda)\, x^2] \leq \lambda f(x^1) + (1-\lambda)\, f(x^2)\ ,$$

for $0 \leq \lambda \leq 1$, and hence f is convex on X. □

If $X \subset \mathbb{R}^n$ is *open* (and convex) we have the following characterization.

Theorem 2.5.2. Let $f : X \to \mathbb{R}$, $X \subset \mathbb{R}^n$ be an open convex set; then f is convex on X if and only if:

viii) for each $x^0 \in X$ there exists a vector $u^0 \in \mathbb{R}^n$ such that

$$f(x) - f(x^0) \geqq u^0(x - x^0)\,, \quad \forall x \in X\,. \tag{2}$$

Proof. Let $X \subset \mathbb{R}^n$ be convex and open and let $x^0 \in X$; we first prove the implication epi f convex $\Rightarrow$ viii). The point $(x^0, f(x^0))$ is a point of the boundary of the convex set epi f. Therefore on the ground of Theorem 2.2.4, i), there exists a vector $(v, v_0) \neq 0$ such that

$$vx + v_0\alpha \geqq vx^0 + v_0 f(x^0) \tag{3}$$

for each $(x, \alpha) \in \operatorname{epi} f$.
If $v_0 = 0$, then $v \cdot (x - x^0) \geqq 0$, $\forall x \in X$, which implies $v = 0$, X being open; therefore we have the absurd result $(v, v_0) = 0$. If $v_0 < 0$, then it is possible to take α sufficiently large in order to have

$$vx + v_0\alpha < vx^0 + v_0 f(x^0)\,,$$

in contradiction to (3).
Therefore it is $v_0 > 0$. Choose $\alpha = f(x)$ and $u^0 = (-v)/v_0$; from relation (3) we obtain viii).
Now let us prove the reverse implication viii) $\Rightarrow$ f convex on X. Let $x^1, x^2 \in X$ and $\lambda \in [0, 1]$; therefore as viii) holds, for each $x^0 \in X$ there exists $u^0 \in \mathbb{R}^n$ such that $f(x^1) - f(x^0) \geqq u^0(x^1 - x^0)$ and $f(x^2) - f(x^0) \geqq u^0(x^2 - x^0)$. Multiplying these inequalities respectively by λ and $(1 - \lambda)$, we obtain

$$\lambda f(x^1) + (1 - \lambda)\, f(x^2) - f(x^0) \geqq u^0[\lambda x^1 + (1 - \lambda)\, x^2 - x^0]\,.$$

Taking $x^0 = \lambda x^1 + (1 - \lambda)\, x^2$, relation i) of Theorem 2.5.1 follows, i.e. f is convex on X. □

The vector u^0 in relation (2) is called a *subgradient* of f at x^0 (see the subsequent Section 2.6). If X is not open and relation (2) holds in $\operatorname{int}(X)$, f is convex in $\operatorname{int}(X)$, but in general not also in X: consider for example

$n = 2$,

$X = \{x \in I\!R^2 \mid 0 \leqq x_i \leqq 1,\ i = 1, 2\}$,

$$f(x) = \begin{cases} 0, & \text{if } 0 \leqq x_1 \leqq 1 \text{ and } 0 < x_2 \leqq 1 \\ 1 - (x_1)^2, & \text{if } 0 \leqq x_1 \leqq 1 \text{ and } x_2 = 0. \end{cases}$$

Relation (2) is satisfied with $u^0 = 0$ for each x belonging to $\operatorname{int}(X)$, but f is not convex on X.

From Theorem 2.5.2 we also have that every convex function f on an open convex set $X \subset I\!R^n$ is the supremum of a family of linear affine functions. More precisely: for each $x^0 \in X$ there exists a linear affine function $l(x) = u^0 x + \alpha$ such that $l(x^0) = f(x^0)$ and $l(x) \leqq f(x)$, $\forall x \in X$.

We have also that the nonempty set (*support set*)

$$U = \{(u, \alpha) \in I\!R^{n+1} \mid ux + \alpha \leqq f(x),\ \forall x \in X\}$$

is convex and

$$f(x) = \sup_{(u,\alpha)\in U} \{ux + \alpha\}, \qquad \forall x \in X.$$

For concave functions the above characterizations hold, with suitable modifications. In particular, if the *hypograph* of f is introduced as the set

$$\operatorname{hypo} f = \{(x, \alpha) \mid (x, \alpha) \in X \times I\!R,\ f(x) \geqq \alpha\},$$

then f is concave on X if and only if $\operatorname{hypo} f$ is a convex set.

Theorem 2.5.3. Let f be defined on the convex set $X \subset I\!R^n$. A necessary condition for f to be convex on X, is that the *lower level set* of f

$$L(f, \alpha) = \{x \mid x \in X,\ f(x) \leqq \alpha,\ \alpha \in I\!R\}$$

is a convex set for each $\alpha \in I\!R$.

Proof. Let f be convex on X and let $x^1, x^2 \in L(f, \alpha)$. Then

$$f[(1-\lambda)\,x^1 + \lambda x^2] \leqq (1-\lambda)\,f(x^1) + \lambda f(x^2) \leqq (1-\lambda)\,\alpha + \lambda\alpha = \alpha.$$

Hence $(1-\lambda)\,x^1 + \lambda x^2 \in L(f,\alpha)$, which is therefore convex. $\square$

We now show that if $L(f,\alpha)$ is convex for each $\alpha \in \mathbb{R}$, it does not follow that f is convex on X. Consider the function f on $\mathbb{R}$, defined by $f(x) = x^3$; f is not convex on $\mathbb{R}$, however the set

$$L(f,\alpha) = \{x \mid x \in \mathbb{R},\ x^3 \leqq \alpha\} = \{x \mid x \in \mathbb{R},\ x \leqq \alpha^{1/3}\}$$

is obviously convex for any $\alpha \in \mathbb{R}$.

Theorem 2.5.4 (Equivalent definitions for differentiable convex functions). Let $X \subset \mathbb{R}^n$ be an open convex set and let $f : X \longrightarrow \mathbb{R}$ be differentiable on X. Then the following assertions are equivalent:

a) f is convex on X.

b) For each $x \in X$ and for each $y \in \mathbb{R}^n$ the function $\varphi'_{x,y}(t) = y\nabla f(x + ty)$ is nondecreasing, with respect to t, on the interval $T_{x,y} = \{t \mid x + ty \in X\}$.

c) For each $x^1, x^2 \in X$ the function $\psi'_{x^1,x^2}(\lambda) = (x^1 - x^2)\,\nabla f(\lambda x^1 + (1-\lambda)\,x^2)$ is nondecreasing, with respect to λ on $[0,1]$.

d) For each $x^1, x^2 \in X$

$$f(x^1) - f(x^2) \geqq (x^1 - x^2)\,\nabla f(x^2)\ .$$

e) For each $x^1, x^2 \in X$

$$f(x^1) - f(x^2) \leqq (x^1 - x^2)\,\nabla f(x^1)\ .$$

f) For each $x^1, x^2 \in X$

$$(x^1 - x^2)\cdot[\nabla f(x^1) - \nabla f(x^2)] \geqq 0\ .$$

Proof.
a) $\Leftrightarrow$ b). On the grounds of Theorem 2.5.1, f is convex on X if and only if the function $\varphi_{x,y}(t) = f(x + ty)$ is convex, $\forall\, x \in X$, $\forall\, y \in \mathbb{R}^n$, on the

interval $T_{x,y} = \{t \mid t \in I\!R,\ x + ty \in X\}$. This is true if and only if the derivative $\varphi'_{x,y}(t)$ is nondecreasing on the same interval $T_{x,y}$.

a) $\Leftrightarrow$ c). On the grounds of Theorem 2.5.1, f is convex on X if and only if the function $\psi_{x^1,x^2}(\lambda) = f(\lambda x^1 + (1-\lambda)\, x^2)$ is convex, $\forall\, x^1, x^2 \in X$, on the interval $[0,1]$ and this is true if and only if the derivative $\psi'_{x^1,x^2}(\lambda) = (x^1 - x^2)\, \nabla f(\lambda x^1 + (1-\lambda)\, x^2)$ is nondecreasing on $[0,1]$.

c) $\Rightarrow$ d). Let $x^1, x^2 \in X$. As the function $\psi'_{x^1,x^2}(\lambda)$ is nondecreasing on the interval $[0,1]$, we have, for each $\lambda \in [0,1]$,

$$\psi'_{x^1,x^2}(\lambda) \geqq \psi'_{x^1,x^2}(0) ;$$

therefore it is

$$(x^1 - x^2)\, \nabla f(\lambda x^1 + (1-\lambda)\, x^2) \geqq (x^1 - x^2)\, \nabla f(x^2) .$$

If $x^1 = x^2$ the implication is trivial, otherwise from Taylor's expansion formula we obtain

$$f(x^1) - f(x^2) = (x^1 - x^2)\, \nabla f(\vartheta x^1 + (1-\vartheta)\, x^2) , \qquad 0 < \vartheta < 1 .$$

The result follows therefore immediately.

d) $\Rightarrow$ e). For each $x^1, x^2 \in X$ it is

$$\begin{aligned} f(x^1) - f(x^2) &= -[f(x^2) - f(x^1)] \leqq -(x^2 - x^1)\, \nabla f(x^1) = \\ &= (x^1 - x^2)\, \nabla f(x^1) . \end{aligned}$$

e) $\Rightarrow$ f). Summing up the two inequalities

$$f(x^1) - f(x^2) \leqq (x^1 - x^2)\, \nabla f(x^1) ;$$

$$f(x^2) - f(x^1) \leqq -(x^1 - x^2)\, \nabla f(x^2) ;$$

we obtain result f).

f) $\Rightarrow$ c). Let $x^1, x^2 \in X$, $\lambda_1, \lambda_2 \in [0,1]$, with $\lambda_1 < \lambda_2$. Since X is convex, $y^1 = \lambda_1 x^1 + (1-\lambda_1)\, x^2 \in X$ and $y^2 = \lambda_2 x^1 + (1-\lambda_2)\, x^2 \in X$. Moreover, $y^1 - y^2 = (\lambda_1 - \lambda_2)(x^1 - x^2)$. We have

$$\psi'_{x^1,x^2}(\lambda_2) - \psi'_{x^1,x^2}(\lambda_1) =$$
$$= (x^1 - x^2)\,\nabla f(\lambda_2 x^1 + (1-\lambda_2)\,x^2) - (x^1 - x^2)\,\nabla f(\lambda_1 x^1 + (1-\lambda_1)\,x^2) =$$
$$= [1/(\lambda_1 - \lambda_2)] \cdot (y^1 - y^2) \cdot (\nabla f(y^2) - \nabla f(y^1)) =$$
$$= [1/(\lambda_2 - \lambda_1)] \cdot (y^2 - y^1) \cdot (\nabla f(y^2) - \nabla f(y^1)) \geqq 0 \; .$$

Hence

$$\psi'_{x^1,x^2}(\lambda_2) \geqq \psi'_{x^1,x^2}(\lambda_1) \; ,$$

i.e. $\psi'_{x^1,x^2}(\lambda)$ is nondecreasing on $[0,1]$. □

For what concerns characterization f) in the previous theorem, we observe that if $X \subset \mathbb{R}^n$, the application or map $g : X \to \mathbb{R}^n$ is said to be *monotone on* X if, for each $x^1, x^2 \in X$,

$$(x^1 - x^2)\,[g(x^1) - g(x^2)] \geqq 0 \; .$$

The equivalence a) ⇔ f) proved above, states therefore that a function, differentiable on the open convex set $X \subset \mathbb{R}^n$, is convex on X if and only if its gradient ∇f is a monotone map on X. See Minty (1964), Rockafellar (1967) and Karamardian (1976) for other applications of monotone operators in mathematical programming and Section 2.12 for generalizations of this concept.

Theorem 2.5.5. Let $f : X \to \mathbb{R}$ be $\mathcal{C}^2(X)$ on the open convex set $X \subset \mathbb{R}^n$; then f is convex on X if and only if its Hessian matrix $Hf(x)$ is positive semidefinite for each $x \in X$.

Proof. Let $x^1, x^2 \in X$ and $Hf(x)$ be positive semidefinite, for each $x \in X$. By Taylor's formula we have

$$f(x^1) - f(x^2) =$$
$$= (x^1 - x^2)\,\nabla f(x^2) + \tfrac{1}{2}\,(x^1 - x^2)\,Hf(\vartheta x^1 + (1-\vartheta)\,x^2)(x^1 - x^2) \; ,$$
$$\vartheta \in (0,1) \; .$$

Therefore, for each $x^1, x^2 \in X$:

$$f(x^1) - f(x^2) \geqq (x^1 - x^2)\,\nabla f(x^2)$$

i.e. f is convex on X.
Conversely, let f be convex on X; then characterization b) of Theorem 2.5.4 holds and since $f \in C^2(X)$,then $\varphi_{x,y}(t)$ is twice differentiable on $T_{x,y}$ and from the said characterization we have $\varphi''_{x,y}(t) \geqq 0$, $\forall\, t \in T_{x,y}$, i.e. for each $x \in X$ and for each $y \in \mathbb{R}^n$ it is $yHf(x+ty)\,y \geqq 0$, $\forall\, t \in T_{x,y}$. In particular for $t = 0 \in T_{x,y}$ we obtain $yHf(x)\,y \geqq 0$, $\forall\, y \in \mathbb{R}^n$, i.e. $Hf(x)$ is positive semidefinite for each $x \in X$. □

When relation (1) holds with a strict inequality and for each $\lambda \in (0,1)$, we speak of *strictly convex functions*:

$$f(\lambda x^1 + (1-\lambda)\,x^2) < \lambda f(x^1) + (1-\lambda)\,f(x^2)\ ,$$
$$\forall\, x^1, x^2 \in X,\ (x^1 \neq x^2),\ \forall\, \lambda \in (0,1)\ .$$

If $X \subset \mathbb{R}^n$ is a convex set and $f : X \to \mathbb{R}$, the following assertions are equivalent:

i) f is strictly convex on X.

ii) For each $x \in X$ and for each $y \in \mathbb{R}^n$, $y \neq 0$, the function $\varphi_{x,y}(t) = f(x+ty)$ is strictly convex on the interval $T_{x,y} = \{t \,|\, t \in \mathbb{R},\ x+ty \in X\}$.

iii) For each $x^1, x^2 \in X$, $x^1 \neq x^2$, the function $\psi_{x^1,x^2}(\lambda) = f(\lambda x^1 + (1-\lambda)\,x^2)$ is strictly convex on $[0,1]$.

If $f : X \to \mathbb{R}$ is differentiable on the open convex set $X \subset \mathbb{R}^n$, the following assertions are equivalent:

i) f is strictly convex on X.

iv) For each $x^1, x^2 \in X$, $x^1 \neq x^2$,

$$f(x^1) - f(x^2) > (x^1 - x^2)\,\nabla f(x^2)\ .$$

v) For each $x^1, x^2 \in X$, $x^1 \neq x^2$,

$$f(x^1) - f(x^2) < (x^1 - x^2)\,\nabla f(x^1)\ .$$

vi) For each $x^1, x^2 \in X$, $x^1 \neq x^2$,

$$(x^1 - x^2) \cdot [\nabla f(x^1) - \nabla f(x^2)] > 0\ ,$$

i.e. the gradient $\nabla f(x)$ is a *strictly monotone map on* X.

The characterization of strictly convex functions, in the case of twice continuous differentiability, is more delicate. We first present the following result.

Theorem 2.5.6. Let $f : X \to I\!R$ be $\mathcal{C}^2(X)$ on the open convex set $X \subset I\!R^n$. A sufficient condition that f be strictly convex on X is that $Hf(x)$ be positive definite on X, i.e. it must be, for each $x \in X$,

$$yHf(x)\,y > 0\,,\quad \forall y \in I\!R^n\,,\quad y \neq 0\ .$$

Proof. The proof is essentially identical to the proof of Theorem 2.5.5 (proof of the "sufficiency"). □

The above condition is only sufficient: indeed consider the function $f(x) = x^4$, defined on $I\!R$; f is strictly convex on $I\!R$, but $Hf(x) = 12x^2$ is not positive definite on $I\!R$, since $Hf(0) = 0$.

Fenchel (1953), pp. 73–88, and Bernstein and Toupin (1962), p. 42, show that if f is twice continuously differentiable on the open convex set $X \subset I\!R^n$, then the following condition is necessary for f to be strictly convex on X: its Hessian matrix is positive definite except on a nowhere dense subset of X.

For the simple case of a function $f : (a,b) \to I\!R$, with $(a,b) \subset I\!R$, the following characterization holds: if $f''(x)$ exists on (a,b), then $f : (a,b) \to I\!R$ is strictly convex on (a,b) if and only if $f''(x)$ is nonnegative on (a,b) and not identically zero on any nontrivial subinterval of (a,b).

The characterizations for concave and strictly concave functions are easily obtainable from the previous results. A function $f : X \to I\!R$ is linear (affine) on the convex set $X \subset I\!R^n$ if and only if it is both convex and concave on X, i.e. if and only if it is

$$f[(1-\lambda)\,x^1 + \lambda x^2] = (1-\lambda)\,f(x^1) + \lambda f(^2)\ ,$$
$$\forall x^1, x^2 \in X\,;\quad \forall \lambda \in [0,1]\ .$$

Let us prove that the above definition is equivalent to the more usual definition of a linear (affine) function on $X = I\!R^n$.

Theorem 2.5.7. The function $f : I\!R^n \to I\!R$ is linear affine if and only if there are a vector $c \in I\!R^n$ and a scalar γ such that $f(x) = cx + \gamma$.

Proof. According to Theorem 2.5.1, i), ii), iii), a function which is convex and concave is characterized by the following relation:

$$f((1-\lambda)\,x^1 + \lambda x^2) = (1-\lambda)\,f(x^1) + \lambda f(x^2)\ ,$$
$$\forall x^1, x^2 \in I\!R^n\,;\quad \forall \lambda \in I\!R\ .$$

Set $\gamma = f(0)$ and $\psi(x) = f(x) - f(0) = f(x) - \gamma$. Since ψ fulfills the same equality and $\psi(0) = 0$, we have

$$\psi(\lambda x) = \lambda \psi(x)\ , \qquad \forall x \in I\!R^n\ ,\ \ \forall \lambda \in I\!R$$
$$\psi(x^1 + x^2) = \psi(x^1) + \psi(x^2)\ , \qquad \forall x^1, x^2 \in I\!R^n\ .$$

Especially, we get for $x = \sum_{i=1}^n x_i e^i$,

$$\psi(x) = \psi\Big(\sum_{i=1}^n x_i e^i\Big) = \sum_{i=1}^n x_i \psi(e^i)\ .$$

Setting $c_i = \psi(e^i)$, we have $\psi(x) = cx$ and $f(x) = cx + \gamma$. □

We note that if $f_1(x), f_2(x), ..., f_n(x)$ are convex functions, all defined on the convex set $X \subset I\!R^n$, then each nonnegative linear combination of these is also a convex function on X. We shall see that this is no longer true for the main generalizations of the convex functions we shall introduce later (Section 2.10).

Moreover, if $\{f_i(x)\}$, $i \in I \subset \mathbb{N}$, is a family (finite or not) of functions which are all convex on a convex set $X \subset \mathbb{R}^n$, then the function

$$\vartheta(x) = \sup_{i \in I} f_i(x)$$

is a convex function on X, since $\operatorname{epi} \vartheta = \bigcap_{i \in I} \operatorname{epi} f_i$ is a convex set.

Regarding the convexity of maps $f : X \to \mathbb{R}^m$, we say that a map (or vector-valued functon) f is convex on the convex set $X \subset \mathbb{R}^n$, if every component f_i, $i = 1, 2, ..., m$, is convex on X.

The following theorem gives results on the composition of convex functions; these results are particularly useful, especially when one has to prove the convexity of a given non-elementary function.

Theorem 2.5.8.

i) If f is convex on $\mathbb{R}^m$, then $f(Ax+b)$ is convex on $\mathbb{R}^n$, for any matrix A of order (m, n) and any vector $b \in \mathbb{R}^m$.

ii) Let $X \subset \mathbb{R}^n$ and $F \subset \mathbb{R}^m$ be convex sets; if $u : F \to \mathbb{R}$ is convex and non-decreasing on F, then for each map $f : X \to F$, convex on X, the composed function $\varphi = u[f(x)]$ is convex on X.

iii) Let $X \subset \mathbb{R}^n$ and $F \subset \mathbb{R}^m$ be convex sets and $J \subset \{1, 2, ..., m\}$. If $u : F \to \mathbb{R}$ is convex [concave] on F, monotone non-decreasing on F with respect to each component f_i, $i \in J$, and monotone non-increasing on F with respect to each component f_i, $i \notin J$, then for each map $f = [f_1, f_2, ..., f_m] : X \to F$ such that f_i is convex [concave] on X for each $i \in J$ and f_i is concave [convex] on X for each $i \notin J$, then the composed function $\varphi = u[f(x)]$ is convex [concave] on X.

Proof. See Bereanu (1969). We shall give the proof of a more general result when we shall introduce the generalized convex functions (see Section 2.15 of the present chapter). □

We have seen that $f : X \to \mathbb{R}$ is convex if and only if $\operatorname{epi} f$ is a convex set; therefore if we consider the function on $\mathbb{R}^n$, defined as $f = +\infty$, it is $\operatorname{epi} f = \emptyset$ and, as the empty set if convex, the function considered is convex too. Similarly, the function on $I\!R^n$, defined as $f = -\infty$ is also convex, since $\operatorname{epi} f = I\!R^{n+1}$. Again: consider a convex function f defined on a proper subset D of $I\!R^n$ and let

$$f_1(x) = \begin{cases} f(x) \ , & \text{if } x \in D \\ +\infty \ , & \text{if } x \notin D \ . \end{cases}$$

The epigraph of f is identical to the epigraph of f_1 where f_1 is defined on all $I\!R^n$. In this way we can always construct convex functions defined throughout $I\!R^n$. We shall speak then of *extended* convex functions. So $f : I\!R^n \to I\!R \cup \{\pm\infty\}$ is an extended convex function if $\operatorname{epi} f$ is convex; the *effective domain* of an extended convex function f is the set

$$\operatorname{dom}(f) = \{x \mid x \in I\!R^n, \ f(x) < +\infty\} \ .$$

It is actually the projection of $\operatorname{epi} f$ on $I\!R^n$. It is easy to see that f is an extended convex function if and only if the inequality

$$f(\lambda x^1 + (1-\lambda)\, x^2) \leqq \lambda f(x^1) + (1-\lambda)\, f(x^2)$$

is true for all $x^1, x^2 \in \operatorname{dom}(f)$ and for all $\lambda \in (0,1)$.

The effective comain of an extended convex function is a convex set in $I\!R^n$. The converse statement generally does not hold, i.e. if $\operatorname{dom}(f)$ is a convex set, f will not necessarily be an extended convex function. Allowing convex functions to take on infinite values requires some caution in arithemtic operations, such as the undefined operations $+\infty + (-\infty)$: see Rockafellar (1970), Section 4. Convex functions that have values at $+\infty$ and $-\infty$ do not arise frequently in applications; however, the *proper* convex functions, i.e. those (extended) convex functions such that

$$f(x) < +\infty \quad \text{for at least one } x, \text{ and}$$

$$f(x) > -\infty \quad \text{for all } x,$$

have a considerable theoretical importance. Extended convex functions that are not proper, are also called *improper*.

An extended (and also a real-valued) convex function can be discontinuous only at some boundary points of its effective domain (see also

Theorem 2.5.16).

Example 2.5.1. Consider the function $f : \mathbb{R} \to \bar{\mathbb{R}}$ given by

$$f(x) = \begin{cases} x^2 - 1 \ , & x < 1 \\ 2 \ , & x = 1 \\ +\infty \ , & x > 1 \end{cases}$$

which is discontinuous at the boundary point $x = 1$ of its effective domain. Some of these discontinuities can be eliminated by means of the *closure operation* for extended convex functions. We recall that an extended real-valued function f on $\mathbb{R}^n$ is *closed* (or lower semicontinuous) if one of the following equivalent conditions holds:

a) The epigraph of f is a closed set in $\mathbb{R}^{n+1}$.

b) All lower level sets of f are closed.

c) $\liminf\limits_{y \to x} f(y) = f(x), \ \forall x.$

By definition the supremum of closed functions is closed and we can introduce the *closure* of f:

$$\mathrm{cl}(f(x)) = \sup \{ g(x) \mid g(y) \leqq f(y), \ g \text{ closed}, \ \forall y \} \ .$$

This condition is equivalent to $\mathrm{epi}(\mathrm{cl}\, f) = \mathrm{cl}(\mathrm{epi}\, f)$ or also to $\mathrm{cl}(f(x)) = \liminf\limits_{y \to x} f(y), \ \forall x.$

Now, by definition, we get: f closed $\Leftrightarrow \mathrm{cl}(f) = f$. It holds the following result.

Theorem 2.5.9. If f is a convex function on $\mathbb{R}^n$, with $f(x) > -\infty$, $\forall x$, then

$$\begin{aligned} \mathrm{cl}(f(x)) &= \sup \{ g(x) \mid g(y) \leqq f(y), \ \forall y, \ g \text{ linear affine} \} = \\ &= \sup \{ ux + \gamma \mid uy + \gamma \leqq f(y), \ \forall y \} \ , \qquad \forall x \ . \end{aligned}$$

Since this result is closely connected with the theory of conjugate functions, we shall give its proof later, in Section 2.7 (see the second remark to Theorem 2.7.4). If $f(x) = -\infty$ for any x, then of course the above equality fails. Since we cannot find a linear affine function which is smaller than f, we have

$$\sup \{ux + \gamma \mid uy + \gamma \leqq f(y),\ \forall y\} = -\infty\,, \qquad \forall x\,,$$

but $\mathrm{cl}(f(x))$ is characterized by the closure of the epigraph of f.

Example 2.5.2. Consider the following (extended) convex function f defined on $I\!R$:

$$f(x) = \begin{cases} 0 \quad , & \text{if } x > 0 \\ K \quad , & \text{if } x = 0, \text{ with } K \in [0, +\infty] \\ +\infty \ , & \text{if } x < 0\,. \end{cases}$$

The closure of f is coinciding with f at each point, except for the origin, where $\mathrm{cl}\, f = 0$.

Example 2.5.3. Consider the (extended) convex function defined as

$$f(x) = \begin{cases} 0 \quad , & \text{if } \|x\| < 1\,; \\ +\infty \ , & \text{if } \|x\| \geqq 1\,. \end{cases}$$

Its closure is given by

$$\mathrm{cl}(f) = \begin{cases} 0 \quad , & \text{if } \|x\| \leqq 1\,; \\ +\infty \ , & \text{if } \|x\| > 1\,. \end{cases}$$

The following results are proved by Rockafellar (1970):

Theorem 2.5.10. If f is a nonproper convex function, then $f(x) = -\infty$, $\forall\, x \in \mathrm{relint}(\mathrm{dom}(f))$.

Therefore a nonproper convex function takes necessarily infinite values except perhaps at boundary points of its effective domain.

Theorem 2.5.11. For any extended convex function f, the set $\text{relint}(\text{epi}\, f)$ is given by all pairs (x, μ) such that $x \in \text{relint}(\text{dom}(f))$ and $f(x) < \mu < +\infty$.

Theorem 2.5.12. If f is a proper convex function on $\mathbb{R}^n$, also $\text{cl}\, f$ is a proper convex function. Moreover, $\text{cl}\, f$ is coinciding with f except at most at relative boundary points of $\text{dom}(f)$, i.e. on $\text{cl}(\text{dom}(f)) \setminus \text{relint}(\text{dom}(f))$.

Corollary 2.5.1. If f is a proper convex function with $\text{dom}(f)$ affine set, then f is closed.

Some important examples of (extended) convex functions which arise in optimization theory and in other contexts are:

a) The *distance function* of a point x from an arbitrary set $X \subset \mathbb{R}^n$:

$$d(x, X) = \inf \{\|x - y\| : y \in X\} .$$

We shall now show that, given a convex set $X \subset \mathbb{R}^n$, then $d(\cdot, X)$ is a convex function on $\mathbb{R}^n$. The reader may recall that given any point $x \notin \bar{X}$ there is a unique minimizing point (see Theorem 2.2.2). In other words

$$d(x, X) = d(x, \bar{X}) = \min \{\|x - y\| : y \in \bar{X}\} = \|x - x^0\| ,$$

where $x^0 \in \bar{X}$ is unique. On the other hand, if $x \in \bar{X}$, then clearly $d(x, X) = 0$ and x^0 is x itself. We will make use of this information in showing that the distance function is convex. Then we need to show that

$$d(\lambda x + (1 - \lambda)\, y, X) \leqq \lambda d(x, X) + (1 - \lambda)\, d(y, X) ,$$

for each $\lambda \in [0, 1]$. But

$$d(\lambda x + (1 - \lambda)\, y, X) = d(\lambda x + (1 - \lambda)\, y, \bar{X}) \leqq$$
$$\leqq \|\lambda x + (1 - \lambda)\, y - \lambda \hat{x} - (1 - \lambda)\, \hat{y}\| ,$$

where $d(x, X) = \|x - \hat{x}\|$ and $d(y, X) = \|y - \hat{y}\|$. But by convexity of X, $\bar{X}$ is convex, and so $\lambda\hat{x} + (1 - \lambda)\,\hat{y} \in \bar{X}$ for each $\lambda \in [0, 1]$. Therefore

$$d(\lambda x + (1 - \lambda)\,y, X) \leqq \|\lambda(x - \hat{x}) + (1 - \lambda)(y - \hat{y})\| \leqq$$
$$\leqq \lambda\,\|x - \hat{x}| + (1 - \lambda)\,\|y - \hat{y}\| = \lambda d(x, X) + (1 - \lambda)\,d(y, X)$$

and convexity of $d(\cdot, X)$ follows.

b) The *indicator function* of a nonempty set $X \subset I\!R^n$:

$$\delta(x, X) = \begin{cases} 0 \quad , & \text{if } x \in X \\ +\infty \; , & \text{if } x \notin X \end{cases}$$

is convex if and only if X is convex.

c) The *support function* of a nonempty convex set $X \subset I\!R^n$:

$$\delta^*(x, X) = \sup_{y \in X} \{x \cdot y\}$$

is a convex function.

d) The function

$$f^*(x) = \sup_{y} \{x \cdot y - f(y)\}$$

is said to be the *conjugate* of the closed convex function f on $I\!R^n$. The conjugate function f^* is a closed convex function; obviously, by definition, the support function is the conjugate of the distance function. More insights on conjugate functions will be given in Section 2.7 of this chapter.
The support function enjoys several important properties, proved mainly by Minkowski for $I\!R^n$ and Hörmander for topological spaces (see Hörmander (1954)). See also Rockafellar (1970), where the following results are proved.

Theorem 2.5.13. The support functions of the nonempty convex sets are the closed proper convex functions which are positively homogeneous. The support functions of the nonempty bounded convex sets are the finite positively homogeneous convex functions.

Theorem 2.5.14. Let C, D be nonempty *closed* convex subsets of $I\!R^n$. Then:

a) $C \subset D$ if and only if $\delta^*(x, C) \leqq \delta^*(x, D)$.

b) C is compact if and only if $\delta^*(x, C)$ is finite-valued.

Theorem 2.5.15. If f is an extended convex function, positively homogeneous and not identically $+\infty$, then $\operatorname{cl} f$ is the support function of the closed convex set

$$C = \{x^* \mid \forall x,\ x \cdot x^* \leqq f(x)\} .$$

e) The *gauge function* or *Minkowski function* of a nonempty set $X \subset I\!R^n$:

$$\gamma(x, X) = \inf \{\lambda \mid \lambda > 0,\ x \in \lambda X\}$$

is convex if X is convex. Note that γ essentially defines a kind of distance from the origin to x, measured with respect to X. It is the factor by which X must be expanded to include x.

An important feature of convex functions is related to their continuity. We have already noted that a function convex on the convex set X is not necessarily continuous on X. Consider, e.g., the function

$$f(x) = \begin{cases} 2 , & x = -1 \\ x^2 , & x > -1 \end{cases}$$

defined on $X = [-1, +\infty)$. It is quick to verify that f is convex on X, but discontinuous at $x = -1$. However, we have the following important result.

Theorem 2.5.16. Let $f : X \to I\!R$ be convex on the convex set $X \subset I\!R^n$; then f is continuous on $\mathrm{relint}(X)$.

Proof. Let $\bar{x} \in \mathrm{relint}(X)$; let us introduce the closed ball

$$\bar{B} = \{x \in I\!R^n : \|x - \bar{x}\| \leqq \delta\} \cap \mathrm{aff}(X)$$

and the hypercube

$$C(\delta) = \{x \in I\!R^n : x_i - \bar{x}_i| \leqq \delta,\ i = 1, 2, ..., n\} \cap \mathrm{aff}(X) \ ,$$

having $\bar{x}$ as center (see Figure 6).

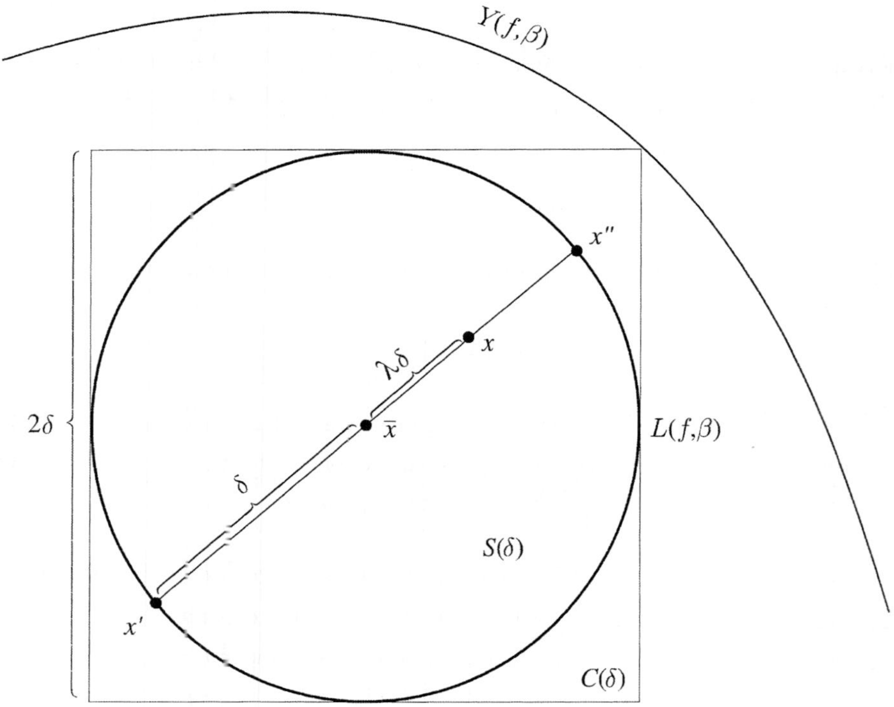

Figure 6.

Without loss of generality, we may suppose that $\exists\,\bar{\delta} > 0$ such that $\forall\,\delta \in [0, \bar{\delta})$, we have $C(\delta) \subset X$, and therefore $\bar{B} \subset C(\delta) \subset X$. Let x^i be an extreme point of $C(\delta)$ and let $\beta = \max_i \{f(x^i)\}$. On the ground of Theorem 2.5.3, $L(f, \beta)$ is convex. We observe moreover that every point $x \in C(\delta)$ can be expressed as a convex combination of the extreme points of $C(\delta)$, extreme points whose number is 2^K, where K is the dimension of $\mathrm{aff}(X)$.

The convexity of f implies

$$f(x) = f\Big(\sum_{i=1}^{2^K} \lambda_i x^i\Big) \leqq \sum_{i=1}^{2^K} \lambda_i f(x^i) \leqq \sum_{i=1}^{2^K} \lambda_i \beta = \beta\ .$$

It follows that $C(\delta) \subset L(f, \beta)$. Let now $x \in \bar{B}$, $x \neq \bar{x}$, and let x', x'' be the intersections of the line of $I\!R^n$, defined by $\bar{x}$ and x, with the boundary of $\bar{B}$. Let $\lambda \in [0, 1]$ and such that, denoted $x(\lambda) = (1 - \lambda)\,\bar{x} + \lambda x''$, it results $x(\lambda) = x$. Then

$$\bar{x} = \tfrac{1}{2}\,(x' + x'') = \tfrac{1}{2}\,x' + \frac{1}{2\lambda}\,x - \frac{1 - \lambda}{2\lambda}\,\bar{x}$$

and hence

$$\bar{x} = \frac{\lambda}{1 + \lambda}\,x' + \frac{1}{1 + \lambda}\,x\ .$$

The convexity of f gives then

$$f(x) = f(x(\lambda)) \leqq (1 - \lambda)\,f(\bar{x}) + \lambda f(x'') \leqq (1 - \lambda)\,f(\bar{x}) + \lambda\beta\ ;$$

$$\begin{aligned} f(\bar{x}) = f\Big(\frac{\lambda}{1 + \lambda}\,x' + \frac{1}{1 + \lambda}\,x\Big) &\leqq \frac{\lambda}{1 + \lambda}\,f(x') + \frac{1}{1 + \lambda}\,f(x) \ \leqq \\ &\leqq \ \frac{\lambda}{1 + \lambda}\,\beta + \frac{1}{1 + \lambda}\,f(x)\ . \end{aligned}$$

Subtracting $f(\bar{x})$ from both members of the first inequality and taking into account also the second inequality, we have

$$-\lambda[\beta - f(\bar{x})] \leqq f(x) - f(\bar{x}) \leqq \lambda[\beta - f(\bar{x})]$$

and therefore

$$|f(x) - f(\bar{x})| \leqq \frac{1}{\delta}\,[\beta - f(\bar{x})] \cdot \|x - \bar{x}\|\ ,$$

as

$$\|x - \bar{x}\| = \|(1 - \lambda)\,\bar{x} + \lambda x'' - \bar{x}\| = \|\lambda(x'' - \bar{x})\| = \lambda\delta\ .$$

It follows that, for each x such that $[\beta - f(\bar{x})] \cdot \|x - \bar{x}\| < \delta\varepsilon$, we have $|f(x) - f(\bar{x})| < \varepsilon$, and therefore f is continuous at x. □

If f is convex on an *open* convex set $X \subset I\!R^n$, it is therefore continuous on X. It can be proved (see, e.g. Rockafellar (1970)) that a convex function $f : X \to I\!R$, where $X \subset I\!R^n$ is an *open* set, is locally Lipschitzian at $x^0 \in X$ and therefore continuous at x^0. If $f : X \to I\!R$ is continuous on the convex set $X \subset I\!R^n$, we have another characterization, due to Jensen, of the convexity of f.

Theorem 2.5.17. Let f be continuous on the convex set $X \subset I\!R^n$; then f is convex on X if and only if

$$f[\tfrac{1}{2}\,(x^1 + x^2)] \leqq \tfrac{1}{2}\,[f(x^1) + f(x^2)]\ . \tag{4}$$

Proof. Necessity: obvious. Sufficiency: we shall prove that it results

$$f(\lambda x^1 + (1 - \lambda)\,x^2) \leqq \lambda f(x^1) + (1 - \lambda)\,f(x^2)\ ,\qquad \forall\,\lambda \in (0,1)\,,\ \ \forall\,x^1, x^2 \in X \tag{5}$$

if this relation is verified for $\lambda = 1/2$. We perform the proof in two stages.

a) For induction upon n we prove that (5) is verified for each number $\lambda = m/2^n$, with $m, n \in I\!N$, $m < 2^n$. Let us suppose that (5) is verified for $\lambda = m/2^n$, with $m = 1, ..., 2^n - 1$; let $\lambda = p/2^{n+1}$ with $p < 2^{n+1}$. If $p = 2k$, then (5) is verified by induction; if $p = 2k + 1$, letting $\lambda_1 = k/2^n$, $\lambda_2 = (k + 1)/2^n$ and $\tilde{x}^i = \lambda_i x^1 + (1 - \lambda_i)\,x^2$, $i = 1, 2$, we have $\lambda = \frac{1}{2}\,(\lambda_1 + \lambda_2)$ and therefore

$$\begin{aligned} f(\lambda x^1 + (1 - \lambda)\,x^2) &= f(\tfrac{1}{2}\,(\tilde{x}^1 + \tilde{x}^2)) \leqq \\ &\leqq \tfrac{1}{2}\,[f(\tilde{x}^1) + f(\tilde{x}^2)] \leqq \tfrac{1}{2}\,[(\lambda_1 + \lambda_2)\,f(x^1) + \\ &\quad + (2 - \lambda_1 - \lambda_2)\,f(x^2)] = \lambda f(x^1) + (1 - \lambda)\,f(x^2)\ . \end{aligned}$$

b) Using the diadic fractions, every $\lambda \in (0,1)$ can be written as limit of a sequence $\{\lambda_n\}$, with λ_n of the form that appears in a). Using then the continuity of f, we get the thesis. □

The functions satisfying inequality (4) are also called *midconvex functions.* Of course the class of midconvex functions is larger than the class of convex functions. Let us now state the boundedness properties of convex functions.

Theorem 2.5.18. If $f : X \to I\!R$ is convex on the convex set $X \subset I\!R^n$, then:

i) f is bounded on every compact subset $Y \subset \text{relint}(X)$;

ii) f is bounded from below on every bounded set $Y \subset X$.

Proof.

i) As f is continuous on $\text{relint}(X)$, it is bounded on every compact subset of X.

ii) Let $x^0 \in \text{relint}(X)$, with x^0 fixed. Then there exists $\delta > 0$ such that the closed ball $\bar{B} = \{y \in \text{aff}(X),\ \|y - x^0\| \leqq \delta\} \subset \text{relint}(X)$. As S is a compact set, then, on the grounds of i), $f(y) \leqq M$, $\forall y \in S$. Now let $x \in X$ be arbitrary and let $y(x) = x^0 + (x^0 - x)\,\delta/\rho$, with $\rho = \|x - x^0\|$; as $\|y(x) - x^0\| = \delta$, we have $y(x) \in S$, therefore $f(y(x)) \leqq M$. From the convexity of f we have

$$f(x^0) = f\Big(\frac{\delta}{\rho+\delta}\,x + \frac{\rho}{\rho+\delta}\,y(x)\Big) \leqq \frac{\delta}{\rho+\delta}\,f(x) + \frac{\rho}{\rho+\delta}\,M\ ,$$

hence

$$\delta f(x) \geqq \rho[f(x^0) - M] + \delta f(x^0) \geqq k[f(x^0) - M] + \delta f(x^0)\ ,$$

if $\rho = \|x - x^0\| \leqq k$. □

We conclude this section with a result that relates positively homogeneous functions and convex functions. We recall that a function f :

$X \to \mathbb{R}$, X cone of $\mathbb{R}^n$, is said to be *positively homogeneous* (*of the first degree*) if, for any number $\alpha \geqq 0$, $f(\alpha x) = \alpha f(x)$, $\forall x \in X$; f is said to be *subadditive* on $X \subset \mathbb{R}$ if, for any $x, y \in X$, $f(x+y) \leqq f(x) + f(y)$.

It is easy to verify the following result: if $f : K \to \mathbb{R}$ is defined on a convex cone $K \subset \mathbb{R}^n$ (with $0 \in K$), then if f is positively homogeneous on K, it is convex on K if and only if it is subadditive on K. This means that $\operatorname{epi} f$ is in this case a closed cone in $\mathbb{R}^{n+1}$. For other results concerning convex functions and homogeneous functions, see Newman (1969).

2.6. Directional Derivatives and Subgradients of Convex Functions

We will now discuss the notion of directional differentiability of a convex function. We first recall the notion of *directional derivative.* Let $f : \mathbb{R}^n \to \mathbb{R} \cup \{\pm\infty\}$ and let x be a point where f is finite; let g be a *direction* of $\mathbb{R}^n$. The *right-sided directional derivative* of f in the direction y is defined as:

$$f'_+(x; y) = \lim_{t \to 0^+} \frac{f(x+ty) - f(x)}{t} ,$$

if the said limits exist, finite or not. Similarly, the *left-sided directional derivative* of f at x in the direction y is defined as

$$f'_-(x; y) = \lim_{t \to 0^-} \frac{f(x+ty) - f(x)}{t} .$$

For $y = 0$, both f'_+ and f'_- are defined to be zero.

The reader can easily verify that

$$-f'_+(x; -y) = f'_-(x; y) ,$$

so that only the right-sided directional derivative could be considered (indeed in Chapter IV only right-sided generalized directional derivatives will be treated and the symbol "+" will be omitted in the related notations).

If $f'_+(x; y) = f'_-(x, y)$, then f admits *directional derivative* $f'(x, y)$ at x; so the unilateral derivative $f'_+(x; y)$ is bilateral if and only if $f'_+(x; -y)$ exists and $f'_+(x; -y) = -f'_+(x; y)$.

In case f is a convex function, the existence of $f'_+(x;y)$ is assured. Indeed we have the following

Theorem 2.6.1. Let f be an extended convex function and let x be a point where f is finite. Then, for each direction y, the ratio $[f(x+ty)-f(x)]/t$ is a nondecreasing function of $t > 0$, so that $f'_+(x;y)$ exists for every direction y and

$$f'_+(x;y) = \inf_{t>0} \frac{f(x+ty)-f(x)}{t} .$$

Moreover, $f'_+(x;y)$ is a convex and positively homogeneous function of y, with $f'_-(x;y) \leqq f'_+(x;y)$.

Proof. Let $x \in \mathbb{R}^n$ be any point such that $f(x)$ is finite. Define $h(y) = f(x+y) - f(x)$; the function h is convex, since $\operatorname{epi} h$ is obtained by translating the convex set $\operatorname{epi} f$ and $h(0) = 0$. Let $\lambda_2 > \lambda_1 > 0$ and $q_1 = \lambda_1/\lambda_2$, $q_2 = (\lambda_2 - \lambda_1)/\lambda_2$. Then $q_1 > 0$, $q_2 > 0$ and $q_1 + q_2 = 1$. Hence

$$h(\lambda_1 y) = h(q_1\lambda_2 y + q_2 0) \leqq q_1 h(\lambda_2 y) + q_2 h(0)$$

and

$$\frac{h(\lambda_1 y)}{\lambda_1} \leqq \frac{h(\lambda_2 y)}{\lambda_2} .$$

The last inequality indicates that $h(ty)/t$ is a nondecreasing function of $t > 0$. Thus for a convex function f we can write that for every direction y

$$f'_+(x;y) = \inf_{t>0} \frac{f(x+ty)-f(x)}{t}$$

and $f'_+(x;y)$ exists, although it may not be finite. Then

$$f'_+(x,\alpha y) = \lim_{\alpha t \to 0^+} \frac{\alpha[f(x+t\alpha y)-f(x)}{\alpha t} = \alpha f'_+(x;y) ,$$

hence f'_+ is positively homogeneous.

Next we show that f'_+ is a convex function of y by proving that f'_+ is a subadditive function with respect to the direction $y \in \operatorname{dom}(f'_+(x;\cdot))$. Let $y^1, y^2 \in \operatorname{dom}(f'_+(x,\cdot))$. Then $x+ty^1$, $x+ty^2 \in \operatorname{dom}(f)$ for $t > 0$ sufficiently small. Now we get

$$f'_+(x\,;\,y^1+y^2) = \lim_{t\to 0+} \frac{f(x+t(y^1+y^2))-f(x)}{t} =$$

$$= \lim_{t\to 0+} \frac{f(\frac{1}{2}(x+2ty^1)+\frac{1}{2}(x+2ty^2))-f(x)}{t} \leq$$

$$\leq \lim_{t\to 0+} \frac{[f(x+2ty^1)-f(x)]+[f(x+2ty^2)-f(x)]}{t} =$$

$$= f'_+(x;y^1)+f'_+(x;y^2)\ .$$

Finally, by subadditivity, for all y with $f'_+(x;y) < +\infty$ and $f'_+(x;-y) < +\infty$, we get

$$f'_+(x;y)+f'_+(x;-y) \geqq f'_+(x;0) = 0$$

and thus

$$f'_+(x;y) \geqq -f'_+(x;-y) = f'_-(x,y)\ .$$

In case of $f'_+(x;y) = +\infty$ or $f'_+(x;-y) = +\infty$ the last inequality holds trivially. □

Other results concerning the differentiability of convex functions are the following ones:

Theorem 2.6.2. If $f : X \to \mathbb{R}$ is convex on an open set $X \subset \mathbb{R}^n$ and all its partial derivatives exist at $x^0 \in X$, then f is differentiable at x^0.

Proof. See, e.g., Roberts and Varberg (1973). □

Theorem 2.6.3 (Rademacher's theorem). A locally Lipschitzian function f on an open set $X \subset \mathbb{R}^n$ is differentiable on X almost everywhere, i.e. except a set of points of measure zero in the Lebesgue sense. It follows that a convex function $f : X \to \mathbb{R}$, where $X \subset \mathbb{R}^n$ is open, is differentiable almost everywhere on X.

Proof. See, e.g., Saks (1937), Roberts and Varberg (1973) and Rockafellar (1970). □

Another important concept to be introduced is the *subgradient* of an extended convex function, a concept related to the ordinary gradient in the case of differentiable convex functions. We have seen in Theorem 2.5.2 that $f : X \to \mathbb{R}$, X open convex set of $\mathbb{R}^n$, is convex on X if and only if there exists a vector $u^0 \in \mathbb{R}^n$ such that for each $x^0 \in X$

$$f(x) - f(x^0) \geqq u^0(x - x^0) , \qquad \forall x \in X .$$

More generally, a vector ξ is called a *subgradient* of a convex function f at a point x, with $f(x)$ finite, if

$$f(y) \geqq f(x) + \xi(y - x) , \qquad \forall y \in \mathbb{R}^n . \tag{1}$$

The set of all subgradients of f at x is called the *subdifferential* of f at x and is denoted by $\partial f(x)$. Relation (1), which we refer to as the *subgradient inequality*, has a simple geometric meaning: it says that the graph of the linear affine function $h(y) = f(x) + \xi(y - x)$ is a non-vertical supporting hyperplane to the convex set epi f at the point $(x, f(x))$.

Clearly $\partial f(x)$ is a closed convex set (maybe void or a singleton) since $\xi \in \partial f(x)$ if and only if a system of (infinite) weak linear inequalities is satisfied. If $\partial f(x) \neq \emptyset$, f is said to be *subdifferentiable at* x.

For example the Euclidean norm $f(x) = \|x\|$, $x \in \mathbb{R}^n$, is not differentiable at $x = 0$, but it is subdifferentiable and $\partial f(0)$ is given by all vectors ξ such that $\|y\| \geq \xi \cdot y$, $\forall y$, i.e. by the closed ball of unitary radius.
From the definition of subgradient we have the following important case.

Proposition 2.6.4. If X is a nonempty convex set of $\mathbb{R}^n$, then $\partial \delta(x, X)$ is the normal cone to X at x.

Proof. By definition $\xi \in \partial \delta(x, X)$ if and only if

$$\delta(y, X) \geqq \delta(x, X) + \xi \cdot (y - x) , \qquad \forall y .$$

This condition means that $x \in X$ and $0 \geqq \xi \cdot (y - x)$, $\forall y \in X$, i.e. that ξ is normal to X at x. □

We must note that ξ is a subgradient of f at x if and only if $(\xi, -1) \in \mathbb{R}^{n+1}$ is a normal vector to a supporting hyperplane for epi f at $(x, f(x))$.

In fact ξ is a subgradient of f at x if and only if (1) holds, i.e. if and only if

$$(\xi, -1) \cdot (y - x, \alpha - f(x)) \leqq 0 \,, \qquad \forall\, \alpha \geqq f(y)$$

i.e. if and only if

$$(\xi, -1) \in N(\mathrm{epi}\, f, (x, f(x))) \,.$$

Subgradients can be characterized by means of directional derivatives, as shown by the next theorem.

Theorem 2.6.5. Let f be an extended convex function and let x be a point where f is finite. Then ξ is a subgradient of f at x if and only if

$$f'_+(x; y) \geqq \xi \cdot y \,, \qquad \forall y \,.$$

In fact the closure of $f'_+(x; y)$ as a convex function of y, is the support function of the closed convex set $\partial f(x)$.

Proof. Setting $z = x + ty$, we can write the subgradient inequality as

$$\frac{[f(x + ty) - f(x)]}{t} \geqq \xi \cdot y \,, \qquad \forall y, \;\; \forall t > 0 \,.$$

Since the difference quotient decreases to $f'_+(x; y)$ as $t \to 0^+$, the first part of the theorem is proved. For the second part we have

$$\begin{aligned} \mathrm{cl}(f'_+(x, y)) &= \sup\{cy + \gamma \mid cy + \gamma \leqq f'_+(x, y), \; \forall y\} = \\ &= \sup\{cy \mid cy \leqq f'_+(x, y), \; \forall y\} = \\ &= \sup\{cy \mid c \in \partial f(x)\} = \delta^*(y, \partial f(x)) \,. \end{aligned}$$

□

Example 2.6.1. The function $f(x_1, x_2, ..., x_n) = \max\{x_i\}$, $i = 1, 2, ..., n$, is a convex function. Let $I(x)$ be the set of indices i such that $x_i = f(x)$. Then

$$f'_+(x; y) = \lim_{t \to 0^+} \max_i \frac{(x_i + ty_i) - f(x)}{t}$$

and since for t sufficiently small every $i \notin I(x)$ can be omitted, we can write

$$f'_+(x;y) = \lim_{t\to 0+} \max_{i\in \bar{I}(x)} \frac{(x_i + ty_i) - f(x)}{t} =$$

$$= \lim_{t\to 0+} \max_{i\in I(x)} \frac{(x_i + ty_i - x_i)}{t} = \max_{i\in I(x)} y_i \ .$$

Hence $\partial f(x)$ consists of all vectors $\xi \in I\!R^n$ such that $\max_{i\in I(x)} y_i \geqq \xi \cdot y$, $\forall y \in I\!R^n$, i.e. of all vectors of components $x_1, x_2, ..., x_n$, such that $x_i \geqq 0$, $\sum_i x_i = 1$, $x_i = 0$ if $i \notin I(x)$.

It must be noted that the assertion of Theorem 2.6.5 is equivalent to $\xi \in \partial(f'_+(x;0))$, where f'_+ is now considered as a function only of the direction (i.e. x is fixed).

Theorem 2.6.6. Let f be a proper convex function; then

i) If $x \notin \mathrm{dom}(f)$, $\partial f(x) = \emptyset$.

ii) If $x \in \mathrm{relint}(\mathrm{dom}(f))$, $\partial f(x) \neq \emptyset$ and $f'_+(x;y)$ is closed and proper as a function of y, with

$$f'_+(x;y) = \sup\{\xi \cdot y \mid \xi \in \partial f(x)\} = \delta^*(x, \partial f(x)) \ .$$

iii) $\partial f(x)$ is nonempty and bounded if and only if $x \in \mathrm{int}(\mathrm{dom}(f))$, in which case $f'_+(x;y)$ is finite for every y.

Proof. Taking $y \in \mathrm{dom}(f)$ in the subgradient inequality, we see that the inequality cannot be satisfied by any ξ if $f(x) = +\infty$. If $x = \mathrm{relint}(\mathrm{dom}(f))$, the effective domain of $f'_+(x;y)$ is the affine set given by the subspace parallel to the affine hull of $\mathrm{dom}\, f$. Since $f'_+(x;y)$ vanishes at the origin, it cannot be identically $-\infty$ on this affine set. Therefore $f'_+(x;y)$ is proper (Theorem 2.5.10) and closed (Corollary 2.5.1); then $f'_+(x;y)$ is itself the support function of $\partial f(x)$ by Theorem 2.6.5, whence the supremum formula and the non-emptiness of $\partial f(x)$. If actually $\mathrm{relint}(\mathrm{dom}(f)) = \mathrm{int}(\mathrm{dom}(f))$, the effective domain of $f'_+(x;y)$ is the whole space, so that the support function $\delta^*(y;\partial f(x))$ is finite everywhere. On the other hand, as $\delta^*(y;\partial f(x))$ is the closure of $f'_+(x;y)$, if

$\delta^*(y; \partial f(x))$ is finite everywhere, $f'_+(x; y)$ must be finite everywhere. As $z \in \text{int}(C)$, with C convex set of $\mathbb{R}^n$, if and only if $\forall y \in \mathbb{R}^n \ \exists \varepsilon > 0$ such that $z + \varepsilon y \in C$ (see Rockafellar (1970), Corollary 6.4.1), then it is $x \in \text{int}(\text{dom}(f))$. As for what regards the boundedness of $\partial f(x)$, assume that $\partial f(x)$ is not bounded; then there exists a sequence $\xi^k \in \partial f(x)$ with $\|\xi^k\| \to \infty$. Set

$$y^k = \xi^k / \|\xi^k\| \ ;$$

then we have $f'_+(x; y^k) \geqq \xi^k \cdot y^k = \|\xi^k\| \to \infty$. Since $f'_+(x; \cdot)$ is continuous (it is convex and finite), it is bounded on the closed unit sphere. Thus we have a contradiction. □

We note that if $x \in \text{int}(\text{dom}(f))$, which is always true when f is a finite convex function on the whole space $\mathbb{R}^n$, then Theorem 2.6.6, iii), implies that

$$f'_+(x; y) = \max\{\xi \cdot y \mid \xi \in \partial f(x)\}$$

with $\partial f(x)$ convex and compact. The next theorem establishes the relationships between gradient and subgradient in case of differentiable functions.

Theorem 2.6.7. Let f be an extended convex function and let x be a point where f is finite. If f is differentiable at x, then $\nabla f(x)$ is the unique subgradient of f at x, so that in particular

$$f(y) \geqq f(x) + (y - x)\,\nabla f(x) \ , \qquad \forall y \ .$$

Conversely, if f has a unique subgradient at x, then f is differentiable at x.

Proof. If f is differentiable at x, we have $f'_+(x; y) = \nabla f(x) \cdot y$, $\forall y \in \mathbb{R}^n$. So, thanks to Theorem 2.6.5 we obtain $\nabla f(x) \cdot y \geqq \xi \cdot y$, $\forall y \in \mathbb{R}^n$, from which necessarily it follows $\partial f(x) = \{\nabla f(x)\}$. Of course we have

$$f(y) \geqq f(x) + (y - x)\,\nabla f(x) \ , \qquad \forall y \ .$$

On the contrary, if $\partial f(x)$ has a unique subgradient ξ, the inequality

$$f'_+(x;y) = \sup_{\xi \in \partial f(x)} \{\xi \cdot y\}$$

becomes

$$f'_+(x;y) = \xi \cdot y \ , \qquad \forall y \in \mathbb{R}^n \ .$$

Then $f'_+(x;y)$ is a continuous and linear function; thanks to the equality $f'_+(x;y) = -f'_+(x;-y)$, we can conclude that the convex function f admits gradient. So f results differentiable at x (Theorem 2.6.2). □

From the definition of subgradient it is immediate to get

$$\partial(\lambda f)(x) = \partial f(x) \ , \qquad \forall x \ , \ \ \forall \lambda > 0 \ .$$

This formula is also valid trivially if $\lambda = 0$, provided $\partial f(x) \neq \emptyset$. The following theorem gives an important calculus rule for subgradients.

<u>Theorem 2.6.8</u>. Let $f_1, ..., f_m$ be proper convex functions on $\mathbb{R}^n$ and let $f = f_1 + ... + f_m$. Then

$$\partial f_1(x) + ... + \partial f_m(x) \subset \partial f(x) \ , \qquad \forall x \ .$$

If the convex sets $\operatorname{relint}(\operatorname{dom} f_i)$, $i = 1, 2, ..., m$, have a common point, then

$$\partial f_1(x) + ... + \partial f_m(x) = \partial f(x) \ , \qquad \forall x \ .$$

<u>Proof</u>. See Rockafellar (1970), Theorem 23.8. □

We conclude this subject with an example of an extended convex function which is not subdifferentiable at a point where it is finite.

<u>Example 2.6.2</u>. The extended convex function, defined as

$$f(x) = \begin{cases} -x^{-\alpha} \ , & x \in \mathbb{R}, \ x \geqq 0 \ , \ \ 0 < \alpha < 1 \\ +\infty \ , & x \in \mathbb{R}, \ x < 0 \end{cases}$$

is differentiable at each point $x > 0$, and therefore also subdifferentiable at these points. At $x = 0$ f is not differentiable and is not subdifferentiable, although it is finite at 0. Indeed it is $f'_+(0; y) = f'(0) = -\infty$ and therefore there exists no vector ξ such that $f'_+(0; y) \geq \xi \cdot y$, $\forall y$.

2.7. Conjugate functions

From Section 2.2 of the present chapter it is known that a closed convex set X is given by the intersection of halfspaces containing X. If $f : \mathbb{R}^n \to \mathbb{R} \cup \{\pm\infty\}$ is a lower semicontinuous convex function, then its epigraph is a closed convex set. The mentioned assertion can be interpreted in such a manner that, assuming f to be proper, f is the supremum of affine functions.

This principle is closely connected with the notion of *conjugate functions*. Let $f : \mathbb{R}^n \to \mathbb{R} \cup \{\pm\infty\}$ be a function and $u \in \mathbb{R}^n$. To find the largest affine function $h(x) = ux - \alpha$, with $h(x) \leqq f(x)$ for all $x \in \mathbb{R}^n$, we get

$$\alpha = \sup_{x \in \mathbb{R}^n} \{ux - f(x)\} .$$

Definition 2.7.1. The function $f^* : \mathbb{R}^n \to \mathbb{R}$, defined by

$$f^*(u) = \sup_x \{ux - f(x)\}$$

is called the (Fenchel) *conjugate* function of f.
Since f is the supremum of affine functions, it is convex and lower semicontinuous.

Examples of conjugate functions

1) If $f(x) = 0$ for all $x \in \mathbb{R}^n$, then

$$f^*(u) = \sup_x \{ux\} = \begin{cases} 0 \quad , & \text{if } u = 0 \\ +\infty \quad , & \text{elsewhere} . \end{cases}$$

2) If $f(x) = ax + \beta$, where $a \in \mathbb{R}^n$, $\beta \in \mathbb{R}$, then

$$f^*(u) = \sup_x \{(u-a)\cdot x - \beta\} = \begin{cases} -\beta \ , & \text{if } u = a \\ +\infty \ , & \text{elsewhere .} \end{cases}$$

3) If $f(x) = \begin{cases} -\beta \ , & \text{if } x = a \\ +\infty \ , & \text{elsewhere} \end{cases}$, then

$$f^*(u) = \sup_x \{ux - f(x)\} = ax + \beta \ .$$

4) If $f(x) = \frac{1}{2}\,(\|x\|)^2$, then

$$f^*(x) = \sup_x \{ux - \tfrac{1}{2}\,(\|x\|)^2\} = \tfrac{1}{2}\,(\|u\|)^2 \ ,$$

since the supremum is attained for $x = u$.

5) If $f(x) = \delta(x, X) = \begin{cases} 0 \ , & \text{if } x \in X \\ +\infty \ , & \text{elsewhere} \end{cases}$, where $X \subset \mathbb{R}^n$ is a nonempty set, then

$$f^*(u) = \delta^*(x, X) = \sup_{x \in X} \{ux\} \ ,$$

the *support function* of X.

The following assertions can be obtained directly from the definition of f^*.

Lemma 2.7.1. We have:

1) $f \leqq g \Rightarrow g^* \leqq f^*$;

2) $g = f + \alpha\,(\alpha \in \mathbb{R}) \Rightarrow g^*(u) = f^*(u) - \alpha\,, \quad \forall\, u \in \mathbb{R}^n$;

3) $g = \lambda f\,(\lambda \in \mathbb{R}) \Rightarrow g^*(u) = \lambda f^*(u/\lambda)\,, \quad \forall\, u \in \mathbb{R}^n$;

4) $g(x) = f(x - x^0)\,(x, x^0 \in \mathbb{R}^n) \Rightarrow g^*(u) = f^*(u) + ux^0$;

5) $g(x) = \inf_{i \in I} f_i(x) \Rightarrow g^*(u) = \sup_{i \in I} f_i^*(u)$;

6) $g(x) = \sup_{i \in I} f_i(x) \Rightarrow g^*(u) \leq \inf_{i \in I} f_i^*(u)$.

If $f(x) = -\infty$ for any $x \in \mathbb{R}^n$, then of course we have

$$f^*(u) = \sup_x \{ux - f(x)\} = +\infty , \qquad \forall u \in \mathbb{R}^n .$$

If $f(x) = +\infty$ for all $x \in \mathbb{R}^n$, then we have

$$f^*(u) = \sup_x \{ux - f(x)\} = -\infty , \qquad \forall u \in \mathbb{R}^n .$$

Thus, it is interesting to discuss only the conjugate of proper functions. For these we have the following result.

Theorem 2.7.1. If $f : \mathbb{R}^n \to \mathbb{R} \cup \{\pm\infty\}$ is a proper convex function, then f^* is proper too.

Proof.

i) Since $\mathrm{dom}(f) \neq \emptyset$, we have

$$f^*(u) = \sup_x \{ux - f(x)\} > -\infty , \qquad \forall u .$$

ii) Let $x^0 \in \mathrm{relint}(\mathrm{dom}\, f) \neq \emptyset$. Since f is a proper convex function, then according to Theorem 2.6.6, we have $\partial f(x^0) \neq \emptyset$. For $u \in \partial f(x^0)$ we have

$$f(x) \geqq f(x^0) + u(x - x^0) , \qquad \forall x \in \mathbb{R}^n$$

and

$$ux^0 - f(x^0) \geqq ux - f(x) , \qquad \forall x \in \mathbb{R}^n .$$

This means

$$f^*(u) = \sup_x \{ux - f(x)\} \leqq ux^0 - f(x^0) < +\infty$$

and $\mathrm{dom}(f^*) \neq \emptyset$. □

In the proof of the previous theorem we have seen that

$$\bigcup_x \partial f(x) \subset \mathrm{dom}(f^*) \ .$$

The above inclusion is in general strict, as the equality may not hold, as shown by the following example.

Example 2.7.1. Let

$$f(x) = \begin{cases} \dfrac{1}{x} \ , & \text{if } x > 0 \\ +\infty \ , & \text{if } x \leqq 0 \ . \end{cases}$$

Then

$$(-\infty, 0) = \bigcup_x \partial f(x) \subset (-\infty, 0] = \mathrm{dom}(f^*) \ .$$

There are other connections between the conjugate functions and the subdifferentials; for this subject see, e.g., Rockafellar (1970). As a consequence of Definition 2.7.1 we have the so-called *Young-Fenchel inequality*:

$$f^*(u) + f(x) \geqq ux \ , \qquad \forall\, x \in \mathbb{R}^n \ , \ \ \forall\, u \in \mathbb{R}^n \ .$$

Theorem 2.7.2. The equality $f^*(u) + f(x^0) = u^0 \cdot x^0$ holds if and only if $u^0 \in \partial f(x^0)$.

Proof. The relation $u^0 \in \partial f(x^0)$ is equivalent to

$$u^0(x - x^0) \leqq f(x) - f(x^0) \ , \qquad \forall\, x \in \mathbb{R}^n \ .$$

This means

$$u^0 x - f(x) \leqq u \cdot x^0 - f(x^0) \ , \qquad \forall\, x \in \mathbb{R}^n$$

or equivalently

$$f^*(u^0) \leqq u \cdot x^0 - f(x^0)$$

i.e.

$$f^*(u^0) + f(x^0) \leqq u \cdot x^0 \ .$$

Taking the Young-Fenchel inequality into account, the thesis is proved. □

We can also introduce the conjugate of f^*, called the *bi-conjugate* of f and denoted f^{**}, according to $f^{**} = (f^*)^*$, i.e.

$$f^{**}(x) = \sup_{u \in \mathbb{R}^n} \{x \cdot u - f^*(u)\} .$$

Obviously f^{**} is a convex lower semi-continuous function, with $f^{**} \leqq f$. The question arises about the equality between f^{**} and f for which we have the following result.

Theorem 2.7.3. If $\partial f(x^0) \neq \emptyset$, then $f^{**}(x^0) = f(x^0)$.

Proof. Let $u^0 \in \partial f(x^0)$, then according to Theorem 2.7.2 we have

$$f^*(u^0) + f(x^0) = ux^0$$

and

$$f(x^0) \geqq f^{**}(x^0) = \sup_{u \in \mathbb{R}^n} \{x \cdot u - f^*(u)\} \geqq f(x^0) . \qquad \square$$

In particular we have the following implications:

$$u^0 \in \partial f(x^0) \Leftrightarrow u^0 \cdot x^0 = f(x^0) + f^*(u^0) ;$$

$$u^0 \in \partial f(x^0) \Rightarrow u^0 \cdot x^0 = f^{**}(x^0) + f^*(u^0) \Leftrightarrow x^0 \in \partial f^*(u^0) .$$

Thus, from $u^0 \in \partial f(x^0)$ we can deduce $x^0 \in \partial f^*(u^0)$. The converse implication is true if $f = f^{**}$.

Theorem 2.7.4 (Fenchel-Moreau theorem). Let $f > -\infty$; then $f = f^{**}$ if and only if f is convex and lower semi-continuous.

Proof. Since the "only if" part of the theorem is obvious, let us assume f convex and lower semi-continuous. If it is identically $f = +\infty$, there is nothing to prove. In other cases epi $f \neq \emptyset$ is a closed convex set.
Assuming the existence of a point $x^0 \in \mathbb{R}^n$, $x^0 \in \mathrm{dom}(f)$, with $f^{**}(x^0) < f(x^0)$, we have $(x^0, f^{**}(x^0)) \notin \mathrm{epi}\, f$. Hence, using the strong separation theorem, we can find a vector $(v, v_0) \in \mathbb{R}^{n+1}$, $(v, v_0) \neq (0, 0)$, such that

$$\sup_{(x,\alpha)\in \text{epi}\, f} \{vx + v_0\alpha\} < v \cdot x^0 + v_0 f^{**}(x^0) \,. \tag{1}$$

If $v_0 > 0$, then with α sufficiently large we get a contradiction. If $v_0 = 0$ we get

$$\sup_{x\in \text{dom}(f)} vx < vx^0 \,.$$

Let $u \in \text{dom}(f^*)$ and $t > 0$. Then

$$\begin{aligned} f^*(u+tv) &= \sup_x \{(u+tv)\,x - f(x)\} \leqq \\ &\leqq \sup_{x\in \text{dom}(f)} \{ux - f(x)\} + t \sup_{x\in \text{dom}(f)} vx = \\ &= f^*(u) + t \sup_{x\in \text{dom}(f)} vx \end{aligned}$$

and

$$\begin{aligned} f^{**}(x^0) &\geqq (u+tv)\,x^0 - f^*(u+tv) \geqq \\ &\geqq (u+tv)\,x^0 - (f^*(u) + t \sup_{x\in \text{dom}(f)} vx) = \\ &= (ux^0 - f^*(u)) + t(vx^0 - \sup_{x\in \text{dom}(f)} vx) \,. \end{aligned}$$

Since this inequality is true for all $t > 0$, then we get $f^{**}(x^0) = +\infty$, which is a contradiction.

Thus $v_0 < 0$. Without loss of generality we can set $v_0 = -1$; then we have $f^*(v) = \sup_x \{vx - f(x)\} < vx^0 - f^{**}(x^0)$, in contradiction to the Young-Fenchel inequality. □

About the previous theorem the following remarks are useful.

a) Because of

$$\begin{aligned} f^{**}(x) &= \sup_x \{ux - f^*(x)\} = \\ &= \sup \{ux + \gamma \mid -\gamma \geqq f^*(u)\} = \\ &= \sup \{ux + \gamma \mid -\gamma \geqq uy - f(y),\ \forall y\} = \\ &= \sup \{ux + \gamma \mid uy - \gamma \leqq f(y),\ \forall y\} \,, \qquad \forall x \,, \end{aligned}$$

the equality $f = f^{**}$ means that f can be represented as a supremum of linear-affine functions which are smaller than f.

b) If f is convex, but not lower semi-continuous (i.e. not closed), then in case of $f > -\infty$ we have $f^{**}(x) = \mathrm{cl}(f(x))$. Together with the former remark we get

$$\mathrm{cl}(f(x)) = \sup\{ux + \gamma \mid uy + \gamma \leqq f(y),\ \forall y\}\ , \qquad \forall x\ ,$$

hence the assertion of Theorem 2.5.9 is proved.

c) Finally, if f is an arbitrary function which admits a linear-affine function smaller than f, then f^{**} provides the so-called *closed convex hull* of f, i.e. the greatest closed convex function which is smaller than f.

In order to give some calculus rules on the conjugate of the sum of functions, we must introduce the notion of *infimal convolution.*

Definition 2.7.2. Let $f_1, f_2, ..., f_m : I\!R^n \to I\!R \cup \{\pm\infty\}$; the function, denoted by

$$f_1 \oplus ... \oplus f_m (f_1 \oplus ... \oplus f_m : I\!R^n \to I\!R \cup \{\pm\infty\})$$

and defined as

$$(f_1 \oplus ... \oplus f_m)(x) = \inf_{\sum x^i = x} \{f_1(x^1) + ... + f_m(x^m)\}\ ,$$

is called the *infimal convolution* of $f_1, ..., f_m$.

Theorem 2.7.5. Let $f_1, ..., f_m : I\!R^n \to I\!R \cup \{\pm\infty\}$; then the following properties hold:

1) $(f_1 \oplus ... \oplus f_m)^* = f_1^* + ... + f_m^*$;

2) $(f_1 + ... + f_m)^* \leqq f_1^* \oplus ... \oplus f_m^*$;

3) If $f_1, ..., f_m$ are proper convex functions, with $\bigcap_i \mathrm{relint}(\mathrm{dom}(f_i)) \neq \emptyset$, then

$$(f_1 + ... + f_m)^* = f_1^* \oplus ... \oplus f_m^*\ .$$

Proof. The first and second assertion can be derived directly from the definition of conjugation and infimal convolution. For the proof of assertion 3) see, e.g., Rockafellar (1970). □

2.8. Extrema of Convex Functions

Convex functions (and their generalizations) play a central role in the analysis of extremum problems. The importance of convex functions in problems of this kind lies mainly in some basic properties, described in the following theorems.

Theorem 2.8.1. Let f be a proper convex function on $\mathbb{R}^n$. Then every local minimum of f in $\mathrm{dom}(f)$ is a global minimum of f on $\mathbb{R}^n$.

Proof. If $x^* \in \mathrm{dom}(f)$ is a local minimum, then

$$f(x) \geqq f(x^*)$$

for all x in a sufficiently small neighbourhood $N(x^*)$. Let z be any point in $\mathbb{R}^n$. Then $((1-\lambda)\, x^* + \lambda z) \in N(x^*)$ for sufficiently small $\lambda \in (0,1)$ and

$$f((1-\lambda)\, x^* + \lambda z) \geqq f(x^*) \ .$$

Since f is a proper convex function,

$$(1-\lambda)\, f(x^*) + \lambda f(z) \geqq f((1-\lambda)\, x^* + \lambda z) \ .$$

Adding the last two inequalities and dividing the result by λ, we obtain

$$f(z) \geqq f(x^*) \ .$$

that is, x^* is a global minimum point. □

Corollary 2.8.1. Let f be a proper convex function on $\mathbb{R}^n$ and let $X \subset \mathrm{dom}(f)$ be a convex set; then every local minimum of f at $x \in X$ is a global minimum of f over X.

Note that generally the minimal value of a convex function can be attained at more than one point. We will now show that the set of the minimizing points of a proper convex function is a convex set.

Theorem 2.8.2. Let f be an extended convex function on $\mathbb{R}^n$. The set of points at which f attains its minimum is convex.

Proof. Let α^* be the value of f at the minimizing points. Then the set $L(f, \alpha^*) = \{x \,|\, x \in \mathbb{R}^n,\ f(x) \leqq \alpha^*\}$ is precisely the set of points at which f attains its minimum, and by Theorem 2.5.3 this set is a convex set. □

Another result, quite important in some applications, is worth mentioning.

Theorem 2.8.3. Let $f : X \to \mathbb{R}$ be a strictly convex function on the convex set $X \subset \mathbb{R}^n$. If f attains its minimum on X, this minimum is attained at a unique point of X.

Proof. Suppose that the minimum is attained at two distinct points $x^1 \in X$, $x^2 \in X$ and let $f(x^1) = f(x^2) = \alpha$. It follows from Theorem 2.8.2 that for every $\lambda \in [0, 1]$ we have $f(\lambda x^1 + (1 - \lambda)\, x^2) = \alpha$, contradicting the fact that f is strictly convex. □

The following proposition generalizes a well-known property of differentiable convex functions.

Theorem 2.8.4. Let f be an extended convex functon and let $f(x^*)$ be finite; a necessary and sufficient condition for x^* to be a minimum point for f is that

$$0 \in \partial f(x^*) \ .$$

Proof. By the definition of subgradient, $0 \in \partial f(x^*)$ if and only if $f(y) \geqq f(x^*)$ for every $y \in \mathbb{R}^n$; that is, x^* is a minimum point for f. □

According to Theorem 2.6.5, one has $0 \in \partial f(x^*)$ if and only if f is finite at x^* and $f'_+(x^*, g) \geqq 0,\ \forall g$. Therefore if f is a differentiable (extended) convex function on $\mathbb{R}^n$, then $\nabla f(x^*) = 0$ if and only if f attains its minimum at x^*. This result remains valid also if we replace $\mathbb{R}^n$ by some open convex subset X of $\mathbb{R}^n$ such that $x^* \in X$. It also indicates the familiar fact that in seeking the unconstrained minimum of a (twice differentiable) convex function, no second-order conditions need to be checked at the stationary points. Other remarkable theorems on the extreme of (extended) convex functions are given in Section 27 of Rockafellar's book on convex analysis (Rockafellar (1970)).

2.9. Systems of Convex Functions and Nonlinear Theorems of the Alternative

We conclude the topics related to convex functions with some results concerning systems of inequalities of convex functions which generate theorems of the alternative for the nonlinear case. The first result is due to Fan, Glicksberg and Hoffman (1957) .

Theorem 2.9.1. Let $f_1, f_2, ..., f_m$ be real-valued convex functions, all defined on the convex set $X \subset \mathbb{R}^n$ and let $g_1, g_2, ..., g_k$ be linear affine functions on $\mathbb{R}^n$. If the system

$$x \in X\,, \quad f_i(x) < 0\,, \quad i = 1, ..., m\,; \qquad g_i(x) = 0\,, \quad i = 1, ..., k$$

admits no solution, then there exist a nonnegative vector $u \in \mathbb{R}^m$ and a vector $v \in \mathbb{R}^k$, with $(u, v) \neq 0$, such that

$$u \cdot f(x) + v \cdot g(x) \geqq 0\,, \qquad \forall x \in X\,.$$

Proof. The set

$$Y = \bigcup_{x \in X} \{(y, z) \mid y \in \mathbb{R}^m,\ z \in \mathbb{R}^k,\ y > f(x),\ z = g(x)\}$$

is convex and $0 \notin Y$. From the first separation Theorem 2.2.5 it follows that there exist a vector $u \in \mathbb{R}^m$ and a vector $v \in \mathbb{R}^k$, with $(u, v) \neq 0$, such that

$$u \cdot y + v \cdot z \geqq 0$$

for each $(y, z) \in Y$.

Since each y_i can be made as large as desired, it is $u \geqq 0$. Let $\varepsilon > 0$;

we then have, for each $x \in X$, $(f(x) + \varepsilon e, g(x)) \in Y$ and therefore, for $x \in X$:

$$u \cdot f(x) + v \cdot g(x) + \varepsilon u \cdot e \geqq 0$$

or

$$u \cdot f(x) + v \cdot g(x) \geqq -\varepsilon u \cdot e , \qquad \text{for } x \in X .$$

This inequality holds for all $\varepsilon > 0$, thus

$$u \cdot f(x) + v \cdot g(x) \geqq 0 , \qquad \forall x \in X .$$ □

The following result may be viewed as a generalization of Gordan's theorem of the alternative (result 11) of Section 2.4).

Corollary 2.9.1. Let $f_1, f_2, ..., f_m$ be real-valued convex functions, all defined on the convex set $X \subset \mathbb{R}^n$. Then either the system

$$x \in X , \quad f_i(x) < 0 , \quad i = 1, ..., m$$

admits solution x or

$$uf(x) \geqq 0 , \quad \forall x \in X , \quad \text{for some } u \geq 0 , \quad u \in \mathbb{R}^m ,$$

but never both.

Proof. Let $\bar{x} \in X$ be a solution of $f(x) < 0$; then for any semipositive vector $u \in \mathbb{R}^m$ we have $uf(\bar{x}) < 0$. Conversely if the system $x \in X$, $f(x) < 0$, has no solution, from Theorem 2.9.1 it follows that there exists a semipositive vector u such that $uf(x) \geqq 0$, $\forall x \in X$. □

It is possible to sharpen Theorem 2.9.1 and make it also a theorem of the alternative if we let $X = \mathbb{R}^n$, $g(x) = Bx - d$, with B a matrix of order k, n and rows linearly independent. Indeed we have the following:

Corollary 2.9.2. Let $f_1, f_2, ..., f_m$ be convex on $\mathbb{R}^n$; let B be a given matrix of order k, n and with linearly independent rows; let d be a given k-dimensional vector. Then either the system

$$f(x) < 0 , \quad Bx = d$$

has a solution $x \in \mathbb{R}^n$ or there exist a vector $u \geq 0$, $u \in \mathbb{R}^m$, and a vector $v \in \mathbb{R}^k$ such that

$$uf(x) + v(Bx - d) \geqq 0 , \qquad \forall\, x \in \mathbb{R}^n$$

but never both.

Proof. See Mangasarian (1969). □

The following theorem, due to Berge (1963), may be considered a nonlinear version of the Farkas-Minkowski theorem of the alternative. We need a previous result.

Lemma 2.9.1. Let $X \subset \mathbb{R}^n$ be a convex set and A a real (m,n) matrix. If the system $Ax \leqq b$, $x \in X$, admits no solution but becomes possible if any one of the inequalities $A_i x \leqq b_i$ $(i = 1, 2, ..., m)$ is eliminated, then there exists $x \in X$ such that $Ax > b$.

Proof. Apply Theorem 2.2.9 with $X_i = \{x \mid A_i x \leqq b_i\}$. □

Theorem 2.9.2. Let $f, f_1, f_2, ..., f_p$ be convex functions on $\mathbb{R}^n$ and A a real (m,n) matrix. If the system

$$f(x) < 0 \; ; \qquad f_k(x) \leqq 0 , \quad k = 1, ..., p \; ,; \qquad Ax \leqq b$$

admits no solution, but there exists a vector $x^0 \in \mathbb{R}^n$ such that the system

$$f_k(x^0) < 0 , \quad k = 1, ..., p \; ; \qquad Ax^0 \leqq b$$

admits solution, then there exist numbers $y_1 \geqq 0, ..., y_p \geqq 0$, $u_1 \geqq 0, ...,$ $u_m \geqq 0$ such that

$$\mathcal{L}(x) = f(x) + \sum_{k=1}^{p} y_k \cdot f_k(x) + \sum_{i=1}^{m} u_i (A_i x - b_i) \geqq 0 , \qquad \forall x \in \mathbb{R}^n .$$

Proof. The proof will be done by induction upon m.

i) For $m = 0$ the property is true: in this case we have to prove that it is

$$\lambda f(x) + \sum_{k=1}^{p} y_k \cdot f_k(x) \geqq 0 , \qquad \forall\, x \in \mathbb{R}^n ,$$

with $\lambda \neq 0$ (i.e. $\lambda > 0$). Indeed, because of the assumptions, also the system

$$f(x) < 0 , \qquad f_k(x) < 0 , \qquad k = 1, ..., p$$

has no solution. Hence, by Theorem 2.9.1 we can find multipliers $\lambda \geqq 0$, $y_k \geqq 0$ (not all vanishing) such that the above inequality holds. Now, of course, if $\lambda = 0$, then the multipliers y_k are also all zero, as if one of them is positive, it will be $\mathcal{L}(x^0) < 0$, which is absurd.

ii) Let us suppose the property true for $m - 1$ and let us prove that it is true for m. We consider two cases:

a) There exists an index q, $1 \leqq q \leqq m$, such that the subsystem obtained by the elimination of the inequation $A_q x \leqq b_q$, remains impossible. Therefore, choosing $u_q = 0$, the property is verified also for m.

b) The elimination of any one of the inequations $A_i x \leqq b_i$, $i = 1, ..., m$, makes the system possible. Therefore, thanks to Lemma 2.9.1, with $X = \{x \mid f(x) < 0,\ f_k(x) \leqq 0,\ k = 1, ..., p\}$, there exists $\bar{x} \in X$ such that $A\bar{x} > b$. On the other hand, as the original system has no solution, on the grounds of Theorem 2.9.1 there exist nonnegative multipliers λ, y_k, u_i ($k = 1, ..., p$; $i = 1, ..., m$), not all zero, such that

$$\mathcal{L}(x) = \lambda f(x) + \sum_{k=1}^{p} y_k f_k(x) + \sum_{i=1}^{m} u_i \cdot (A_i x - b_i) \geqq 0 ,$$
$$\forall x \in I\!R^n .$$

Now we prove that the multiplier λ is positive. Suppose the contrary: then $y_1 = y_2 = ... = y_p = 0$, as if $y_k > 0$, then it is $\mathcal{L}(x^0) < 0$. Therefore $\mathcal{L}(x) = u \cdot (Ax - b) \geqq 0$, $\forall x \in I\!R^n$, i.e. $\mathcal{L}(x) = 0$, $\forall x \in I\!R^n$, being $\mathcal{L}(x^0) = 0$, as $\mathcal{L}(x^0) \geqq 0$ and $\mathcal{L}(x^0) \leqq 0$ by assumption. On the other hand, from $A\bar{x} - b > 0$ and from the fact that $u_1 \geqq 0, ..., u_m \geqq 0$, $u \neq 0$, it results $\mathcal{L}(\bar{x}) > 0$, which is absurd. □

Theorem 2.9.2 holds also if the functions involved are defined on a solid (i.e. with a nonempty interior) convex set $X \subset \mathbb{R}^n$ and $x^0 \in \text{int}(X)$.

Let us now deduce the theorem of the alternative of Farkas-Minkowski from Theorem 2.9.2. It is quite evident the systems

$$Ax = b \ , \qquad x \geqq 0 \tag{S_1}$$

and

$$A^T y \geqq 0 \ , \qquad by < 0 \tag{S_2}$$

cannot admit simultaneously solution. It remains to prove that if (S_2) does not admit solution, then (S_1) admits solution. Let us write (S_2) as

$$f(y) = by < 0 \ , \qquad -A^T y \leqq 0$$

and let (S_2) be impossible. Then, thanks to Theorem 2.9.2, there exists a nonnegative vector x such that

$$by - x^T A^T y = y^T (b - Ax) \geqq 0 \ , \qquad \forall\, y \in \mathbb{R}^m \ ,$$

hence it is $b - Ax = 0$, $x \geqq 0$. □

Another important theorem for systems of convex and linear functions is due to Bohnenblust, Karlin and Shapley (1950):

Theorem 2.9.3. Let $X \subset \mathbb{R}^n$ be a nonempty compact convex set and let f_i, $i \in M$, be a finite or infinite family $\mathcal{M}$ of real-valued convex functions, lower semicontinuous on X; let h_i, $i \in K$, be a finite or infinite family $\mathcal{K}$ of linear (affine) functions on $\mathbb{R}^n$. If the system

$$f_i(x) \leqq 0 \, , \ \ i \in M \ ; \qquad h_i(x) = 0 \, , \ \ i \in K \ ; \qquad x \in X$$

has no solution, then for some finite subfamily $f_1, f_2, ..., f_m$ of $\mathcal{M}$ and some finite subfamily $h_1, h_2, ..., h_k$ of $\mathcal{K}$, there exist vectors $p \in \mathbb{R}^m$ and $q \in \mathbb{R}^k$ such that

$$p \geqq 0 \ \ (p, q) \neq 0 \, , \ \ pf(x) + qh(x) \geqq 0 \ , \qquad \forall\, x \in X \ .$$

If K is empty, the last inequality becomes strict.

Proof. See Berge and Ghouila-Houri (1965) or Mangasarian (1969). □

A more general approach to the theorems of the alternative involving nonlinear systems has been proposed by Giannessi (1984) and Cambini (1986). Assume we are given the positive integers n and v, the nonempty sets $\mathcal{H} \subset \mathbb{R}^v$, $X \subset \mathbb{R}^n$, $Z \subset \mathbb{R}$ and the real-valued function $F : X \to \mathbb{R}^v$. We want to study the conditions for the generalized system

$$F(x) \in \mathcal{H} , \qquad x \in X$$

to have (or not to have) solution.

Definition 2.9.1. The function $w : \mathbb{R}^v \to \mathbb{R}$ is called *weak separation function* if

$$\mathcal{H}^w = \{h \in \mathbb{R}^v \mid w(h) \notin Z\} \supset \mathcal{H} ;$$

$S : \mathbb{R}^v \to \mathbb{R}$ is called *strong separation function* if

$$\mathcal{H}^s = \{h \in \mathbb{R}^n \mid s(h) \notin Z\} \subset \mathcal{H} .$$

The following theorems hold:

Theorem 2.9.4. Let the sets $\mathcal{H}$, X and the function F be given.

i) The systems

$$F(x) \in \mathcal{H}, \quad x \in X \text{ and } w(F(x)) \in Z , \qquad \forall x \in X$$

are not simultaneously possible whatever the weak separation function w might be.

ii) The systems

$$F(x) \in \mathcal{H}, \quad x \in X \text{ and } s(F(x)) \in Z , \qquad \forall x \in X$$

are not simultaneously impossible whatever the strong separation function s might be.

Proof.

i) If $F(x) \in \mathcal{H}$, $x \in X$, is possible, i.e. if $\exists \bar{x} \in X$ such that

$$\bar{h} = F(\bar{x}) \in \mathcal{H} ,$$

then we have $w(F(\bar{x})) = w(\bar{h}) \notin Z$, so that the second system of i) is false.

ii) If $F(x)$ admits no solution, i.e. if $h = F(x) \notin \mathcal{H}$, $\forall x \in X$, then we have $s(F(x)) = s(h) \in Z$, $\forall x \in X$, so that the second system of ii) is true. This completes the proof. □

Many applications of the previous general theorem to nonlinear theorems of the alternative, optimality, regularity and duality conditions for mathematical programming problems are given in Cambini (1986), Giannessi (1984, 1987), Martein (1985), Pellegrini (1991), Tardella (1989). Here we report only the following one: consider the particular case where $v = l+m$; $\mathcal{H} = \{(u,w) \in I\!R^l \times I\!R^m : u \in \text{int}(U),\ w \in V\}$; $Z = (-\infty, 0]$; $f : X \to I\!R^l$; $g : X \to I\!R^m$; $F(x) = f(x), g(x))$; $h = (u,w)$, where the positive integers l and m, the closed convex cones $U \subset I\!R^l$, $V \subset I\!R^m$, with $\text{int}(U) \neq \emptyset$ (otherwise $\mathcal{H} = \emptyset$), and the functions f, g are given.

Definition 2.9.2. Let C be a convex cone. A function $F : X \to I\!R^s$ is said to be *C-convex-like* on the convex set $X \subset I\!R^n$, if $\forall x, y \in X$ there exists $z \in X$ such that $F(z) - (1-\alpha)\, F(x) - \alpha F(y) \in C$, $\forall \alpha \in [0,1]$.

Theorem 2.9.5. Let $F(x) = f(x), g(x))$ be $\text{cl}(\mathcal{H})$-convex-like.

i) If the system

$$f(x) \in \text{int}(U) \ ; \qquad g(x) \in V \ ; \qquad x \in X \tag{1}$$

is impossible, then:

$$\exists \bar{\vartheta} \in U^* , \quad \bar{\lambda} \in V^* , \qquad \text{with } (\bar{\vartheta}, \bar{\lambda}) \neq 0 ,$$

such that

$$\bar{\vartheta} f(x) + \bar{\lambda} g(x) \leqq 0 \,, \qquad \forall x \in X \,.$$

ii) If the previous inequality holds and moreover

$$\{x \in X \mid f(x) \in \operatorname{int}(U),\ g(x) \in V,\ \bar{\lambda} g(x) = 0\} = \emptyset \,, \quad \text{when } \bar{\vartheta} = 0 \,,$$

then system (1) is impossible.

Proof. See Cambini (1986), Giannessi (1984). □

2.10. Generalized Convex Functions

Generalized convex functions play a very important role in optimization theory; historically the first type of generalized convex function was considered by De Finetti (1949) who first introduced the *quasiconvex functions* (a name given later by Fenchel (1953)).

We have seen in Theorem 2.5.3 that a *necessary*, but not sufficient, condition for a function $f : X \to \mathbb{R}$ to be convex on the convex set $X \subset \mathbb{R}^n$ is that its lower-level sets

$$L(f, \alpha) = \{x \mid x \in X,\ f(x) \leqq \alpha,\ \alpha \in \mathbb{R}\}$$

are convex for any real number α. A quasiconvex function on a convex set $X \subset \mathbb{R}^n$ is just a function characterized by the convexity of its lower-level sets. We have therefore the following definition.

Definition 2.10.1. The function $f : X \to \mathbb{R}$ is *quasiconvex* on the convex set $X \subset \mathbb{R}^n$ if $L(f, \alpha)$ is convex for any $\alpha \in \mathbb{R}$.

We shall see that an equivalent definition of a quasiconvex function $f : X \to \mathbb{R}$, where $X \subset \mathbb{R}^n$ is convex, is the following one:

$$x^1, x^2 \in X \,, \quad f(x^1) \leqq f(x^2) \Rightarrow f(\lambda x^1 + (1 - \lambda)\, x^2) \leqq f(x^2) \,, \qquad \forall \lambda \in [0, 1]$$

or, equivalently, in a symmetric form:

$$f(\lambda x^1 + (1-\lambda)\, x^2) \leqq \max\{f(x^1), f(x^2)\} \,,$$
$$\forall\, x^1, x^2 \in X \,, \quad \forall\, \lambda \in [0,1] \,.$$

Theorem 2.10.1. If $X \subset \mathbb{R}^n$ is a convex set and $f : X \to \mathbb{R}$, then the following conditions are equivalent:

i) f is quasiconvex on X, i.e.

$$f(\lambda x^1 + (1-\lambda)\, x^2) \leqq \max\{f(x^1), f(x^2)\} \,,$$
$$\forall\, x^1, x^2 \in X \,, \quad \forall\, \lambda \in [0,1] \,.$$

ii) For each $x \in X$ and each $y \in \mathbb{R}^n$ the function $g_{x,y}(t) = f(x+ty)$ is quasiconvex on the interval

$$T_{x,y} = \{t \mid t \in \mathbb{R},\ x+ty \in X\} \,.$$

iii) For each $x^1, x^2 \in X$ the function $h_{x^1,x^2}(\lambda) = f(\lambda x^1 + (1-\lambda)\, x^2)$ is quasiconvex on the interval $[0,1]$.

iv) For each $\alpha \in \mathbb{R}$ the lower-level set

$$L(f,\alpha) = \{x \mid x \in X,\ f(x) \leqq \alpha\}$$

is convex (recall that the empty set $\emptyset$ is convex by definition).

v) For each $\alpha \in \mathbb{R}$ the strict lower-level set

$$SL(f,\alpha) = \{x \mid x \in X,\ f(x) < \alpha\}$$

is convex.

Proof.
i) $\Rightarrow$ ii) $\Rightarrow$ iii): the proof is quite similar to the one given in Theorem 2.5.1 for the corresponding implications regarding convex functions.

iii) $\Rightarrow$ iv). Let $x^1, x^2 \in L(f,\alpha)$ and $\lambda \in [0,1]$. From the quasiconvexity of h_{x^1,x^2}, it results

$$h_{x^1,x^2}(\lambda) \leqq \max\{h_{x^1,x^2}(1), h_{x^1,x^2}(0)\}$$

that is

$$f(\lambda x^1 + (1-\lambda)\,x^2) \leqq \max\{f(x^1), f(x^2)\} \leqq \alpha$$

and therefore $\lambda x^1 + (1-\lambda)\,x^2 \in L(f,\alpha)$, i.e. $L(f,\alpha)$ is a convex set.

iv) $\Rightarrow$ i). Let $x^1, x^2 \in X$ and $\lambda \in [0,1]$ and let $\alpha = \max\{f(x^1), f(x^2)\}$. As $x^1, x^2 \in L(f,\alpha)$ and $L(f,\alpha)$ is convex, it results that $\lambda x^1 + (1-\lambda)\,x^2 \in L(f,\alpha)$, i.e. $f(\lambda x^1 + (1-\lambda)\,x^2) \leqq \alpha$ and hence $f(x)$ is quasiconvex on X.

iv) $\Rightarrow$ v). Let $x^1, x^2 \in SL(f,\alpha)$, $\lambda \in [0,1]$ and let $\alpha_0 = \max\{f(x^1), f(x^2)\} < \alpha$; then it is $x^1, x^2 \in L(f,\alpha_0)$ and $\lambda x^1 + (1-\lambda)\,x^2 \in L(f,\alpha_0)$. Hence $f(\lambda x^1 + (1-\lambda)\,x^2) \leqq \alpha_0 < \alpha$, i.e. $\lambda x^1 + (1-\lambda)\,x^2 \in SL(f,\alpha)$.

v) $\Rightarrow$ iv). Let $x^1, x^2 \in L(f,\alpha)$, $\lambda \in [0,1]$; then $x^1, x^2 \in SL(f,\alpha+\varepsilon)$, $\forall\,\varepsilon > 0$ and $\lambda x^1 + (1-\lambda)\,x^2 \in SL(f,\alpha+\varepsilon)$, $\forall\,\varepsilon > 0$. Hence $f(\lambda x^1 + (1-\lambda)\,x^2) < \alpha + \varepsilon$, $\forall\,\varepsilon > 0$, that is $f(\lambda x^1 + (1-\lambda)\,x^2 \leqq \alpha$. □

Another characterization of quasiconvexity for functions on one variable, in terms of unimodal functions, is given in Avriel, Diewert, Schaible and Zang (1987) and in Martos (1975). The following theorem gives the characterization of differentiable quasiconvex functions, defined on an open convex set. This result is due to Arrow and Enthoven (1961).

Theorem 2.10.2. Let $X \subset \mathbb{R}^n$ be an open convex set and $f: X \to \mathbb{R}$ be differentiable on X; then f is quasiconvex on X if and only if

$$x^1, x^2 \in X\,, \quad f(x^1) \leqq f(x^2) \Rightarrow (x^1 - x^2)\,\nabla f(x^2) \leqq 0\,.$$

Proof. Let f be quasiconvex and $f(x^1) \leqq f(x^2)$; then

$$f(x^2 + \lambda(x^1 - x^2) = f(\lambda x^1 + (1-\lambda)\,x^2) \leqq f(x^2)\,, \qquad \forall\,\lambda \in [0,1]$$

and thus,

$$\lim_{\lambda \downarrow 0} \frac{f(x^2 + \lambda(x^1 - x^2)) - f(x^2)}{\lambda} = (x^1 - x^2)\,\nabla f(x^2) \leqq 0\,.$$

Conversely, let the condition of the theorem be fulfilled and $f(x^1) \leqq f(x^2)$. We absurdly assume that there exists a $\bar{\lambda} \in (0,1)$ with $f(\bar{\lambda} x^1 + (1 - \bar{\lambda})\, x^2) > f(x^2)$. This means that for the function

$$h_{x^1,x^2}(\lambda) = f(\lambda x^1 + (1-\lambda)\, x^2)$$

it holds

$$h_{x^1,x^2}(\bar{\lambda}) > h_{x^1,x^2}(0) \geqq h_{x^1,x^2}(1)\ .$$

Then, because of the differentiability of h, it follows that h is continuous and that the set $\{\lambda \in [0,\bar{\lambda}] | h(\lambda) = h(0)\}$ is closed and admits a maximum. Therefore there exists $\hat{\lambda} \in [0,\bar{\lambda})$ such that $h(\hat{\lambda}) = h(0)$ and $h(\lambda) > h(0)$, $\forall\, \lambda \in (\hat{\lambda}, \bar{\lambda}]$. By the mean value theorem we have

$$h'_{x^1,x^2}(\lambda_0) > 0\ , \qquad \lambda_0 \in (\hat{\lambda}, \bar{\lambda})$$

and

$$h_{x^1,x^2}(0) < h_{x^1,x^2}(\lambda_0) < h_{x^1,x^2}(\bar{\lambda})\ .$$

But this means that there exists a point $x^0 = \lambda_0 x^1 + (1-\lambda_0)\, x^2$ between x^1 and x^2, with

$$f(x^0) = h_{x^1,x^2}(\lambda_0) > h_{x^1,x^2}(1) = f(x^1)\ .$$

From the assumptions of the theorem from $f(x^1) < f(x^0)$ it follows

$$(x^1 - x^0)\, \nabla f(x^0) \leqq 0\ ,$$

i.e.

$$(1-\lambda_0)(x^1 - x^2)\, \nabla f(x^0) \leqq 0\ ,$$

i.e. $h'_{x^1,x^2}(\lambda_0) \leqq 0$, in contradiction with the previous result. □

By the contraposition rule it follows that the implication of Theorem 2.10.2 can be rewritten as

$$x^1, x^2 \in X\ , \quad (x^1 - x^2)\, \nabla f(x^2) > 0 \Rightarrow f(x^1) > f(x^2)\ .$$

A function $f : X \to I\!R$ is defined to be *quasiconcave* on the convex set $X \subset I\!R^n$ if and only if $-f$ is quasiconvex on the same set X (this kind of definition also holds for the other classes of generalized convex functions which will be introduced later). Another result on the characterization of continuous and respectively of differentiable quasiconvex functions is contained in the following

Theorem 2.10.3.

a) Let $f : X \to I\!R$ be continuous on the convex set $X \subset I\!R^n$; then f is quasiconvex on X if and only if

$$x^1, x^2 \in X\,, \quad f(x^1) < f(x^2) \Rightarrow f(\lambda x^1 + (1-\lambda)\,x^2) \leqq f(x^2),$$
$$\forall\,\lambda \in [0,1]\,.$$

b) Let $f : X \to I\!R$ be differentiable on the open convex set $X \subset I\!R^n$; then f is quasiconvex on X if and only if

$$x^1, x^2 \in X\,, \quad f(x^1) < f(x^2) \Rightarrow (x^1 - x^2)\,\nabla f(x^2) \leqq 0\,.$$

Proof. Obviously a) and b) are necessary for quasiconvexity; moreover, with a reasoning analogous to the proof of Theorem 2.10.2, we see that, under differentiability, a) and b) are equivalent. Now we show that a) is sufficient for the quasiconvexity of f; this means that the implication of a) holds even if $f(x^1) = f(x^2)$. Assuming there exists a $\bar{\lambda} \in (0,1)$ with $f(\bar{\lambda}x^1 + (1-\bar{\lambda})\,x^2) > f(x^1) = f(x^2)$, because of the continuity of f, we can also find a point $x^0 = \lambda_0 x^1 + (1-\lambda_0)\,x^2$ between x^1 and $\bar{\lambda}x^1 + (1-\bar{\lambda})\,x^2$, i.e. $\lambda_0 \in (\bar{\lambda}, 1)$, such that

$$f(\bar{\lambda}x^1 + (1-\bar{\lambda})\,x^2) > f(x^0) > f(x^1) = f(x^2)\,.$$

But this is a contradiction to a). In fact, since $\bar{\lambda}x^1 + (1-\bar{\lambda})\,x^2$ is between x^0 and x^2, according to

$$\bar{\lambda}x^1 + (1-\bar{\lambda})\,x^2 = \frac{\bar{\lambda}}{\lambda_0}\,x^0 + \left(1 - \frac{\bar{\lambda}}{\lambda_0}\right)x^2\,, \qquad \frac{\bar{\lambda}}{\lambda_0} \in [0,1]$$

and $f(x^2) < f(x^0)$, we should have

$$f(\bar{\lambda}x^1 + (1-\bar{\lambda})\,x^2) \leqq f(x^0)\ .$$ □

All the given definitions of quasiconvexity are in general rather difficult to verify, as they imply conditions involving infinitely many inequalities. However, if the function is twice continuously differentiable on X, we have necessary and sufficient conditions for the quasiconvexity, which is easier to verify. We give separately the necessary conditions and the sufficient conditions for the quasiconvexity of a twice continuously differentiable function.

Theorem 2.10.4. Let $f : X \to \mathbb{R}$ be twice continuously differentiable on the open convex set $X \subset \mathbb{R}^n$. The following condition is necessary for f to be quasiconvex on X:

$$y \in \mathbb{R}^n,\ x \in X,\ y\nabla f(x) = 0 \Rightarrow yHf(x)\,y \geqq 0\ .$$

Proof. Let f be quasiconvex and assume that there exist x^0, y^0 with $y^0\nabla f(x^0) = 0$ and $y^0 Hf(x^0)\,y^0 < 0$. Without loss of generality we can set $\|y\| = 1$. Because of the continuity of the elements of the Hessian, there exists a neighbourhood $U_\delta(x^0)$ such that $y^0Hf(x)\,y^0 < 0$ for each $x \in U_\delta(x^0)$. We have especially $y^0Hf(x^0 \pm \lambda y^0)\,y^0 < 0$, $\forall\,\lambda \in [0,\delta]$. Using Taylor's expansion formula, this allows us to obtain

$$f(x^0 + \delta y^0) - f(x^0) = \tfrac{1}{2}\,y^0Hf(x^0 + \vartheta_1\delta y^0)\,y^0 < 0$$
$$f(x^0 - \delta y^0) - f(x^0) = \tfrac{1}{2}\,y^0Hf(x^0 - \vartheta_2\delta y^0)\,y^0 < 0$$

where $\vartheta_1, \vartheta_2 \in (0,1)$.
Thus, it follows that $f(x^0 \pm \delta y^0) < f(x^0)$, but because of the quasiconvexity we have

$$f(x^0) = f(\tfrac{1}{2}\,(x^0 - \delta y^0) + \tfrac{1}{2}\,(x^0 + \delta y^0)) \ \leqq$$
$$\leqq\ \max\{f(x^0 - \delta y^0), f(x^0 + \delta y^0)\} < f(x^0)\ ,$$

which is absurd. □

Note that the conditions given by Theorem 2.10.4 are not sufficient for the quasiconvexity of $f(x)$; indeed, consider, e.g., $f(x) = -x^4$, $x \in I\!R$. The conditions of Theorem 2.10.4 are satisfied, but obviously f is not quasiconvex on $I\!R$.

Let us now give a sufficient condition for the quasiconvexity of a twice continuously differentiable function.

Theorem 2.10.5 (Otani (1983)). Let f be twice continuously differentiable on the open convex set $X \subset I\!R^n$; suppose $\nabla f(x) \neq 0$ for every $x \in X$. If

$$y \in I\!R,\ x \in X,\ y \nabla f(x) = 0 \Rightarrow y Hf(x)\, y \geqq 0$$

then f is quasiconvex on X.

Proof. Let $f(x^1) \leqq f(x^2)$; assume that there exists a number $\bar{\lambda} \in (0,1)$ with $f(\bar{\lambda} x^1 + (1-\bar{\lambda})\, x^2) > f(x^2)$. With $h(\lambda) = h_{x^1 x^2}(\lambda) = f(\lambda x^1 + (1-\lambda)\, x^2)$ we have

$$h(\bar{\lambda}) > h(0) \geqq h(1)\ . \tag{1}$$

Let λ_0 be the maximum point of h on the interval $(0,1)$. Then with

$$x^0 = \lambda_0 x^1 + (1-\lambda_0)\, x^2 \tag{2}$$

we have

$$h'(\lambda_0) = (x^1 - x^2)\, \nabla f(x^0) = 0\ . \tag{3}$$

Now we consider the equation

$$f(x^0 + \alpha(x^1 - x^2) + \beta \nabla f(x^0)) = f(x^0)\ .$$

Since $[\nabla f(x^0)]^T \nabla f(x^0) \neq 0$, using the implicit function theorem, the above equation is solvable in a neighbourhood of $\alpha = 0$, $\beta = 0$ by $\beta = \beta(\alpha)$.

Setting $x(\alpha) = x^0 + \alpha(x^1 - x^2) + \beta(\alpha)\, \nabla f(x^0)$ we have $\beta(0) = 0$ and by differentiation of the equality

$$f(x(\alpha)) = f(x^0 + \alpha(x^1 - x^2) + \beta(\alpha)\, \nabla f(x^0)) = f(x^0) \tag{4}$$

for small $|\alpha|$ we obtain

$$(x^1 - x^2 + \beta'(\alpha)\,\nabla f(x^0))\,\nabla f(x(\alpha)) = 0 \ .$$

In particular, with (3) we get $\beta'(0) = 0$. Again differentiation provides

$$\begin{aligned}&(x^1 - x^2 + \beta'(\alpha)\,\nabla f(x^0))\,Hf(x(\alpha)) \,\cdot \\ &\cdot\,(x^1 - x^2 + \beta'(\alpha)\,\nabla f(x^0)) + [\nabla f(x(\alpha))]^T \,\cdot \\ &\cdot\,\nabla f(x^0)\cdot\beta''(\alpha) = 0 \ .\end{aligned}$$

According to the assumption of our theorem, the first part of the above sum is nonnegative and thus we have

$$[\nabla f(x(\alpha))]^T\,\nabla f(x^0)\,\beta''(\alpha) \leqq 0$$

for α sufficiently small.
Since $[\nabla f(x(\alpha))]^T\,\nabla f(x^0)$ is close to $[\nabla f(x^0)]^T\,\nabla f(x^0) > 0$, when $|\alpha|$ is close to zero, we can see that $\beta''(\alpha) \leqq 0$ for $|\alpha|$ sufficiently small and therefore (since $\beta'(0) = 0$) $\beta(\alpha) \leqq \beta(0) = 0$ for $|\alpha|$ sufficiently small.
Now, since $[\nabla f(x^0)]^T\,\nabla f(x^0) > 0$, for $\lambda \in [0, 1]$ with $|\lambda - \lambda_0|$ sufficiently small, we have

$$[\nabla f(x^0 + (\lambda - \lambda_0)(x^1 - x^2))](-\nabla f(x^0)) < 0$$

and, taking (2) and (4) into account and with $\beta(\lambda - \lambda_0) \leqq 0$,

$$\begin{aligned}h(\lambda) &= f(\lambda x^1 + (1-\lambda)\,x^2) = f(x^0 + (\lambda - \lambda_0)\cdot(x^1 - x^2)) \ \geqq \\ &\geqq \ f(x^0 + (\lambda - \lambda_0)(x^1 - x^2) + \beta(\lambda - \lambda_0)\,\nabla f(x^0)) = \\ &= f(x^0) = h(\lambda_0) \ .\end{aligned}$$

However, this holds for each maximum point λ_0 of h, thus h is constant on $[0, 1]$ and it holds $h(\lambda) = h(\lambda_0)$ for each $\lambda \in [0, 1]$, which is in contradiction to (1). Thus f is quasiconvex (see also Figure 7). □

We note that the conditions expressed by Theorem 2.10.5 only require the positive semidefiniteness of the Hessian matrix of $f(x)$ on the hyperplane $\{y \mid y\nabla f(x) = 0\}$. The first results for the quasiconvexity of twice continuously differentiable functions are due to Arrow and Enthoven

(1961); Theorem 2.10.5 generalizes a result of Katzner (1970). For other papers on quadratic forms subject to a system of linear equalities see, e.g., Debreu (1952), Bellman (1960), Mann (1943), Diewert, Avriel and Zang (1981), Crouzeix and Ferland (1982). We list the main results. The following theorem gives an equivalent condition for the semipositiveness of a quadratic form on a hyperplane.

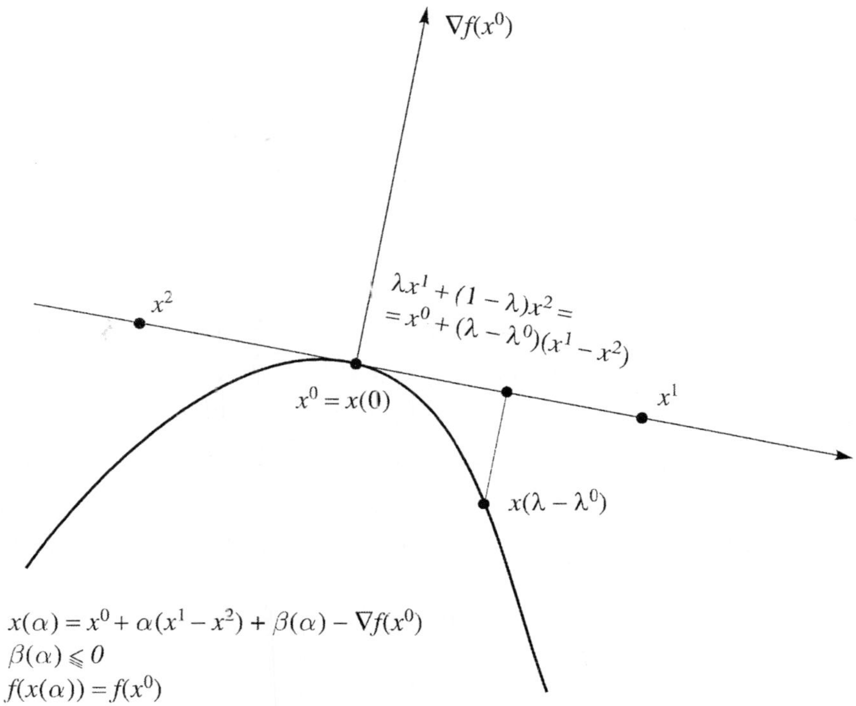

Figure 7.

Theorem 2.10.6. Let A be a symmetric matrix and $a \neq 0$ a vector; then the following conditions are equivalent:

(1) $yAy \geqq 0$ for each y such that $ya = 0$.

(2) The equation (of degree $(n-1)$ in λ)

$$\begin{vmatrix} A - \lambda I & a \\ a^T & 0 \end{vmatrix} = 0$$

admits only nonnegative roots.

Proof. See, e.g., Hancock (1960) or the mathematical appendix of Samuelson (1947). □

So we have the following

Corollary 2.10.1. Let $f : X \to I\!R$ be twice continuously differentiable on the open convex set $X \subset I\!R^n$ and let $\nabla f(x) \neq 0$, $\forall x \in X$; then f is quasiconvex on X, if and only if the equation (of degree $(n-1)$ in λ)

$$\begin{vmatrix} Hf(x) - \lambda I & \nabla f(x) \\ [\nabla f(x)]^T & 0 \end{vmatrix} = 0$$

admits only nonnegative roots for each $x \in X$.

Other interesting results on the positive semidefiniteness of a symmetric matrix A, of order n, on the hyperplane $\{y \,|\, ya = 0\}$, $a, y \in I\!R^n$, $a \neq 0$, are given in the quoted paper of Crouzeix and Ferland who prove that the following conditions are equivalent:

C1. $ya = 0$ implies $yAy \geqq 0$.

C2. Either A is positive semidefinite or A has one simple negative eigenvalue and there exists a vector $b \in I\!R^n$ such that $Ab = a$ and $ab \leqq 0$.

C3. The bordered matrix

$$A_a = \begin{bmatrix} A & a \\ a^T & 0 \end{bmatrix}$$

has one simple negative eigenvalue.

C4. For all nonempty subsets $Q \subset \{1, 2, ..., n\}$

$$\det \begin{bmatrix} A_Q & a_Q \\ a_Q^T & 0 \end{bmatrix} \leqq 0$$

where A_Q is obtained from A by deleting rows and columns whose indices are not in Q and a_Q is obtained analogously from a (i.e. the so-called "bordered principal minors" of the matrix A are all nonpositive).

The same authors show that the following conditions are equivalent.

D1. There exists a scalar λ, $0 \leqq \lambda < +\infty$ such that $A + \lambda aa^T$ is positive semidefinite.

D2. Either A is positive semidefinite or A has one simple negative eigenvalue and there exists a vector $b \in \mathbb{R}^n$ such that $Ab = a$ and $ab < 0$.

D3. For all nonempty subsets $Q \subset \{1, 2, ..., n\}$ we have

$$\det \begin{bmatrix} A_Q & a_Q \\ a_Q^T & 0 \end{bmatrix} \leqq 0$$

and if equality holds, then $\det(A_Q) \geqq 0$.

Obviously the D-conditions are sufficient for the C-conditions.

We shall be more precise on the result of Theorem 2.10.5 when we introduce the convex and generalized convex functions *at a point* x^0 (Section 2.14).

Under the assumption $\nabla f(x) \neq 0$, $\forall x \in X$, $X \subset \mathbb{R}^n$ open and convex, the conditions imposed by Theorem 2.10.5 are therefore necessary and sufficient for the quasiconvexity of $f \in C^2(X)$. Obviously we may have quasiconvex functions with a zero gradient at some points. Consider, e.g., the function $f : \mathbb{R}^2 \to \mathbb{R}$ defined as $f(x_1, x_2) = (x_1)^4 + (x_2)^4$, which is quasiconvex on $\mathbb{R}^2$ but with $\nabla f(x^0) = 0$ at $x^0 = (0, 0)$. A rather strong sufficient condition (including also the case $\nabla f(x) = 0$) for the quasiconvexity of a twice continuously differentiable function is contained in the following result.

<u>Theorem 2.10.7</u>. If $f : X \to \mathbb{R}$ is twice continuously differentiable on the open convex set $X \subset \mathbb{R}^n$, then f is quasiconvex on X if

$$x \in X\,, \quad y \neq 0\,, \quad y\nabla f(x) = 0 \quad \Rightarrow \quad yHf(x)\,y > 0\,. \tag{5}$$

Proof. Let $f(x^1) \leqq f(x^2)$ and assume on the contrary that $(x^1 - x^2)\,\nabla f(x^2) > 0$. Again we consider the function

$$h(\lambda) = h_{x^1,x^2}(\lambda) = f(\lambda x^1 + (1-\lambda)\,x^2) \;.$$

By the Weierstrass theorem, there exists $\lambda_0 \in [0,1]$ such that $h(\lambda_0)$ is the maximum of h on $[0,1]$. We cannot have $\lambda_0 = 0$, as $(x^1 - x^2)\,\nabla f(x^2) = h'(0) \leqq 0$, in contrast with the assumption and we cannot have $\lambda_0 = 1$ since $h(1) = f(x^1) \leqq f(x^2) = h(0)$ and $\lambda_0 = 0$ cannot be the maximum point. So we have proved that there exists $\lambda_0 \in (0,1)$ for which $(x^1 - x^2)\,\nabla f(x^0) = 0$. Taking the condition of the theorem into account, we get

$$(x^1 - x^2)\,Hf(x^0)(x^1 - x^2) > 0 \;.$$

Obviously, there exists a neighbourhood $U(x^0)$ such that

$$(x^1 - x^2)\,Hf(x)(x^1 - x^2) > 0 \;,$$

for each $x \in U(x^0)$. In particular we have, for $|t|$ sufficiently small,

$$\begin{aligned} h(\lambda_0 + t) - h(\lambda_0) &= f(x^0 + t(x^1 - x^2)) - f(x^0) = \\ &= \tfrac{1}{2}\,t^2 (x^1 - x^2)\,Hf(x^0 + \delta t(x^1 - x^2))(x^1 - x^2) > 0 \;, \end{aligned}$$

where $\delta \in (0,1)$.
This is a contradiction since λ_0 was assumed to be a maximum of h on $[0,1]$. □

It must be noted that implication (5) is taken by Diewert, Avriel and Zang (1981) as a characterization of another class of generalized convex functions: the *strongly pseudoconvex functions* (twice continuously differentiable); we shall see later that implication (5) is also sufficient for the *strict pseudo-convexity* of $f(x) \in C^2$. So any condition assuring the positive definiteness of $Hf(x)$ in the subspace orthogonal to $\nabla f(x) \neq 0$ is a sufficient condition, not only for the quasiconvexity of f, but also for other classes of generalized convex functions contained in the class of quasiconvex functions, such as the pseudoconvex functions and the strict pseudoconvex functions. Among these conditions we quote the following, expressed in terms of *bordered determinants* and due to Arrow and Enthoven (1961), perhaps historically the first sufficient conditions for the quasiconvexity of a twice continuously differentiable function:

For each $x \in X$ it is $\Delta_k(x) < 0$, $k = 1, 2, ..., n$, where

$$\Delta_k(x) = \begin{vmatrix} \dfrac{\partial^2 f}{\partial x_1 \partial x_1} & \dfrac{\partial^2 f}{\partial x_1 \partial x_2} & \cdots & \dfrac{\partial^2 f}{\partial x_1 \partial x_k} & \dfrac{\partial f}{\partial x_1} \\ \vdots & & & & \vdots \\ \dfrac{\partial^2 f}{\partial x_k \partial x_1} & \dfrac{\partial^2 f}{\partial x_k \partial x_2} & \cdots & \dfrac{\partial^2 f}{\partial x_k \partial x_k} & \dfrac{\partial f}{\partial x_k} \\ \dfrac{\partial f}{\partial x_1} & \dfrac{\partial f}{\partial x_2} & \cdots & \dfrac{\partial f}{\partial x_k} & 0 \end{vmatrix} .$$

See also Ferland (1972b).

Another characterization of twice continuously differentiable quasiconvex functions is given by Diewert, Avriel and Zang (1981), who introduce the concepts of *semistrict local maximum*.

Following Martos (1967, 1975) we say that a function $f : X \to I\!R$ is *quasimonotone* or *quasilinear* on the convex set $X \subset I\!R^n$ if f is both quasiconvex and quasiconcave on X. The term "quasilinear" is preferable, in order to avoid confusion with another concept (see Section 2.12 of this chapter). Equivalently f is quasilinear on X if for any points $x^1, x^2 \in X$, $x^0 \in (\lambda x^1 + (1 - \lambda)\, x^2)$, $0 < \lambda < 1$, we have

$$\min\{f(x^1), f(x^2)\} \leqq f(x^0) \leqq \max\{f(x^1), f(x^2)\ .$$

Again we have that $f : X \to I\!R$ is quasilinear on the convex set $X \subset I\!R^n$ if and only if all its lower and upper level sets $L(f, \alpha)$ and $U(f, \alpha)$ (equivalently: all its strict lower and upper level sets $SL(f, \alpha)$, $SU(f, \alpha)$) are convex for every $\alpha \in I\!R$. This result, which follows directly from the definition of quasilinear functions, implies that the level sets, i.e. the sets

$$Y(f, \alpha) = \{x \mid x \in X,\ f(x) = \alpha\}$$

of a quasilinear function are convex, for all $\alpha \in I\!R$. The converse of this statement, however, is not true: a function with convex level sets need not be quasilinear. Consider the following example:

$$f(x) = \begin{cases} 0\,, & \text{if } 0 \leqq x \leqq 1 \\ 2\,, & \text{if } 1 < x \leqq 2 \\ 1\,, & \text{if } 2 < x \leqq 3\,. \end{cases}$$

The set $Y(f,\alpha)$ is empty for $\alpha \neq 0,1,2$ and for $\alpha = 0$ it is $Y(f,0) = [0,1]$; for $\alpha = 1$ we have $Y(f,1) = (2,3]$; for $\alpha = 2$ we have $Y(f,2) = (1,2]$. Thus $Y(f,\alpha)$ is convex for all α, but f is not quasiconvex and hence not quasilinear. Note that f is not continuous. In fact, continuity assures the converse statement, as shown by the following result, due to Greenberg and Pierskalla (1971).

Theorem 2.10.8. If the level sets $Y(f,\alpha)$ are convex for every $\alpha \in \mathbb{R}$ and f is continuous on the convex set $X \subset \mathbb{R}^n$, then f is quasilinear on X.

Proof. We show the quasiconvexity of f. Let $f(x^1) \leqq f(x^2)$ and absurdly assume that there exists a number $\bar{\lambda} \in (0,1)$ such that

$$f(\bar{\lambda}x^1 + (1-\bar{\lambda})\,x^2) > f(x^2)\,.$$

Because of the continuity of f we can find a point $x^0 = \lambda_0 x^1 + (1-\lambda_0)\,x^2$ between x^1 and $\bar{\lambda}x^1 + (1-\bar{\lambda})\,x^2$, i.e. a number $\lambda_0 \in (\bar{\lambda},1]$ such that

$$f(x^0) = f(x^2) < f(\bar{\lambda}x^1 + (1-\bar{\lambda})\,x^2\,.$$

But this provides a contradiction to the condition of the theorem. In fact, since $\bar{\lambda}x^1 + (1-\bar{\lambda})\,x^2$ is between x^0 and x^2, according to

$$\bar{\lambda}x^1 + (1-\bar{\lambda})\,x^2 = \frac{\bar{\lambda}}{\lambda_0}\,x^0 + \Big(1 - \frac{\bar{\lambda}}{\lambda_0}\Big)\,x^2\,, \qquad \frac{\bar{\lambda}}{\lambda_0} \in [0,1]$$

and $f(x^0) = f(x^2)$, we should have, taking into account the assumption on the convexity of the level sets of f,

$$f(\bar{\lambda}x^1 + (1-\bar{\lambda})\,x^2) = f(x^0) = f(x^2)\,.$$

Thus f is quasiconvex. Since the condition is fulfilled also for $-f$ we can state that $-f$ is quasiconvex, too, i.e. f is quasilinear. □

If f is differentiable, quasilinearity is characterized as follows:

Theorem 2.10.9. Let f be differentiable on the open convex set $X \subset \mathbb{R}^n$; then f is quasilinear on X if and only if

$$x^1, x^2 \in X\,, \quad f(x^1) = f(x^2) \Rightarrow (x^1 - x^2)\, \nabla f(x^2) = 0\ .$$

Proof.

1) Let f be quasilinear and $f(x^1) = f(x^2)$; then by Theorem 2.10.2 we have $(x^1 - x^2)\, \nabla f(x^2) = 0$.

2) We show now quasiconvexity. Let $f(x^1) \leqq f(x^2)$ and assume $(x^1 - x^2)\, \nabla f(x^2) > 0$. Even if $f(x^1) < f(x^2)$, thanks to Darboux's theorem applied to the function $h(\lambda) = f(\lambda x^1 + (1-\lambda)\, x^2)$ on the interval $[0,1]$, we can find $x^0 \in [x^1, x^2)$ such that $f(x^0) = f(x^2)$. By the assumption of the theorem, we have then $(x^0 - x^2)\, \nabla f(x^2) = 0$. It follows that, for $\lambda_0 \in (0,1]$:

$$\begin{aligned} 0 &= [\lambda_0 x^1 + (1-\lambda_0)\, x^2 - x^2]\, \nabla f(x^2) = \\ &= [x^1 - (1-\lambda_0)\, x^1 + (1-\lambda_0)\, x^2 - x^2]\, \nabla f(x^2) = \\ &= \lambda_0 (x^1 - x^2)\, \nabla f(x^2) \end{aligned}$$

in contradiction with the assumption. This means that f is quasiconvex, but this means that $-f$ is also quasiconvex, and therefore f is quasilinear. □

For other insights on quasilinear functions, see Martos (1975), Thompson and Parke (1973).

Other types of generalized convex functions we shall examine are the *strictly quasiconvex functions*, the *semistrictly quasiconvex functions*, the *pseudoconvex functions* and the *strictly pseudoconvex functions*.

Definition 2.10.2 (Ponstein (1967), Katzner (1970)). A function $f : X \to \mathbb{R}$, defined on the convex set $X \subset \mathbb{R}^n$ is called *strictly quasiconvex* on X if

$$x^1, x^2 \in X\,, \quad x^1 \neq x^2\,,$$

$$f(x^1) \leqq f(x^2) \Rightarrow f(\lambda x^1 + (1-\lambda)\, x^2) < f(x^2)\,, \qquad \forall\, \lambda \in (0,1)$$

or, equivalently:

$$x^1, x^2 \in X\,, \quad x^1 \neq x^2\,, \quad \lambda \in (0,1)\ :$$
$$f(\lambda x^1 + (1-\lambda)\,x^2) < \max\{f(x^1), f(x^2)\}\ .$$

Strictly quasiconvex functions are also called "strongly quasiconvex functions" by Avriel (1976), "unnamed convex functions" by Ponstein (1967) and "X-convex functions" by Thompson and Parke (1973). It is easy to see that, as a consequence of the definitions, a strictly quasiconvex function is quasiconvex and that a strictly convex functions is strictly quasiconvex. However, not every convex function is strictly quasiconvex; for example a convex function with a "flat region" cannot be strictly quasiconvex. For differentiable functions we have the following characterization, due to Diewert, Avriel and Zang (1981).

Theorem 2.10.10. Let $f : X \to I\!R$ be differentiable on the open convex set $X \subset I\!R^n$; then f is strictly quasiconvex on X if and only if

$$x \in X\,, \quad y \in I\!R^n\,, \quad y \neq 0\,, \quad y\nabla f(x) = 0 \Rightarrow g_{x,y}(t) = f(x+ty)\ ,$$

defined for $t \geqq 0$, does not attain a local maximum at $t = 0$.

Proof.

i) Let f be strictly quasiconvex, $y \neq 0$ and $y\nabla f(x) = 0$; we assume that the conditions of the theorem are not fulfilled. Then with t sufficiently small we have $x \pm 2ty \in X$, $f(x \pm 2ty) \leqq f(x)$ and, according to the definition of strictly quasiconvexity we have $f(x \pm ty) < f(x)$, but $f(x) < \max\{f(x-ty), f(x+ty)\} < f(x)$ which is absurd.

ii) Let $x^1, x^2 \in X$, $x^1 \neq x^2$ and the conditions of the theorem be fulfilled and let $f(x^1) \leqq f(x^2)$. Assuming there exists a number $\bar{\lambda} \in (0,1)$ with $f(\bar{\lambda}x^1 + (1-\bar{\lambda})\,x^2) \geqq f(x^2)$ then, thanks to the Weierstrass theorem, we can also find a number $\lambda_0 \in (0,1)$ for which the function

$$g(\lambda) = f(\lambda x^1 + (1-\lambda)\,x^2) = f(x^2 + \lambda(x^1 - x^2))$$

admits a maximum on the interval $[0,1]$. Hence

$$g'(\lambda_0) = (x^1 - x^2)\, \nabla f(x^2 + \lambda_0(x^1 - x^2)) = 0$$

and so we get a contradiction to the conditions of the theorem. □

Definition 2.10.3 (Elkin (1968), Ginsberg (1973), Diewert, Avriel and Zang (1981)). A function $f : X \to I\!R$, defined on the convex set $X \subset I\!R^n$, is called *semistrictly quasiconvex* on X if

$$x^1, x^2 \in X, \quad f(x^1) < f(x^2) \Rightarrow f(\lambda x^1 + (1-\lambda)\, x^2) < f(x^2) ,$$
$$\forall \lambda \in (0,1)$$

or equivalently if

$$x^1, x^2 \in X, \quad f(x^1) \neq f(x^2) : f(\lambda x^1 + (1-\lambda)\, x^2) < \max\{f(x^1), f(x^2)\} ,$$
$$\forall \lambda \in (0,1) .$$

Beware! Semistrictly quasiconvex functions are called "strictly quasiconvex" by Mangasarian (1969), Karamardian (1967), Thompson and Parke (1973) and Avriel (1976). They are called "functionally convex" by Hanson (1964); moreover, Martos (1965, 1975) calls "explicitly quasiconvex" the quasiconvex functions satisfying Definition 2.10.2.

We have adopted a terminology widely used and which is motivated by the fact that those semistrictly quasiconvex functions, which satisfy certain continuity properties, have an intermediate position between strictly quasiconvex and quasiconvex functions (see the subsequent Theorem 2.10.11). As a consequence of the definitions we can immediately conclude that a strictly quasiconvex function is also semistrictly quasiconvex; similarly a convex function is semistrictly quasiconvex.

For other characterizations of differentiable semistrictly quasiconvex functions, see Diewert, Avriel and Zang (1981). We have already observed that, under certain continuity properties, semistrictly quasiconvex functions have an intermediate position between strictly quasiconvex and quasiconvex functions. Lacking these continuity properties, semistrictly quasiconvex functions need not be quasiconvex. Consider for example the following function, defined on $I\!R$:

$$f(x) = \begin{cases} 1 , & \text{for } x = 0 \\ 0 , & \text{for } x \neq 0 . \end{cases}$$

This function is semistrictly quasiconvex on $I\!R$, but not quasiconvex. If we require that f be lower semicontinuous, then we have the following result, established by Karamardian (1967).

Theorem 2.10.11. Let $f : X \to I\!R$ be a lower semicontinuous function on the convex set $X \subset I\!R^n$; if f is semistrictly quasiconvex on X, then f is quasiconvex on X, but not conversely.

Proof. By Definition 2.10.3 we have

$$f(x^1) < f(x^2), \quad \lambda \in (0,1) \Rightarrow f(\lambda x^1 + (1-\lambda)\, x^2) < f(x^2) .$$

Hence if $f(x^1) < f(x^2)$, there is nothing to prove. Now assume that $f(x^1) = f(x^2)$; we will show (by contradiction) that there exists no $\hat{x} \in (\lambda x^1 + (1-\lambda)\, x^2)$, $\lambda \in (0,1)$ such that $f(x^2) < f(\hat{x})$. This will then establish the quasiconvexity of f. Let $\hat{x}$ belong to the open line segment (x^1, x^2) such that $f(x^2) < f(\hat{x})$. Then $\hat{x} \in \Omega = \{x \mid f(x^2) < f(x),\ x \in (x^1, x^2)\}$.
Since f is lower semicontinuous on X, Ω is open relative to (x^1, x^2); hence there exists an $\bar{x} \in (\hat{x}, x^2) \cap \Omega$. By the semistrict quasiconvexity of f we have (since $\hat{x} \in \Omega$ and $\bar{x} \in \Omega$)

$$f(x^2) < f(\hat{x}) \Rightarrow f(\bar{x}) < f(\hat{x})$$

and

$$f(x^1) < f(\bar{x}) \Rightarrow f(\hat{x}) < f(\bar{x}) ,$$

a contradiction.
Hence no such $\hat{x}$ exists and f is therefore quasiconvex on X. That the converse is not true follows from the following example:

$$f(x) = \begin{cases} x \quad , & \text{for } x \leqq 0 \\ 0 \quad , & \text{for } 0 < x < 1 \\ x - 1 \ , & \text{for } x \geqq 1 \end{cases}$$

is quasiconvex on $I\!R$, but if we take $x^1 = -\frac{1}{2}$, $x^2 = \frac{1}{2}$, $\lambda = \frac{1}{10}$, then $f(x^1) < f(x^2)$, but $f(\lambda x^1 + (1-\lambda)\, x^2) = f(x^2)$. $\square$

An important property of a differentiable convex function is that a stationary point is also a global minimum point; however, this useful property is not restricted to (differentiable) convex functions only. The family of *pseudoconvex functions*, introduced by Mangasarian (1965) and, under the name of *semiconvex functions*, by Tuy (1964), strictly includes the family of differentiable convex functions and has the above-mentioned property.

Definition 2.10.4. Let $f : X \to I\!R$ be differentiable on the open set $X \subset I\!R^n$; then f is *pseudoconvex* on X if:

$$x^1, x^2 \in X\,, \quad f(x^1) < f(x^2) \Rightarrow (x^1 - x^2)\,\nabla f(x^2) < 0$$

or equivalently if

$$x^1, x^2 \in X\,, \quad (x^1 - x^2)\,\nabla f(x^2) \geqq 0 \Rightarrow f(x^1) \geqq f(x^2)\ .$$

From this definition it appears obvious that, if f is pseudoconvex and $\nabla f(x^2) = 0$, then x^2 is a point of global minimum of f over X. Pseudoconvexity plays a key role in obtaining sufficient optimality conditions for a nonlinear programming problem as, if a differentiable objective function can be shown or assumed to be pseudoconvex, then the usual first-order stationarity conditions are able to produce a global minimum point.

Other characterizations of pseudoconvex functions on open convex sets are due to Diewert, Avriel and Zang (1981) and to Crouzeix and Ferland (1982).

Theorem 2.10.12. Let $f : X \to I\!R$ be differentiable on the open convex set $X \subset I\!R^n$; then f is pseudoconvex on X if and only if

$$x \in X\,, \quad y \neq 0\,, \quad y \nabla f(x) = 0 \Rightarrow g(t) = f(x + ty)\ ,$$

defined for $t \geqq 0$, attains a local minimum at $t = 0$.

Proof.

i) Let f be pseudoconvex and $ty\nabla f(x) \geqq 0$, $\forall t \geqq 0$; then we have $f(x + ty) \geqq f(x)$, i.e. $g(t) = f(x + ty)$ admits a minimum at $t = 0$.

ii) Let the condition of the theorem be fulfilled and let $(x^1 - x^2)\,\nabla f(x^2) \geqq 0$. We have to show that $f(x^2) \leqq f(x^1)$. Consider the function $g(t) = f(x^2 + t(x^1 - x^2))$, $t \in [0,1]$, and absurdly suppose $f(x^1) < f(x^2)$, i.e. $g(1) < g(0)$.
If $g'(0) > 0$, i.e. $(x^1 - x^2)\,\nabla f(x^2) > 0$, then there exists a local maximum point t_0 for g, with $t_0 \in (0,1)$, and therefore $g'(t_0) = 0$, i.e. $(x^1 - x^2)\,\nabla f(x^0) = 0$. However, t_0 is not a local minimizer for g, which contradicts the assumptions.
If $g'(0) = 0$, i.e. $(x^1 - x^2)\,\nabla f(x^2) = 0$, then $t = 0$ is, by assumption, a local minimum point for g, i.e. x^2 is a local minimum point for f. But this, together with the absurd assumption $g(1) < g(0)$, implies the existence of a local maximum point $t_0 \in (0,1)$ for g. Therefore $g'(t_0) = 0$ and again we have a contradiction with the assumptions of the theorem. □

Theorem 2.10.13. Let $f : X \to I\!R$ be twice-continuously differentiable on the open convex set $X \subset I\!R^n$; then f is pseudoconvex on X if and only if for all $x \in X$:

i) $y\nabla f(x) = 0 \Rightarrow yHf(x)\,y \geqq 0$, and

ii) whenever $\nabla f(x) = 0$, then f has a local minimum at x.

Proof. See Crouzeix and Ferland (1982). □

We shall see that differentiable convex functions are pseudoconvex, and pseudoconvex functions are, in turn, semistrictly quasiconvex and therefore quasiconvex.

We shall revert to the notion of pseudoconvexity after the introduction of generalized convex functions at a point. We shall prove that a quasidifferentiable quasiconvex function (defined on the open convex set $X \subset I\!R^n$) is pseudoconvex at any point $x^0 \in X$, where $\nabla f(x^0) \neq 0$. Ortega and Rheinboldt (1970) and Thompson and Parke (1973) have introduced a definition of pseudoconvexity for nondifferentiable functions.

Definition 2.10.5. An arbitrary function $f : X \to \mathbb{R}$ is pseudoconvex on X if

$$x^1, x^2 \in X\,,\quad \lambda \in (0,1)\,,\quad f(x^1) < f(x^2) \Rightarrow$$

$$f(\lambda x^1 + (1-\lambda)\,x^2) \leqq f(x^2) - \lambda(1-\lambda)\,\beta(x^1, x^2)\ ,$$

where $\beta(x^1, x^2)$ is a positive number, depending on x^1 and x^2.

This definition is motivated by the fact that, when applied to differentiable functions, it is equivalent to Definition 2.10.4: see Ortega and Rheinboldt (1970), Thompson and Parke (1973) and Avriel, Diewert, Schaible and Zang (1988).

Functions that are both pseudoconvex and pseudoconcave are called *pseudomonotone* of *pseudolinear*. The latter term is preferable in order to avoid confusion with another concept: see Section 2.12 of this chapter. Pseudolinear functions are particularly important in certain optimization problems, both in the scalar case and in the vector case; see Martos (1975), Chew and Choo (1984), Kortanek and Evans (1967), Komlosi (1993), Rapcsak (1991), Thompson and Parke (1973).

The following theorem, due to Kortanek and Evans (condition ii)) and to Chew and Choo (condition iii)), characterizes the class of pseudolinear functions.

Theorem 2.10.14. Let $f : X \to \mathbb{R}$, where $X \subset \mathbb{R}^n$ is an open convex set. Then the following statements are equivalent:

i) f is pseudolinear;

ii) for any $x^1, x^2 \in X$ it is $(x^1 - x^2)\,\nabla f(x^2) = 0$ if and only if $f(x^1) = f(x^2)$;

iii) there exists a function $p : X \times X \to \mathbb{R}_+$ such that

$$f(x^1) = f(x^2) + p(x^1, x^2) \cdot (x^1 - x^2)\,\nabla f(x^2)\ .$$

Proof. We prove only the equivalence between i) and ii).

1) Let f be pseudolinear; then from the definition we get

$$(x^1 - x^2)\,\nabla f(x^2) = 0 \Rightarrow f(x^1) = f(x^2)\ .$$

Since f is also quasilinear, then

$$f(x^1) = f(x^2) \Rightarrow (x^1 - x^2)\,\nabla f(x^2) = 0\ .$$

2) Condition ii) can be written in the form

$$y\nabla f(x) = 0 \Leftrightarrow f(x + ty) = f(x)\ , \qquad \forall t\ :\ x + ty \in X\ ,$$

which means that whenever $y\nabla f(x) = 0$, the function $g(t) = f(x+ty)$ admits a minimum at $t = 0$. Thus f, but also $-f$, is pseudoconvex and therefore pseudolinear. □

The class of pseudolinear functions includes many functions useful for applications, e.g. the class of linear fractional functions (see, e.g. Martos (1975)).

An example of a pseudolinear function is given by $f(x) = x + x^3$, $x \in I\!R$. More generally Kortanek and Evans (1967) observe that if f is pseudolinear on the convex set $X \subset I\!R^n$, then the function $F(x) = f(x) + [f(x)]^3$ is also pseudolinear on X.

The notion of *strictly pseudoconvex function* was introduced by Ponstein (1967) for differentiable functions.

Definition 2.10.6. A function $f : X \to I\!R$, differentiable on the open set $X \subset I\!R^n$, is *strictly pseudoconvex* on X if

$$x^1, x^2 \in X\,, \quad x^1 \neq x^2\,, \quad f(x^1) \leqq f(x^2) \Rightarrow (x^1 - x^2)\,\nabla f(x^2) < 0$$

or equivalently if

$$x^1, x^2 \in X\,, \quad x^1 \neq x^2\,, \quad (x^1 - x^2)\,\nabla f(x^2) \geqq 0 \Rightarrow f(x^1) > f(x^2)\ .$$

The comparison of Definitions 2.10.4 and 2.10.6 shows that strictly pseudoconvexity implies pseudoconvexity. Ponstein (1967, p. 118) shows that pseudoconvexity plus strict quasiconvexity implies strict pseudoconvexity and that strict pseudoconvexity implies strict quasiconvexity. If in a minimization problem the strict pseudoconvexity of the objective function can be assumed or shown, then the solution to the first-order optimality conditions affords a unique global minimum.

Other characterizations of strictly pseudoconvex functions are given by Diewert, Avriel and Zang (1981):

Theorem 2.10.15. Let $f : X \to I\!R$ be differentiable on the open convex set $X \subset I\!R^n$; then f is strictly pseudoconvex on X if and only if

$$x \in X\,, \quad y \neq 0\,, \quad y\nabla f(x) = 0 \Rightarrow g(t) = f(x+ty)\,,$$

defined for $t \geqq 0$, attains a strict local minimum at $t = 0$.

Proof. Analogous to the proof of Theorem 2.10.12. □

Theorem 2.10.16. Let $f : X \to I\!R$ be a twice-continuously differentiable function on the open convex set $X \subset I\!R^n$; then f is strictly pseudoconvex on X if and only if:

$$x \in X\,, \quad y \neq 0\,, \quad y\nabla f(x) = 0 \Rightarrow$$

either $yHf(x)\,y > 0$ or $yHf(x)\,y = 0$

and $g(t) = f(x+ty)$ attains a strict local minimum at $t = 0$.

Proof. See Diewert, Avriel and Zang (1981). □

2.11. Relationships Between the Various Classes of Generalized Convex Functions. Properties in Optimization Problems

We have already given some relationships between the generalized convex functions introduced; now we shall complete the subject and shall prove the less trivial implications. We now assume that f is defined on an open convex set $X \subset I\!R^n$, where it is also differentiable if and when this property is required by the related implication. First we illustrate the various relationships by a diagram.

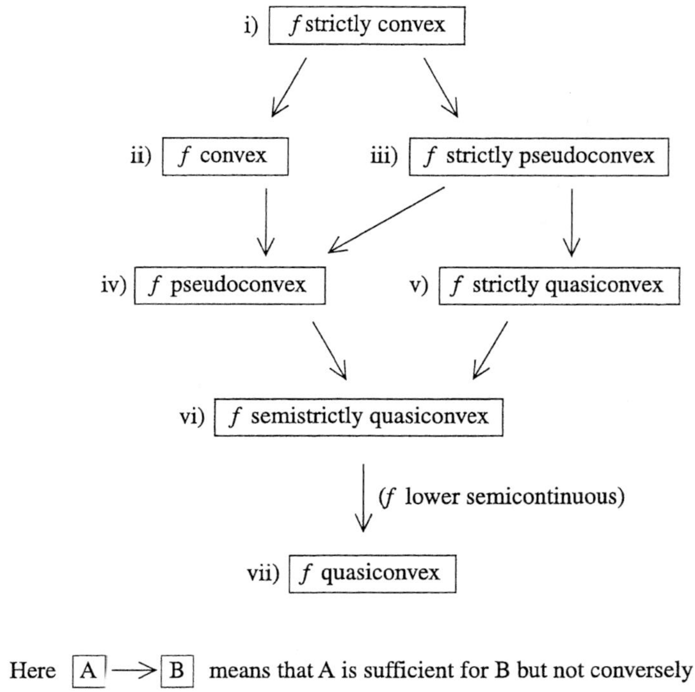

Figure 8.

Before proving the nontrivial implications of the above diagram, it is useful to point out, for functions $f : I\!R \to I\!R$ and from a geometric point of view, the differences between the several types of generalized convex functions. Take into consideration the following functions:

$f_1(x) = x^3 + x$

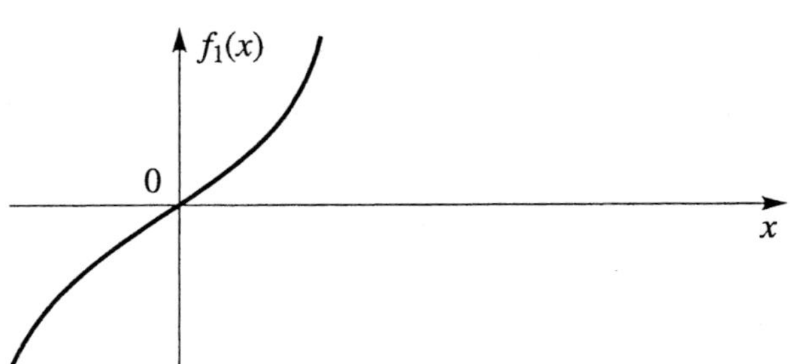

Figure 9.

$$f_2(x) = x^3$$

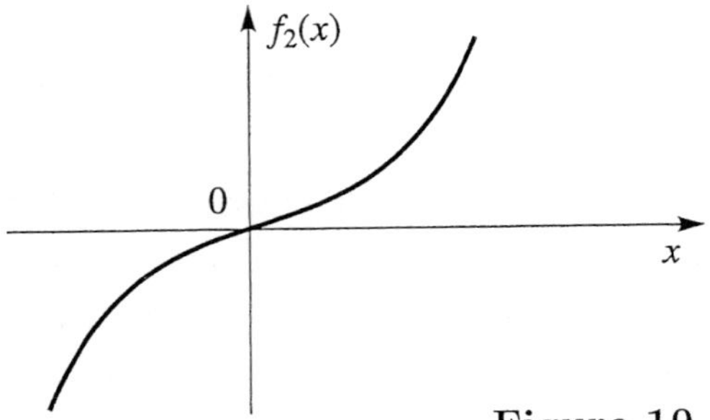

Figure 10.

$$f_3(x) = \begin{cases} 0 \quad , & \text{for } x < 0 \\ x^2 \quad , & \text{for } x \geqq 0 \end{cases}$$

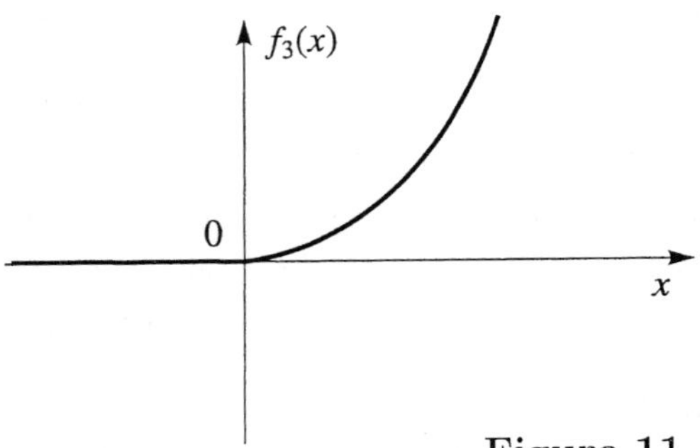

Figure 11.

$$f_4(x) = \begin{cases} -x^2 \quad , & \text{for } x < 0 \\ 0 \quad , & \text{for } x \in [0, 1] \\ (x-1)^2 \quad , & \text{for } x > 1 \end{cases}$$

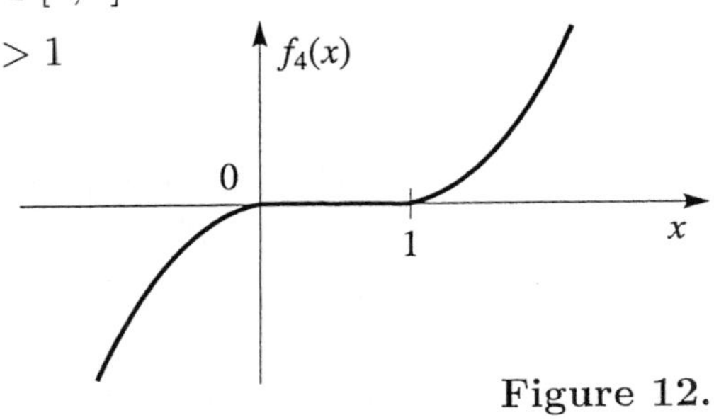

Figure 12.

Then it holds that

f_1 is pseudoconvex (even strictly pseudoconvex), but not convex (or strictly convex).

f_2 is semistrictly quasiconvex (even strictly quasiconvex), but not pseudoconvex (or strictly pseudoconvex).

f_3 is semistrictly quasiconvex (even pseudoconvex and convex), but not strictly quasiconvex (or strictly pseudoconvex of strictly convex).

f_4 is quasiconvex but not semistrictly quasiconvex.

The non trivial implications shown in Figure 8 are the following ones:

1) f pseudoconvex $\Rightarrow$ f semistrictly quasiconvex;

2) f semistrictly quasiconvex and lower semicontinuous $\Rightarrow$ f quasiconvex.

Let us prove implication 1). Let f be pseudoconvex on X, $x^1, x^2 \in X$, where X is an open convex subset of $\mathbb{R}^n$, with $f(x^1) < f(x^2)$. Assume that there exists a number $\bar{\lambda} \in (0,1)$ with $f(\bar{\lambda}x^1 + (1-\bar{\lambda})\, x^2) \geqq f(x^2)$. Let us denote by λ_0 the maximum point of $h(\lambda) = f(\lambda x^1 (1-\lambda)\, x^2)$, $\lambda \in (0,1)$. Then we have, with $x^0 = \lambda_0 x^1 + (1-\lambda_0)\, x^2$,

$$h'(\lambda_0) = (x^1 - x^2)\, \nabla f(x^0) = 0 \ .$$

However, because of $f(x^0) > f(x^1)$, from the definition of pseudoconvexity we have

$$(x^1 - x^0)\, \nabla f(x^0) < 0$$

and thus, since $x^1 - x^0 = (1-\lambda_0)(x^1 - x^2)$, $\lambda_0 \in (0,1)$,

$$(x^1 - x^2)\, \nabla f(x^0) < 0 \ ,$$

in contradiction with the previous result. □

For the implication 2), see Theorem 2.10.11.

We call the reader's attention on the fact that, contrary to what holds for convex functions, each nonnegative linear combination of quasiconvex, semistrictly quasiconvex, strictly quasiconvex and pseudoconvex functions need not be, respectively, a quasiconvex, a semistrictly quasiconvex, a strictly quasiconvex, a pseudoconvex function. For example, in the case of pseudoconvex functions, if f and ϑ are two pseudoconvex functions on $\mathbb{R}^n$ and we define $\psi(x) = f(x) + \vartheta(x)$, then the requirement that $(x^1 - x^2)\,\nabla\psi(x^2) \geqq 0$ does not ensure that $(x^1 - x^2)\,\nabla f(x^2) \geqq 0$ and $(x^1 - x^2)\,\nabla\vartheta(x^2) \geqq 0$ and hence we cannot conclude that $f(x^1) \geqq f(x^2)$, $\vartheta(x^2) \geqq f(x^2)$ and hence $\psi(x^1) \geqq \psi(x^2)$ and so ψ may not be pseudoconvex on $\mathbb{R}^n$. For instance $f(x) = -x$ and $\vartheta(x) = x + x^3$ are functions both pseudoconvex on $\mathbb{R}$, but their sum $\psi(x) = x^3$ is not pseudoconvex on $\mathbb{R}$. See Debreu and Koopmans (1982) and Crouzeix and Lindberg (1986) for an analysis on the linear combination of quasiconvex functions.

Let us now review the most important properties of generalized convex functions with respect to extremum problems.

Theorem 2.11.1. Let $f : X \to \mathbb{R}$ be defined on the convex set $X \subset \mathbb{R}^n$ and let $x^* \in X$ be a point of local minimum of f over X. If f is semistrictly quasiconvex, then x^* is a point of global minimum of f over X.

Proof. If x^* is a local minimum point, then there exists $\varepsilon > 0$ such that $f(x^*) \leqq f(x)$ for each $x \in X \cap N_\varepsilon(x^*)$, where $N_\varepsilon(x^*)$ is an open ball centered in x^* and of radius ε.
Assume now that there exists an $\bar{x} \in X$, but not in $N_\varepsilon(x^*)$, such that $f(\bar{x}) < f(x^*)$. From the strict quasiconvexity of f it follows that $f(\lambda\bar{x} + (1-\lambda)\,x^*) < f(x^*)$, $\forall\,\lambda \in (0,1)$. But for $\lambda < \delta/\|\bar{x} - x^*\|$ we have that $\lambda\bar{x} + (1-\lambda)\,x^* \in X \cap N_\delta(x^*)$ and hence $f(x^*) \leqq f(\lambda\bar{x} + (1-\lambda)\,x^*)$ for $0 < \lambda < \delta/\|\bar{x} - x^*\|$, which contradicts a previous inequality. Therefore x^* is a point of global minimum of f on X. □

Note that Theorem 2.11.1 is not in general true for quasiconvex functions. For instance, consider the function $f : \mathbb{R} \to \mathbb{R}$ defined by

$$f(x) = \begin{cases} x \quad , & \text{for } x \leqq 0 \\ 0 \quad , & \text{for } 0 < x < 1 \\ x - 1 \ , & \text{for } x \geqq 1 \ . \end{cases}$$

This function is quasiconvex on $\mathbb{R}$. Every point $x \in (0,1)$ is a point of local minimum, but not a point of global minimum. Moreover, we note that if f is a semistrictly quasiconvex function on a convex set $X \subset \mathbb{R}^n$, then the global minimum can be attained at more than one point: consider, e.g., the semistrictly quasiconvex function $f : \mathbb{R} \to \mathbb{R}$ defined by

$$f(x) = \begin{cases} 0 \ , & \text{for } x < 0 \\ x^2 \ , & \text{for } x \geqq 0 \ . \end{cases}$$

However, if f is strictly quasiconvex, the following result holds.

Theorem 2.11.2. If f is strictly quasiconvex on the convex set $X \subset \mathbb{R}^n$, then any local minimum point is unique and global.

Proof. Assume $\bar{x}$ and $\bar{\bar{x}}$ are two distinct local minimum points, with $f(\bar{\bar{x}}) \geqq f(\bar{x})$; then, thanks to strict quasiconvexity, $f(\lambda \bar{x} + (1-\lambda)\,\bar{\bar{x}}) < f(\bar{\bar{x}})$, $\forall\, \lambda \in (0,1)$ and $\bar{\bar{x}}$ is not a local minimum point. Thus there is at most one local minimum point; but since f is also semistrictly quasiconvex, the local minimum is also global. □

We have seen that for quasiconvex functions not every local minimum is a global one. Local minima, however, that are not global cannot be strict minima, as shown by the following result, due to Wolfe (1967) and Luenberger (1968).

Theorem 2.11.3. Let f be a quasiconvex function on the convex set $X \subset \mathbb{R}^n$; if x^* is a strict local minimum point of f, then x^* is also the unique global minimum point of f on X.

Proof. Suppose x^* is a strict local minimum point for f, i.e. $f(x^*) < f(x)$, $\forall\, x \in N_\delta(x^*)$, $x \neq x^*$, and suppose that x^* is not a strict global minimum. Therefore there exists an $\bar{x} \in X$, $\bar{x} \neq x^*$, such that

$$f(\bar{x}) \leqq f(x^*) \ .$$

Then by quasiconvexity we have

$$f(\lambda \bar{x} + (1-\lambda)\, x^*) \leqq f(x^*) \ , \qquad \forall\, \lambda \in [0,1] \ .$$

But for sufficiently small λ, it follows that $(\lambda\bar{x}+(1-\lambda)\,x^*) \in X \cap N_\delta(x^*)$, contradicting the fact that x^* is a strict local minimum point. □

By adding some further hypotheses, the implication of Theorem 2.11.2 can be reversed as shown in the following

Theorem 2.11.4. If f assumes its greatest lower bound on the convex set $X \subset \mathbb{R}^n$ and has a unique local minimum over every closed interval, then f is strictly quasiconvex on X.

Proof. See Thompson and Parke (1973). □

We give now some results on pseudoconvex functions and on the set of minimum points of convex and generalized convex functions.

Theorem 2.11.5. Let $f : D \to \mathbb{R}$ be pseudoconvex on the open set $D \subset \mathbb{R}^n$; then for any convex set $X \subset D$ we have

$$\min_{x \in X} f(x) = f(x^*) \Leftrightarrow (x - x^*)\,\nabla f(x^*) \geqq 0\ , \qquad \forall x \in X\ .$$

Proof. If x^* is a minimum point of $f(x)$ over X, the function $\varphi(\lambda) = f(x^* + \lambda(x - x^*))$ has at $\lambda = 0$ a minimum point on the interval $[0, 1]$ and hence $(x - x^*)\,\nabla f(x^*) = \varphi'(0) \geqq 0$. The second part of the theorem follows immediately from Definition 2.10.4. □

Corollary 2.11.1. Let f be a pseudoconvex function on the open convex set $X \subset \mathbb{R}^n$; then we have

$$\min_{x \in X} f(x) = f(x^*) \Leftrightarrow \nabla f(x^*) = 0\ .$$

Let $X \subset \mathbb{R}^n$ be convex and let f be differentiable on an open set D containing X. Let

$$X_f^\Delta = \{x^* \in X \mid f(x^*) = \min_{x \in X} f(x)\}\ ,$$

i.e. X_f^Δ is the set of global minimizers of f over X. (Some authors denote this set by "$\arg\min f(x)$, $x \in X$".)

Theorem 2.11.6.

i) If f is quasiconvex on X, then X_f^Δ is convex.

ii) If f is strictly convex or strictly pseudoconvex or strictly quasiconvex on X, then X_f^Δ contains at most one point, i.e. $x^* \in X_f^\Delta \Rightarrow X_f^\Delta = \{x^*\}$ (in other words, if $X_f^\Delta \neq \emptyset$, then x^* is the unique global minimum point of f over X).

Proof.

i) Being that $X_f^\Delta = L(f, \alpha^*)$, with $\alpha^* = \min\{f(x) \mid x \in X\}$, the convexity of X_f^Δ is ensured by the definition of quasiconvex functions.

ii) For strictly convex functions, see Theorem 2.8.3; for strictly quasiconvex function, see Theorem 2.11.2. Finally remember that strictly pseudoconvex functions are "intermediate" between these two classes. □

Corollary 2.11.2. Let f be strictly pseudoconvex on the open convex set $X \subset \mathbb{R}^n$; then we have

$$X_f^\Delta = \{x^*\} \Leftrightarrow \nabla f(x^*) = 0 \ .$$

Let us now consider the following nonlinear programming problem

$$\min_x \{f(x) \mid x \in X,\ g_i(x) \leqq 0,\ i = 1, 2, ..., m\} \tag{P}$$

where $X \subset \mathbb{R}^n$ is convex. Denote by x^* a solution of (P) and let $I(x^*) = \{i \mid g_i(x^*) = 0\}$ be the so-called set of *effective* or *active* constraints of (P).

Theorem 2.11.7. Let the function f in (P) be semistrictly quasiconvex on X, every g_i, $i \in I(x^*)$, be quasiconvex on X and every g_i, $i \notin I(x^*)$, be continuous at x^*. Then

$$\min_x \{f(x) \mid x \in X,\ g_i(x) \leqq 0,\ i \in I(x^*)\} = f(x^*) \ .$$

Proof. Let us suppose aburdly that there exists a vector $\bar{x} \in X$ such that $g_i(\bar{x}) \leqq 0$, $\forall i \in I(x^*)$, and $f(\bar{x}) < f(x^*)$. Then for each $\lambda \in (0, 1)$ we have $x_\lambda = \lambda\bar{x} + (1 - \lambda)\, x^* \in X$ and

$$f(x_\lambda) < f(x^*) \tag{1}$$

$$g_i(x_\lambda) \leqq \max\{g_i(\bar{x}),\ g_i(x^*) = 0\} \leqq 0\ , \qquad \forall i \in I(x^*)\ .$$

For each $i \notin I(x^*)$, $g_i(x^*) < 0$ and, thanks to the continuity of g_i at x^*, we have $g_i(x_\lambda) < 0$ for λ sufficiently small. Then x_λ satisfies the feasible set of (P) and (1) contradicts the optimality of x^*. □

In the following table we sum up the local/global minimum properties of the generalized convex functions introduced (see also Avriel and Zang (1980)). An entry (x) in the table indicates that the column property is satisfied by the row class. It must be noted that the converse relations of the table (i.e. that a column property would imply a row class) need not hold.

Avriel (1976) and Avriel and Zang (1980) have introduced the "arcwise connected functions" and have shown that, under more mild regularity assumptions, a local/global property of a function implies its belonging to a class of these generalized convex functions. See also Section 2.17 of this chapter.

We conclude this section with some remarks on the *radial* minimum points (see also Thompson and Parke (1973)). A necessary and sufficient

Table 1.

Necessary local/global properties of generalized convex functions.

Functions	strict local min ⇒ strict global min	local min ⇒ global min	local min ⇒ unique global min	stationary point ⇒ global min	stationary point ⇒ unique global min
Quasiconvex	(x)				
semistrictly quasiconvex	(x)	(x)			
strictly quasiconvex	(x)	(x)	(x)		
pseudoconvex	(x)	(x)		(x)	
strictly pseudoconvex	(x)	(x)	(x)	(x)	(x)

condition that x^0 be a global minimum of f over a set $X \subset \mathbb{R}^n$ is that x^0 be a *global radial minimum* of f over X, i.e. for every vector y and every $\lambda \geqq 0$, x^0 must be a global minimum point of f on the set $X \cap \{x \mid x = x^0 + \lambda y,\ \lambda \in \mathbb{R}\}$.

For a local minimum things run differently: a *necessary* condition that x^0 be a local minimum point of f over X is that x^0 be a local radial minimum point of f, i.e. for every vector y there exists a scalar $\lambda_0(y)$ such that $f(x^0) \leqq f(x^0 + \lambda y)$, whenever $\lambda \in (0, \lambda_0(y))$ and $x^0 + \lambda y \in X$. But a local radial minimum point need not be a local minimum point, as the following famous example, due to G. Peano, shows. The function $f : \mathbb{R}^2 \to \mathbb{R}$, defined as $f(x, y) = (y - x^2)(y - 2x^2)$, has a local radial minimum at $x^0 = (0, 0)$, but $x^0 = (0, 0)$ is not a local minimum for $f(x, y)$. See also related comments in Section 3.2 of the next chapter.

However, for a quasiconvex function, defined on a convex set $X \subset \mathbb{R}^n$, the above necessary condition becomes also sufficient. We have therefore the following results.

Theorem 2.11.8. If f is quasiconvex on the convex set $X \subset \mathbb{R}^n$, then any local radial minimum point of f over X is also a local minimum point of f over X.

Proof. See Thompson and Parke (1973). □

Theorem 2.11.9. If f is semistrictly quasiconvex on the convex set $X \subset \mathbb{R}^n$, then $x^0 \in X$ is a global minimum point of f over X if and only if x^0 is a local radial minimum point of f over X.

Proof. See Thompson and Parke (1973). □

2.12. Generalized Monotonicity and Generalized Convexity

We have seen in Theorem 2.5.4, point f), the characterization, due to Minty (1964), of a differentiable convex function by means of the notion of monotonicity of its gradient. In general a vector-valued function (or map) $F : X \to \mathbb{R}^n$, $X \subset \mathbb{R}^n$, is *monotone* on X if for every pair of distinct points $x^1, x^2 \in X$, we have

$$(x^1 - x^2)\,[F(x^1) - F(x^2)] \geqq 0 \; .$$

This notion has been generalized by Karamardian (1976) and Karamardian and Schaible (1990) to other notions of *generalized monotone maps.*

Definition 2.12.1. Let $X \subset \mathbb{R}^n$; a map $F : X \to \mathbb{R}^n$ is *pseudomonotone* on X if, for every pair of distinct points, $x^1, x^2 \in X$, we have

$$(x^1 - x^2)\,F(x^2) \geqq 0 \Rightarrow (x^1 - x^2)\,F(x^1) \geqq 0 \; .$$

Obviously every monotone map is pseudomonotone; the converse is not in general true. It is now clear why it is better to adopt the term "pseudolinear" (instead of "pseudomonotone") when we are dealing with functions both pseudoconvex and pseudoconcave. The following lemma is proved by Karamardian and Schaible (1990).

Lemma 2.12.1. A map $F : X \to \mathbb{R}^n$ is pseudomonotone on X if and only if for every pair of distinct points $x^1, x^2 \in X$, we have

$$(x^1 - x^2)\,F(x^2) > 0 \Rightarrow (x^1 - x^2)\,F(x^1) > 0 \; .$$

Proof. First note that the pseudomonotonicity of F is equivalent to

$$(x^1 - x^2)\,F(x^1) < 0 \Rightarrow (x^1 - x^2)\,F(x^2) < 0 \; ;$$

therefore it is equivalent to

$$(x^2 - x^1)\,F(x^1) > 0 \Rightarrow (x^2 - x^1)\,F(x^2) > 0 \; . \qquad \square$$

The following result shows that pseudoconvex functions over a convex set are characterized by the pseudomonotonicity of their gradients, similarly to the characterization of differentiable convex functions via monotonicity of their gradients.

Theorem 2.12.1. Let $f : X \to \mathbb{R}$ be differentiable on the open convex set $X \subset \mathbb{R}^n$; then f is pseudoconvex on X if and only if its gradient ∇f is pseudomonotone on X.

Proof (Karamardian (1976)). Suppose that f is pseudoconvex on X; let x^1 and x^2 be two arbitrary distinct points in X with

$$(x^1 - x^2)\,\nabla f(x^2) \geqq 0 \; . \tag{1}$$

We want to show that

$$(x^1 - x^2)\,\nabla f(x^1) \geqq 0 \; .$$

Assume the contrary, i.e.

$$(x^1 - x^2)\,\nabla f(x^1) < 0 \; . \tag{2}$$

From the pseudoconvexity of f, we have $f(x^1) \geqq f(x^2)$. Since every pseudoconvex function on a convex set is also quasiconvex, from the last inequality it follows that

$$(x^2 - x^1)\,\nabla f(x^1) \leqq 0 \; ,$$

which contradicts (2). Therefore ∇f is pseudomonotone.
Now, suppose that ∇f is pseudomonotone on X; let x^1 and x^2 be two arbitrary distinct points in X, with

$$(x^1 - x^2)\,\nabla f(x^2) \geqq 0 \; .$$

We want to show that $f(x^1) \geqq f(x^2)$. Assume the contrary, i.e.

$$f(x^1) < f(x^2) \; . \tag{3}$$

From the mean-value theorem, we have

$$f(x^1) - f(x^2) = (x^1 - x^2)\,\nabla f(\bar{x}) \tag{4}$$

where $\bar{x} = \bar{\lambda} x^1 + (1 - \bar{\lambda})\, x^2$, for some $0 < \bar{\lambda} < 1$. Now from (3) and (4) we have $(x^2 - \bar{x})\,\nabla f(\bar{x}) > 0$. Since ∇f is pseudomonotone, from the last inequality and Lemma 2.12.1, it follows that

$$(x^2 - \bar{x})\,\nabla f(x^2) > 0 \; .$$

But then we have $(x^1 - x^2)\,\nabla f(x^2) < 0$ which contradicts (1). Hence f is pseudoconvex. □

Definition 2.12.2. A map $F : X \to \mathbb{R}^n$ is said to be *quasimonotone* on $X \subset \mathbb{R}^n$ if, for every pair of distinct points $x^1, x^2 \in X$, we have

$$(x^1 - x^2)\, F(x^2) > 0 \Rightarrow (x^1 - x^2)\, F(x^1) \geqq 0 \, .$$

Obviously, by Lemma 2.12.1, every pseudomonotone map is quasimonotone, but the converse is not true. (Take, for example, $f(x) = x^2$, $X = \mathbb{R}$.) Also for the case of quasimonotonicity there exists a link with quasiconvexity. More precisely, we have the following result, due to Karamardian and Schaible (1990).

Theorem 2.12.2. Let $f : X \to \mathbb{R}$ be differentiable on the open convex set $X \subset \mathbb{R}^n$; then f is quasiconvex on X if and only if ∇f is quasimonotone on X.

Proof. This is quite similar to the proof of Theorem 2.12.1. In fact

a) If f is quasiconvex, then

$$(x^1 - x^2)\, \nabla f(x^2) > 0 \Rightarrow f(x^1) > f(x^2) \Rightarrow$$
$$\Rightarrow (x^2 - x^1)\, \nabla f(x^1) \leqq 0 \Rightarrow (x^1 - x^2)\, \nabla f(x^1) \geqq 0 \, ,$$

i.e. ∇f is quasimonotone.

b) Let ∇f be quasimonotone and assume that f is not quasiconvex. Then there exist $x^1, x^2 \in X$ such that $f(x^1) \leqq f(x^2)$ and $\bar{\lambda} \in (0,1)$ such that, for $\bar{x} = x^2 + \bar{\lambda}(x^1 - x^2)$, we have

$$f(\bar{x}) > f(x^2) \geqq f(x^1) \, . \tag{5}$$

The mean-value theorem implies the existence of $\hat{x}$ and x^* such that

$$f(\bar{x}) - f(x^1) = (\bar{x} - x^1)\, \nabla f(\hat{x})$$
$$f(\bar{x}) - f(x^2) = (\bar{x} - x^2)\, \nabla f(x^*)$$

where

$$\hat{x} = x^2 + \hat{\lambda}(x^1 - x^2) , \qquad x^* = x^2 + \lambda^*(x^1 - x^2) ,$$
$$0 < \lambda^* < \bar{\lambda} < \hat{\lambda} < 1 . \tag{6}$$

Then (5) implies that

$$(\bar{x} - x^1)\, \nabla f(\hat{x}) > 0$$
$$(\bar{x} - x^2)\, \nabla f(x^*) > 0 .$$

This yields

$$(x^* - \hat{x})\, \nabla f(\hat{x}) > 0 \tag{7}$$
$$(\hat{x} - x^*)\, \nabla f(x^*) > 0 \tag{8}$$

in view of (6). From (8) we obtain

$$(x^* - \hat{x})\, \nabla f(x^*) < 0$$

which together with (7) contradicts the quasimonotonicity of ∇f. □

It must be noted that Theorems 2.12.1 and 2.12.2 supply pure first-order characterizations of pseudoconvex and quasiconvex functions. Karamardian and Schaible (1990) discuss seven kinds of monotone and generalized monotone maps and give the characterizations of the corresponding generalized convex functions in terms of generalized monotonicity of their gradients. See also the survey paper of Schaible (1994).

2.13. Comparison Between Convex and Generalized Convex Functions

We have seen that quasiconvex functions form the widest class of generalized convex functions introduced in Section 2.10. It may be interesting to compare the most relevant properties of convex functions and quasiconvex functions. The following tables are taken from Greenberg and Pierskalla (1971). As usual, X is a convex subset of $\mathbb{R}^n$.

Table 2.
Analogous (or generalized) properties.

Convex functions	Quasiconvex functions
1a) f is convex if and only if epi f is a convex set.	**1b)** f is quasiconvex if and only if its lower level set $L(f, \alpha)$ is convex for any $\alpha \in \mathbb{R}$.
2a) f is linear if and only if epi f and hypo f are convex sets.	**2b)** f is quasilinear if and only if its lower level set $L(f, \alpha)$ and its upper level set $U(f, \alpha)$ are convex for any $\alpha \in \mathbb{R}$.
3a) $f : X \to \mathbb{R}$ is continuous on $\text{relint}(X)$.	**3b)** $f : X \to \mathbb{R}$ is continuous almost everywhere on $\text{relint}(X)$ (Deak (1962)).
4a) One-sided partial derivatives exist everywhere on $\text{relint}(X)$.	**4b)** One-sided partial derivatives exist almost everywhere on $\text{relint}(X)$ (Deak (1962)).
5a) If f is $\mathcal{C}^2(X)$, then f is convex on X if and only if $Hf(x)$ is positive semidefinite on X.	**5b)** If f is $\mathcal{C}^2(X)$ and $\nabla f(x) \neq 0$, $\forall x \in X$, then f is quasiconvex on X if and only if all the *bordered* principal minors of $Hf(x)$ are non-positive for each $x \in X$.
6a) If f is differentiable on X, then f is convex on X if and only if $x^1, x^2 \in X$, $f(x^1) - f(x^2) \geqq (x^1 - x^2)\nabla f(x^2)$.	**6b)** If f is differentiable on X, then f is quasiconvex on X if and only if $x^1, x^2 \in X$, $f(x^1) \leqq f(x^2) \Rightarrow (x^1 - x^2)\nabla f(x^2) \leqq 0$.
7a) $\sup_{x \in \text{relint}(X)} f(x) < +\infty$ if X is compact (Fenchel (1953)).	**7b)** $\sup_{x \in \text{relint}(X)} f(x) < +\infty$ if X is compact (Greenberg and Pierskalla (1971)).

Convex functions	Quasiconvex functions
8a) Every local minimum point is a global minimum point.	**8b)** A strict local minimum is also a strict global minimum.
9a) The set of global minimum points X_f^{Δ} is a convex set.	**9b)** The set of global minimum points X_f^{Δ} is a convex set.
10a) $f(\lambda x) \leqq \lambda f(x), \forall \lambda \in [0,1]$, if $f(0) \leqq 0$.	**10b)** $f(\lambda x) \leqq f(x), \forall \lambda \in [0,1]$, if $f(x) \geqq f(0)$.
11a) $\varphi(\lambda) = f\left[\lambda x^1 + (1-\lambda)\, x^2\right]$ is convex on $[0,1]$, $\forall x^1, x^2 \in X$, if and only if f is convex on X.	**11b)** $\varphi(\lambda) = f\left[\lambda x^1 + (1-\lambda)\, x^2\right]$ is quasiconvex on $[0,1]$, $\forall x^1, x^2 \in X$, if and only if f is quasiconvex on X.
12a) $g(x) = \sup_{i \in I} \{f_i(x)\}$ is convex, where I is any index set.	**12b)** $g(x) = \sup_{i \in I} \{f_i(x)\}$ is quasiconvex, where I is any index set.
13a) $g(x) = F\left[f(x)\right]$ is convex if F is convex and nondecreasing.	**13b)** $g(x) = F\left[f(x)\right]$ is quasiconvex if F is nondecreasing (see also Section 2.15).

Table 3.
Properties of convex functions with no analogue for quasiconvex functions

1) $g(x) = \sum_{i=1}^{n} \lambda_i g_i(x)$, $\lambda_i \geqq 0$, $i = 1, ..., n$ is a convex function if g_i, $i = 1, ..., n$, is convex. In case g_i is a quasiconvex function, then the property is no more valid. Take, e.g., the functions

$$f_1(x) = \begin{cases} 0 \quad , & \text{if } x \leqq 0 \\ -x^2 \quad , & \text{if } x > 0 \end{cases} \quad ; \qquad f_2(x) = \begin{cases} -x^2 \quad , & \text{if } x \leqq 0 \\ 0 \quad , & \text{if } x > 0 \end{cases} .$$

Then $f_1(x) + f_2(x) = -x^2$, which is not quasiconvex.

2) f^{**}, the biconjugate of f, equals f if f is convex and closed (f proper). If f is quasiconvex the property is no longer valid. Take, e.g., the function $f(x) = -e^{-x^2}$ for which we have $f^{**}(x) = -1$.

3) If f is convex on the bounded set $X \subset \mathbb{R}^n$, it is $\inf_{x \in \text{relint}(X)} f(x) > -\infty$ (see Fenchel (1953)). This property does not hold for quasiconvex functions, as shown by the following example:

$$f(x) = \begin{cases} 1/(x-1) \ , & \text{for } x \in [0,1) \\ 0 \ , & \text{for } x = 1 \ . \end{cases}$$

f is quasiconvex on $X = [0,1]$ but $\inf_{x \in \text{relint}(X)} f(x) = -\infty$.

4) The theorems of Berge (Theorem 2.9.2), Fan-Glicksberg-Hoffman (Theorem 2.9.1), and Bohnenblust-Karlin-Shapley (Theorem 2.9.3) are not extensible, in their original forms, by substituting convexity with quasiconvexity.

2.14. Generalized Convexity at a Point

We now present another generalization of convex sets and convex functions, introduced by Mangasarian (1969), namely the convexity of a set at a point x^0 and the (generalized) convexity of a function at a point x^0. So far we have discussed convexity and generalized convexity of a function on a given (convex) set $X \subset \mathbb{R}^n$. In the various definitions two points of X, x^1 and x^2, must vary on X. If we keep fixed one point x^0 and let vary only the second point x over X, we obtain local definitions (or better: "pointwise definitions") of, e.g., convex, pseudoconvex, quasiconvex functions *at a point* $x^0 \in X$, with respect to X. This concept is useful because in many cases "global" generalized convexity is really not needed, especially in obtaining sufficient optimality conditions in nonlinear programming.

Definition 2.14.1. Let $X \subset \mathbb{R}^n$ be a nonempty set and $x^0 \in X$; then X is said to be *convex at* x^0 or *star-shaped at* x^0 if

$$\lambda x^0 + (1-\lambda)\, x \in X\,, \quad \forall\, x \in X\,, \quad \forall\, \lambda \in [0,1]\,.$$

Definition 2.14.2. Let $f : X \to \mathbb{R}$, $X \subset \mathbb{R}^n$ be star-shaped at $x^0 \in X$; the function f is said to be *convex at* x^0, with respect to X if

$$f(\lambda x^0 + (1-\lambda)\, x) \leqq \lambda f(x^0) + (1-\lambda)\, f(x)\,, \qquad \forall\, x \in X\,, \quad \forall\, \lambda \in [0,1]\,;$$

f is said to be *strictly convex at* x^0, with respect to X, if

$$x \in X\,, \quad x \neq x^0\,, \quad f(\lambda x^0 + (1-\lambda)\, x) < \lambda f(x^0) + (1-\lambda)\, f(x)\,, \qquad \forall\, \lambda \in (0,1)\,.$$

Definition 2.14.3. Let $f : X \to \mathbb{R}$ be differentiable at $x^0 \in X \subset \mathbb{R}^n$; the function f is said to be *pseudoconvex at* x^0, with respect to X, if

$$x \in X\,, \quad (x - x^0)\, \nabla f(x^0) \geqq 0 \Rightarrow f(x) \geqq f(x^0)$$

or equivalently if

$$x \in X\,, \quad f(x) < f(x^0) \Rightarrow (x - x^0)\, \nabla f(x^0) < 0\,.$$

Definition 2.14.4. Let $f : X \to \mathbb{R}$, $X \subset \mathbb{R}^n$ be star-shaped at $x^0 \in X$; the function f is said to be *quasiconvex at* x^0, with respect to X, if

$$x \in X\,, \quad f(x) \leqq f(x^0) \Rightarrow f(\lambda x^0 + (1-\lambda)\, x) \leqq f(x^0)\,, \qquad \forall\, \lambda \in [0,1]\,.$$

Obviously if a function is convex, pseudoconvex, quasiconvex at *each point* of a convex set $X \subset \mathbb{R}^n$, it will be, respectively, convex, pseudoconvex, quasiconvex on X. In the same manner we can introduce functions which are semistrictly quasiconvex at x^0, strictly quasiconvex at x^0 and strictly pseudoconvex at x^0.

The following results are proved by Mangasarian (1969) and Martos (1975).

1) $f : X \to \mathbb{R}$, $X \subset \mathbb{R}^n$, is convex at $x^0 \in X$ if and only if $\operatorname{epi} f$ is star-shaped at $(x^0, f(x^0))$.

2) Let $f : X \to \mathbb{R}$ be differentiable and convex at $x^0 \in X \subset \mathbb{R}^n$, X open and star-shaped set at x^0. Then we have

$$f(x) - f(x^0) \geqq (x - x^0)\,\nabla f(x^0)\,, \qquad \forall\, x \in X\,.$$

If f is strictly convex at $x^0 \in X$, then we have

$$f(x) - f(x^0) > (x - x^0)\,\nabla f(x^0)\,, \qquad \forall\, x \in X\,, \quad x \neq x^0\,.$$

3) Let $f : X \to \mathbb{R}$ be twice-continuously differentiable and convex at $x^0 \in X \subset \mathbb{R}^n$, X open and star-shaped at x^0; then $Hf(x^0)$ is positive semidefinite (i.e. $yHf(x^0)\,y \geqq 0$, $\forall\, y \in \mathbb{R}^n$).

4) Let $f : X \to \mathbb{R}$ be differentiable and quasiconvex at $x^0 \in X \subset \mathbb{R}^n$, X open and star-shaped at x^0; then we have

$$x \in X\,, \quad f(x) \leqq f(x^0) \Rightarrow (x - x^0)\,\nabla f(x^0) \leqq 0$$

or equivalently

$$x \in X\,, \quad (x - x^0)\,\nabla f(x^0) > 0 \Rightarrow f(x) > f(x^0)\,.$$

We point out that the converse implications of 2), 3) and 4) do not hold (the reader is invited to build numerical or graphical examples). We must note that if $f : X \to \mathbb{R}$ is differentiable and convex at $x^0 \in X \subset \mathbb{R}^n$, X open and star-shaped at x^0, then f is pseudoconvex at x^0, but not conversely. However, if f is pseudoconvex at x^0, it is not necessarily also quasiconvex at x^0. Obviously a differentiable function on the open set $X \subset \mathbb{R}^n$ and with $f(x) \geqq f(x^0)$, $\forall\, x \in X$, is pseudoconvex at x^0; thus, e.g., $f(x) = -\cos x$ is pseudoconvex at $x^0 = 0$, with respect to $\mathbb{R}$, but not quasiconvex at $x^0 = 0$, with respect to $\mathbb{R}$. See also Mangasarian (1969), Martos (1975), Lopez Cerda and Valls Verdejo (1976), Giorgi (1987). We point out also that some properties, related to minimization problems, continue to hold under local convexity or local generalized convexity. For example, if f is strictly convex at $x^0 \in X$, then if x^0 is a minimum point of f over X, it is the unique minimum point. If x^0 is a local minimum point of f over X and f is convex at x^0, then x^0 is also a global minimum point. The same holds even if f is semistrictly quasiconvex at x^0. Theorem 2.11.5 and Corollary 2.11.1 hold also under the assumption that f is pseudoconvex at $x^* \in X$, with respect to X. Similarly, Corollary 2.11.2 also holds under the assumption that f is strictly pseudoconvex at $x^* \in X$ with respect to X.

The following result has been proved independently by Ferland (1971) and Giorgi (1987); see also Crouzeix and Ferland (1982).

Theorem 2.14.1. Let $f : X \to \mathbb{R}$ be continuous on the open set $X \subset \mathbb{R}^n$ and differentiable at $x^0 \in X$. If f is quasiconvex at x^0, with respect to X, and if $\nabla f(x^0) \neq 0$, then f is pseudoconvex at x^0.

Proof. Consider a point $x^1 \in X$ such that

$$(x^1 - x^0)\,\nabla f(x^0) \geqq 0 \tag{1}$$

but for which

$$f(x^1) < f(x^0) \ . \tag{2}$$

Then x^1 belongs to the nonempty set

$$X_0 = \{x \mid x \in X,\ f(x) \leqq f(x^0)\} \ ,$$

whose elements, thanks to the quasiconvexity of $f(x)$ at x^0, verify the relation

$$x \in X_0 \Rightarrow (x - x^0)\,\nabla f(x^0) \leqq 0 \ . \tag{3}$$

Let us now consider the sets, both nonempty,

$$W = \{x \mid x \in X,\ (x - x^0)\,\nabla f(x^0) \geqq 0\} \ ,$$

$$X_{00} = X_0 \cap W \ .$$

The following implication holds:

$$x \in X_{00} \Rightarrow x \in H_0 = \{x \mid x \in X,\ (x - x^0)\,\nabla f(x^0) = 0\} .$$

It is therefore evicent that X_{00} is included in the hyperplane (recall that $\nabla f(x^0) \neq 0$)

$$H = \{x \mid x \in \mathbb{R}^n,\ (x - x^0)\,\nabla f(x^0) = 0\} ,$$

supporting X_0, owing to (3). Relations (1) and (2) point out that x^1 belongs to W and X_0 and hence to X_{00}, H_0 and H. Moreover, thanks to the continuity of f on X, relation (2) says that x^1 lies in the interior of X_0: therefore x^1 at the same time belongs to the interior of a set and to a hyperplane supporting the same set, which is absurd. So relation (2) is false and (1) implies $f(x^1) \geqq f(x^0)$. □

Corollary 2.14.1. Let $f : X \to \mathbb{R}$ be differentiable and quasiconvex on the open convex set $X \subset \mathbb{R}^n$; then f is pseudoconvex on X if and only if f has a minimum point at $x^0 \in X$, whenever $\nabla f(x^0) = 0$.

Proof. The necessity part follows at once from the definition of pseudoconvex functions. As for sufficiency, let $x^0 \in X$, $\nabla f(x^0) = 0$; the point x^0 is then a (global) minimum point of f over X and we have

$$(x - x^0)\,\nabla f(x^0) = 0 \Rightarrow f(x) \geqq f(x^0) , \qquad \forall x \in X .$$

Thus f is pseudoconvex at x^0 with respect to X. On the other hand, on the grounds of Theorem 2.14.1, the quasiconvex function f is pseudoconvex at every point $x \in X$ where $\nabla f(x) \neq 0$. □

From Theorem 2.14.1 it also follows that all those criteria used to test the quasiconvexity of a differentiable or twice-continuously differentiable function $f : X \to \mathbb{R}$, X open subset of $\mathbb{R}^n$, and that imply $\nabla f(x) \neq 0$, $\forall x \in X$, in effect identify the class of pseudoconvex functions. This is the case for Theorem 2.10.5 and for the other criteria equivalent to the conditions expressed by this same theorem. This is also the case for the classical conditions on the bordered Hessian, given by Arrow and Enthoven (1961) and generalized by Ferland (1971, 1972b).

2.15. Convexity, Pseudoconvexity and Quasiconvexity of Composite Functions

In spite of the numerous characterizations available for establishing the convexity or the generalized convexity of a function, it may be difficult to decide whether a given function belongs to a particular class of generalized convex functions. Under suitable conditions, however, it is possible to identify the class to which a composite function belongs, after having established the convexity or the generalized convexity of the (simpler) components of the same function.

The following results are due mainly to Mangasarian (1970); however, see also Bereanu (1969, 1972), Bector (1973), Martos (1975) and Schaible (1971, 1972).

Let $\Phi : A \to \mathbb{R}$, where $A \subseteq \mathbb{R}^m \times \mathbb{R}^k$; Φ is said to be *increasing-decreasing* (incr-decr) on A if and only if for every (y^1, z^1) and (y^2, z^2) in A

$$y^2 \geqq y^1 \text{ and } z^2 \leqq z^1 \Rightarrow \Phi(y^2, z^2) \geqq \Phi(y^1, z^1) .$$

Φ is said to be y-*increasing* (y-incr) on A if and only if for each (y^1, z) and (y^2, z) in A

$$y^2 \geqq y^1 \Rightarrow \Phi(y^2, z) \geqq \Phi(y^1, z)$$

and Φ is said to be y-*decreasing* (y-decr) on A if and only if for each (y^1, z) and (y^2, z) in A

$$y^2 \leqq y^1 \Rightarrow \Phi(y^2, z) \geqq \Phi(y^1, z) .$$

The following lemma follows directly from the definition of differentiability and the mean value theorem.

Lemma 2.15.1. Let $\Phi : A \to \mathbb{R}$ be differentiable on the open convex set $A \subset \mathbb{R}^m \times \mathbb{R}^k$. Then Φ is incr-decr on A if and only if $\nabla_y \Phi(y, z) \geqq 0$ and $\nabla_z \Phi(y, z) \leqq 0$ for all $(y, z) \in A$.

The following theorem represents the principal result on the generalized convexity of composite functions.

<u>Theorem 2.15.1</u>. Let $X \subset I\!R^n$ be a convex set, let $f(x) = (f_1(x), ..., f_m(x))$, $g(x) = (g_1(x), ..., g_k(x))$, both defined on X, and let Φ be a real-valued function defined on $I\!R^m \times I\!R^k$. Let $\psi(x) = \Phi(f(x), g(x))$ and let any one of the following four assumptions hold:

i) f is convex; g is concave; Φ is incr-decr;

ii) f is linear; g is linear;

iii) f is convex; g is linear; Φ is y-incr;

iv) f is concave; g is linear; Φ is y-decr.

Then the following results hold:

I) If Φ is convex on $I\!R^m \times I\!R^k$, then ψ is convex on X.

II) If X is open, f and g are differentiable on X and Φ is pseudoconvex on $I\!R^m \times I\!R^k$, then ψ is pseudoconvex on X.

III) If Φ is quasiconvex, then ψ is quasiconvex.

<u>Proof</u>.

I) We shall first prove this part of the theorem under assumption i). Let $x^1, x^2 \in X$ and let $0 \leqq \lambda \leqq 1$. Then

$$\psi((1-\lambda)\,x^1 + \lambda x^2) =$$

$$= \Phi(f((1-\lambda)\,x^1 + \lambda x^2), g((1-\lambda)\,x^1 + \lambda x^2)) \leqq$$

$$\leqq \Phi((1-\lambda)\,f(x^1) + \lambda f(x^2), (1-\lambda)\,g(x^1) + \lambda g(x^2))$$

(since f is convex, g is concave and φ is incr-decr) $\leqq$

$$\leqq (1-\lambda)\,\Phi(f(x^1), g(x^1)) + \lambda\Phi(f(x^2), g(x^2))$$

(since Φ is convex) $=$

$$= (1-\lambda)\,\psi(x^1) + \lambda\psi(x^2)$$

and hence ψ is convex.
Under assumption ii) the first inequality above is an equality and under assumption iii) or iv) it remains an inequality.

II) We again prove first this part of the theorem under assumption i). Let $x^1, x^2 \in X$; then

$$(x^2 - x^1)\,\nabla\psi(x^1) =$$
$$= (x^2 - x^1) \cdot [\nabla_f \Phi(f(x^1), g(x^1)) \cdot \nabla f(x^1) +$$
$$+ \nabla_g \Phi(f(x^1), g(x^1)) \cdot \nabla g(x^1)] \ .$$

By the convexity of f, concavity of g, incr-decr property of Φ and Lemma 2.15.1, we obtain

$$(x^2 - x^1)\,\nabla\psi(x^1) \ \leqq$$
$$\leqq \ [f(x^2) - f(x^1)] \cdot \nabla_f \Phi(f(x^1), g(x^1)) + [g(x^2) - g(x^1)] \cdot$$
$$\cdot \nabla_g \Phi(f(x^1), g(x^1)) \ .$$

Hence

$$(x^2 - x^1)\,\nabla\psi(x^1) \geqq 0 \Rightarrow [f(x^2) - f(x^1)] \cdot \nabla_f \Phi(f(x^1), g(x^1)) +$$
$$+ \ [g(x^2) - g(x^1)] \cdot \nabla_g \Phi(f(x^1), g(x^1)) \geqq 0$$

(by the above inequality)

$$\Rightarrow \Phi(f(x^2), g(x^2)) \geqq \Phi(f(x^1), g(x^1))$$

(by the pseudoconvexity of Φ)

$$\Rightarrow \psi(x^2) \geqq \psi(x^1) \ .$$

Hence ψ is pseudoconvex. Under assumption ii) the first inequality in the above proof becomes an equality and under assumption iii) or iv) it remains an inequality.

III) Again we prove this part of the theorem under assumption i). Let $x^1, x^2 \in X$ and $0 \leqq \lambda \leqq 1$. If $\psi(x^2) \leqq \psi(x^1)$, then

$$\Phi(f(x^2), g(x^2)) \leqq \Phi(f(x^1), g(x^1)) .$$

Since Φ is quasiconvex we have

$$\Phi((1-\lambda) f(x^1) + \lambda f(x^2), (1-\lambda) g(x^1) + \lambda g(x^2)) \leqq$$
$$\leqq \Phi(f(x^1), g(x^1))$$

and by the assumptions it follows

$$\Phi(f((1-\lambda) x^1 + \lambda x^2), g((1-\lambda) x^1 + \lambda x^2)) \leqq \Phi(f(x^1), g(x^1)) ,$$

i.e. $\psi((1-\lambda) x^1 + \lambda x^2) \leqq \psi(x^1)$ and hence ψ is quasiconvex. The rest of the proof is similar. □

Theorem 2.15.1 can be easily reformulated with reference to the concave and generalized concave case. Moreover, if we set $k = 0$, i.e. $\psi(x) = \Phi(f(x))$, the result I) i) is a well-known result on the convexity of composite functions and III) i) is a similar known result on the quasiconvexity of composite functions (see Berge (1963), Bereanu (1969, 1972), Fenchel (1953), Greenberg and Pierskalla (1971), Martos (1975); see, moreover, Theorem 2.5.8 and results 13a) and 13b) of Table 1 of Section 2.13). Theorem 2.15.1 can be applied to a large class of functions; particularly important in view of applications to nonlinear programming problems is the case of *nonlinear fractional functions*.

Let X be a convex set in $\mathbb{R}^n$, let $\rho : X \to \mathbb{R}$ and $\sigma : X \to \mathbb{R}$ and consider $\psi(x) = \rho(x)/\sigma(x)$. Suppose that any one of the following assumptions hold on X (Table 4).

Table 4.

	1	2	3	4	5	6	7	8	9
ρ	convex $\geqq 0$	concave $\leqq 0$	convex $\leqq 0$	concave $\geqq 0$	linear	linear $\leqq 0$	linear $\geqq 0$	convex	concave
σ	concave > 0	convex < 0	convex > 0	concave < 0	linear $\neq 0$	convex $\neq 0$	concave $\neq 0$	linear > 0	linear < 0

Then ψ is pseudoconvex on X if X is open and ρ and σ are differentiable on X; otherwise ψ is quasiconvex on X. In Bector (1973) and Martos (1975) it is proved that, under the above assumptions, ψ is indeed semistrictly quasiconvex on X.

The above results follow from Theorem 2.15.1 by observing that the real-valued function $\Phi(y, z) = y/z$, $(y, z) \in I\!R \times I\!R$ is pseudoconvex and hence also quasiconvex on either of the convex sets

$$\{(y, z) \mid (y, z) \in I\!R \times I\!R, \ z > 0\}$$

or

$$\{(y, z) \mid (y, z) \in I\!R \times I\!R, \ z < 0\} ,$$

and by making the identifications

$$f(x) = \rho(x) \ ; \qquad g(x) = \sigma(x)$$

for the cases 1, 5, 8 and 9; and

$$f(x) = \sigma(x) \ ; \qquad g(x) = \rho(x)$$

for the cases 2, 6 and 7; and

$$f(x) = [\rho(x) \, ; \, \sigma(x)]$$

for the case 3; and

$$g(x) = [\rho(x) \, ; \, \sigma(x)]$$

for the case 4.

Note moreover that, according to the above assertions, the linear fractional function

$$f(x) = \frac{cx + \alpha}{dx + \beta} ,$$

with $c, d \in I\!R^n$; $\alpha, \beta \in I\!R$, is pseudolinear (and hence semistrictly quasilinear) on each convex set $X \subset I\!R^n$ on which we have $dx + \beta \neq 0$.

Theorem 2.15.1 can also be applied to *reciprocal functions*, i.e. to functions of the type $h(x) = 1/f(x)$, where f is defined on the convex set $X \subset I\!R^n$. From the above theorem we obtain that:

i) If f is positive and concave on X, then h is convex (and positive) on X.

ii) If f is negative and convex on X, then h is concave (and negative) on X.

Theorem 2.15.1 can also be applied to *bi-nonlinear functions*, i.e. to functions of the type $\varphi(x) = \rho(x) \cdot \mu(x)$, where ρ and μ are differentiable on $X \subset \mathbb{R}^n$. Then the following implications hold on X:

1) ρ convex $\leqq 0$, μ concave $> 0 \Rightarrow \varphi$ pseudoconvex $\leqq 0$.

2) ρ convex < 0, μ concave $\geqq 0 \Rightarrow \varphi$ pseudoconvex $\leqq 0$.

3) ρ convex < 0, μ convex $\leqq 0 \Rightarrow \varphi$ pseudoconvex $\geqq 0$.

4) ρ concave $\geqq 0$, μ convex $> 0 \Rightarrow \varphi$ pseudoconcave $\geqq 0$.

For other applications of Theorem 2.15.1 to special composite functions see Avriel, Diewert, Schaible and Zang (1987), Mangasarian (1970), Martos (1975), Schaible (1971, 1972).

Another class of convex composite functions which has some importance, especially in geometric programming, is given by the *logarithmic convex functions* or *L-convex functions*, examined in detail by Klinger and Mangasarian (1968).

Definition 2.15.1. A function $f : X \to \mathbb{R}$, defined and positive on the convex set $X \subset \mathbb{R}^n$, is *logarithmic convex* or *L-convex* on X if $\log f$ is convex on X. It is L-concave on X if $\log f$ is concave on X.

From the previous definition it can be proved that the following properties are equivalent:

i) f is L-convex on X;

ii) $1/f$ is L-concave on X;

iii) $f = e^h$, with h convex on X;

iv) f^α is convex on X for each $\alpha \in (0, 1)$;

v) for each $x^i \in X$ and for each $t \in \mathbb{R}^m$, $t \geq 0$, $\sum_{i=1}^m t_i = 1$, we have

$$f\left(\sum_{i=1}^{m} t_i x^i\right) \leqq \prod_{i=1}^{m} (f(x^i))^{t_i} \ ;$$

vi) $x^1, x^2 \in X$, $\lambda \in [0, 1]$:

$$f(\lambda x^1 + (1-\lambda)\, x^2) \leqq [f(x^1)]^\lambda \, [f(x^2)]^{1-\lambda} \ .$$

If f is also differentiable on the open convex set X, the following properties are equivalent:

i) f is L-convex on X;

ii) $x^1, x^2 \in X$:

$$(x^1 - x^2)\, \frac{\nabla f(x^2)}{f(x^2)} \leqq \log \frac{f(x^1)}{f(x^2)} \leqq (x^1 - x^2)\, \frac{\nabla f(x^1)}{f(x^1)} \ ;$$

iii) $x^1, x^2 \in X$:

$$(x^1 - x^2) \left[\frac{\nabla f(x^1)}{f(x^1)} - \frac{\nabla f(x^2)}{f(x^2)}\right] \geqq 0 \ .$$

If f is twice-continuously differentiable on the open convex set X, the following properties are equivalent:

i) f is L-convex on X;

ii) for each $x \in X$ the matrix

$$f(x) \cdot Hf(x) - \nabla f(x) \cdot [\nabla f(x)]^T$$

is positive semidefinite.

For proofs of the above equivalences, see Klinger and Mangasarian (1968) and Stancu-Minasian (1992). Moreover, it can be shown that if f is L-convex on the convex set X, then it is convex on X, but not conversely and that if f is concave on X, then it is L-concave on X, but not conversely.

2.16. Convexity, Pseudoconvexity and Quasiconvexity of Quadratic Functions

Quadratic functions consist of the sum of a quadratic form and a linear function; they are generally expressed as:

$$Q(x) = \frac{1}{2}\, x^T A x + bx \ ,$$

where A is a real symmetric matrix of order n and $b \in \mathbb{R}^n$. First we note that if the quadratic form $F(x) = x^T A x$ is convex on some convex set $X \subset \mathbb{R}^n$, the above sum $Q(x)$ will also be convex on the same set; this does not occur if $x^T A x$ is pseudoconvex or quasiconvex on some convex set, as the sum of a pseudoconvex or quasiconvex function and a linear function is not necessarily pseudoconvex or quasiconvex.

Generalized convexity of quadratic forms and functions has been studied by many authors with different techniques. One of the first approaches is due to Martos (1969, 1971, 1975) who characterized quasiconvex and pseudoconvex quadratic functions on the nonnegative orthant by means of the concept of *positive subdefinite matrices*. This approach was followed also by Cottle and Ferland (1972) who derived additional criteria. Ferland (1971) and Schaible (1971) independently obtained a characterization of quasiconvex and pseudoconvex quadratic functions on arbitrary solid (i.e. with a nonempty interior) convex sets. For a survey on the main results on this subject, see Avriel, Diewert, Schaible and Zang (1987), Schaible (1981).

The following results are taken from the quoted references. First we note that if $X \subset \mathbb{R}^n$ is a convex set with a nonempty interior, then the quadratic function $Q(x) = \frac{1}{2}\, x^T A x + bx$ is convex (concave) on X if and only if A is positive (negative) semidefinite (the proof is obtained taking Theorem 2.5.5 into account). Similarly $Q(x)$ is strictly convex (strictly concave) on X if A is positive (negative) definite. The following result is due to Cottle (1967).

Theorem 2.16.1. Let $X \subset \mathbb{R}^n$ be a nonempty convex set; then $Q(x)$ is convex on X if and only if $Q(x)$ is convex on every translation $X + a$ of X.

The following result on the quasiconvexity of quadratic functions is perhaps more interesting and is due to Martos (1975).

Theorem 2.16.2. The quadratic function $Q(x)$ is quasiconvex on $\mathbb{R}^n$ if and only if it is convex on $\mathbb{R}^n$.

Proof. Let y be any vector of $\mathbb{R}^n$ and $\alpha > 0$ a number such that

$$Q(\alpha y) \leqq Q(-\alpha y) \ . \tag{1}$$

(Change the sign of y, not of α, if necessary.)
Then

$$\tfrac{1}{2}\alpha^2 y^T Ay + \alpha by \leqq \tfrac{1}{2}\alpha^2 y^T Ay - \alpha by$$

i.e.

$$2\alpha by \leqq 0 \ .$$

If, however, this holds for some $\alpha > 0$, it also holds for any $\alpha > 0$, thus also (1) holds for any $\alpha > 0$. Now if $Q(x)$ is quasiconvex on $\mathbb{R}^n$, then (1) implies that for all $\alpha > 0$

$$[\alpha y - (-\alpha y)]^T [A(-\alpha y) + b] = -2\alpha^2 y^T Ay + 2\alpha by \leqq 0$$

i.e.

$$by \leqq \alpha y^T Ay \ .$$

The last inequality holds for all $\alpha > 0$ only if $y^T Ay \geqq 0$ (or $(-y)^T A(-y) \geqq 0$ if the sign of y has been changed); as y has been chosen arbitrarily, thus $Q(x)$ is convex on $\mathbb{R}^n$. The converse of the theorem is obvious. □

The previous theorem shows that there is no reason to study the generalized convexity of quadratic functions on $\mathbb{R}^n$. However, there may be quadratic functions or quadratic forms that are pseudoconvex or quasiconvex on a convex subset of $\mathbb{R}^n$ (e.g. $\mathbb{R}^n_+$), but not convex on that subset. For example the quadratic form of two variables $F(x) = -x_1 x_2$ is quasiconvex on $\mathbb{R}^2_+$, but not convex there. However, Martos has observed that for quadratic functions we do not have to distinguish between semistrict quasiconvexity and quasiconvexity, i.e. $Q(x)$ is semistrictly quasiconvex on the convex set $X \subset I\!R^n$ if and only if it is quasiconvex on X.

When $X = I\!R^n_+$ we have several interesting results due to Martos (1969, 1975), Cottle and Ferland (1971, 1972), Ferland (1978, 1981). We need some definitions concerning certain classes of matrices (see Martos (1969)).

Definition 2.16.1. A real symmetric matrix A of order n and its corresponding quadratic form $x^T Ax$ are called *positive subdefinite* if for all $x \in I\!R^n$

$x^T Ax < 0 \Rightarrow Ax \geq 0$ or $Ax \leq 0$

and *strictly positive subdefinite* if for all $x \in I\!R^n$

$x^T Ax < 0 \Rightarrow Ax > 0$ or $Ax < 0$.

We note that if A is positive semidefinite, it is also strictly positive subdefinite and positive subdefinite, but not conversely.

Definition 2.16.2. A real symmetric matrix A of order n and its corresponding quadratic form $x^T Ax$ are said to be (strictly) *merely* positive subdefinite if they are (strictly) positive subdefinite, but not positive semidefinite.

Similarly we call a function *merely quasiconvex* (merely pseudoconvex) on some convex set if it is quasiconvex (pseudoconvex) but not convex on that set. We have the following basic results.

Theorem 2.16.3. The symmetric matrix A is merely positive subdefinite if and only if

i) A has one (simple) negative eigenvalue, and

ii) $A \leq 0$.

Proof. See Martos (1969, 1975). □

A more useful criterion, in view of its applications, is contained in the following

Theorem 2.16.4. The symmetric matrix A is merely positive subdefinite if and only if

i) $A \leq 0$, and

ii) all the principal minors of A are nonpositive.

Proof. See Cottle and Ferland (1972). □

Theorem 2.16.5. The quadratic function $Q(x) = x^T Ax + bx$ is quasiconvex on the nonnegative orthant $\mathbb{R}^n_+$ if and only if the following bordered matrix

$$\bar{A} = \begin{bmatrix} A & b \\ b^T & 0 \end{bmatrix}$$

is merely positive subdefinite. If $Q(x)$ is quasiconvex on $\mathbb{R}^n_+$ and $b \neq 0$, then $Q(x)$ is also pseudoconvex on $\mathbb{R}^n_+$.

Proof. See Cottle and Ferland (1972) and Ferland (1971). □

For what concerns quadratic forms $F(x) = x^T Ax$, we have the following results of Martos (1969).

Theorem 2.16.6. The quadratic form $x^T Ax$ is quasiconvex on $\mathbb{R}^n_+$ if and only if it is positive subdefinite . It is pseudoconvex on $\mathbb{R}^n_+$ if and only if it is strictly positive subdefinite.

A simple way to test whether a merely quasiconvex quadratic form, quasiconvex on $\mathbb{R}^n_+$, is also pseudoconvex on $\mathbb{R}^n_+$ is given in the following result, again due to Martos (1969).

Theorem 2.16.7. A merely quasiconvex quadratic form $x^T Ax$ on $\mathbb{R}^n_+$ is merely pseudoconvex on $\overset{o}{\mathbb{R}}{}^n_+$ if and only if A does not contain a row of zeros.

Generalizations of the above results for arbitrary solid convex sets were obtained by the authors cited at the beginning of this section. Here we give the following results, due to Ferland (1972a, 1981). Let A be a real symmetric matrix having exactly one negative eigenvalue; let v be either one of the two possible choices of normalized eigenvector associated with the unique negative eigenvalue. Let us consider the following two sets

$$T_A^1 = \{y \in \mathbb{R}^n \mid y^T A y \leqq 0 \text{ and } vy \geqq 0\}$$

$$T_A^2 = \{y \in \mathbb{R}^n \mid y^T A y \leqq 0 \text{ and } vy \leqq 0\} .$$

Consider the quadratic function $Q(x) = \frac{1}{2}\, x^T A x + bx$, with A a real symmetric matrix of order n and $b \in \mathbb{R}^n$. Associate with Q the set

$$M = \{x \in \mathbb{R}^n \mid Ax + b = 0\} .$$

Theorem 2.16.8. The quadratic function $Q(x) = \frac{1}{2}\, x^T A x + bx$ is merely quasiconvex on the solid convex set $X \subset \mathbb{R}^n$ if and only if M is nonempty, A has exactly one (single) negative eigenvalue and $X \subset T_A^1 + M$ or $X \subset T_A^2 + M$.

If M is nonempty and A has one (single) negative eigenvalue, the sets $T_A^1 + M$ and $T_A^2 + M$ are the *maximal domain of quasiconvexity* for the quadratic function $Q(x) = \frac{1}{2}\, x^T A x + bx$ (Ferland (1972a)). Note that M is nonempty if and only if $\operatorname{rank}(A, b) = \operatorname{rank}(A)$. Another criterion for the quasiconvexity of $Q(x)$ on a solid convex set is contained in the following theorem, again due to Ferland (1981).

Theorem 2.16.9. The quadratic function $Q(x) = \frac{1}{2}\, x^T A x + bx$ is merely quasiconvex on the solid convex set $X \subset \mathbb{R}^n$ if and only if A and the bordered Hessian matrix

$$\begin{bmatrix} A & Ax+b \\ (Ax+b)^T & 0 \end{bmatrix}$$

has exactly one negative eigenvalue for all $x \in X$.

2.17. Other Types of Generalized Convex Functions

A generalization of the class of convex functions can be obtained via generalized means. Note that the right-hand side of the inequality

$$f(\lambda x^1 + (1-\lambda)\, x^2) \leqq \lambda f(x^1) + (1-\lambda)\, f(x^2) \,, \qquad \lambda \in [0,1] \,, \tag{1}$$

which defines convex functions, is just the weighted arithmetic mean of $f(x^1)$ and $f(x^2)$; so a first generalization is obtained by substituting other weighted means in the right-hand side of (1). This approach has been investigated by Avriel (1972), Avriel and Zang (1974), Martos (1975), Ben-Tal (1977), Castagnoli and Mazzoleni (1989b), Mond (1983).

Before stating this generalized type of convexity, we review some classical definitions and results on generalized means (see Hardy, Littlewood and Polya (1934)).

Let α and β be positive numbers, $\lambda \in [0,1]$ and $r \neq 0$ any real number. Then the positive number

$$M_r(\alpha, \beta, \lambda) = \{\lambda \cdot \alpha^r + (1-\lambda)\, \beta^r\}^{\frac{1}{r}}$$

is called the *generalized* r*-th mean* of α and β. This notion of generalized mean easily extends to more than two positive numbers $\alpha_1, \alpha_2, ..., \alpha_n$, keeping the sum of the nonnegative weights $\lambda_1, \lambda_2, ..., \lambda_n$ equal to one. The same remark extends to vectors.

For $r = 0$ we define

$$M_0(\alpha, \beta, \lambda) = \lim_{r \to 0} M_r(\alpha, \beta, \lambda) = \alpha^\lambda \cdot \beta^{1-\lambda} \,.$$

This is the well-known *geometric mean* of α and β. For $r = +\infty$ and $r = -\infty$, respectively, we define

$$M_{+\infty}(\alpha, \beta, \lambda) = \lim_{r \to +\infty} M_r(\alpha, \beta, \lambda) = \max\{\alpha, \beta\} \,;$$

$$M_{-\infty}(\alpha, \beta, \lambda) = \lim_{r \to -\infty} M_r(\alpha, \beta, \lambda) = \min\{\alpha, \beta\} \,.$$

Theorem 2.17.1.

a) If $\alpha = \beta$, then $M_r(\alpha, \alpha, \lambda) = \alpha = \beta$, $-\infty \leqq r \leqq +\infty$.

b) If $\alpha > \beta$, then $\alpha > M_r(\alpha, \beta, \lambda) > \beta$ for any finite r and $\lambda \in (0,1)$.

c) If $s > r$, then $M_s(\alpha, \beta, \lambda) \geqq M_r(\alpha, \beta, \lambda)$ and the inequality is strict if and only if $\alpha \neq \beta$ and $\lambda \in (0,1)$.

Proof. The assertions a) and b) are easy conclusions of the definitions. Assertion c) can be proved by differentiation of M_r.
Since the derivative of M_r with respect to r is positive for $r \neq 0$ and M_r is (by definition) continuous at $r = 0$, we can conclude that M_r is strictly increasing with respect to r. □

Following the terminology of Avriel (1972), we now extend the definition of convex functions as follows:

Definition 2.17.1. Let $f : X \to \mathbb{R}$ be *positive* on the convex set $X \subset \mathbb{R}^n$; then f is said to be r^+*-convex* on X if

$$f(\lambda x^1 + (1-\lambda)\, x^2) \leqq M_r(f(x^1), f(x^2), \lambda)\ , \qquad \forall\, x^1, x^2 \in X\,,\quad \forall\, \lambda \in [0,1]\ . \tag{2}$$

Note that (2) gives the usual definition of (positive) convex functions for $r = 1$ and of (positive) quasiconvex functions for $r = +\infty$. In general, for $r > 1$ we obtain from (2) generalized convex functions and for $r < 1$ we obtain special convex functions. In particular for $r = 0$ we get the logarithmic convex functions:

$$f(\lambda x^1 + (1-\lambda)\, x^2) \leqq [f(x^1)]^\lambda \cdot [f(x^2)]^{1-\lambda}\ , \qquad \forall\, x^1, x^2 \in X\,,\quad \forall\, \lambda \in [0,1]\ .$$

Since $M_s(f(x^1), f(x^2), \lambda) \geqq M_r(f(x^1), f(x^2), \lambda)$ for $s > r$, it follows that a function that is r^+-convex will also be s^+-convex for all $s > r$. Thus we have a continuous transition from the class of positive convex functions ($r = 1$) to the class of positive quasiconvex functions ($r = +\infty$), via the intermediate class of r^+-convex functions, with $1 < r < +\infty$.

Recall that we restricted our definition of $M_r(f(x^1), f(x^2), \lambda)$ to a positive f in order to allow zero and negative values of r. Avriel (1972) and Martos (1975) independently define *r-convex functions* as follows.

Definition 2.17.2. The function $f : X \to \mathbb{R}$, defined on the convex set $X \subset \mathbb{R}$, is said to be *r-convex* on X if

$$f(\lambda x^1 + (1-\lambda)\, x^2) \leqq \log M_r\{e^{f(x^1)}, e^{f(x^2)}, \lambda\} \,, \quad \forall\, x^1, x^2 \in X \,, \quad \forall\, \lambda \in [0,1] \,.$$

By the definitions of the weighted arithmetic mean, r-convex functions are thus also defined equivalently as those functions satisfying the inequalities (for all $x^1, x^2 \in X$ and for all $\lambda \in [0,1]$):

$$f(\lambda x^1 + (1-\lambda)\, x^2) \leqq \begin{cases} \log\,[\lambda e^{rf(x^1)} + (1-\lambda)\, e^{rf(x^2)}]^{\frac{1}{r}} & , \text{ if } r \neq 0 \,, \\ \lambda f(x^1) + (1-\lambda)\, f(x^2) & , \text{ if } r = 0 \,, \\ \max\{f(x^1), f(x^2)\} & , \text{ if } r = +\infty \,, \\ \min\{f(x^1), f(x^2)\} & , \text{ if } r = -\infty \,. \end{cases}$$

Note that r-convexity, which is no longer restricted to positive values of f, reduces to ordinary convexity when $r = 0$ and to quasiconvex functions when $r = +\infty$. As noted above, r-convexity implies s-convexity for all $s > r$ (being $\log \xi$ a strictly increasing function for $\xi \in (0, +\infty)$). So r-convex functions represent a continuous parametric transition from the class of convex functions ($r = 0$) to the class of quasiconvex functions ($r = +\infty$), via the intermediate class of r-convex functions, with $0 < r < +\infty$. Avriel (1972) calls the functions that satisfy Definition 2.17.2 with $r < 0$, *superconvex* and with $r > 0$, *subconvex*. This superconvexity implies convexity which implies subconvexity which in turn implies quasiconvexity.

It is often difficult, from an algebraic point of view, to deal with r-convex functions. However, for r-convex functions with a finite r we have the following useful results due to Avriel (1972).

Theorem 2.17.2. Let $f : X \to \mathbb{R}$, with $X \subset \mathbb{R}^n$ convex; then f is r-convex on X, with $r \neq 0$, if and only if the function

$$\hat{f} = \exp(rf(x))$$

is convex for $r > 0$ and concave for $r < 0$.

Proof. The proof follows easily from the definition of r-convex functions.□

Theorem 2.17.3. Let $f : X \to \mathbb{R}$ be a twice-continuously differentiable function on the open convex set $X \subset \mathbb{R}^n$. Then f is r-convex on X if and only if the matrix Q, given by

$$Q(x) = r\nabla f(x)(\nabla f(x))^T + Hf(x) ,$$

is positive semidefinite for all $x \in X$.

Theorem 2.17.4. Let $f : X \to \mathbb{R}$ be a twice-continuously differentiable quasiconvex function on the open convex set $X \subset \mathbb{R}^n$. If there exists a real number r^* satisfying

$$r^* = \sup_{\substack{x \in X \\ \|z\|=1}} \frac{-z^T Hf(x)\, z}{(z^T \nabla f(x))^2}$$

whenever $z^T \nabla f(x) \neq 0$, then f is r^*-convex.

A further extension of convexity by the use of more general means is possible (see Avriel (1976), Ben-Tal (1977), Castagnoli and Mazzoleni (1989b), Mond (1983)).

Definition 2.17.3. Let $f : X \to \mathbb{R}$ be defined on the convex set $X \subset \mathbb{R}^n$ and let ϑ be a continuous strictly increasing scalar function that includes $f(x^1)$ and $f(x^2)$ in its domain, for any $x^1, x^2 \in X$. Then f is said to be *ϑ-convex* on X if

$$f(\lambda x^1 + (1-\lambda)\, x^2) \leqq \vartheta^{-1}\{\lambda\vartheta[f(x^1)] + (1-\lambda)\,\vartheta[f(x^2)]\} ,$$
$$\forall\, x^1, x^2 \in X , \quad \forall\, \lambda \in [0,1] . \tag{3}$$

Note that if $\vartheta(x) = x$, then (3) reduces to the usual definition of convexity. If $\vartheta(x) = x^r$ for $r \neq 0$ and $\vartheta(x) = \log x$ for $r = 0$, then (3) reduces to r^+-convexity. If $\vartheta(x) = e^{rx}$ for $r \neq 0$ and $\vartheta(x) = x$ for $r = 0$, then (3) reduces to r-convexity.

So far we have extended the definition of convexity by generalizing the right-hand side of the usual definition of convex function. A further extension is possible by generalizing the left-hand side of (3) as well. This has been done by Avriel and Zang (1980) who introduced the *arcwise connected functions*: in essence, since $f(\lambda x^1 + (1-\lambda)\, x^2)$, $\lambda \in [0,1]$, consists of the values of f at all points on the straight line between x^1 and x^2, we can consider, instead of lines, a more general path from x^1 to x^2.

Definition 2.17.4. The set $X \subset \mathbb{R}^n$ is said to be *arcwise connected* if for every pair of points $x^1, x^2 \in X$, there exists a continuous vector-valued function $H_{x^1,x^2}(\lambda)$ called an *arc* or a *path*, defined on the unit interval $[0,1]$ and with values in X such that

$$H_{x^1,x^2}(0) = x^1 \; ; \qquad H_{x^1,x^2}(1) = x^2 \; .$$

Obviously convex sets are arcwise connected and every arcwise connected set is also connected. On the other hand, a nonconvex set may be arcwise connected. Consider, e.g., the following example, taken from Avriel (1976). Let $X \subset \mathbb{R}^2$ be the points lying outside a circle centered at the origin and with radius r:

$$X = \{(x_1, x_2) \mid x_1, x_2 \in \mathbb{R},\ (x_1)^2 + (x_2)^2 - r^2 \geqq 0\} \; .$$

X is obviously not convex; now the function, expressed in polar coordinates

$$H_{x^1,x^2}(\lambda) = \begin{bmatrix} [(1-\lambda)\, r_1 + \lambda r_2] \cos((1-\lambda)\, \alpha_1 + \lambda \alpha_2) \\ [(1-\lambda)\, r_1 + \lambda r_2] \sin((1-\lambda)\, \alpha_1 + \lambda \alpha_2) \end{bmatrix} ,$$

$$0 \leqq \lambda \leqq 1 , \quad x^1, x^2 \in X \; ,$$

where

$$x_1^i = r_i \cos \alpha_i \qquad x_2^i = r_i \sin \alpha_i \; , \qquad i = 1, 2$$

is an arc and X is an arcwise connected set.

Definition 2.17.5. A function $f : X \to \mathbb{R}$, defined on the arcwise connected set $X \subset \mathbb{R}^n$, is called *ϑ-arcwise convex* on X if there exists an arc $H_{x^1,x^2}(\lambda)$ with values in X such that

$$f[H_{x^1,x^2}(\lambda)] \leqq \vartheta^{-1}\{\lambda\vartheta[f(x^1)] + (1-\lambda)\,\vartheta[f(x^2)]\,,$$
$$\forall x^1, x^2 \in X\,, \quad \forall \lambda \in [0,1]\,,$$

where ϑ is a continuous strictly increasing scalar function including $f(x^1)$ and $f(x^2)$ in its domain.

Avriel and Zang (1980) study extensively the subclass of ϑ-arcwise convex functions where $\vartheta(x) = x$ and call these functions *arcwise connected functions*. Similarly these same authors introduce *arcwise quasiconnected* and *arcwise pseudoconnected* functions. These classes of functions have, under some mild regularity conditions, some interesting local/global minimum properties; for details the reader is referred to Avriel and Zang (1980) and to Avriel, Diewert, Schaible and Zang (1987). Another subclass of ϑ-arcwise convex functions is obtained assuming that in Definition 2.17.5 the arc $H_{x^1,x^2}(\lambda)$ is an *h-mean value function* given by

$$H_{x^1,x^2}(\lambda) = h^{-1}\left[\lambda h(x^1) + (1-\lambda)\,h(x^2)\right]\,, \qquad \lambda \in [0,1]\,,$$

where h is a continuous one-to-one and onto function defined on a subset of $\mathbb{R}$ including the range of f and with values in $\mathbb{R}^n$. This type of function was explored in a most elegant way by Ben-Tal (1977); see also Castagnoli and Mazzoleni (1989b).

<u>Definition 2.17.6</u>. Let f be defined on the arcwise connected set $X \subset \mathbb{R}^n$ and let h and ϑ be the functions previously defined; then f is said to be *$(h-\vartheta)$-convex* on X if

$$f[h^{-1}(\lambda h(x^1)) + (1-\lambda)\,h(x^2)] \leqq$$
$$\leqq \vartheta^{-1}\{\lambda \cdot \vartheta[f(x^1)] + (1-\lambda)\,\vartheta[f(x^2)]\}\,,$$
$$\forall x^1, x^2 \in X\,, \quad \forall \lambda \in [0,1]\,.$$

Note that if in Definition 2.17.6 we take $h(x) = x$, $x \in \mathbb{R}^n$, and $\vartheta = e^{r\alpha}$, $r \neq 0$, we obtain as a special case the r-convex functions. Avriel (1976) and Avriel and Zang (1980) point out the existence of functions that are $(h-\vartheta)$-convex but that do not belong to any other class of generalized convex functions previously described in the present chapter. This is the case, for example, of the "Rosenbrock's curved valley function", defined by

$$f(x_1, x_2) = 100\,[x_2 - (x_1)^2]^2 + (1 - x_1)^2\ ,$$

which is a continuously differentiable nonconvex function on $\mathbb{R}^2$, having a unique minimum at $x^0 = (1, 1)$.

Some of the results obtained for $(h - \vartheta)$-convex functions include the following theorems (see Avriel (1976), Ben-Tal (1977)).

Theorem 2.17.5.

a) If f and g are $(h - \vartheta)$-convex, then $\vartheta^{-1}\,[\vartheta(f) + \vartheta(g)]$ is also $(h - \vartheta)$-convex.

b) If f, h and ϑ are differentiable, then f is $(h - \vartheta)$-convex if and only if

$$\begin{aligned} \vartheta[f(x^2)] \geqq\ & \vartheta[f(x^1)] + \vartheta'[f(x^1)] \cdot \\ & \cdot \sum_{i=1}^{n} \sum_{j=1}^{n} \frac{\partial f(x^1)}{\partial x_j} \cdot \frac{\partial h_j^{-1}(h(x^1))}{\partial x_i} \cdot \\ & \cdot (h_i(x^2) - h_i(x^1))\ , \qquad \forall\, x^1, x^2\ . \end{aligned}$$

The result under b) reduces to the usual definition of differentiable convex functions if $\vartheta[f(x)] = f(x)$ and $h(x) = x$.

The following result is an immediate consequence of the inequality sub b) in the previous theorem.

Theorem 2.17.6. Let f be a differentiable $(h - \vartheta)$-convex function, with h differentiable on $X \subset \mathbb{R}^n$, ϑ differentiable in the range of f, and let $x^0 \in X$ satisfy $\nabla f(x^0) = 0$. Then x^0 is a global minimum of f over X.

Moreover, $(h - \vartheta)$-convexity is related to ordinary convexity by the following result, again due to Avriel (1976) and Ben-Tal (1977)

<u>Theorem 2.17.7</u>. The function f is $(h - \vartheta)$-convex if and only if the function $\hat{f}$, given by

$$\hat{f} = \vartheta[f(h^{-1}(y))] \, ,$$

is convex.

Applications of $(h - \vartheta)$-convexity to nonlinear programming can be found in Avriel (1976) and Ben-Tal (1977).

Another approach in extending the notion of convex function is concerned with differentiable functions. Hanson (1981) noted that the (generalized) convexity requirement, utilized to prove sufficient optimality conditions for a differentiable mathematical programming problem (see the next chapter), can be further weakened by substituting the linear term $(x^1 - x^2)$, appearing in the definition of differentiable convex, pseudoconvex, quasiconvex function, with an arbitrary vector-valued function.

<u>Definition 2.17.6</u>. The differentiable function $f : X \to I\!R$, X open subset of $I\!R^n$, is said to be *invex* on X if there exists a vector-valued function $\eta : X \times X \to I\!R^n$ such that

$$f(x^1) - f(x^2) \geqq \eta(x^1, x^2)\, \nabla f(x^2) \, , \qquad \forall\, x^1, x^2 \in X \, .$$

The name "invex" was given by Craven (1981b) and stands for "invariant convex", since $f = g \circ \vartheta$ will be invex if the differentiable function $g : I\!R^n \to I\!R$ is convex and the differentiable surjective function $\vartheta : I\!R^r \to I\!R^n$ has Jacobian matrix $\nabla\vartheta$ of full rank. Indeed, for any $y, z \in I\!R^r$ with $u = \vartheta(y)$ and $x = \vartheta(z)$, from the chain rule for differentiation we have:

$$\nabla f(y)\, \eta(z, y) = \nabla g(u)\, \nabla\vartheta(y)\, \eta(z, y) = \nabla g(u)(x - u) \, ,$$

where $\eta(z, y)$ is now chosen in order to satisfy the equality $\nabla\vartheta(y) \cdot \eta(z, y) = (x - u)$ (the hypothesis on the rank of $\nabla\vartheta$ assures the existence of such a solution). Immediately we can deduce the invexity of f:

$$\begin{aligned} \nabla f(y) \cdot \eta(z, y) &= \nabla g(u)(x - u) \leqq g(x) - g(u) = \\ &= g[\vartheta(z)] - g[\vartheta(y)] = f(z) - f(y) \, . \end{aligned}$$

We can note that the same property does not hold for convex functions; take, e.g., the function $y = e^x$, $x \in I\!R$, which is convex. Its convexity is destroyed for the transformed function $y = \exp \sqrt[3]{u}$.

Similarly f is said to be *pseudo-invex* on X if for some vector-valued function $\eta(x^1, x^2)$ and all $x^1, x^2 \in X$, we have

$$\eta(x^1, x^2)\, \nabla f(x^2) \geqq 0 \Rightarrow f(x^1) \geqq f(x^2) \ .$$

The function $f : X \to I\!R$ is said to be *quasi-invex* on X if there exists a vector-valued function $\eta(x^1, x^2)$, not identically equal to zero, such that, for all $x^1, x^2 \in X$, we have

$$f(x^1) \leqq f(x^2) \Rightarrow \eta(x^1, x^2)\, \nabla f(x^2) \leqq 0 \ .$$

Local definitions of invex (pseudo-invex, quasi-invex) functions at a point $x^0 \in X$ have also been given: see Kaul and Kaur (1985). It follows that, by taking $\eta(x^1, x^2) = (x^1 - x^2)$, convex functions are invex, pseudoconvex functions are pseudoinvex and quasiconvex functions are quasiinvex. Moreover, invex functions are both pseudoinvex and quasiinvex and the sum of two (or more) functions that are invex with respect to the same function $\eta(x^1, x^2)$ is also invex.

Ben Israel and Mond (1986) and Kaul and Kaur (1985) have studied the relationships between the various classes of (generalized) invex functions and (generalized) convex functions. Let us complete their results (here and also in the sequel, for similar comparisons, $X \subset I\!R^n$ is assumed to be open and convex):

I) A differentiable convex function is also invex, but not conversely.

II) A differentiable pseudoconvex function is also pseudoinvex, but not conversely.

III) A differentiable quasiconvex function is also quasiinvex, but not conversely.

IV) Every invex function is also pseudoinvex with respect to the *same* function $\eta(x^1, x^2)$, but not conversely.

V) Every pseudoinvex function is also quasiinvex, but not conversely.

Further insights into these relationships can be deduced by means of the following interesting characterization of invex functions (Craven and Glover (1985), Ben-Israel and Mond (1986)).

Theorem 2.17.8. Let $f : X \to I\!R$ be differentiable on the open set $X \subset I\!R^n$; then f is invex on X if and only if every stationary point of f is a global minimum point over X.

Proof. Clearly if f is invex, then $\nabla f(x^0) = 0$ implies $f(x) \geqq f(x^0)$, $\forall x \in X$. Assume now that

$$\nabla f(x^0) = 0 \Rightarrow f(x) \geqq f(x^0) , \qquad \forall x \in X .$$

If $\nabla f(x^0) = 0$, take any $\eta(x, x^0)$. If $\nabla f(x^0) \neq 0$, take

$$\eta(x, x^0) = \frac{[f(x) - f(x^0)]}{[\nabla f(x^0)]^T \nabla f(x^0)} \cdot \nabla f(x^0) . \qquad \square$$

It follows from Theorem 2.17.8 that if f has no stationary points, then f is invex and that both pseudoconvex and pseudoinvex functions are invex. Thus (unlike pseudoconvex and convex functions) there is no distinction between pseudoinvex and invex functions. This is not in contrast with the previous property IV), which is established with respect to a *same* function η. We note that some authors (see, e.g., Hanson and Mond (1987), Kim (1988)) still consider pseudoinvexity as a generalization of invexity.

Another result, useful in detecting the relationships between the various classes of generalized convex functions considered in this chapter, is Corollary 2.14.1: a differentiable quasiconvex function in the open convex set $X \subset I\!R^n$ is pseudoconvex on X if and only if $f(x)$ has a global minimum at $x \in X$, whenever $\nabla f(x) = 0$.

Thus, under the assumption of quasiconvexity, invexity and pseudoconvexity coincide. So for an invex function not to be pseudoconvex, it must also not be quasiconvex.

Another result concerning the above-mentioned relationships is given by the following theorem, due to Pini (1991).

Theorem 2.17.9. The class of pseudoconvex functions on $X \subset \mathbb{R}^n$ is strictly included in the class of invex functions if $n > 1$; if $n = 1$ (functions of one real variable) the two classes coincide.

We can therefore complete the previous results I)–V):

VI) The class of invex and pseudoinvex functions coincides.

VII) The classes of quasiconvex and invex functions have only a partial overlapping; for example $f(x) = x^3$ is quasiconvex on $\mathbb{R}$, but not invex since the stationary point $x_0 = 0$ is not a global minimum point for f. The function $f(x_1, x_2) = (x_1)^3 + x_1 - 10(x_2)^3 - x_2$ is invex, as it lacks stationary points, but not quasiconvex, as $x^0 = (0,0)$, $x_1 = 2$, $x_2 = 1$, gives $f(x) - f(x^0) < 0$ but $(x - x^0)\nabla f(x^0) > 0$.

The relationship between these notions of generalized convexity of differentiable functions, defined on an open convex set $X \subset \mathbb{R}^n$, can be represented in the diagram of Figure 13 (an incorrect diagram appears in Ben-Israel and Mond (1986); the flaw was corrected by Giorgi (1990)).

Another approach to characterize invex functions is through some associated sets; more precisely, Zang, Choo and Avriel (1977) characterized by means of the lower level sets $L(f, \alpha) = \{x \,|\, x \in X,\ f(x) \leqq \alpha\}$, $\alpha \in \mathbb{R}$, those functions whose stationary points are global minimum points, i.e. invex functions.

Definition 2.17.7. If $L(f, \alpha)$ is nonempty, then it is *strictly lower semicontinuous* if, for every $x \in L(f, \alpha)$ and sequences $\{\alpha_i\}$, with $\alpha_i \rightarrow \alpha$, $L(f, \alpha_i)$ nonempty, there exists $k \in \mathbb{N}$, a sequence $\{x^i\} \rightarrow x$ and $\beta(x) \in \mathbb{R}$, $\beta(x) > 0$, such that $x^i \in L(f, \alpha_i - \beta(x)\, \|x^i - x\|)$, $i = k, k+1, \ldots$.

The cited authors prove the following result:

Theorem 2.17.10. A function $f : X \rightarrow \mathbb{R}$, differentiable on the open set $X \subset \mathbb{R}^n$ is invex on X if and only if $L(f, \alpha)$ is strictly lower semicontinuous for every α such that $L(f, \alpha)$ is nonempty.

Proof. See Zang, Choo and Avriel (1977). □

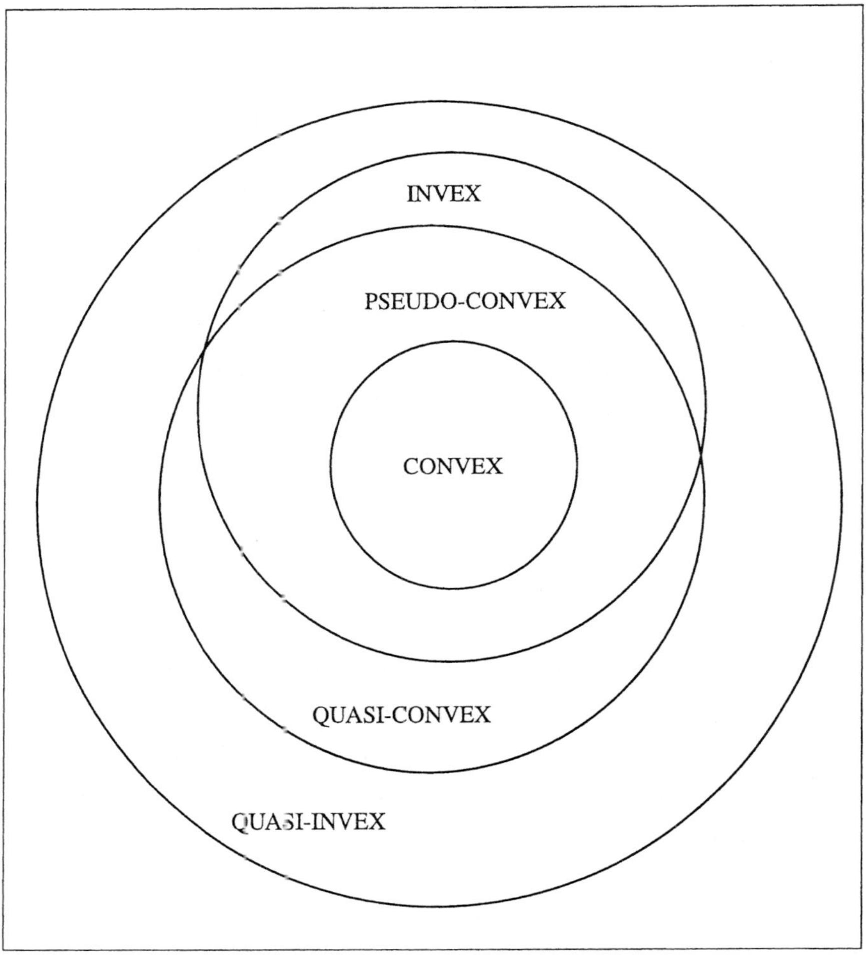

Figure 13.

In order to consider also some type of invexity for nondifferentiable functions, Ben-Israel and Mond (1986), Weir and Mond (1988) and Weir and Jeyakumar (1988) introduced the following definition.

Definition 2.17.8. A function $f : X \to I\!R$ is said to be *pre-invex* on X if there exists a vector-valued function $\eta : X \times X \to I\!R^n$ such that

$$[x^2 + \lambda\eta(x^1, x^2)] \in X \,, \qquad \forall\, \lambda \in [0,1]\,, \;\; \forall\, x^1, x^2 \in X$$

and

$$f[x^2 + \lambda\eta(x^1, x^2)] \; \leqq \; \lambda f(x^1) + (1-\lambda)\, f(x^2) \,, \qquad \forall\, x^1, x^2 \in X\,, \;\; \forall\, \lambda \in [0,1] \,.$$

Of course pre-invexity is a generalization of convexity; Weir and Mond (1988) have given the following example of a pre-invex function which is not convex: $f(x) = -|x|$, $x \in I\!R$. Then f is pre-invex with η given by

$$\eta(x_1, x_2) = \begin{cases} x_1 - x_2 \,, & \text{if } x_2 \leqq 0 \text{ and } x_1 \leqq 0 \\ x_1 - x_2 \,, & \text{if } x_2 \geqq 0 \text{ and } x_1 \geqq 0 \\ x_2 - x_1 \,, & \text{if } x_2 > 0 \text{ and } x_1 < 0 \\ x_2 - x_1 \,, & \text{if } x_2 < 0 \text{ and } x_1 > 0 \,. \end{cases}$$

As for convex functions, every local minimum of a pre-invex function is a global minimum and nonnegative linear combinations of pre-invex functions are pre-invex. Pre-invex functions are utilized by Weir and Mond (1988) to establish proper efficiency results in multiple objective optimization (see Chapter VI). For differentiable functions it appears that pre-invexity is a sufficient condition for invexity.

Theorem 2.17.11. If f is differentiable and pre-invex on the open set $X \subset I\!R^n$, then it is invex on X.

Proof. If f is pre-invex on $X \subset I\!R^n$, we have, for each $x^1, x^2 \in X$, and for each $\lambda \in [0,1]$,

$$f[x^2 + \lambda\eta(x^1, x^2)] - f(x^2) \leqq \lambda[f(x^1) - f(x^2)] \,.$$

Assume $\lambda > 0$ and divide by λ to obtain

$$\frac{1}{\lambda}\,[f(x^2 + \lambda\eta(x^1, x^2)) - f(x^2)] \leqq f(x^1) - f(x^2) \,.$$

Taking the limit as $\lambda \to 0^+$, we have the definition of invex functions. □

For other characterizations of nondifferentiable invex functions by means of subgradients and directional derivatives, see Craven and Glover (1985), Reiland (1989, 1990), Giorgi and Guerraggio (1996), Jeyakumar (1987).

References to Chapter II

K.J. ARROW and A.C. ENTHOVEN (1961), Quasiconcave programming, Econometrica, 29, 779–800.

K.J. ARROW and L. HURWICZ and H. UZAWA (1961), Constraint qualifications in maximization problems, Naval Res. Logistics Quart., 8, 175–191.

M. AVRIEL (1972), R-convex functions, Math. Programming, 2, 309–323.

M. AVRIEL (1973), Solutions of certain nonlinear programs involving r-convex functions, J.O.T.A., 11, 159–174.

M. AVRIEL (1976), Nonlinear programming – Analysis and Methods, Prentice Hall, Englewood Cliffs, N.J.

M. AVRIEL, W.E. DIEWERT, S. SCHAIBLE and I. ZANG (1987), Generalized Concavity, Plenum Publ. Corp., New York.

M. AVRIEL, W.E. DIEWERT, S. SCHAIBLE and W.T. ZIEMBA (1981), Introduction to concave and generalized concave functions; in S. Schaible and W.T. Ziemba (Eds.), Generalized Concavity in Optimization and Economics, Academic Press, New York, 21–50.

M. AVRIEL and S. SCHAIBLE (1978), Second order characterization of pseudoconvex functions, Math. Programming, 14, 170–185.

M. AVRIEL and I. ZANG (1974), Generalized convex functions with applications to nonlinear programming; in P.A. Van Moeseke (Ed.), Mathematical Programs for Activity Analysis, North Holland, Amsterdam, 23–33.

M. AVRIEL and I. ZANG (1980), Generalized arcwise connected functions and characterizations of local-global minimum properties, J.O.T.A., 32, 407–425.

M.S. BAZARAA, J.J. JARVIS and H.D. SHERALI (1990), Linear Programming and Network Flows, J. Wiley, New York

M.S. BAZARAA and C.M. SHETTY (1976), Foundations of Optimization, Springer Verlag, Berlin.

E.F. BECKENBACK (1937), Generalized convex functions, Bull. Amer. Math. Soc., 43, 363–371.

C.R. BECTOR (1970), Some aspects of quasiconvex programming, Z. Angew. Math. Mech., 50, 495–497.

C.R. BECTOR (1973), On convexity, pseudo-convexity and quasi-convexity of composite functions, Cahiers Centre Etudes Res. Opér., 15, 411–428.

C.R. BECTOR and S. CHANDRA (1986), ρ-convexity and first order duality for a nonlinear programming problem, Congressus Numerantium, 52, 53–62.

R. BELLMANN (1960), Introduction to Matrix Analysis, McGraw-Hill, New York.

A. BEN-ISRAEL (1969), Linear equations and inequalities on finite dimensional, real or complex, vector spaces: a unified theory, J. Math. Anal. Appl., 27, 367–389.

A. BEN-ISRAEL and B. MOND (1986), What is invexity?, J. Austral. Math. Soc., 28 (B), 1–9.

A. BEN-TAL (1977), On generalized means and generalized convex functions, J.O.T.A., 21, 1–13.

A. BEN-TAL and A. BEN-ISRAEL (1976), A generalization of convex functions via support properties, J. Austral. Math. Soc., 21 (A), 341–361.

A. BEN-TAL and A. BEN-ISRAEL (1981), F-convex functions: properties and applications; in S. Schaible and W.T. Ziemba (Eds.), Generalized Concavity in Optimization and Economics, Academic Press, New York, 301–314.

B. BEREANU (1969), On the composition of convex functions, Revue Roumaine de Math. Pures et Appl., 14, 1077–1084.

B. BEREANU (1972), Quasi-convexity, strict quasi-convexity and pseudo-convexity of composite objective functions, Revue Française d'Autom. Inform. Rech. Opér., 6 R, 15–26.

C. BERGE (1963), Topological Spaces, Oliver and Boyd, Edinburgh.

C. BERGE and A. GHOUILA-HOURI (1965), Programming, Games and Transportation Networks, Methuen, London.

B. BERNSTEIN and R.A. TOUPIN (1962), Some aspects of the Hessian matrix of a strictly convex function, J. Reine und Angew. Math., 210, 65–72.

H.F. BOHNENBLUST, S. KARLIN and L.S. SHAPLEY (1950), Solutions of discrete two-persons games; in H.W. Kuhn and A.W. Tucker (Eds.), Contributions to the Theory of Games, Vol. I, Annals of Mathematics Studies N. 24, Princeton Univ. Press, Princeton, 51–72.

J.M. BORWEIN (1977), Multivalued coonvexity: a unified approach to equality and inequality constraints, Math. Programming, 13, 163–180.

A. BRØNDSTED (1964) , Conjugate convex functions in topological vector spaces, Mat. Fys. Medd. Dan. Vid. Selsk., 34, 1–27.

A. CAMBINI (1986), Nonlinear separation theorems, duality and optimality conditions; in R. Conti, E. De Giorgi and F. Giannessi (Eds.), Optimization and Related Fields, Springer Verlag, Lecture Notes in Mathematics, N. 1190, Berlin, 57–93.

E. CASTAGNOLI and P. MAZZOLENI (1986), Generalized convexity for functions and multifunctions and optimality conditions, Technical Rep., Dep. Oper. Res., Univ. of Pisa, N. 134.

E. CASTAGNOLI and P. MAZZOLENI (1989a), About derivatives of some generalized concave functions; in C. Singh and B.K. Dass (Eds.), Continuous-time, Fractional and Multiobjective Programming, Analytic Publishing Co., Delhi, 53–64.

E. CASTAGNOLI and P. MAZZOLENI (1989b), Towards a unified type of concavity; in C. Singh and B.K. Dass (Eds.), Continuous-time, Fractional and Multiobjective Programming, Analytic Publishing Co., Delhi, 225–240.

K.L. CHEW and E.U. CHOO (1984), Pseudolinearity and efficiency, Math. Programming, 28, 226–239.

F.H. CLARKE (1983), A new approach to Lagrange multipliers, Math. Oper. Res., 1, 165–174.

F.H. CLARKE (1983), Optimization and Nonsmooth Analysis, J. Wiley, New York.

L. COLLATZ and W. WETTERLING (1975), Optimization Problems, Springer Verlag, Berlin.

R.W. COTTLE (1967), On the convexity of quadratic forms over convex sets, Op. Res., 15, 170–172.

R.W. COTTLE and J.A. FERLAND (1971), On pseudo-convex functions of nonnegative variables, Math. Programming, 1, 95–101.

R.W. COTTLE and J.A. FERLAND (1972), Matrix-theoretic criteria for the quasi-convexity and pseudo-convexity of quadratic functions, Linear Algebra and Its Applications, 5, 123–136.

B.D. CRAVEN (1981a), Invex functions and constrained local minima, Bull. Austral. Math. Soc., 24, 357–366.

B.D. CRAVEN (1981b), Duality for generalized convex fractional problems; in S. Schaible and W.T. Ziemba (Eds.), Generalized Concavity in Optimization and Economics, Academic Press, New York, 473–489.

B.D. CRAVEN and B.M. GLOVER (1985), Invex functions and duality, J. Austral. Math. Soc., 39 (A), 1–20.

J.P. CROUZEIX (1980), A second order condition for quasiconvexity, Math. Programming, 18, 349–352.

J.P. CROUZEIX (1981), Continuity and differentiability properties of quasiconvex functions on R^n; in S. Schaible and W.T. Ziemba (Eds.), Generalized Concavity in Optimization and Economics, Academic Press, New York, 109–130.

J.P. CROUZEIX and J.A. FERLAND (1982), Criteria for quasiconvexity and pseudoconvexity: relationships and comparisons, Math. Programming, 23, 193–205.

J.P. CROUZEIX and P.O. LINDBERG (1986), Additively decomposed quasiconvex functions, Math. Programming, 35, 42–57.

E. DEAK (1962), Ueber konvexe und interne Funktionen, sovie eine gemeinsame Verallgemeinerung von beiden, Ann. Univ. Sci. Budapest Sect. Math., 5, 109–154.

G. DEBREU (1952), Definite and semidefinite quadratic forms, Econometrica, 20, 295–300.

G. DEBREU and T.C. KOOPMANS (1982), Additively decomposed quasiconvex functions, Math. Programming, 14, 1–38.

B. DE FINETTI (1949), Sulle stratificazioni convesse, Ann. Mat. Pura e Appl., 30, 173–183.

W.E. DIEWERT (1981a), Generalized concavity and economics; in S. Schaible and W.T. Ziemba (Eds.), Generalized Concavity in Optimization and Economics, Academic Press, New York, 511–541.

W.E. DIEWERT (1981b), Alternative characterizations of six kinds of quasiconcavity in the nondifferentiable case with applications to nonsmooth programming; in S. Schaible and W.T. Ziemba (Eds.), Generalized Concavity in Optimization and Economics, Academic Press, New York, 51–93.

W.E. DIEWERT, M. AVRIEL and I. ZANG (1981), Nine kinds of quasiconcavity and concavity, J. Econ. Theory, 25, 397–420.

P. DOMBI (1985), On extremal points of quasiconvex functions, Math. Programming, 33, 115–119.

H.D. EGGLESTON (1958), Convexity, Cambridge Univ. Press, Cambridge.

R.M. ELKIN (1968), Convergence theorems for the Gauss-Seidel and other minimization algorithms, Ph.D. Dissertation, Univ. of Maryland, College Park.

K.-H. ELSTER and R. NEHSE (1980), Optimality conditions for some nonconvex problems; in K. Iracki, K. Malanowski and S. Walukiewicz (Eds.), Optimization Techniques, Part 2, Springer Verlag, Berlin, 1–9.

G.M. EWING (1977), Sufficient conditions for global minima of suitable convex functionals from variational and control theory, SIAM Review, 19, 202–220.

K. FAN, I. GLICKSBERG and A.J. HOFFMAN (1957), Systems of inequalities involving convex functions, Amer. Math. Soc. Proc., 8, 617–622.

W. FENCHEL (1953), Convex Cones, Sets and Functions, Lecture Notes, Princeton Univ., Princeton.

J.A. FERLAND (1971), Quasi-convex and pseudo-convex functions on solid convex sets, Technical Report N. 71-4, Dept. of Operations Research, Stanford Univ., Stanford.

J.A. FERLAND (1972a), Maximal domains of quasiconvexity and pseudoconvexity for quadratic functions, Math. Programming, 2, 178–192.

J.A. FERLAND (1972b), Mathematical programming problems with quasiconvex objective functions, Math. Programming, 3, 296–301.

J.A. FERLAND (1978), Matrix criteria for pseudoconvex functions in the class C^2, Linear Algebra and Its Applications, 21, 47–57.

J.A. FERLAND (1981), Matrix-theoretic criteria for the quasiconvexity of twice continuous differentiable functions, Linear Algebra and Its Appl., 38, 51–63.

D. GALE (1951), Convex polyhedral cones and linear inequalities; in T.C. Koopmans (Ed.), Activity Analysis of Production and Allocation, J. Wiley & Sons, New York, 287–297.

D. GALE (1960), The Theory of Linear Economic Models, McGraw-Hill, New York.

L. GERENCSER (1973), On a close relation between quasiconvex and convex functions and related investigations, Mathematische Operationsforsch. und Statistik, 4, 201–211.

F. GIANNESSI (1982), Metodi Matematici della Programmazione. Problemi Lineari e non Lineari, Pitagora Editrice, Bologna.

F. GIANNESSI (1984), Theorems of the alternative and optimality conditions, J.O.T.A., 42, 331–365. Errata coorige in J.O.T.A., 44, 1984, 363–364.

F. GIANNESSI (1987), Theorems of the alternative for multifunctions with applications to optimization: general results, J.O.T.A., 55, 233–256.

W. GINSBERG (1973), Concavity and quasiconcavity in economics, J. Ec. Theory, 6, 596–605.

G. GIORGI (1984), Quasiconvex programming revisited, Calcolo, 21, 307–316.

G. GIORGI (1987), A note on quasiconvex functions that are pseudoconvex, Trabajos de Investigacion Oper., 2, 80–83.

G. GIORGI (1990), A note on the relationships between convexity and invexity, J. Austral. Math. Soc., 32 (B), 97–99.

G. GIORGI and A. GUERRAGGIO (1996), Various types of nonsmooth invex functions, J. Inf. Optim. Sciences, 17, 137–150.

G. GIORGI and E. MOHLO (1992), Generalized invexity: relationships with generalized convexity and applications to optimality and duality conditions; in P. Mazzoleni (Ed.), Generalized Concavity for Economic Applications, Proceedings of the Workshop held in Pisa, April, 2, 1992, Tecnoprint, Bologna, 53–70.

B.M. GLOVER (1984), Generalized convexity in nondifferentiable programming, Bull. Austral. Math. Soc., 30, 193–218.

A.J. GOLDMAN (1956), Resolution and separation theorems for polyhedral convex sets; in H.W. Kuhn and A.W. Tucker (Eds.), Linear Inequalities and Related Systems, Princeton Univ. Press, Princeton, 41–51.

H.J. GREENBERG and W. PIERSKALLA (1971), A review of quasiconvex functions, Op. Res., 19, 1553–1570.

M. GUIGNARD (1969), Generalized Kuhn-Tucker conditions for mathematical programming problems in a Banach space, SIAM on Control, 7, 232–241.

H. HANCOCK (1960), Theory of Maxima and Minima, Dover Publications, New York (original publication: 1917).

M.A. HANSON (1964), Bounds for functionally convex optimal control problems, J. Math. Anal. Appl., 8, 84–89.

M.A. HANSON (1981), On sufficiency of the Kuhn-Tucker conditions, J. Math. Anal. Appl., 80, 545–550.

M.A. HANSON and B. MOND (1987), Convex transformable programming problems and invexity, J. Inf. Optim. Sciences, 8, 201–207.

M.A. HANSON and N.G. RUEDA (1989), A sufficient condition for invexity, J. Math. Anal. Appl., 138, 193–198.

G.H. HARDY, J.E. LITTLEWOOD and G. POLYA (1934), Inequalities, Cambridge Univ. Press, Cambridge.

H. HARTWIG (1983), On generalized convex functions, Optimization, 14, 49–60.

L. HÖRMANDER (1954), Sur la fonction d'appui des ensembles convexes dans un espace localement convexe, Ark. Math., 3, 181–186.

R. HORST (1984), On the convexification of nonlinear programming problems: an applications-oriented survey, European J. of Oper. Res., 15, 382–392.

A.D. IOFFE (1986), On the theory of subdifferentials; in J.B. Hiriart-Urruty (Ed.), Fermat Days 85: Mathematics for Optimization, North Holland, Amsterdam, 183–200.

A.D. IOFFE and V.L. LEVIN (1972), Subdifferentials of convex functions, Trans. Moscow Math. Soc., 26, 1–72.

A.D. IOFFE and V.M. TIHOMIROV (1979), Theory of Extremal Problems, North Holland, Amsterdam.

J.L.W.V. JENSEN (1906), Sur les fonctions convexes et les inégalités entre les valeurs moyennes, Acta Mathematica, 30, 175–193.

V. JEYAKUMAR (1985), Strong and weak invexity in mathematical programming, Methods Oper. Res., 55, 109–125.

V. JEYAKUMAR (1986), ρ-convexity and second-order duality, Utilitas Math., 29, 71–85.

V. JEYAKUMAR (1987) On optimality conditions in nonsmooth inequality constrained minimization, Numer. Funct. Anal. and Optim., 9, 535–546.

Y. KANNAI (1977), Concavifiability and constructions of concave utility functions, J. Math. Econ., 4, 1–56.

S. KARAMARDIAN (1967), Strictly quasi-convex (concave) functions and duality in mathematical programming, J. Math. Anal. Appl., 20, 344–358.

S. KARAMARDIAN (1975), Complementarity over cones with monotone and pseudomonotone maps, J.O.T.A., 18, 445–454.

S. KARAMARDIAN and S. SCHAIBLE (1990), Seven kinds of monotone maps, J.O.T.A., 66, 37–46.

S. KARAMARDIAN, S. SCHAIBLE and J.P. CROUZEIX (1993), Characterizations of generalized monotone maps, J.O.T.A., 76, 399–413.

S. KARLIN (1959), Mathematical Methods and Theory in Games, Programming and Economics, I and II, Addison-Wesley, Reading, Mass.

D.M. KATZNER (1970), Static Demand Theory, The MacMillan Company, New York.

R.N. KAUL and S. KAUR (1982), Generalizations of convex and related functions, European J. of Oper. Research, 9, 369–377.

R.N. KAUL and S. KAUR (1985), Optimality criteria in nonlinear programming involving nonconvex functions, J. Math. Anal. Appl., 105, 104–112.

D.S. KIM (1988), Pseudo-invexity in mathematical programming, Atti Accademia Peloritana dei Pericolanti, Classe I di Scienze Fisiche, Mat. e Naturali, 66, 347–355.

A.P. KIRMAN and L.M. TOMASINI (1986), A note on convexity, Metroeconomica, 20, 136–144.

A. KLINGER and O.L. MANGASARIAN (1968), Logarithmic convexity and geometric programming, J. Math. Anal. Appl., 24, 388–408.

S. KOMLOSI (1983), Some properties of nondifferentiable pseudoconvex functions, Math. Programming, 26, 232–237.

S. KOMLOSI (1993), First and second order characterizations of pseudolinear functions, European J. of Oper. Research, 67, 278–286.

K.O. KORTANEK and J.P. EVANS (1967), Pseudoconcave programming and Lagrange regularity, Op. Res., 15, 882–891.

H.W. KUHN and A.W. TUCKER (Eds.) (1956), Linear Inequalities and Related Systems, Annals of Mathematics Studies N. 38, Princeton Univ. Press, Princeton.

J. KYPARISIS and A.V. FIACCO (1987), Generalized convexity and concavity of the optimal value function in nonlinear programming, Math. Programming, 39, 285–304.

A. LEROUX (1984), Other determinantal conditions for concavity and quasiconcavity, J. Math. Economics, 13, 43–49.

M.A. LOPEZ CERDA and V. VALLS VERDEJO (1976), Propiedades de la funciones cuasiconvexas, Trabajos de Estat. y de Invest. Operativa, 27, 107–114.

D.G. LUENBERGER (1968), Quasi-convex programming, SIAM J. Appl. Math., 16, 1090–1095.

O.L. MANGASARIAN (1965), Pseudo-convex functions, S.I.A.M. J. on Control, 3, 281–290.

O.L. MANGASARIAN (1969), Nonlinear Programming, McGraw-Hill, New York.

O.L. MANGASARIAN (1970), Convexity, pseudoconvexity and quasiconvexity of composite functions, Cahiers du Centre d'Etudes de Recherche Opér., 12, 114–122.

H.B. MANN (1943), Quadratic forms with linear constraints, American Math. Monthly, 50, 430–433.

L. MARTEIN (1985), Regularity conditions for constrained extremum problems, J.O.T.A., 47, 217–233.

D.H. MARTIN (1985), The essence of invexity, J.O.T.A., 47, 65–76.

B. MARTOS (1965), The direct power of adjacent vertex programming methods, Management Science, 12, 241–255.

B. MARTOS (1967), Quasi-convexity and quasi-monotonicity in nonlinear programming, Studia Scientiarum Mathematicarum Hungarica, 2, 265–273.

B. MARTOS (1969), Subdefinite matrices and quadratic forms, S.I.A.M. J. Appl. Math., 17, 1215–1223.

B. MARTOS (1971), Quadratic programming with a quasiconvex objective function, Op. Res., 19, 87–97.

B. MARTOS (1975), Nonlinear Programming. Theory and Methods, North Holland, Amsterdam.

D. MCFADDEN (1978), Convex analysis; in M. Fuss and D. McFadden (Eds.), Production Economics: A Dual Approach to Theory and Applications, Vol. 1, North Holland, Amsterdam, 383–408.

P. MEREAU and J.C. PAQUET (1974), Second order conditions for pseudoconvex functions, S.I.A.M. J. Appl. Math., 27, 131–137.

G.J. MINTY (1964), On the monotonicity of the gradient of a convex function, Pacific J. of Mathematics, 14, 243–247.

B. MOND (1983), Generalized convexity in mathematical programming, Bull. Austral. Math. Soc., 27, 185–202.

P. NEWMAN (1969), Some properties of concave functions, J. Econ. Theory, 1, 291–314.

H. NIKAIDO (1954), On Von Neumann's minimax theorem, Pacific J. of Mathematics, 4, 65–72.

H. NIKAIDO (1968), Convex Structures and Economic Theory, Academic Press, New York.

J.M. ORTEGA and W.C. RHEINBOLDT (1970), Iterative Solutions of Nonlinear Equations in Several Variables, Academic Press, New York.

K. OTANI (1983), A characterization of quasi-convex functions, J. of Econ. Theory, 31, 194–196.

L. PELLEGRINI (1991), On a Lagrangian sufficient optimality condition, J.O.T.A., 68, 19–33.

A.L. PERESSINI, F.E. SULLIVAN and J.J. UHL (1988), The Mathematics of Nonlinear Programming, Springer Verlag, Berlin.

R. PINI (1991), Invexity and generalized convexity, Optimization, 22, 513–525.

J. PONSTEIN (1967), Seven kinds of convexity, S.I.A.M. Review, 9, 115–119.

T. RADO (1935), On convex functions, Trans. Amer. Math. Soc., 37, 266–285.

T. RAPCSAC (1991), On pseudolinear functions, European J. of Op. Res., 50, 353–360.

T.W. REILAND (1989), Generalized invexity for nonsmooth vector-valued mappings, Numer. Funct. Anal. and Optim., 10, 1191–1202.

T.W. REILAND (1990), Nonsmooth invexity, Bull. Austral. Math. Soc., 42, 437–446.

A.W. ROBERTS and D.E. VARBERG (1973), Convex Functions, Academic Press, New York.

R.T. ROCKAFELLAR (1967), Convex programming and systems of elementary monotonic relations, J. Math. Anal. Appl., 19, 543–564.

R.T. ROCKAFELLAR (1970), Convex Analysis, Princeton Univ. Press, Princeton.

R.T. ROCKAFELLAR (1974), Conjugate Duality and Optimization, C.B.M.S. Series N. 16, S.I.A.M. Publications, Philadelphia.

R.T. ROCKAFELLAR (1981), The Theory of Subgradients and its Applications to Problems of Optimization: Convex and Nonconvex Functions, Heldermann Verlag, Berlin.

N.G. RUEDA (1989), Generalized convexity in nonlinear Programming, J. of Information & Optimization Sciences, 10, 395–400.

S. SAKS (1937), Theory of the Integral, Hafner Publ. Co., New York.

P.A. SAMUELSON (1947), Foundations of Economic Analysis, Harvard Univ. Press, Cambridge, Mass.

H.H. SCHAEFER (1966), Topological Vector Spaces, MacMillan, New York.

S. SCHAIBLE (1971), Beiträge zur Quasikonvexen Programmierung, Doctoral Dissertation, Universität Köln.

S. SCHAIBLE (1972), Quasi-convex optimization in general real linear spaces, Zeitschrift für Operations Research, 16, 205–213.

S. SCHAIBLE (1973a), Quasiconvexity and pseudoconvexity of cubic functions, Math. Programming, 5, 243–247.

S. SCHAIBLE (1973b), Quasi-concave, strictly quasi-concave and pseudoconcave functions; in R. Henn, H.P. Künzi and H. Schubert (Eds.), Methods of Op. Res., 17, 308–316.

S. SCHAIBLE (1981), Quasiconvex, pseudoconvex and strictly pseudoconvex quadratic functions, J.O.T.A., 35, 303–338.

S. SCHAIBLE (1994), Generalized monotonicity – A survey; in S. Komlosi, T. Rapcsak and S. Schaible (Eds.), Generalized Convexity – Proceedings, Pécs, Hungary, 1992, Springer Verlag, Berlin, 229–249.

S. SCHAIBLE and W.T. ZIEMBA (Eds.) (1981), Generalized Concavity and Optimization in Economics, Academic Press, New York.

C. SINGH (1983), Elementary properties of arcwise connected sets and functions, J.O.T.A., 41, 377–387.

I.M. STANCU-MINASIAN (1992), Metode de Rezolvare a Problemelor de Programare Fractionara, Editura Academiei Romane, Bucharest.

J. STOER and C. WITZGALL (1970), Convexity and Optimization in Finite Dimensions – I, Springer Verlag, Berlin.

Y. TANAKA (1990), Note on generalized convex functions, J.O.T.A., 66, 345–349.

Y. TANAKA, M. FUKUSHIMA and T. IBARAKI (1989), On generalized pseudoconvex functions, J. Math. Anal. Appl., 144, 342–355.

F. TARDELLA (1989), On the image of a constrained extremum problem and some applications to the existence of a minimum, J.O.T.A., 60, 93–104.

W.A. THOMPSON and D.W. PARKE (1973), Some properties of generalized concave functions, Op. Res., 21, 305–313.

A.W. TUCKER (1956), Dual systems of homogeneous linear relations; in H.W. Kuhn and A.W. Tucker (Eds.), Linear Inequalities and Related Systems, Princeton Univ. Press, Princeton, 3–18.

H. TUY (1964), Sur les inégalités linéaires, Colloquium Math., 13, 107–123.

F.A. VALENTINE (1964), Convex Sets, McGraw-Hill, New York.

J.P. VIAL (1982), Strong convexity of sets and functions, J. Math. Economics, 9, 187–205.

J.P. VIAL (1983), Strong and weak convexity of sets and functions, Math. Oper. Res., 8, 231–259.

T. WEIR and V. JEYAKUMAR (1988), A class of nonconvex functions and mathematical programming, Bull. Austral. Math. Soc., 38, 177–189.

T. WEIR and B. MOND (1988), Pre-invex functions in multiple objective optimization, J. Math. Anal. Appl., 136, 29–38.

H. WEYL (1935), Elementare Theorie der Konvexen Polyeder, Comm. Math. Helv., 7, 290–306. Translated into English in H.W. Kuhn and A.W. Tucker (Eds.), Contributions to the Theory of Games, Vol. I, Annals of Mathematics Studies N. 24, Princeton Univ. Press, Princeton, 1950, 3–18.

P. WOLFE (1967), Methods of nonlinear programming; in J. Abadie (Ed.), Nonlinear Programming, North Holland, Amsterdam, 99–131.

I. ZANG, E.U. CHOO and M. AVRIEL (1977), On functions whose stationary points are global minima, J.O.T.A., 22, 195–208.

CHAPTER III. SMOOTH OPTIMIZATION PROBLEMS. SADDLE POINT CONDITIONS

3.1. Introduction

In this chapter we shall mainly analyse the optimality conditions for various types of extremum problems, under differentiability assumptions of the functions involved in these problems. An exception is found in the last section of the chapter. We shall treat separately necessary and sufficient optimality conditions. In some cases, e.g. for the unconstrained extremum problem or for the extremum problem with constraints expressed by equalities, these topics go back to the past two or three centuries. Other cases have been treated more recently: the basic "modern" starting articles are the papers of Fritz John (1948) and Kuhn and Tucker (1951). Prior to these papers is the unpublished thesis of W. Karush (1939); for an interesting account on the history of optimization problems, see Lenstra, Rinnooy Kan and Schrijver (1991)). See also Chapter 1 of Fiacco and McCormick (1968), Pourciau (1980), Prékopa (1980).

We shall be mainly concerned with the following types of extremum problems (or mathematical programming problems):

$$\min_{x \in X} f(x) \; ; \qquad (P_0)$$

$$\min_{x \in S} f(x)$$

$$S = \{x \mid x \in X,\ g_i(x) \leqq 0,\ i = 1, \ldots, m\} \; ; \qquad (P)$$

$$\min_{x \in S_1} f(x)$$

$$S_1 = \{x \mid x \in X,\ g_i(x) \leqq 0,\ i = 1, \ldots, m \,;\ h_j(x) = 0,\ j = 1, \ldots, r\} \,, \qquad (P_1)$$

where $X \subset \mathbb{R}^n$ is any set, f, g_i $(i = 1, \ldots, m)$ are real-valued functions, all defined and differentiable on an open set $D \subset \mathbb{R}^n$, with $X \subset D$; h_j

$(j = 1, ..., r < n)$ are real-valued functions, all defined and continuously differentiable on D.

3.2. Unconstrained Extremum Problems and Extremum Problems with a Set Constraint

In this section we discuss necessary and sufficient conditions for the existence of extremum points of differentiable functions, in absence of explicit functional constraints. In other words, we shall be concerned with problem (P_0). When $X \subset I\!R^n$ is open or, more generally, when the optimal point x^0 is interior to X, (P_0) is a *free* or *unconstrained* minimization problem. Obviously this last is the oldest among the various types of extremum problems. In the other cases we have a minimization problem with a *set constraint.*

Definition 3.2.1. Given $x^0 \in X$, the vector $y \in I\!R^n$ is said to be a *feasible direction from* x^0 for (P_0) if $\exists \bar{\alpha} > 0$ such that $x^0 + \alpha y \in X$, $\forall \alpha \in [0, \bar{\alpha}]$.

From a geometric point of view a feasible direction y is a vector such that if $x^0 \in X$, then every point $x^0 + \alpha y$ of the segment joining x^0 and $x^0 + \bar{\alpha} y$, belongs to X. The following theorem gives a necessary condition for the existence of local solutions for (P_0).

Theorem 3.2.1. Let $x^0 \in X$ be a point of local minimum for (P_0); then we have $y \nabla f(x^0) \geqq 0$ for any feasible direction y from x^0.

Proof. Being that y is a feasible direction, then $x^0 + \alpha y \in X$, $\forall \alpha \in [0, \bar{\alpha}]$. As f is differentiable on $D \supset X$, we have

$$f(x^0 + \alpha y) = f(x^0) + \alpha y \nabla f(x^0) + o(\|\alpha y\|) ,$$

where $\lim_{\alpha \to 0} \{o(\|\alpha y\|)/\|\alpha y\|\} = 0$.
Therefore, for $\alpha > 0$ suitably small, the sign of $y \nabla f(x^0)$ coincides with the sign of

$$f(x^0 + \alpha y) - f(x^0) .$$

As $f(x^0 + \alpha y) - f(x^0) \geqq 0$ we get our assertion. □

Theorem 3.2.1 holds similarly for a maximum problem, for which the necessary optimality condition becomes $y\nabla f(x^0) \leqq 0$, y feasible direction from x^0. If $x^0 \in \operatorname{int}(X)$, every direction from x^0 is feasible and therefore it must be

$$y\nabla f(x^0) \geqq 0 \ \text{ and } \ -y\nabla f(x^0) \geqq 0 \,, \qquad \forall\, y \in \mathbb{R}^n \,.$$

So, if $x^0 \in \operatorname{int}(X)$, it must be $y\nabla f(x^0) = 0$ and, as this is true for every direction y, it must be $\nabla f(x^0) = 0$. This is the well-known generalization to the case of n real variables of the classical Fermat theorem:

<u>Theorem 3.2.2</u>. Let $x^0 \in \operatorname{int}(X)$ be a point of local minimum (or local maximum) of f on $X \subset \mathbb{R}^n$; if $f : D \to \mathbb{R}$ is differentiable at x^0, then $\nabla f(x^0) = 0$.

In the free extremum problems the points x^0, for which we have $\nabla f(x^0) = 0$, are called *stationary* or *critical* points for f.

The proof of Theorem 3.2.1 is based on a first-order approximation of f in a neighbourhood of the minimum point x^0; if f is twice continuously differentiable in a neighbourhood of x^0, we obtain the following:

<u>Theorem 3.2.3</u>. In (P_0) let f be twice continuously differentiable in a neighbourhood of $x^0 \in X$; if x^0 is a local minimum of f on X, then, for each feasible vector y from x^0, we have

i) $y\nabla f(x^0) \geqq 0$;

ii) $y\nabla f(x^0) = 0 \Rightarrow yHf(x^0)\,y \geqq 0$.

<u>Proof</u>. Condition i) has already been proven in Theorem 3.2.1. To prove ii), let us consider Taylor's expansion formula in Peano's form:

$$f(x^0 + \alpha y) =$$
$$= f(x^0) + \alpha y\nabla f(x^0) + \tfrac{1}{2}\,\alpha^2 yHf(x^0)\,y + o(\|\alpha y\|^2) \,,$$

where $\lim_{\alpha \to 0} \{o(\|\alpha y\|^2)/\|\alpha y\|^2\} = 0$.
If $y\nabla f(x^0) = 0$, we have

$$f(x^0 + \alpha y) - f(x^0) = \frac{1}{2}\alpha^2 y Hf(x^0)\, y + o(\|\alpha y\|^2) ,$$

and, for $\alpha > 0$ suitably small, the sign of $yHf(x^0)\,y$ coincides with the sign of $f(x^0 + \alpha y) - f(x^0)$. Being $f(x^0 + \alpha y) - f(x^0) \geqq 0$ we get our assertion. □

If x^0 is a local maximum of f on X, then, for each feasible y from x^0, we have

i) $y\nabla f(x^0) \leqq 0$;

ii) $y\nabla f(x^0) = 0 \Rightarrow yHf(x^0)\, y \leqq 0$.

If $x^0 \in \operatorname{int}(X)$, Theorem 3.2.3 is substituted by the following:

Theorem 3.2.4. Let $x^0 \in \operatorname{int}(X)$ be a point of local minimum (or local maximum) of f on X. If f is twice continuously differentiable in a neighbourhood of x^0, then we have

$$yHf(x^0)\, y \geqq 0 \ (\leqq \ 0) , \qquad \forall\, y \in \mathbb{R}^n ,$$

i.e. the Hessian matrix $Hf(x^0)$ is positive (negative) semidefinite.

The next theorem provides sufficient optimality conditions for a free extremum problem, i.e. for (P_0).

Theorem 3.2.5.

i) Let $x^0 \in \operatorname{int}(X)$ and suppose f is twice continuously differentiable at x^0. If $\nabla f(x^0) = 0$ and $yHf(x^0)\, y > 0$, $\forall\, y \in \mathbb{R}^n$, $y \neq 0$, then f has a strict local minimum at x^0.

ii) Let $x^0 \in \operatorname{int}(X)$ and suppose f is twice continuously differentiable in some neighbourhood $N_\delta(x^0)$ of x^0. If $\nabla f(x^0) = 0$ and for every $x \in N_\delta(x^0)$ and for every $y \in \mathbb{R}^n$ we have $yHf(x)\, y \geqq 0$, then x^0 is a point of local minimum of f.

iii) Let $x^0 \in \text{int}(X)$ and f be pseudoconvex at x^0; then $\nabla f(x^0) = 0$ is a necessary and sufficient condition such that f has a global minimum at x^0.

iv) Let $x^0 \in \text{int}(X)$ and f be differentiable on $D \supset X$; then f admits a global minimum at every stationary point x^0 if f is invex on D.

Proof.

i) This part of the theorem is an easy consequence of Taylor's expansion formula and is left to the reader.

ii) Suppose that $\nabla f(x^0) = 0$ and $yHf(x)\, y \geqq 0$, $\forall y \in \mathbb{R}^n$, $\forall x \in N_\delta(x^0)$, but that x^0 is not a local minimum for f. Then there exists a vector $w \in N_\delta(x^0)$ such that $f(x^0) > f(w)$. Let $w = x^0 + \vartheta y$, where $\|y\| = 1$ and $\vartheta > 0$. By Taylor's expansion formula in the Lagrange form, we have

$$f(w) = f(x^0) + \vartheta y \nabla f(x^0) + \tfrac{1}{2}\vartheta^2 y \cdot Hf(x^0 + \lambda\vartheta y) \cdot y \ ,$$

where $\lambda \in (0, 1)$. Our assumptions lead to

$$yHf(x^0 + \lambda\vartheta y)\, y < 0 \ ,$$

and since $x^0 + \lambda\vartheta y \in N_\delta(x^0)$, we get a contradiction.

To prove iii) and iv) recall the definitions and properties of pseudoconvex and invex functions (especially Theorems 2.11.5, 2.17.8 and Corollary 2.11.1). □

It is easy to prove that if in condition ii) of the previous theorem $zHf(x)\, z > 0$, $\forall x \in N_\delta(x^0)$, $x \neq x^0$, and $\forall z \in \mathbb{R}^n$, $z \neq 0$, then f has a strict local minimum at x^0. Obviously for a maximum, the sign of the above inequalities must be reversed.

We have seen in the previous results some second-order conditions that involve the nature (i.e. the sign) of quadratic forms. We recall a useful criterion to state the sign of a quadratic form. Given an n real matrix $A = [a_{ij}]$, the following determinants

$$\Delta_k = \begin{vmatrix} a_{11} & a_{12} & \dots & a_{1k} \\ a_{21} & a_{22} & \dots & a_{2k} \\ \dots & \dots & \dots & \dots \\ a_{k1} & a_{k2} & \dots & a_{kk} \end{vmatrix}, \qquad k = 1, 2, ..., n$$

are called *leading principal minors* or *north west (N.W.) principal minors*, whereas the determinants

$$\tilde{\Delta}_k = \begin{vmatrix} a_{ii} & a_{ij} & \dots & a_{is} \\ a_{ji} & a_{jj} & \dots & a_{js} \\ \dots & \dots & \dots & \dots \\ a_{si} & a_{sj} & \dots & a_{ss} \end{vmatrix},$$

where $(i, j, ..., s)$ is any selection of k integers from the sets of integers $\{1, 2, ..., s\}$, are called *principal minors* of A of order k. Namely, the principal minors of order k are formed by taking k rows and the *same* k columns of A. Then we have the following:

Theorem 3.2.6. Let $Q(x) = x^T Ax$, $x \in I\!\!R^n$, be a real quadratic form. Then:

i) $Q(x)$ is positive definite if and only if $\Delta_1 > 0, \Delta_2 > 0, ..., \Delta_n > 0$, i.e. if and only if all the leading principal minors are positive;

ii) $Q(x)$ is negative definite if and only if $\Delta_1 < 0, \Delta_2 > 0, ..., (-1)^n \Delta_n > 0$, i.e. if and only if the leading principal minors alternate in sign, being Δ_1 negative;

iii) $Q(x)$ is positive semidefinite if and only if all $\tilde{\Delta}_1 \geqq 0, \tilde{\Delta}_2 \geqq 0, ..., \tilde{\Delta}_n \geqq 0$;

iv) $Q(x)$ is negative semidefinite if and only if all $\tilde{\Delta}_1 \leqq 0, \tilde{\Delta}_2 \geqq 0, ..., (-1)^n \tilde{\Delta}_n \geqq 0$;

v) a positive (negative) semidefinite quadratic form $Q(x)$ is positive (negative) definite if and only if $|A| \neq 0$.

Proof. See Gantmacher (1959) or Hestenes (1966). □

One could believe that if x^0 is a stationary point for f and also a minimum (maximum) point with respect to *every* direction starting from x^0, i.e. it is a minimum (maximum) point for $g(t) = f(x^0 + tv)$, $\forall\, v \neq 0$, then x^0 is a minimum (maximum) point for f. The following example shows that this is not true (see, however, also Theorems 2.11.8 and 2.11.9).

Example 3.2.1. Let $f(x, y) = y^2 - 3x^2y + 2x^4 = (y - x^2)(y - 2x^2)$. The function has a stationary point in $x^0 = (0, 0)$ and has a minimum point at the origin, with respect to every line starting from the origin. On the other hand, the origin is *not* a minimum point for f, as this function changes sign in every neighbourhood of the origin, i.e. in every neighbourhood of the origin there are points where f is positive and points where f is negative.

Stationary points such as the origin for the function of Example 3.2.1 are also called *saddle points*. More precisely we have the following:

Definition 3.2.2. A stationary point x^0 is a *saddle point* for f if in every neighbourhood of x^0 there are points x such that $f(x) > f(x^0)$ and points x such that $f(x) < f(x^0)$.

We shall see in the last section of the present chapter another definition of saddle point of the so-called "Lagrangian function". There the words "saddle point" makes reference to points where f has a minimum with respect to a direction and a maximum with respect to another direction. In this second meaning the origin is not a saddle point for the function of Example 3.2.1, whereas a saddle point according to this second definition is also a saddle point according to Definition 3.2.2. Always with reference to the second meaning, a typical saddle point is the origin for the function $f(x, y) = x^2 - y^2$ (see Figure 1), whose surface just looks like a saddle of a horse. Also on the grounds of what is observed, one must note that a saddle point does not always have such a representation (see Takayama (1985)).

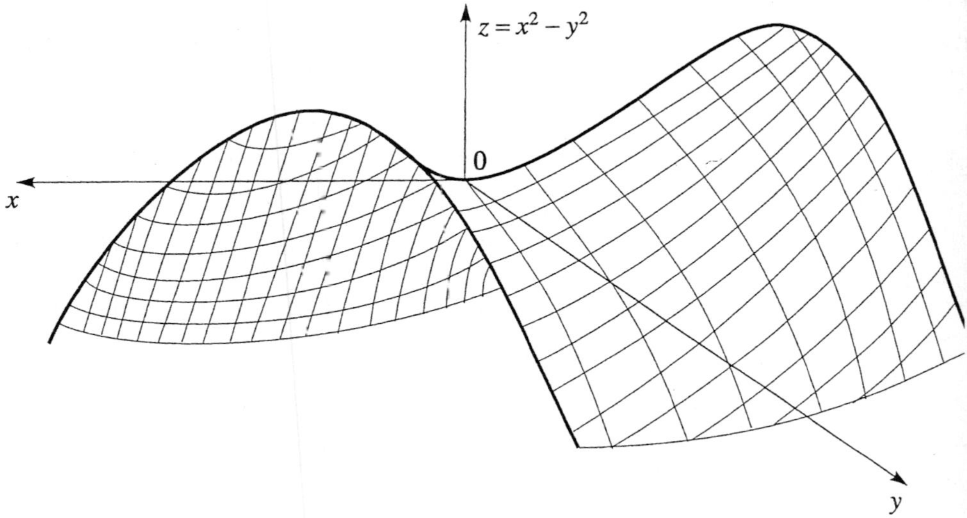

Figure 1.

We now give a sufficient condition for the existence of saddle points for a twice differentiable function.

<u>Theorem 3.2.7</u>. In problem (P_0) let $x^0 \in \mathrm{int}(X)$ and f be twice continuously differentiable in a neighbourhood of x^0; if

i) $\nabla f(x^0) = 0$, and

ii) $Hf(x^0)$ is indefinite,

then x^0 is a saddle point for f, i.e. is not a minimum nor a maximum point for f.

Proof. Let us apply Taylor's expansion formula in Peano's form:

$$f(x^0 + \alpha y) - f(x^0) = \alpha y \nabla f(x^0) + \frac{\alpha^2}{2}\, yHf(x^0)\, y + o(\|\alpha y\|^2)\ ,$$

with $\lim_{\alpha \to 0} \{o(\|\alpha y\|^2 / \|\alpha y\|\} = 0$.
As $\nabla f(x^0) = 0$, then

$$f(x^0 + \alpha y) - f(x^0) = \frac{\alpha^2}{2}\, yHf(x^0)\, y + o(\|\alpha y\|^2)\ .$$

Being $Hf(x^0)$ indefinite, then

$$f(x^0 + \alpha y) - f(x^0) \begin{cases} > 0 & \texttt{for some } y \in \mathbb{R}^n \\ < 0 & \texttt{for other } y \in \mathbb{R}^n\ , \end{cases}$$

i.e. x^0 is a saddle point for f. □

Other results on optimality conditions for (P_0) are obtained by imposing various convexity assumptions on f (see Section 2.11 of Chapter II).

3.3. Equality Constrained Extremum Problems

Another classical extremum problem which goes back to J.L. Lagrange (1762) is problem (P_1), where $X \subset \mathbb{R}^n$ is open and there are only constraints of the type $h_j(x) = 0$, $j = 1, 2, ..., r$. That is, we consider the problem

$$\min_{x \in S_e} f(x)$$

$$S_e = \{x \mid x \in X \subset \mathbb{R}^n\,;\ h_j(x) = 0,\ j = 1, 2, ..., r\}\ , \qquad (P_e)$$

where X is open, f and each h_j are continuously differentiable on X and $r < n$.
If in (P_e) we have $h : X \to \mathbb{R}^r$ continuously differentiable on X and rank $(\nabla h(x)) = r$, the relation $h(x) = 0$ generates a *smooth hypersurface* of $\mathbb{R}^n$. Let us try to characterize the feasible movements from the optimal point x^0 for (P_e). As the constraints are not necessarily linear, if $x^0 \in S_e$, there may not exist a feasible vector y such that $h(x^0 + \alpha y) = 0$ for every $\alpha > 0$ sufficiently small. Therefore we must generalize the concept of

feasible direction to the one of *feasible arc* (or *contained path*, as it is called by Arrow, Hurwicz, Uzawa (1961)) from point x^0, i.e. a continuous curve described by an r-vector-valued function $x(t) \to \mathbb{R}^n$ such that:

a) $x(t)$ is continuous on $[0, \bar{t}]$;

b) $x(0) = x^0$; $x(t) \in S_e$, $\forall\, t \in [0, \bar{t}]$;

c) $x(t)$ is differentiable at $t = 0$ and $x'(0) = y$ (here the prime denotes differentiation).

Under these assumptions, the vector y appearing in c) is a *tangent* or *attainable* direction at x^0 for S_e. The set of all these attainable vectors for S_e at x^0 is also called *tangent space* of S_e at x^0 or *cone of attainable directions* to S_e at x^0 (see also the next section of the present chapter). Let us denote this cone by $A(S_e, x^0)$. We now state the necessary optimality conditions for (P_e) using some previous results.

Lemma 3.3.1. In problem (P_e) let the Jacobian matrix $\nabla h(x)$ have full rank at least at x^0 and let h be continuously differentiable in a neighbourhood of x^0. Then:

i) For every vector $y \in \mathbb{R}^n$ such that

$$\nabla h(x^0) \cdot y = 0 \tag{1}$$

there exists a feasible arc $x(t)$, with $x'(0) = y$.

ii) On the contrary, if there exists a feasible arc at x^0 for S_e, then $y = x'(0)$ verifies (1).

Proof.

i) Let $y \in \mathbb{R}^n$ be a vector, with $\nabla h(x^0)\, y = 0$; since $r < n$, without loss of generality, we can assume that the determinant formed by the first r columns of $\nabla h(x)$ is nonzero. Then, being $\nabla h(x)$ continuous, the implicit function theorem assures the existence (in the local sense) of r functions

$$x_i = \hat{x}_i(x_{r+1}, ..., x_n) \ , \qquad i = 1, ..., r \ , \tag{2}$$

such that

$$h_i(\hat{x}_1(x_{r+1}, ..., x_n), ..., \hat{x}_r(x_{r+1}, ..., x_n), x_{r+1}, ..., x_n) = 0 \ ,$$
$$i = 1, ..., r$$

i.e. $x = (\hat{x}_1(...), ..., \hat{x}_r(...), x_{r+1}, ..., x_n)$ belongs to S_e for each $x_{r+1}, ...,$ x_n. Let $x = x(t)$, $0 \leqq t \leqq \bar{t}$ be a feasible arc, with the last $n-r$ components satisfying the conditions

$$x_j(0) = x_j^0 \ ; \qquad x_j'(0) = y_j \ , \qquad j = r+1, ..., n \tag{3}$$

and with the first r components expressed by relation (2), in terms of the remaining ones. Then $x_j(0) = x_j^0$, $j = 1, ..., r$. To obtain the thesis it is sufficient to show that $x_j'(0) = y_j$, $j = 1, ..., r$. Since $x(t) \in S_e$, i.e. $h(x(t)) = 0$ for $0 \leqq t \leqq \bar{t}$, then

$$\nabla h(x^0)\, x'(0) = 0 \ . \tag{4}$$

On the other hand, relation (1) holds. If in (1) and in relation (4) we consider as fixed the last $n-r$ components of y and $x'(0)$, we obtain, respectively, two normal systems, which, thanks to the second relation of (3), imply $x_j'(0) = y_j$, $j = 1, ..., r$.

ii) $x(t) \in S_e$, $t \in [0, \bar{t}]$, implies $h(x(t)) = 0$; deriving this last relation with respect to t, and setting $t = 0$, we obtain relation (4) and therefore the thesis follows, by setting $y = x'(0)$. □

If we denote by $D(x^0)$ the *null space* of $\nabla h(x^0)$ (or *linearizing cone at* x^0 for (P_e)):

$$D(x^0) = \{y \in I\!R^n \mid \nabla h(x^0) \cdot y = 0\} \ ,$$

we have just stated that, under the assumptions of Lemma 3.3.1, the equality $D(x^0) = A(S_e, x^0)$.

Lemma 3.3.2. Let $x^0 \in S_e$ be a local minimum point of f on S_e; then

$$y\nabla f(x^0) \geqq 0 , \qquad \forall y \in A(S_e, x^0) .$$

Proof. Let $y \in A(S_e, x^0)$; then there exists a feasible arc $x(t) : [0, \bar{t}] \to \mathbb{R}^n$, $x(t) \in S_e$, $\forall t \in [0, \bar{t}]$, $x(0) = x^0$ and $x'(0) = y$. If we define $\varphi(t) = f(x(t))$, this function admits for $t = 0$ a local minimum on $[0, \bar{t}]$. Therefore $\varphi'(0) = y\nabla f(x^0) \geqq 0$. □

Before giving the main theorem, let us again consider problem (P_e). Sometimes it is possible to "incorporate" the relation $h(x) = 0$ into f so as to obtain an equivalent but unconstrained problem in a new objective function. We borrow the following example from Marlow (1978). Consider the problem in $\mathbb{R}^3$ of dividing a length $a > 0$ into three parts so that the product of the three lengths will be maximum:

$$\max x_1 x_2 x_3$$

subject to

$$x_1 + x_2 + x_3 = a \qquad x_1, x_2, x_3 > 0 .$$

If we solve for $x_3 = a - x_1 - x_2$ and recall that $\max f(x) = \min(-f(x))$ we obtain the equivalent unconstrained minimization problem

$$\min -a x_1 x_2 + x_1^2 x_2 + x_1 x_2^2$$

to be solved in $\mathbb{R}^2_+ \backslash \{0\}$. The new problem can readily be solved by considering the critical points that belong to $\mathbb{R}^2_+ \backslash \{0\}$. What if we are not able, or we do not choose, to incorporate the equality constraints into the objective function? Based on what follows, one answer is that we can consider the related problem

$$\min \{f(x) - \lambda h(x)\}$$

where minimization is with respect to x alone and the vector $\lambda \in \mathbb{R}^r$ is a parameter. The objective function $\psi(x, \lambda) = f(x) - \lambda h(x)$ of this new problem is called *Lagrangian function* or simply *Lagrangian* of (P_e).

We are now in a position to state the following classical result, attributed to J.L. Lagrange (1762).

Theorem 3.3.1. In (P_e) let the Jacobian matrix $\nabla h(x)$ have full rank at $x^0 \in S_e$ and let h be continuously differentiable at least in a neighbourhood of x^0. If x^0 is a local minimum of f on S_e, then there exists a vector $\lambda \in \mathbb{R}^r$ such that

$$\nabla f(x^0) = \lambda \nabla h(x^0) . \tag{5}$$

If, moreover, f and h are twice continuously differentiable in a neighbourhood of x^0, then, too,

$$yH_x(f(x^0) - \lambda h(x^0))\, y \geqq 0 \tag{6}$$

for each vector y satisfying relation (1).

Proof. Thanks to Lemmas 3.3.1 and 3.3.2 $y\nabla f(x^0) \geqq 0$ for all vectors $y \in D(x^0) = \{y \in \mathbb{R}^n \mid \nabla h(x^0)\, y = 0\}$. Apply to these relations the theorem of the alternative of Farkas-Minkowski (Theorem 2.4.1 bis), identifying A with $\begin{bmatrix} \nabla h(x^0) \\ -\nabla h(x^0) \end{bmatrix}$ and b with $\nabla f(x^0)$, in order to obtain that these relations are equivalent to (5).

Now, let y satisfy (1) and let $x(t)$ be the associated feasible arc; if x^0 is a local minimum point for (P_e), this implies that $t = 0$ is a local minimum point of $\varphi(t) = f(x(t)) - \lambda h(x(t)) = f(x(t))$ on $[0, \bar{t}]$. It results, thanks to (5), $\varphi'(0) = 0$ and moreover $\varphi''(0) \geqq 0$. Then relation (6) is proved if we show that its first member equals $\varphi''(0)$. This is true, as it results

$$\begin{aligned} \varphi''(t) = {} & x'(t) \cdot Hf(x(t)) \cdot x'(t) + \\ & + \nabla f(x(t)) \cdot x''(t) - \sum_{i=1}^{r} \lambda_i [x'(t) \cdot Hh_i(x(t)) \cdot x'(t)] - \\ & - \sum_{i=1}^{r} \lambda_i \, [\nabla h_i(x(t)) \cdot x''(t)] . \end{aligned}$$

Therefore we have

$$\varphi''(0) = y \cdot [Hf(x^0) - \sum_{i=1}^{r} \lambda_i Hh_i(x^0)] \cdot y + (\nabla f(x^0) - \lambda \nabla h(x^0)) \cdot x''(0)$$

and the thesis follows from relation (5). □

Theorem 3.3.1 says that a first-order necessary condition for x^0 to be a local solution for (P_e) is that x^0 be a critical point for the Lagrangian function $\psi(x, \lambda)$:

$$\nabla_x \psi(x, \lambda) = \nabla f(x) - \lambda \nabla h(x) = 0$$

$$\nabla_\lambda \psi(x, \lambda) = -h(x) = 0 , \qquad \text{i.e. } h(x) = 0 .$$

Obviously, being that $\lambda \in \mathbb{R}^n$, we can equivalently consider the Lagrangian function in the form $\psi(x, \lambda) = f(x) + \lambda h(x)$. The elements of λ in (5) are called *Lagrangian multipliers*; the famous method of the Lagrangian multipliers consists in the search for solutions of system (5), but this provides only necessary conditions for the solution of (P_e). A point $x^0 \in S_e$ and such that $\nabla h(x^0)$ has full rank is also called a *regular point*. Let us now consider some sufficient optimality conditions for (P_e).

Theorem 3.3.2.

i) In (P_e) let f, h be twice continuously differentiable in a neighbourhood of $x^0 \in S_e$. If there exists a vector $\lambda \in \mathbb{R}^r$ such that

$$\nabla f(x^0) = \lambda \nabla h(x^0)$$

and such that

$$y H_x(f(x^0) - \lambda h(x^0))\, y > 0$$

for all vectors $y \neq 0$ such that $\nabla h(x^0)\, y = 0$, then x^0 is a point of strict local minimum of $f(x)$ on S_e.

ii) If the Lagrangian function $\psi(x, \lambda)$ is pseudoconvex at x^0, with respect to S_e, then condition (5) is sufficient for x^0 to be a point of global minimum of $f(x)$ on S_e, i.e. a solution for (P_e).

Proof.

i) Assume that x^0 is not a strict local minimum. Then there exist a neighbourhood $N_\delta(x^0)$ and a sequence $\{z^k\}$, $z^k \in N_\delta(x^0)$, $z^k \neq x^0$, converging to x^0 such that for every $z^k \in \{z^k\}$ we have $h_i(z^k) = 0$,

$i = 1, ..., r$, and $f(x^0) \geqq f(z^k)$. Let $z^k = x^0 + \vartheta_k y^k$, where $\vartheta_k > 0$ and $\|y^k\| = 1$. The sequence $\{\vartheta_k, y^k\}$ has a subsequence that converges to $(0, \bar{y})$, where $\|\bar{y}\| = 1$. By the Mean Value Theorem we get for each k in this subsequence

$$h_i(z^k) - h_i(x^0) = \vartheta_k y^k \nabla h_i(x^0 + \eta_{i,k} \vartheta_k y^k) = 0 \ , \qquad i = 1, ..., r$$

where $\eta_{i,k} \in (0, 1)$ and

$$f(z^k) - f(x^0) = \vartheta_k y^k \nabla f(x^0 + \xi_k \vartheta_k y^k) \leqq 0 \ ,$$

where $\xi_k \in (0, 1)$.
Dividing the two last expressions by ϑ_k and taking limits as $k \to +\infty$, we get $\bar{y} \nabla h_i(x^0) = 0$, $i = 1, ..., r$ and $\bar{y} \nabla f(x^0) \leqq 0$. From Taylor's expansion formula we obtain

$$\psi(z^k, \lambda) = \psi(x^0, \lambda) + \vartheta_k \cdot y^k \nabla_x \psi(x^0, \lambda) + + \tfrac{1}{2} (\vartheta_k)^2 \cdot y^k \cdot H_x \psi(x^0 + \eta_k \vartheta_k y^k, \lambda) \cdot y^k \ ,$$

with $\eta_k \in (0, 1)$.
Dividing this expression by $\frac{1}{2} (\vartheta_k)^2$ and taking into account that $\nabla_x \psi(x^0, \lambda) = 0$ and

$$\psi(z^k, \lambda) - \psi(x^0, \lambda) = f(z^k) - f(x^0) \leqq 0 \ ,$$

we obtain

$$y^k \cdot H_x \psi(x^0 + \eta_k \vartheta_k y^k, \lambda) \, y^k \leqq 0 \ .$$

Letting $k \to -\infty$ we obtain

$$\bar{y} H_x \psi(x^0, \lambda) \, \bar{y} \leqq 0$$

and, being $\bar{y} \neq 0$ and satisfying relation (1), this completes the proof of i).

ii) If (5) holds, i.e. $\nabla_x \psi(x^0, \lambda) = 0$ and $\psi(x, \lambda)$ is pseudoconvex at x^0, then x^0 is a point of global minimum of $\psi(x, \lambda)$ on S_e:

$$\psi(x^0, \lambda) \leqq \psi(x, \lambda) , \qquad \forall x \in S_e ,$$

i.e.

$$f(x^0) - \lambda h(x^0) \leqq f(x) - \lambda h(x) , \qquad \forall x \in S_e ,$$

i.e.

$$f(x^0) \leqq f(x) , \qquad \forall x \in S_e . \qquad \square$$

In Theorem 3.3.1 we assumed that the Jacobian matrix $\nabla h(x^0)$ has full rank r ($< n$). We now give a generalization of that theorem, in the sense that no regularity condition on the rank of the Jacobian is required. The following theorem is a particular case of a more general result, due to Fritz John (1948), and may be referred to as the *Lagrange-Fritz John necessary conditions* for the optimality in (P_e).

Theorem 3.3.3. In (P_e) let f be differentiable and h be continuously differentiable in a neighbourhood of x^0; if x^0 is a local minimum of $f(x)$ on S_e, then there exist multipliers $\lambda_0, \lambda_1, ..., \lambda_r$, not all zero, such that

$$\lambda_0 \nabla f(x^0) - \sum_{i=1}^{r} \lambda_i \nabla h_i(x^0) = 0 . \tag{7}$$

As we have already proved the classical theorem of Lagrange for (P_e), we omit the proof of the previous theorem, which is a particular case of a more general theorem, which will be proved in Section 3.8 of this chapter. Obviously, if in (7) it is $\lambda_0 \neq 0$, then we obtain relation (5). It can be proved (see Hadley (1964)) that if the rank of the augmented Jacobian matrix $G = \begin{pmatrix} \nabla h(x^0) \\ \nabla f(x^0) \end{pmatrix}$, a rank which is less than or equal to $r + 1$, is equal to the rank of $\nabla h(x^0)$, then $\lambda_0 \neq 0$ in (7). If, however, $\mathrm{rank}(G) > \mathrm{rank}(\nabla h(x^0))$, then in (7) $\lambda_0 = 0$. Moreover, if

$\text{rank}(G) = r + 1$, then f does not take one local minimum at x^0 on S_e. Then if $f(x)$ takes on a local minimum at $x^0 \in S_e$, then $\text{rank}(G) < r+1$.

Example 3.3.1. The point $x^0 = (0,0)$ is a minimum point for the function $f(x) = x_1 + x_2$, subject to $h(x) = x_1^2 + x_2^2 = 0$. Condition (5) is not verified at x^0 and therefore the method of the Lagrangian multipliers is not applicable. Indeed for $\psi(x_1, x_2, \lambda) = x_1 + x_2 + \lambda(x_1^2 + x_2^2)$, the system

$$\begin{cases} 1 + 2\lambda x_1 &= 0 \\ 1 + 2\lambda x_2 &= 0 \\ x_1^2 + x_2^2 &= 0 \end{cases}$$

admits no solution. On the contrary, the Lagrange-Fritz John conditions (7) are verified at x^0, as for $\tilde{\psi}(x_1, x_2, \lambda) = \lambda_0(x_1 + x_2) + \lambda_1(x_1^2 + x_2^2)$, the system

$$\begin{cases} \lambda_0 + 2\lambda_1 x_1 &= 0 \\ \lambda_0 + 2\lambda_1 x_2 &= 0 \\ x_1^2 + x_2^2 &= 0 \end{cases}$$

has the solution $x_1 = x_2 = 0$, with $\lambda_0 = 0$ and λ_1 arbitrary.

The sufficient conditions stated in Theorem 3.3.2, i), involve determining the sign of a quadratic form subject to linear constraints. This problem has been extensively treated e.g. by Bellmann (1960), Debreu (1952), Mann (1943), Chabrillac and Crouzeix (1984). The subsequent Corollary 3.3.1 to Theorem 3.3.2 is nothing but an application to point i) of the theorem stating conditions for the positive definiteness of a quadratic form subject to a system of linear constraints (the case of one linear constraint has also been discussed in Section 2.10 of Chapter II).

We recall first, for the reader's convenience, the said theorem. Let A be a symmetric matrix of order n and B a $m \times n$ matrix ($m < n$). Denote by M_{kq} the matrix obtained from a matrix M by keeping only the elements of the *first* k rows and q columns.

Theorem 3.3.4. Let $|B_{mm}| \neq 0$; then the quadratic form $x^T Ax$ is positive definite for all nonzero x satisfying $Bx = 0$ if and only if

$$(-1)^m \begin{vmatrix} A_{kk} & B_{mk}^T \\ B_{mk} & 0 \end{vmatrix} > 0$$

for $k = m+1, ..., n$. The same quadratic form is negative definite for all nonzero x satisfying $Bx = 0$ if and only if

$$(-1)^k \begin{vmatrix} A_{kk} & B_{mk}^T \\ B_{mk} & 0 \end{vmatrix} > 0$$

for $k = m+1, ..., n$.

We note that if we do not assume $|B_{mm}| \neq 0$, the above conditions are only sufficient.

Corollary 3.3.1. In problem (P_e) let f and h be twice continuously differentiable at least in a neighbourhood of $x^0 \in S_e$; if there exists a vector $\lambda \in I\!R^r$ such that

$$\nabla_x \psi(x^0, \lambda) = 0$$

and if the following determinant ("bordered Hessian")

$$(-1)^r \cdot \begin{vmatrix} \dfrac{\partial^2 \psi(x^0,\lambda)}{\partial x_1 \partial x_1} & \cdots & \dfrac{\partial^2 \psi(x^0,\lambda)}{\partial x_1 \partial x_p} & \dfrac{\partial h_1(x^0)}{\partial x_1} & \cdots & \dfrac{\partial h_r(x^0)}{\partial x_1} \\ \vdots & & \vdots & \vdots & & \\ \dfrac{\partial^2 \psi(x^0,\lambda)}{\partial x_p \partial x_1} & \cdots & \dfrac{\partial^2 \psi(x^0,\lambda)}{\partial x_p \partial x_p} & \dfrac{\partial h_1(x^0)}{\partial x_p} & \cdots & \dfrac{\partial h_r(x^0)}{\partial x_p} \\ \dfrac{\partial h_1(x^0)}{\partial x_1} & \cdots & \dfrac{\partial h_1(x^0)}{\partial x_p} & 0 & \cdots & 0 \\ \vdots & & \vdots & \vdots & & \vdots \\ \dfrac{\partial h_r(x^0)}{\partial x_1} & \cdots & \dfrac{\partial h_r(x^0)}{\partial x_p} & 0 & \cdots & 0 \end{vmatrix}$$

is positive for $p = r+1, ..., n$, then f has a strict local minimum at x^0 on S_e.
A similar result for the strict local maximum is obtained by changing $(-1)^r$ in $(-1)^p$.

We conclude this section with a short discussion on an interpretation of the Lagrange multipliers. So far we have discussed the problem

$\min f(x)$

subject to

$h(x) = 0$,

with $f : \mathbb{R}^n \to \mathbb{R}$, $h : \mathbb{R}^n \to \mathbb{R}^r$, $r < n$. Let us suppose that this problem has a solution at the regular point x^0 and let λ^0 be the corresponding Lagrange multipliers vector. Consider now the family of problems

$$\min f(x) \qquad (P_b)$$

subject to:

$h(x) = b$,

with $f : \mathbb{R}^n \to \mathbb{R}$, $h : \mathbb{R}^n \to \mathbb{R}^r$, $b \in \mathbb{R}^r$, $r < n$.

It is rather intuitive to think that, under suitable conditions, for a sufficiently small range of b near the zero vector, (P_b) will have a solution $x^0(b)$ near $x^0(0) = x^0$. For each of these solutions there will be therefore a corresponding optimal value $f(x^0(b))$ and this value can be regarded as a function of b. The function $F(b) = f(x^0(b))$ is also called the *value function* for problem (P_b). We have the following *sensitivity theorem*:

<u>Theorem 3.3.5</u>. Let f and h be twice continuously differentiable on an open set $X \subset \mathbb{R}^n$ and consider problem (P_b). Suppose that for $b = 0$ there is a local solution x^0 that is a regular point and that, together with its associated Lagrange multipliers vector λ, satisfies the second-order sufficient optimality conditions for a strict local minimum (Theorem 3.3.2). Then for every $b \in \mathbb{R}^r$ in a region containing 0 there is an $x^0(b)$, depending

continuously on b, such that $x^0(0) = x^0$ and such that $x^0(b)$ is a local minimum point for (P_b). Furthermore

$$\nabla_b f(x^0(b))|_{b=0} = \lambda \ .$$

<u>Proof</u>. Consider the system of equations

$$\nabla f(x) - \lambda \nabla h(x) = 0$$

$$b - h(x) = 0 \ .$$

By hypothesis there is a solution x^0, λ to this system when $b = 0$. The Jacobian matrix of this system at x^0, λ is

$$\left[\begin{array}{c|c} H_x \psi(x^0, \lambda) & (-\nabla h(x^0))^T \\ \hline -\nabla h(x^0) & 0 \end{array} \right] .$$

Because of the assumptions, the second-order sufficient optimality conditions are satisfied at x^0 and so it follows that this Jacobian matrix is nonsingular (Corollary 3.3.1). Thus, by the Implicit Function Theorem, there is a solution $x^0(b), \lambda(b)$ to the system, which is in fact continuously differentiable for sufficiently small $||b||$ and such that $x^0(0) = x^0$ and $\lambda(0) = \lambda$. Obviously for sufficiently small $||b||$, $x^0(b)$ will be also a local minimum point for (P_b). Moreover, by the chain rule we have

$$\nabla_b f(x^0(b))|_{b=0} = \nabla_x f(x^0) \cdot \nabla_b x^0(0)$$

and

$$\nabla_b h(x^0(b))|_{b=0} = \nabla_x h(x^0) \cdot \nabla_b x^0(0) \ .$$

But in view of the relation $h(x^0) = 0$, we have $\nabla_x h(x^0) \, \nabla_b x^0(0) = I$, the identity matrix of order r, and, in view of the relation $\nabla f(x^0) - \lambda \nabla h(x^0) = 0$, we obtain

$$\nabla_b f(x^0(b))|_{b=0} = \lambda \ . \qquad \square$$

A possible interpretation of this result is the following: the Lagrange multiplier λ_i for the i-th constraint is the *rate* at which the optimal value of the objective function of (P_e) changes with respect to changes in the i-th constraint constant b_i. For example, if $\lambda_i = 0$ at the optimum, then small changes in the constraint constant b_i will not affect the optimal value of the objective function. In economic problems b_i often denotes the stock available of some resource, and the objective function denotes revenue, profits or costs. Then the Lagrange multiplier λ_i is referred to as a *shadow price* (or *marginal value*) imputed to a unit of that resource.

3.4. Local Cone Approximations of Sets

In several questions of optimization theory, both in the smooth and nonsmooth case, it is useful to substitute, at least locally, a given set (e.g. the feasible set) with another set with a simpler structure, usually a cone. We speak, in this case, of *local cone approximation* of a given set. The development of the theory of mathematical optimization since the 1950s was closely connected with cone approximations of sets. Thus Kuhn and Tucker (1951) used in their classical paper on nonlinear programming a local cone approximation of the feasible set to establish optimality properties and regularity conditions. Other cones were introduced subsequently, also to develop numerical methods (e.g. the so-called method of "feasible directions"; see Zoutendijk (1960), Ben-Israel, Ben-Tal and Szlobec (1981)). A new direction in these developments started in the middle of the seventies when Clarke (1975) introduced a convex cone approximation (*Clarke tangent cone*) which gives, by its convexity, the possibility to treat nonconvex problems.

We distinguish between *classical* cone approximations and *modified* cone approximations, the latter of which are formulated on the basis of the Clarke tangent cone.

I) Classical cone approximations

Historically the first definition of a local cone approximation of a set is due to Bouligand (1932); see, however, also Severi (1930, 1935).

Definition 3.4.1. Given $S \subset \mathbb{R}^n$, $S \neq \emptyset$, and $x^0 \in \bar{S}$, the set

$$T(S,x^0) = \{y \in \mathbb{R}^n \mid \exists \{x^k\} \subset S, \lim_{k \to +\infty} x^k = x^0,$$

$$\exists \{\lambda_k\}, \lambda_k > 0 : y = \lim_{k \to +\infty} \lambda_k (x^k - x^0)\}$$

is called (*Bouligand*) *tangent cone to S at x^0* or *contingent cone to S at x^0*.

This cone has been used first in optimization theory by Abadie (1967), Varaiya (1967) and Dubovitskij and Miljutin (1963, 1965). Many authors have given various equivalent definitions of such a cone. We consider only the following ones:

$$T_1(S, x^0) = \{y \in I\!R^n \mid \exists \{x^k\} \subset S, \ \exists \{\mu_k\} \subset I\!R_+, \\ \lim_{k \to +\infty} \mu_k = 0 : x^k = x^0 + \mu_k \cdot y + \mu_k \cdot o(1)\} \ ;$$

see Bazaraa, Goode and Nashed (1974).

$$T_2(S, x^0) = \{y \in I\!R^n \mid \forall N(y), \ \forall \lambda > 0, \ \exists t \in (0, \lambda), \\ \exists \bar{y} \in N(y) \ : \ x^0 + t\bar{y} \in S\} \ ;$$

see Bazaraa, Goode and Nashed (1974).

$$T_3(S, x^0) = \bigcap_{N(x^0)} \operatorname{cl}(\operatorname{cone}((S \cap N(x^0)) - x^0)) \ ;$$

see Varaiya (1967).

$$T_4(S, x^0) = \Big\{y \in I\!R^n \mid \exists \{t^k\} \subset I\!R_+, \ \exists \{x^k\} \subset S : \lim_{k \to +\infty} t_k = 0, \\ \lim_{k \to +\infty} \frac{x^k - x^0}{t_k} = y\Big\} \ ;$$

see Hestenes (1966, 1975), Kurcyusz (1976).

$$T_5(S, x^0) = \{y \in I\!R^n \mid \exists \{t_k\} \subset I\!R_+, \ t_k \to 0, \\ \exists \{y^k\} \to y : x^0 + t_k y^k \in S\} \ ;$$

see Rockafellar (1981), Saks (1937).

$$T_6(S, x^0) = \Big\{y \in I\!R^n \mid \exists \lambda \geqq 0, \\ \exists \{x^k\} \subset S \backslash \{x^0\} : x^k \to x^0, \ \lambda \frac{x^k - x^0}{\|x^k - x^0\|} \to y\Big\} \ ;$$

see Hestenes (1966, 1975), Rogak and Scott-Thomas (1973).

$$T_7(S,x^0) = \{y \in I\!R^n \mid \exists \varphi : I\!R_+ \to I\!R^n, \lim_{t \to 0^+} \varphi(t) = y ;$$
$$\forall \lambda > 0 \; \exists t \in (0,\lambda) : x^0 + t\varphi(t) \in S\} ;$$

see Elster and Thierfelder (1988a, 1988b).

For a complete proof of the equivalence between $T(S,x^0)$ and any one of the above cones $T_i(S,x^0)$, $i = 1, ..., 7$, see Giorgi and Guerraggio (1992a, 1992b), Elster and Thierfelder (1985). Most of the above-mentioned equivalences are straightforward consequences of the definitions. Here we prove only the not immediate equivalence $T(S,x^0) = T_3(S,x^0)$.

Theorem 3.4.1. Let $S \subset I\!R^n$ and $x^0 \in \bar{S}$; it is $T(S,x^0) = T_3(S,x^0)$.

Proof. If $y \in T(S,x^0)$, then

$$y = \lim_{k \to +\infty} \lambda_k(x^k - x^0) ,$$

with $x^k \in S$, $\lambda_k > 0$ and $x^k \to x^0$. Since $x^k \to x^0$, then given any open ball $N(x^0)$ there exists a positive integer $\bar{k}$ such that $x^k \in S \cap N(x^0)$ for all $k > \bar{k}$. Therefore $\lambda_k(x^k - x^0) \in \text{cone}((S \cap N(x^0)) - x^0)$ for all $k > \bar{k}$ and hence

$$y = \lim_{k \to +\infty} \lambda_k(x^k - x^0) \in \text{cl}(\text{cone}((S \cap N(x^0)) - x^0)) .$$

But since this is true for any $N(x^0)$, then

$$y \in \bigcap_{N(x^0)} \text{cl}(\text{cone}((S \cap N(x^0)) - x^0)) ,$$

i.e.

$$T(S,x^0) \subset T_3(S,x^0) .$$

Conversely, let $y \in T_3(S,x^0)$; given any positive integer k, choose an open ball $N_{1/k}(x^0)$. If $y \in T_3(S,x^0)$ this implies that y is the limit of vectors of the form $\lambda_l(x^l - x^0)$, where $\lambda_l > 0$ and $x^l \in S \cap N_{1/k}(x^0)$. Now choose

l_k such that $\|y - \lambda_{l_k}(x^{l_k} - x^0)\| < 1/k$. By varying k we thus generate the sequence $\{\lambda_{l_k}\}$ and $\{x^{l_k}\}$. Note that $\lambda_{l_k} > 0$, $x^{l_k} \in S$, $x^{l_k} \to x^0$ and $\lambda_{l_k}(x^{l_k} - x^0) \to y$, i.e. $x \in T(S, x^0)$. Therefore $T_3(S, x^0) \subset T(S, x^0)$ and the proof is complete. □

The various definitions of Boulingand tangent cone can be reformulated for a topological vector space, even if for the equivalence of some definitions we must suppose that the space is metrizable.

Theorem 3.4.2.

i) $T(S, x^0)$ is a nonempty closed cone with $0 \in T(S, x^0)$;

ii) if x^0 is an isolated point of S, then $T(S, x^0) = \{0\}$; if $x^0 \in \text{int}(S)$, then $T(S, x^0) = \mathbb{R}^n$;

iii) $T(\bar{S}, x^0) = T(S, x^0)$;

iv) $S_1 \subset S_2 \Rightarrow T(S_1, x^0) \subset T(S_2, x^0)$ (isotony property).

v) $T(S_1 \cap S_2, x^0) \subset T(S_1, x^0) \cap T(S_2, x^0)$
$T(S_1 \cup S_2, x^0) = T(S_1, x^0) \cup T(S_2, x^0)$;

vi) $T(S_1 \times S_2, (x^1, x^2)) \subset T(S_1, x^1) \times T(S_2, x^2)$.

For the proof of this theorem (some assertions are a trivial consequence of the definition), as well as of the other theorems of this section, we refer to Bazaraa and Shetty (1976), Giorgi and Guerraggio (1992a, 1992b), Elster and Thierfelder (1984, 1988a, 1988b), Ward (1988). We note that i) follows immediately from characterization $T_3(S, x^0)$.

Some examples of a contingent cone to a set at a point $x^0 = 0$ are given in the figures below. In Figure 8 S is given by the union of the edges of an infinite sequence of squares, every one with a side which is twice that of the previous square. $T(S, x^0)$ is the acute angle with vertex at x^0 and with edges r' and r''. In Figure 9 $T(S, x^0) = \mathbb{R}^2$.

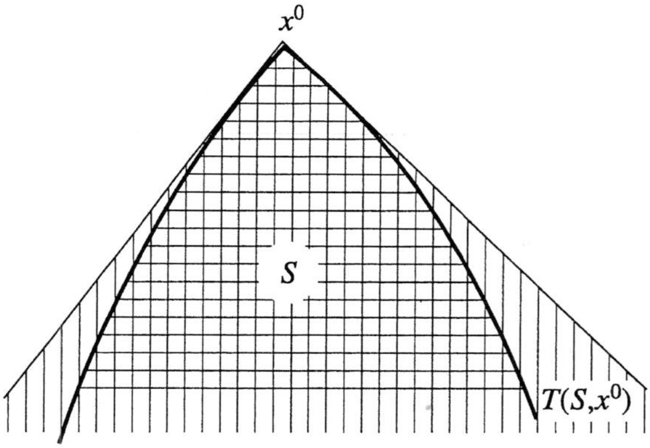

Figure 2.

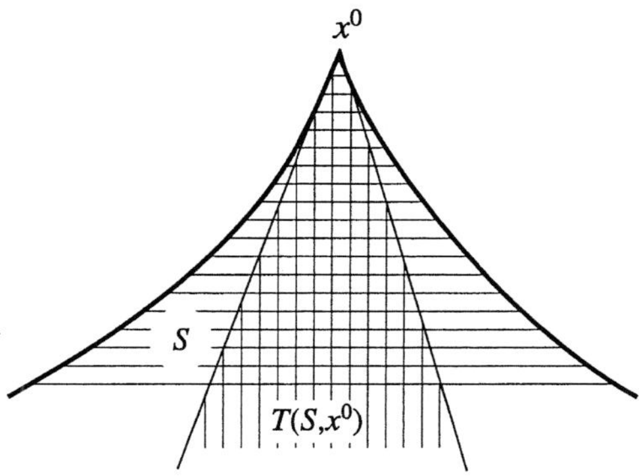

Figure 3.

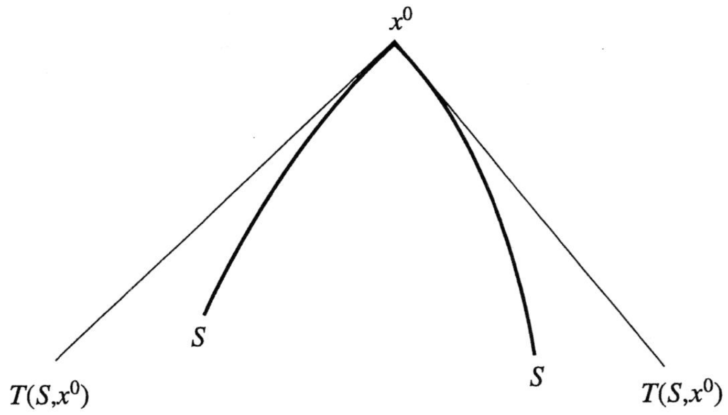

Figure 4.

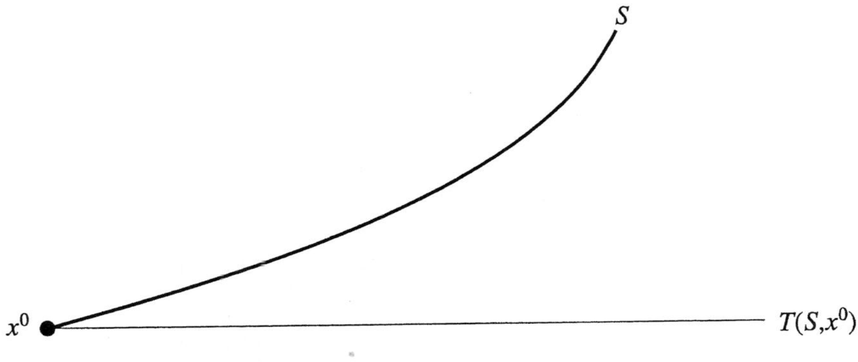

Figure 5.

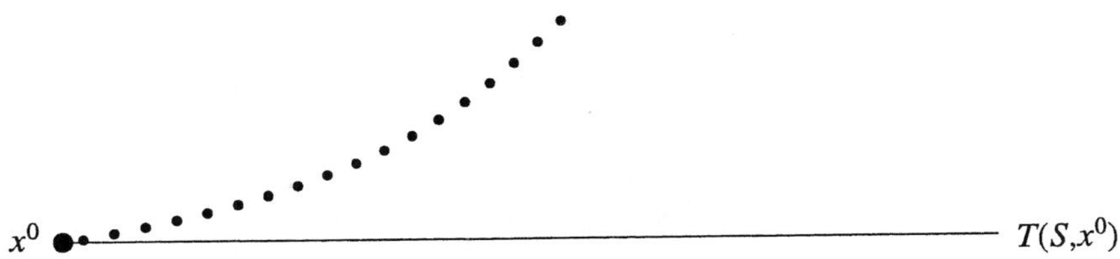

Figure 6.

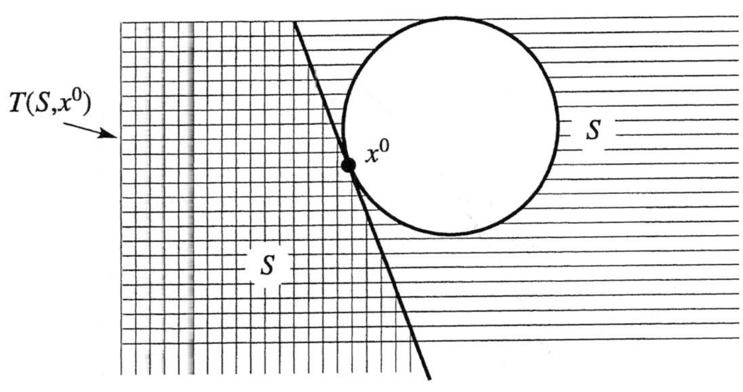

Figure 7.

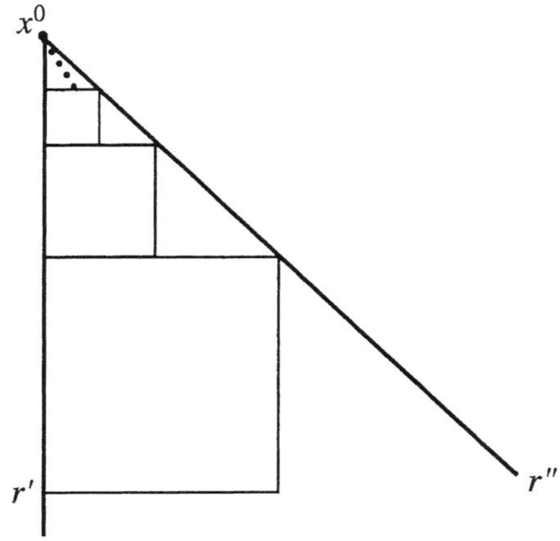

Figure 8.

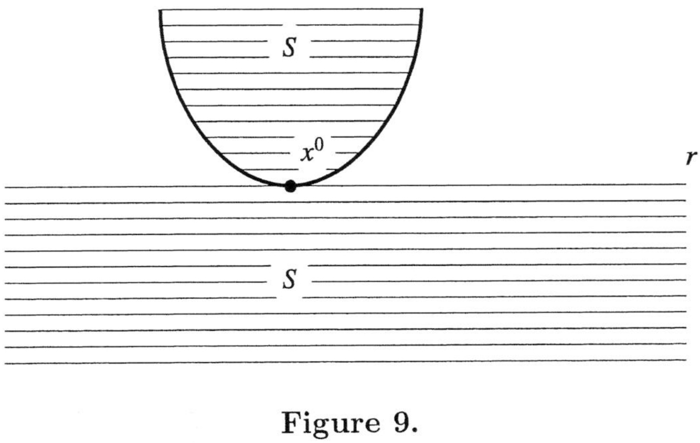

Figure 9.

We shall see that if S is convex, then $\bar{S} - x^0 \subset T(S, x^0)$ and $T(S, x^0)$ is a convex set (see Theorem 3.4.10).

Other examples of contingent cone are the following: $T(\mathbb{Q}^n, x^0) = \mathbb{R}^n$; $T(\mathbb{Z}^n, x^0) = \{0\}$, if $x^0 \in \mathbb{Z}^n$. Vectors $y \in T(S, x^0)$ are also called "sequential tangent vectors" to S at x^0. M. Guignard (1969) introduced (for Banach spaces) the notion of *pseudotangent cone*, in order to generalize the optimality conditions for a smooth mathematical programming problem.

Definition 3.4.2. Let $S \subset \mathbb{R}^n$, $S \neq \emptyset$, $x^0 \in \bar{S}$; the set

$$P(S, x^0) = \overline{\text{conv}(T(S, x^0))}$$

is called *pseudotangent cone* to S at x^0.

Obviously $P(S, x^0)$ is a convex cone and satisfies all properties listed in Theorem 3.4.2 with the exception of the second relation of v). Obviously $T(S, x^0) \subset P(S, x^0)$.

It may occur (especially if S is not convex) that $T(S, x^0)$ is a rough local approximation of S around x^0. This is due to the fact that on the ground of Definition 3.4.1, the requirement that $y \in T(S, x^0)$ is quite weak. A stronger requirement is obtained by substituting the sequence $\{x^k\}$ with an "arc". This cone has been introduced by Kuhn and Tucker (1951) and generalized by Ursescu (1982).

<u>Definition 3.4.3</u>. Let $S \subset \mathbb{R}^n$, $S \neq \emptyset$, $x^0 \in \bar{S}$; the set

$$A(S, x^0) = \{y \in \mathbb{R}^n \mid \exists \varphi : \mathbb{R}_+ \to \mathbb{R}^n,\ \varphi(0) = x^0,$$
$$\varphi'(0) = y,\ \exists \lambda > 0,\ \forall t \in (0, \lambda) : \varphi(t) \in S\}$$

is called *cone of attainable directions* to S at x^0.

Note that in the above definition the function φ is not required to be continuous and that the differentiability is required only at zero (Kuhn and Tucker require the arc to be differentiable, albeit in their proof only differentiability at zero is used; the cone of Kuhn and Tucker is therefore not the same as $A(S, x^0)$; see Palata (1989) and Peterson (1973)).

Other equivalent definitions of this cone are the following:

$$A_1(S, x^0) = \{y \in \mathbb{R}^n \mid \exists \varphi : \mathbb{R}_+ \to \mathbb{R}^n,\ \lim_{x \to 0} \varphi(t) = y,$$
$$\exists \lambda > 0,\ \forall t \in (0, \lambda) : x^0 + t\varphi(t) \in S\} ;$$

$$A_2(S, x^0) = \{y \in \mathbb{R}^n \mid \forall \{t_k\} \subset \mathbb{R}_+,\ t_k \to 0,$$
$$\exists \{y^k\} \to y : x^0 + t_k y^k \in S \text{ for large } k\} ;$$

$$A_3(S,x^0)=\{y\in \mathbb{R}^n \mid \forall\, N(y),\ \exists\,\lambda>0,\ \forall\, t\in(0,\lambda),\\ \exists\,\bar{y}\in N(y): x^0+t\bar{y}\in S\}\ .$$

This last definition is due to Ursescu (1982).

Vectors $x\in A(S,x^0)$ are said to be "curvilinear tangents"; it is immediate to note that $A(S,x^0)\subset T(S,x^0)$. In Figures 2, 3, 4, 5, 7, 9 we have $A(S,x^0)=T(S,x^0)$; $A(S,x^0)=\{0\}$ in Figure 6 and $A(S,x^0)=r'$ in Figure 8. Moreover $A(\mathbb{Q}^n,x^0)=\mathbb{R}^n$, $\forall x^0$; $A(\mathbb{Z}^n,x^0)=\{0\}$, $\forall\, x^0\in\mathbb{Z}^n$.

Theorem 3.4.3.

i) $A(S,x^0)$ is a nonempty closed cone with $0\in A$.

ii) If x^0 is an isolated point of S, then $A(S,x^0)=\{0\}$ and $A(S,x^0)=\mathbb{R}^n$ if $x^0\in \operatorname{int}(S)$.

iii) $A(\bar{S},x^0)=A(S,x^0)$.

iv) $S_1\subset S_2 \Rightarrow A(S_1,x^0)\subset A(S_2,x^0)$.

v) $A(S_1\cap S_2,x^0)\subset A(S_1,x^0)\cap A(S_2,x^0)$;
$A(S_1\cup S_2,x^0)\supset A(S_1,x^0)\cup A(S_2,x^0)$.

vi) $A(S_1\times S_2,(x^1,x^2))=A(S_1,x^1)\times A(S_2,x^2)$.

We note that in Arrow-Hurwicz and Uzawa (1961) and in Bazaraa, Goode and Shetty (1972) the cone $A(S,x^0)$ is not recognized as a closed set. A further strengthening of Definition 3.4.3 requires that the arc contained in the set is "linear". We obtain in this case the *cone of feasible directions*, specified in the following

Definition 3.4.4. Let $S\subset\mathbb{R}^n$, $S\neq\emptyset$, $x^0\in\bar{S}$; the set

$$Z(S,x^0)=\{y\in\mathbb{R}^n \mid \exists\,\lambda>0,\ \forall\, t\in(0,\lambda): x^0+ty\in S\}$$

is called *cone of feasible directions to* S *at* x^0.

Vectors $y \in Z(S, x^0)$ are called *linear tangent vectors*; we have $Z(S, x^0) \subset A(S, x^0)$.

In Figures 3, 7 and 9 $Z(S, x^0) = A(S, x^0) = T(S, x^0)$; in Figure 2 $Z(S, x^0) = \text{int}(A(S, x^0)) = \text{int}(T(S, x^0))$; in Figures 4 and 5 $Z(S, x^0) = \{0\}$; in Figure 6 $Z(S, x^0) = A(S, x^0) = \{0\}$ and in Figure 8 $Z(S, x^0) = A(S, x^0)$.

Moreover, $Z(\mathbb{Q}^n, x^0) = \{0\}$, $\forall x^0 \in \mathbb{Q}^n$; $Z(\mathbb{Q}^n, x^0) = \emptyset$, $\forall x^0 \notin \mathbb{Q}^n$; $Z(\mathbb{Z}^n, x^0) = \{0\}$, $\forall x^0 \in \mathbb{Z}^n$. Another cone, quite similar to $Z(S, x^0)$ is given by the following

Definition 3.4.5. Let $S \subset \mathbb{R}^n$, $S \neq \emptyset$, $x^0 \in \bar{S}$; the set

$$F(S, x^0) = \{y \in \mathbb{R}^n \mid \forall \lambda > 0,\ \exists t \in (0, \lambda) : x^0 + ty \in S\}$$

is called *radial tangent cone to* S *at* x^0; see Vlach (1970, 1981).

Theorem 3.4.4.

i) $0 \in Z(S, x^0)$ if and only if $0 \in F(S, x^0)$ and if and only if $x^0 \in S$.

ii) If x^0 is an isolated point of S, it is $F(S, x^0) = Z(S, x^0) = \{0\}$; if $x^0 \in \text{int}(S)$, it is $F(S, x^0) = Z(S, x^0) = \mathbb{R}^n$.

iii) $S_1 \subset S_2 \Rightarrow Z(S_1, x^0) \subset Z(S_2, x^0)$ and $F(S_1, x^0) \subset F(S_2, x^0)$.

iv) $Z(S_1 \cap S_2, x^0) = Z(S_1, x^0) \cap Z(S_2, x^0)$;
$Z(S_1 \cup S_2, x^0) \supset Z(S_1, x^0) \cup Z(S_2, x^0)$;
$F(S_1 \cup S_2, x^0) = F(S_1, x^0) \cup F(S_2, x^0)$;
$F(S_1 \cap S_2, x^0) \subset F(S_1, x^0) \cap F(S_2, x^0)$.

v) $Z(S_1 \times S_2, (x^1, x^2)) = Z(S_1, x^1) \times Z(S_2, x^2)$;
$F(S_1 \times S_2, (x^1, x^2)) \subset F(S_1, x^1) \times F(S_2, x^2)$.

We have $F(\mathbb{Q}^n, x^0) = \mathbb{R}^n$, $\forall x^0 \in \mathbb{Q}^n$; $F(\mathbb{Q}^n, x^0) = \mathbb{R}^n \backslash \{0\}$, $\forall x^0 \notin \mathbb{Q}^n$; $F(\mathbb{Z}^n, x^0) = \{0\}$, $\forall x^0 \in \mathbb{Z}^n$. We remark also that in Figures 3, 7 and 9 $F(S, x^0) = Z(S, x^0) = A(S, x^0) = T(S, x^0)$; in Figure 2 $F(S, x^0) = Z(S, x^0) = \text{int}(A(S, x^0)) = \text{int}(T(S, x^0))$; in Figures 4, 5 $F(S, x^0) =$

$Z(S,x^0)=\{0\}$; in Figure 6 $F(S,x^0)=Z(S,x^0)=A(S,x^0)=\{0\}$ and in Figure 8 $F(S,x^0)=T(S,x^0)$.

In any case $Z(S,x^0)\subset F(S,x^0)$ and if $x^0\in\bar{S}\cap(\overline{\mathbb{R}^n\backslash S})$ then the following "duality properties" hold between the cones $F(S,x^0)$ and $Z(S,x^0)$.

<u>Theorem 3.4.5</u>. Let $x^0\in\bar{S}\cap(\overline{\mathbb{R}^n\backslash S})$; it results that:

i) $Z(S,x^0)=\mathbb{R}^n\backslash F(\mathbb{R}^n\backslash S,x^0)$;

ii) $F(S,x^0)=\mathbb{R}^n\backslash Z(\mathbb{R}^n\backslash S,x^0)$.

A further restriction of the previous concepts is given by the *cone of interior directions to S at x^0* or *internal cone* or *cone of interior displacements*; this cone was introduced by Dubovitskij and Miljutin (1963, 1965) with the name of "cone of permissible variations" and was subsequently utilized by Bazaraa and Goode (1972) for obtaining general necessary optimality conditions for a smooth programming problem.

<u>Definition 3.4.6</u>. Let $S\subset\mathbb{R}^n$, $S\neq\emptyset$, $x^0\in\bar{S}$; the set

$$I(S,x^0)=\{y\in\mathbb{R}^n\mid\exists N(y),\ \exists\lambda>0,\ \forall t\in(0,\lambda),\\ \forall\bar{y}\in N(y):x^0+t\bar{y}\in S\}$$

is called *cone of interior directions to S at x^0*.

Other equivalent characterizations of this cone are:

$$I_1(S,x^0)=\{y\in\mathbb{R}^n\mid\forall\varphi:\mathbb{R}_+\to\mathbb{R}^n,\ \text{with}\ \lim_{t\to 0}\varphi(t)=y,\\ \exists\lambda>0,\ \forall t\in(0,\lambda):x^0+t\varphi(t)\in S\}\ ;$$

$$I_2(S,x^0)=\Big\{y\in\mathbb{R}^n\mid\exists N(y),\\ \exists N(0):x^0+\Big(\bigcup_{t>0}tN(y)\Big)\cap N(0)\subset S\Big\}\ .$$

(This characterization is due to Laurent (1972); see also Sachs (1978).)

$$I_3(S,x^0) = \{y \in I\!R^n \mid \forall \{y^k\} \to y,\ \forall \{\lambda_k\} \subset I\!R_+, \lim_{k\to+\infty} \lambda_k = 0 : x^0 + \lambda_k y^k \in S \text{ for large } k\} ;$$

$$I_4(S,x^0) = \{y \in I\!R^n \mid \forall x \in I\!R^n,\ \forall \varphi : I\!R \to I\!R, \text{ with } \varphi(\lambda) = o(\lambda) \text{ for } \lambda \to 0^+ : x^0 + \lambda y + \varphi(\lambda)\, x \in S \text{ for small } \lambda\} .$$

From the definitions it appears that

$$I(S,x^0) \subset Z(S,x^0) \subset A(S,x^0) \subset T(S,x^0) \subset P(S,x^0) .$$

In Figures 2 and 3, $I(S,x^0) = \mathrm{int}(T(S,x^0))$; in Figures 4, 5, 6, 7 and 8 $I(S,x^0) = \emptyset$; in Figure 9 $I(S,x^0)$ is the whole space $I\!R^n$, with exclusion of the line r. Moreover, $I(Q^n,x^0) = I(Z\!\!Z^n,x^0) = \emptyset$.

<u>Theorem 3.4.6.</u>

i) $I(S,x^0)$ is an open cone;

ii) If x^0 is an isolated point of S, then $I(S,x^0) = \emptyset$; if $x^0 \in \mathrm{int}(S)$, then $I(S,x^0) = I\!R^n$;

iii) $I(S,x^0) = I(\mathrm{int}(S,x^0))$;

iv) $S_1 \subset S_2 \Rightarrow I(S_1,x^0) \subset I(S_2,x^0)$;

v) $I(S_1 \cap S_2,x^0) = I(S_1,x^0) \cap I(S_2,x^0)$;
$I(S_1 \cup S_2,x^0) \supset I(S_1,x^0) \cup I(S_2,x^0)$;

vi) $I(S_1 \times S_2,(x^1,x^2)) = I(S_1,x^1) \times I(S_2,x^2)$.

Moreover, it holds the following coimplication: $0 \in I(S,x^0) \Leftrightarrow x^0 \in \mathrm{int}(S)$. The cone $I(S,x^0)$ has the following dual properties, with respect to the contingent cone $T(S,x^0)$:

<u>Theorem 3.4.7.</u> Let $x^0 \in \bar{S} \cap (\overline{I\!R^n \backslash S})$; then:

i) $I(S,x^0) = I\!R^n \backslash T(I\!R^n \backslash S, x^0)$;

ii) $T(S,x^0) = I\!R^n \backslash I(I\!R^n \backslash S, x^0)$.

A weaker form of the internal cone is given by the *cone of quasi-interior directions*, introduced by Ursescu (1982) and utilized (see Section 3.8) in obtaining general necessary optimality conditions for a smooth programming problem.

Definition 3.4.7. Let $S \subset I\!R^n$, $S \neq \emptyset$, $x^0 \in \bar{S}$; the set

$$Q(S,x^0) = \{y \in I\!R^n \mid \exists\, N(y),\ \forall\, \lambda > 0,\ \exists\, t \in (0,\lambda),\ \forall\, \bar{y} \in N(y) : x^0 + t\bar{y} \in S\}$$

is called *cone of quasi-interior directions to S at x^0.*

This cone can be equivalently characterized as follows (see Elster and Thierfelder (1988b), Giorgi and Guerraggio (1992b)):

$$Q_1(S,x^0) = \{y \in I\!R^n \mid \exists\, \{t_k\} \subset I\!R_+,\ t_k \to 0,\ \forall\, \{y^k\} \to y : x^0 + t_k y^k \in S \text{ for large } k\}\ ;$$

$$Q_2(S,x^0) = \{y \in I\!R^n \mid \forall\, \varphi : I\!R_+ \to I\!R^n, \text{ with } \lim_{t \to 0} \varphi(t) = y,\ \forall\, \lambda > 0\ \exists\, t \in (0,\lambda) : x^0 + ty \in S\}\ .$$

The cone $Q(S,x^0)$ has the properties specified by the following theorem (again the proofs can be found in Elster and Thierfelder (1984), Giorgi and Guerraggio (1992b); see also Ursescu (1982)).

Theorem 3.4.8.

i) $Q(S,x^0)$ is an open cone;

ii) If x^0 is an isolated point of S, then $Q(S,x^0) = \emptyset$; if $x^0 \in \operatorname{int}(S)$, then $Q(S,x^0) = I\!R^n$;

iii) $Q(\operatorname{int}(S), x^0) = Q(S,x^0)$;

iv) $S_1 \subset S_2 \Rightarrow Q(S_1, x^0) \subset Q(S_2, x^0)$;

v) $Q(S_1 \cap S_2, x^0) \subset Q(S_1, x^0) \cap Q(S_2, x^0)$;
$Q(S_1 \cup S_2, x^0) \supset Q(S_1, x^0) \cup Q(S_2, x^0)$;

vi) $Q(S_1 \times S_2, (x^1, x^2)) \subset Q(S_1, x^1) \times Q(S_2, x^2)$.

We note that in Figures 2 to 9 of this section, we have $I(S, x^0) = Q(S, x^0)$, with $x^0 = 0$. If we take S as the set of points belonging to the sequence of circles in Figure 10, it is quite immediate to see that $I(S, x^0) = \emptyset$ and $Q(S, x^0) \neq \emptyset$ (given by the cone depicted in the figure; again we have, for simplicity, $x^0 = 0$).

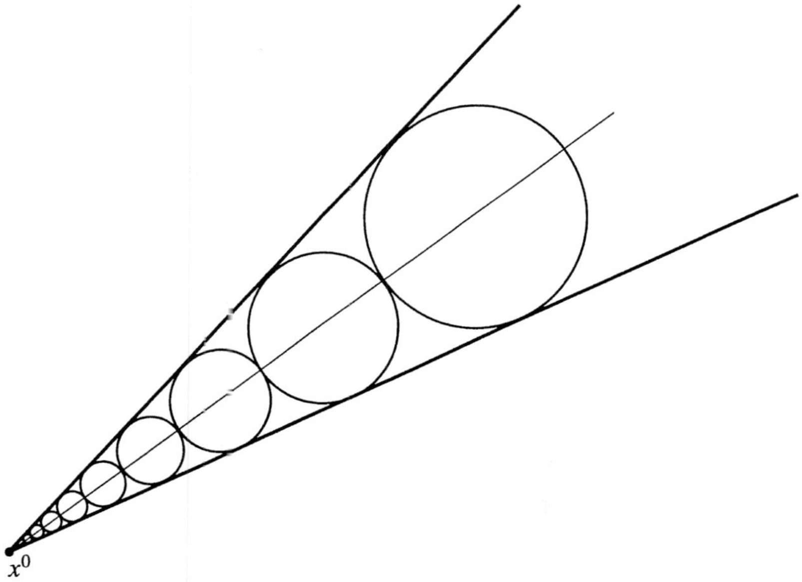

Figure 10.

Also for the cone of quasi-interior directions we have $0 \in Q(S, x^0) \Leftrightarrow x^0 \in \text{int}(S)$. Moreover, $Q(\mathbb{Q}^n, x^0) = Q(\mathbb{Z}^n, x^0) = \emptyset$. The cone $Q(S, x^0)$ has the following duality properties with respect to the cone of attainable directions $A(S, x^0)$:

Theorem 3.4.9. Let $x^0 \in \bar{S} \cap (\overline{\mathbb{R}^n \backslash S})$; then

i) $A(S, x^0) = \mathbb{R}^n \backslash Q(\mathbb{R}^n \backslash S, x^0)$;

ii) $Q(S, x^0) = \mathbb{R}^n \backslash A(\mathbb{R}^n \backslash S, x^0)$.

From the definitions of cones $I(\cdot, \cdot)$, $Q(\cdot, \cdot)$ and $F(\cdot, \cdot)$ it appears that

$$I(S, x^0) \subset Q(S, x^0) \subset F(S, x^0) \ .$$

Numerical examples show that cones $Q(\cdot, \cdot)$ and $Z(\cdot, \cdot)$ are not comparable with an inclusion relation; the same holds for $Q(\cdot, \cdot)$ and $A(\cdot, \cdot)$ and for $A(\cdot, \cdot)$ and $F(\cdot, \cdot)$.

Thus, with reference to the introduced cones, we have the following situation:

$$\begin{array}{ccccc} I(S, x^0) & \subset & Q(S, x^0) & & \\ \cap & & \cap & & \\ Z(S, x^0) & \subset & F(S, x^0) & & \\ \cap & & \cap & & \\ A(S, x^0) & \subset & T(S, x^0) & \subset & P(S, x^0) \end{array}$$

Passing from the tangent cone to the subsequent ones we have a local cone approximation representing, with increasing "fidelity", the set around the point x^0. However, in these "passages" the cones lose some properties. For example, under the assumption that $x^0 \in \bar{S} \cap (\overline{\mathbb{R}^n \backslash S})$, it is easy to prove that $T(S, x^0) \cup T(\mathbb{R}^n \backslash S, x^0) = \mathbb{R}^n$, whereas it is not necessarily $A(S, x^0) \cup A(\mathbb{R}^n \backslash S, x^0) = \mathbb{R}^n$. This loss of properties is less relevant if S is a convex set. In this case we have the following results.

Theorem 3.4.10. Let $S \subset \mathbb{R}^n$ be a nonempty convex set and let $x^0 \in \bar{S}$; then

i) $I(S, x^0) = Q(S, x^0) = \text{cone}(\text{int}(S) - x^0)$;

ii) $Z(S, x^0) = F(S, x^0)$; if $x^0 \in S$, then $Z(S, x^0) = F(S, x^0) = \text{cone}(S - x^0)$;

iii) $A(S,x^0) = T(S,x^0) = P(S,x^0) = \mathrm{cl}(\mathrm{cone}(S - x^0))$;

iv) $I(S,x^0) = \mathrm{int}(T(S,x^0))$.

If, moreover, $\mathrm{int}(S) \neq \emptyset$, it is

v) $T(S,x^0) = \overline{I(S,x^0)}$.

Proof. i) From the definition of $I(S,x^0)$ and $Q(S,x^0)$ and from the fact that (by Theorem 2.1.7 ii)) from $x^0 + \lambda N(y) \subset S$ for some $\lambda > 0$ and $x^0 \in \bar{S}$ it follows $x^0 + tN(y) \subset S$ for all $t \in (0,\lambda)$, we have the equality $I(S,x^0) = \mathrm{cone}(\mathrm{int}(S) - x^0)$ and therefore the convexity of the cones $I(S,x^0)$ and $Q(S,x^0)$.
To prove ii), first note that the inclusions $Z(S,x^0) \subset F(S,x^0) \subset \mathrm{cone}(S - x^0)$ are simple consequences of the definitions (they hold without convexity assumptions). Moreover, from the convexity we get immediately $Z(S,x^0) = F(S,x^0)$. Now, let $x^0 \in S$ and $y \in \mathrm{cone}(S - x^0)$; then $x^0 + \lambda y \in S$ for a scalar $\lambda > 0$. Thus $x^0 + ty \in S$, $\forall t \in (0,\lambda)$, i.e. $y \in Z(S,x^0)$.
Let us now prove iii). Let $y \in T(S,x^0)$; then for every $N(y)$ there exists a scalar $\lambda > 0$ such that $(x^0 + \lambda N(y)) \cap S \neq \emptyset$. Hence it is $T(S,x^0) \subset \mathrm{cl}(\mathrm{cone}(S - x^0))$. Moreover, from the convexity of S (and of $\bar{S}$) it follows: $(x^0 + tN(y)) \cap \bar{S} \neq \emptyset$, $\forall t \in (0,\lambda)$. Therefore $y \in A(\bar{S},x^0) = A(S,x^0)$ and $A(S,x^0) = T(S,x^0) \subset \mathrm{cl}(\mathrm{cone}(S - x^0))$. Finally from ii) it follows $\mathrm{cl}(\mathrm{cone}(S - x^0)) \subset \mathrm{cl}(\mathrm{cone}(\bar{S} - x^0)) = \bar{Z}(\bar{S},x^0) \subset T(\bar{S},x^0) = T(S,x^0)$ and all the cones are equal. So, obviously, $T(S,x^0)$ and $A(S,x^0)$ are convex cones and we have also the equality $T(S,x^0) = P(S,x^0)$.
To get relation iv), first recall that

$$T(S,x^0) = \mathrm{cl}(\mathrm{cone}(S - x^0)) \ ;$$

$$I(S,x^0) = \mathrm{cone}(\mathrm{int}(S) - x^0) \ .$$

By Corollary 2.1.3 we get

$$\mathrm{int}(T(S,x^0)) = \mathrm{int}(\mathrm{cl}(\mathrm{cone}(S - x^0))) = \mathrm{int}(\mathrm{cone}(S - x^0)) \ .$$

So we have to show that

$$\mathrm{int}(\mathrm{cone}(S - x^0)) = \mathrm{cone}(\mathrm{int}(S) - x^0) \ .$$

Without loss of generality we assume $x^0 = 0$. The inclusion

$$\text{cone}(\text{int}(S)) \subset \text{int}(\text{cone}(S))$$

is trivial, since $\text{cone}(\text{int}(S)) \subset \text{cone}(S)$ and $\text{cone}(\text{int}(S))$ is an open set. For the converse inclusion, let $y \in \text{int}(\text{cone}(S))$; then there exists a neighbourhood of y which is contained in $\text{cone}(S)$. In the same manner there exists an n-dimensional simplex, i.e. $n+1$ vectors $y^1, ..., y^{n+1} \in \text{cone}(S)$, with $y \in \text{int}(\text{conv}\,\{y^1, ..., y^{n+1}\}) \subset \text{cone}(S)$. Now we can find vectors $x^1, ..., x^{n+1} \in S$ and numbers $\lambda_1, ..., \lambda_{n+1} > 0$ such that $y^i = \lambda_i x^i$, $i = 1, ..., n+1$. Moreover, with $\lambda = \max\{\lambda_1, ..., \lambda_{n+1}\}$ we get (since $0 \in \bar{S}$) $\dfrac{1}{\lambda}\, y^i = \dfrac{\lambda_i}{\lambda}\, x^i \in \bar{S}$ for all $i = 1, ..., n+1$ and hence

$$\frac{1}{\lambda}\, y \in \text{int}\Big(\text{conv}\Big\{\frac{1}{\lambda}\, y^1, ..., \frac{1}{\lambda}\, y^{n+1}\Big\}\Big) \subset \text{int}(\bar{S}) = \text{int}(S)\ ,$$

i.e. $y \in \text{cone}(\text{int}(S))$.

In the last relation we have used Corollary 2.1.3.

Assertion v) is straightforward: if $\text{int}(S) \neq \emptyset$, then also $I(S, x^0) = \text{cone}(\text{int}(S) - x^0) \neq \emptyset$ and by Corollary 2.1.3 again we have

$$T(S, x^0) = \overline{\text{int}(T(S, x^0))} = \overline{I(S, x^0)}\ . \qquad \square$$

From the previous results it appears that, even if S is convex, there are three distinct levels of local cone approximations. The first approximation is an open cone; the second approximation has no topological properties with respect to openness and closedness; the third approximation is a closed cone. Always for the case of $S \subset I\!R^n$ convex, we formulate the following two properties of the contingent cone $T(S, x^0)$. Proofs can be found in Aubin and Frankowska (1990) and in Bazaraa and Shetty (1976).

Theorem 3.4.11. If S_1 and S_2 are closed convex sets, with $0 \in \text{int}(S_1 - S_2)$ or if S_1 and S_2 are convex, with $\text{relint}(S_1) \cap \text{relint}(S_2) \neq \emptyset$, then $T(S_1 \cap S_2, x^0) = T(S_1, x^0) \cap T(S_2, x^0)$.

Theorem 3.4.12. Let $S \subset I\!R^n$ be convex; we have

i) $T(AS, Ax^0) = AT(S, x^0)$, where A is a matrix of dimension m, n.

ii) $T(S_1 + S_2, (x^1 + x^2)) = T(S_1, x^1) + T(S_2, x^2)$.

II) Modified cone approximations

Some applications of the classical cone approximations introduced above, need additional assumptions concerning either the convexity of S or of the cone. To this extent, Clarke (1975) introduced a new cone approximation which obtained great interest owing to its convexity property. Originally Clarke used for this cone another representation with respect to the one given below, due to Rockafellar (1980, 1981), for which the connection to the other classical cones introduced above is more transparent. See also Aubin and Ekeland (1984), Aubin and Frankowska (1990), Elster and Thierfelder (1984, 1988a, 1988b), Giorgi and Guerraggio (1992a, 1992b), Hiriart-Urruty (1979) Ward (1987, 1988). In the following the cones $T(\cdot,\cdot)$, $A(\cdot,\cdot)$, $Z(\cdot,\cdot)$, $F(\cdot,\cdot)$, $I(\cdot,\cdot)$, $Q(\cdot,\cdot)$ will be modified in such a way that the point x^0 can be varied, too. If we do so, then the convexity behaviour of the cones will be influenced. With such a modification of the classical cones, however, the isotony property is lost.

Definition 3.4.8. Let $S \subset I\!R^n$, $S \neq \emptyset$ and $x^0 \in \bar{S}$; then

$$T_m(S, x^0) = \{y \in I\!R^n \mid \forall N(y)\ \forall V(x^0)\ \forall \lambda > 0\ \exists t \in (0, \lambda)$$
$$\exists \bar{x} \in V(x^0) \cap I\!R^n \backslash S \cup \{x^0\}\ \exists \bar{y} \in N(y) : \bar{x} + t\bar{y} \in S\}$$

is the *modified contingent cone to* S *at* x^0.

$$A_m(S, x^0) = \{y \in I\!R^n \mid \forall N(y)\ \exists V(x^0)\ \exists \lambda > 0\ \forall t \in (0, \lambda)$$
$$\forall \bar{x} \in V(x^0) \cap S \cup \{x^0\}\ \exists \bar{y} \in N(y) : \bar{x} + t\bar{y} \in S\}$$

is the *modified cone of attainable directions to* S *at* x^0 or *Clarke tangent cone to* S *at* x^0. This cone will be denoted in the sequel by $TC(S, x^0)$.

$$Z_m(S, x^0) = \{y \in I\!R^n \mid \exists V(x^0)\ \exists \lambda > 0\ \forall t \in (0, \lambda)$$
$$\forall \bar{x} \in V(x^0) \cap S \cup \{x^0\} : \bar{x} + ty \in S\}$$

is the *modified cone of feasible directions to* S *at* x^0 or *hypertangent cone to* S *at* x^0; this cone will be denoted in the sequel by $H(S, x^0)$.

$$F_m(S, x^0) = \{y \in I\!R^n \mid \forall V(x^0)\ \forall \lambda > 0\ \exists t \in (0, \lambda)$$
$$\exists \bar{x} \in V(x^0) \cap I\!R^n \backslash S \cup \{x^0\}\ : \bar{x} + ty \in S\}$$

is the *modified radial tangent cone to* S *at* x^0.

$$I_m(S, x^0) = \{y \in I\!R^n \mid \exists N(y)\ \exists V(x^0)\ \exists \lambda > 0\ \forall t \in (0, \lambda)$$
$$\forall \bar{x} \in V(x^0) \cap S \cup \{x^0\}\ \forall \bar{y} \in N(y) : \bar{x} + t\bar{y} \in S\}$$

is the *modified cone of interior directions to* S *at* x^0 or *cone of epi-Lipschitzian directions to* S *at* x^0. In the sequel this cone will be denoted by $E(S, x^0)$.

$$Q_m(S, x^0) = \{y \in I\!R^n \mid \exists N(y)\ \forall V(x^0)\ \forall \lambda > 0\ \exists t \in (0, \lambda)$$
$$\exists \bar{x} \in V(x^0) \cap I\!R^n \backslash S \cup \{x^0\}\ \forall \bar{y} \in N(y) : \bar{x} + t\bar{y} \in S\}$$

is the *modified cone of quasi-interior directions to* S *at* x^0.

The definitions of the cones $TC(\cdot, \cdot)$, $H(\cdot, \cdot)$ and $E(\cdot, \cdot)$ can be found in the literature (see especially Rockafellar (1980, 1981), Clarke (1975, 1976, 1983), Penot (1979)); the other cones $F_m(\cdot, \cdot)$, $T_m(\cdot, \cdot)$, $Q_m(\cdot, \cdot)$ have been given here for the purpose of completeness.

The cone $TC(S, x^0)$ was the first "modified" cone to be introduced by Clarke. The most current definitions of $TC(S, x^0)$ consider only the set operation $V(x^0) \cap S$ and not, as in our definition, the union of the point x^0 with the set $V(x^0) \cap S$ (in other words, our definition allows point $\bar{x}$ to coincide with x^0). We have chosen this second possibility for the characterization of the modified cones, in order to avoid some troubles with cones $T_m(\cdot, \cdot)$, $F_m(\cdot, \cdot)$ and $Q_m(\cdot, \cdot)$. In Giorgi and Guerraggio (1992c) it is, however, proved that the two mentioned definitions of Clarke tangent cone coincide when, as in our assumptions, $x^0 \in \bar{S}$.

All modified cones just obtained are not isotonic, but they are convex or complements of convex cones. Similar to the previous theorem, we shall formulate the basic properties of the modified cones. For the proofs, again we refer the reader to the quoted authors (see especially G. Giorgi and A. Guerraggio (1992a, 1992b), Elster and Thierfelder (1984, 1988a, 1988b),

Clarke (1983), Ward (1988)). We first note that Clarke tangent cone to S at x^0 can be also characterized as follows:

$$TC_1(S,x^0) = \{y \in \mathbb{R}^n \mid \forall \{x^k\} \subset S, \lim_{k\to+\infty} x^k = x^0, \forall \{\lambda_k\} \subset \mathbb{R}_+,$$
$$\lim_{k\to+\infty} \lambda_k = 0, \exists \{y^k\} \to y : x^k + \lambda_k\, y^k \in S \text{ for large } k\} ;$$

$$TC_2(S,x^0) = \{\forall N(y), \exists V(x^0), \exists \lambda > 0, \forall \bar{x} \in S \cap V(x^0)$$
$$\forall t \in (0,\lambda) : \bar{x} + tN(y) \cap S \neq \emptyset\} ;$$

$$TC_3(S,x^0) = \Big\{ \bigcap_{N(0)} \; \bigcup_{\substack{N(x^0)\\ \lambda>0}} \; \bigcap_{\substack{x\in S\cap N(x^0)\\ t\in(0,\lambda)}} \Big[\frac{S-x}{t} + N(0)\Big]\Big\} .$$

The last characterization, in terms of the so-called "Painlevé-Kuratowski limits of sets", was first given by Rockafellar (1980, 1981). Thus we have, by definition,

$$TC_3(S,x^0) = \liminf_{\substack{x\to x^0\\ t\to 0^+}} t^{-1}(S-x) .$$

The Clarke tangent cone can also be described in terms of the distance function. We have then the following characterization:

$$TC_4(S,x^0) = \Big\{y \in \mathbb{R}^n \mid \limsup_{\substack{x\to x^0\\ \lambda\to 0^+}} \frac{d_S(x+\lambda y) - d_S(x)}{\lambda} = 0\Big\} ,$$

where $d_S(x) = \inf_{v\in S} \|x - v\|$.

The above definition is the original one due to Clarke (1975).

<u>Theorem 3.4.13</u>.

i) $H(S,x^0)$ is a convex cone. $F_m(S,x^0)$ is the complement of a convex cone.

ii) $0 \in H(S,x^0) \Leftrightarrow 0 \in F_m(S,x^0) \Leftrightarrow x^0 \in S$.

iii) $H(S, x^0) \subset Z(S, x^0)$;
$F_m(S, x^0) \supset F(M, x^0)$.

iv) $H(S_1 \cap S_2, x^0) \supset H(S_1, x^0) \cap H(S_2, x^0)$;
$F_m(S_1 \cup S_2, x^0) \subset F_m(S_1, x^0) \cup F_m(S_2, x^0)$.

v) $H(S_1 \times S_2, (x^1, x^2)) = H(S_1, x^1) \times H(S_2, x^2)$;
$F_m(S_1 \times S_2, (x^1, x^2)) \subset F_m(S_1, x^1) \times F_m(S_2, x^2)$.

Theorem 3.4.14.

i) $E(S, x^0)$ is a convex cone. $T_m(S, x^0)$ is the complement of a convex cone.

ii) $E(S, x^0)$ is an open convex cone and $x^0 \in \text{int}(S) \Leftrightarrow 0 \in E(S, x^0)$;
$T_m(S, x^0)$ is a closed cone with $0 \in T_m(S, x^0)$.

iii) $E(S, x^0) \subset H(S, x^0)$;
$T_m(S, x^0) \supset F_m(S, x^0)$;
$E(S, x^0) \subset I(S, x^0)$;
$T_m(S, x^0) \supset T(S, x^0)$.

iv) $E(S_1 \cap S_2, x^0) \supset E(S_1, x^0) \cap E(S_2, x^0)$;
$T_m(S_1 \cup S_2, x^0) \subset T_m(S_1, x^0) \cup T_m(S_2, x^0)$.

v) $E(S_1 \times S_2, (x^1, x^2)) = E(S_1, x^1) \times E(S_2, x^2)$;
$T_m(S_1 \times S_2, (x^1, x^2)) \subset T_m(S_1, x^1) \times T_m(S_2, x^2)$.

vi) $E(\text{int}(S), x^0) \supset E(S, x^0)$;
$T_m(\bar{S}, x^0) \subset T_m(S, x^0)$.

Theorem 3.4.15.

i) $TC(S, x^0)$ is a convex cone. $Q_m(S, x^0)$ is the complement of a convex cone.

ii) $TC(S, x^0)$ is a closed cone with $0 \in TC(S, x^0)$;
$Q_m(S, x^0)$ is an open cone and $x^0 \in \text{int}(S) \Leftrightarrow 0 \in Q_m(S, x^0)$.

iii) $H(S, x^0) \subset TC(S, x^0) \subset A(S, x^0)$;
$Q(S, x^0) \subset Q_m(S, x^0) \subset F_m(S, x^0)$.

iv) $TC(S_1 \times S_2, (x^1, x^2)) = TC(S_1, x^1) \times TC(S_2, x^2)$;
$Q_m(S_1 \times S_2, (x^1, x^2)) \subset Q_m(S_1, x^1) \times Q_m(S_2, x^2)$.

v) $TC(\bar{S}, x^0) = TC(S, x^0)$;
$Q_m(\text{int}(S), x^0) = Q_m(S, x^0)$.

Note that for $TC(S, x^0)$ and $Q_m(S, x^0)$ there are no properties concerning union and intersection. Owing to the importance of Clarke tangent cone in optimization and nonsmooth analysis, here we prove the convexity of the same (closedness follows easily from the definition, e.g. from characterization $TC_2(\cdot, \cdot)$). Proving the convexity of $TC(S, x^0)$ is equivalent to proving that $y^1 + y^2 \in TC(S, x^0)$ if $y^1, y^2 \in TC(S, x^0)$. Let us choose an arbitrary neighbourhood $N(y^1 + y^2)$; it is always possible to find $N_1(y^1)$ and $N_2(y^2)$ such that $N_1(y^1) + N_2(y^2) \subset N(y^1 + y^2)$. As $y^1, y^2 \in TC(S, x^0)$, it follows the existence of two neighbourhoods of x^0, $V_1(x^0)$ and $V_2(x^0)$, and of two real numbers $\lambda_1, \lambda_2 > 0$ such that $(S - \bar{x}) \cap tN_1(y^1) \neq \emptyset$, $\forall \bar{x} \in V_1(x^0) \cap S \cup \{x^0\}$ and $\forall t \in (0, \lambda_1)$, and $(S - \bar{x}) \cap tN_2(y^2) \neq \emptyset$ $\forall \bar{x} \in V_2(x^0) \cap S \cup \{x^0\}$ and $\forall t \in (0, \lambda_2)$. If we choose $V(x^0)$ and $\lambda > 0$ such that $\lambda \leqq \min\{\lambda_1, \lambda_2\}$, $V(x^0) \subset V_1(x^0)$, $V(x^0) + tN_1(y^1) \subset V_2(x^0)$, $\forall t \in (0, \lambda)$, we have, $\forall \bar{x} \in V(x^0) \cap S \cup \{x^0\}$ and $\forall t \in (0, \lambda)$, that there exists a vector $\bar{y} \in N_1(y^1)$ such that $\bar{x} + t\bar{y} \in S$ and $\bar{x} + t\bar{y} \in V_2(x^0)$. We have also the existence of a vector $\tilde{y} \in N_2(y^2)$ such that $\bar{x} + t\bar{y} + t\tilde{y} \in S$, i.e. $t(\bar{y} + \tilde{y}) \in S - \bar{x}$. The vector $y^1 + y^2 \in TC_3(S, x^0)$, as $\bar{y} + \tilde{y} \in N(y^1 + y^2)$. □

Note, moreover, that if x^0 is an isolated point of S, then $TC(S, x^0) = \{0\}$; if $x^0 \in \text{int}(S)$, then $TC(S, x^0) = \mathbb{R}^n$. Sharper than Theorem 3.4.10, iv) and v), is the following assertion, due to Rockafellar (1980, 1981).

<u>Theorem 3.4.16</u>.

$E(S, x^0) = \text{int}(TC(S, x^0))$

and if $E(S, x^0) \neq \emptyset$, then

$$TC(S, x^0) = \bar{E}(S, x^0) \ .$$

If we assume that the set S is convex, then we can prove the following:

Theorem 3.4.17. Let $S \subset I\!R^n$ be convex; then

i) $E(S, x^0) = I(S, x^0)$.

ii) $TC(S, x^0) = A(S, x^0)$.

For the proof see Giorgi and Guerraggio (1992b) or Elster and Thierfelder (1984).

Taking into account the results given in I) and II), we obtain the following inclusion diagram:

$$\begin{array}{ccccccc}
E(S,x^0) & \subset & I(S,x^0) & \subset & Q(S,x^0) & \subset & Q_m(S,x^0) \\
\cap & & \cap & & \cap & & \cap \\
H(S,x^0) & \subset & Z(S,x^0) & \subset & F(S,x^0) & \subset & F_m(S,x^0) \\
\cap & & \cap & & \cap & & \cap \\
TC(S,x^0) & \subset & A(S,x^0) & \subset & T(S,x^0) & \subset & T_m(S,x^0)
\end{array}$$

Other relationships between the modified cone approximations are contained in the following proposition (see Elster and Thierfelder (1988b)).

Theorem 3.4.18. If x^0 is a boundary point of S, then it holds that

$$E(S, x^0) = -E(I\!R^n \backslash S, x^0) \ ;$$

$$TC(S, x^0) = -TC(I\!R^n \backslash S, x^0) \ .$$

Also for the modified local cone approximations, under the assumption $x^0 \in \bar{S} \cap \overline{(I\!R^n \backslash S)}$, we have the following duality properties:

Theorem 3.4.19. Let $x^0 \in \bar{S} \cap \overline{(I\!R^n \backslash S)}$. Then we have

$$\begin{aligned}
H(S,x^0) &= \mathbb{R}^n \backslash F_m(\mathbb{R}^n \backslash S, x^0) \\
F_m(S,x^0) &= \mathbb{R}^n \backslash H(\mathbb{R}^n \backslash S, x^0) \\
E(S,x^0) &= \mathbb{R}^n \backslash T_m(\mathbb{R}^n \backslash S, x^0) \\
T_m(S,x^0) &= \mathbb{R}^n \backslash E(\mathbb{R}^n \backslash S, x^0) \\
TC(S,x^0) &= \mathbb{R}^n \backslash Q_m(\mathbb{R}^n \backslash S, x^0) \\
Q_m(S,x^0) &= \mathbb{R}^n \backslash TC(\mathbb{R}^n \backslash S, x^0) \ .
\end{aligned}$$

Thus, under the assumptions of the theorem, cones $F_m(\cdot,\cdot)$, $T_m(\cdot,\cdot)$ and $Q_m(\cdot,\cdot)$ are complements of a convex cone.

With reference to the previous diagram, showing the various inclusions between the considered cone approximations, it is useful to make also the following remarks:

1) Under the assumption $x^0 \in \bar{S} \cap \overline{(\mathbb{R}^n \backslash S)}$, the cones lying opposite each other with respect to a fictitious midpoint of the scheme are "dual" according to the relation

$$K_1(S,x^0) = \mathbb{R}^n \backslash K_2(\mathbb{R}^n \backslash S, x^0) \ ,$$

where K_1 and K_2 are the two said cone approximations.

2) For the cones $K(\cdot,\cdot)$ lying in the first column of the scheme

$$K(S_1 \times S_2, (x^1, x^2)) = K(S_1, x^1) \times K(S_2, x^2) \ ;$$

for the cones $K(\cdot,\cdot)$ lying in the last column of the scheme

$$K(S_1 \times S_2, (x^1, x^2)) \subset K(S_1, x^1) \times K(S_2, x^2) \ .$$

3) The cones lying in the first row of the scheme are open and

$$x^0 \in \operatorname{int}(S) \Leftrightarrow 0 \in K(S,x^0) \ ;$$

$$K(\operatorname{int}(S), x^0) \supset K(S,x^0) \ .$$

The cones lying in the third row of the scheme are closed and

$$0 \in K(S, x^0) \ ;$$

$$K(\bar{S}, x^0) \subset K(S, x^0) \ .$$

For the cones lying in the second row

$$x^0 \in S \Leftrightarrow 0 \in K(S, x^0)$$

and these cones are in general neither open nor closed.

4) The cones lying in the first column of the scheme are convex; the cones lying in the last column of the scheme are complements of convex cones (under the assumption $x^0 \in \bar{S} \cap \overline{(\mathbb{R}^n \backslash S)}$). The cones in the second and third column are isotonic, according to

$$S_1 \subset S_2 \Rightarrow K(S_1, x^0) \subset K(S_2, x^0) \ .$$

The cones investigated in this section are the most important cone approximations used in optimization theory. However, we can find numerous other cones in the literature, sometimes differing only slightly. Again we observe that some cones can be changed if we refer to a topological space X: by varying the topology, different cones can arise.

Some authors (Dolecki (1982), Penot (1979), Rockafellar (1980, 1981)) introduce the cone approximations making use of the notion of generalized limit of set-valued maps, in the sense of Painlevé-Kuratowski (see also Aubin and Ekeland (1984) and Aubin and Frankowska (1990)). This is the case of characterization $TC_5(S, x^0)$. Since the corresponding limit sets are always closed, in this way only cones $A(S, x^0)$, $TC(S, x^0)$, $T(S, x^0)$ and $T_m(S, x^0)$ can be obtained by means of this operation (the cone $P(S, x^0)$ is excluded from the said possibility).

The characterizations of cones $A(S, x^0)$ and $T(S, x^0)$ in terms of the Painlevé-Kuratowski limits are then:

$$A(S, x^0) = \bigcap_{N(0)} \bigcup_{\lambda > 0} \bigcap_{t \in (0, \lambda)} \left\{ \frac{S - x^0}{t} + N(0) \right\} = \liminf_{t \to 0^+} \ t^{-1}(S - x^0) \ ;$$

$$T(S, x^0) = \bigcap_{N(0)} \bigcap_{\lambda > 0} \bigcup_{t \in (0, \lambda)} \left\{ \frac{S - x^0}{t} + N(0) \right\} = \limsup_{t \to 0^+} \ t^{-1}(S - x^0) \ .$$

These two cones can also be described by means of the distance function $d_S(x)$. We have already given in $TC_4(S,x^0)$ such a characterization of the Clarke tangent cone. For cones $A(S,x^0)$ and $T(S,x^0)$ we have

$$A(S,x^0) = \left\{ y \in I\!R^n \mid \limsup_{\lambda \to 0^+} \frac{d_S(x^0+\lambda y) - d_S(x^0)}{\lambda} = 0 \right\} ;$$

$$T(S,x^0) = \left\{ y \in I\!R^n \mid \liminf_{\lambda \to 0^+} \frac{d_S(x^0+\lambda y) - d_S(x^0)}{\lambda} = 0 \right\} .$$

A cone which has received some attention in recent years was introduced by Michel and Penot (1984) for normed spaces. It is the *prototangent cone to S at x^0*, defined as

$$PMP(S,x^0) = \Big\{ y \in I\!R^n \mid \forall\, z \in I\!R^n,\ \forall\, \{t_k\} \subset I\!R_+,\ \lim_{k \to +\infty} t_k = 0,$$
$$\forall\, \{x^k\} \in S,\ \lim_{k \to +\infty} x^k = x^0,\ \lim_{k \to +\infty} \frac{x^k - x^0}{t_k} = z,$$
$$\exists\, \{y^k\} \to y : x^k + t_k y^k \in S \text{ for large } k \Big\} .$$

The prototangent cone is convex and closed and it is

$$TC(S,x^0) \subset PMP(S,x^0) \subset A(S,x^0) .$$

The local cone approximations reveal their utility in the smooth optimization problems but especially in the nonsmooth problems. A general axiomatic definition of an abstract cone approximation, which generates all the cones presented sub I) and II) will be presented in Chapter IV. See also Vlach (1981), Ioffe (1985) and especially Ward (1988).

All the previous cones are associated with a set in order to obtain a local approximation of that set around a point; there exists also the necessity to associate a cone with a set in a global form. We have already introduced the conical hull of a set and the convex conical hull; another cone, useful for this purpose, was introduced by Rockafellar (1970).

<u>Definition 3.4.9</u>. Given a nonempty set $S \subset I\!R^n$, the cone

$$0^+S = \{ y \in I\!R^n \mid S + ty \subset S,\ \forall\, t > 0 \}$$

is called *recession cone of* S. If $S = \emptyset$, then by definition $0^+\emptyset = \mathbb{R}^n$.

We note that 0^+S is a convex cone since $S + t(\lambda_1 y^1 + \lambda_2 y^2) = (S + t\lambda_1 y^1) + t\lambda_2 y^2 \subset S + t\lambda_2 y^2 \subset S$ for each $t > 0$, $\lambda_1, \lambda_2 \geqq 0$ and $y^1, y^2 \in 0^+S$. Moreover, 0^+S is the largest (convex) cone $K \subset \mathbb{R}^n$ such that $S + K \subset S$. If S is convex and nonempty, then 0^+S can be represented as:

$$0^+S = \{y \in \mathbb{R}^n \mid S + y \subset S\} .$$

It is intuitive that 0^+S is useful to state the boundedness of a set $S \neq \emptyset$. Indeed, if 0^+S is not a singleton formed by the origin, then S is unbounded, but the converse does not hold; e.g. $0^+\mathbb{Q}^n = \{0\}$. If S is closed and convex, then S is bounded if and only if $0^+S = \{0\}$.

As the recession cone is not a local cone approximation, we do not insist further on its algebraic and topological properties. We only remark that $0^+S \subset 0^+\bar{S}$. In fact, let $y \in 0^+S$ and let $\bar{s} \in \bar{S}$; then, $\forall t > 0$, we have

$$\bar{s} + ty = (\lim_{k \to +\infty} s^k) + ty = \lim_{k \to +\infty} (s^k + ty) = \lim_{k' \to +\infty} s^{k'} \in \bar{S}$$

where $s^k, s^{k'} \in S$. The converse relation between 0^+S and $0^+\bar{S}$ does not hold, as shown by the following simple example: $S = \mathbb{R}^2 \backslash \{x^0\}$; we have $0^+S = \{0\}$, whereas $0^+\bar{S} = \mathbb{R}^2$. Moreover, the equality $0^+S = 0^+\bar{S}$ generally does not hold even if S is convex (take, e.g., $S = \{(x_1, x_2) \mid x_1 \in (-1, 1),\ x_2 \geqq 0\} \cup \{(-1, 0), (1, 0)\}$; we have $\{0\} = 0^+S \subset 0^+\bar{S} = \{(0, x_2) \mid x_2 \geqq 0\}$).

With $S + 0^+S \subset S$ we even get $\bar{S} + \overline{0^+S} \subset \overline{S + 0^+S} \subset \bar{S}$, i.e. $\overline{0^+S} \subset 0^+\bar{S}$.

Example 3.4.1. Consider the following sets of $\mathbb{R}^2$:

$$S_1 = \{(x_1, x_2) \mid x_1 > 0,\ x_2 > x_1^{-\alpha},\ \alpha > 0\} ;$$

$$S_2 = \{(x_1, x_2) \mid x_2 \geqq x_1^2\} ;$$

$$S_3 = \{(x_1, x_2) \mid x_1^{\frac{1}{2}} + x_2^{\frac{2}{3}} \leqq 1\} ;$$

$S_4 = \{(x_1, x_2) \mid x_1 > 0,\ x_2 > 0\} \cup \{(0,0)\}$;

$S_5 = \{(x_1, x_2) \mid x_1 > 0,\ x_2 > 0\}$.

The corresponding recession cones are:

$0^+S_1 = \{(x_1, x_2) \mid x_1 \geqq 0,\ x_2 \geqq 0\}$;

$0^+S_2 = \{(x_1, x_2) \mid x_1 = 0,\ x_2 \geqq 0\}$;

$0^+S_3 = \{(x_1, x_2) \mid x_1 = 0 = x_2\} = \{(0,0)\}$;

$0^+S_4 = \{(s_1, x_2) \mid x_1 > 0,\ x_2 > 0\} \cup \{(0,0)\} = S_4$;

$0^+S_5 = \{(x_1, x_2) \mid x_1 \geqq 0,\ x_2 \geqq 0\} = \bar{S}_5$.

3.5. Necessary Optimality Conditions for Problem (P) Where the Optimal Point is Interior to X

Theorem 3.3.1 gives a condition, necessary under a regularity assumption, for the local optimality of point x^0 for problem (P_e), i.e. for problem (P_1) where there are no inequality constraints and X is an open set. One could remark that it should not be restrictive to study only problem (P_e), as an inequality of the type $g_i(x) \leqq 0$ can be equivalently substituted by the equality

$$g_i(x_1, ..., x_n) - x_{n+i}^2 = 0 \tag{1}$$

where the variable x_{n+i} is called "slack variable".

By means of transformation (1) one could think to have removed any question due to the presence of inequalities and therefore to transfer to problem (P_e) the study of problem (P).

However, transformation (1) does not remove the specific difficulties of problem (P), such as the validation of the regularity conditions on the new constraints. Moreover, the "transformed problem" may be complicated, due to the increase of the number of variables and to possible changes of "structure". For example if g_i is an affine function, the new transformed constraint (1) is no more affine. Transformation (1), indeed advantageous in some cases, can cause considerable complications for other cases (see, e.g., Mangasarian and Fromovitz (1967)).

Let us therefore consider problem (P):

$$\min_{x \in S} f(x)$$

$$S = \{x \mid x \in X,\ g_1(x) \leqq 0, ..., g_m(x) \leqq 0\} \ , \tag{P}$$

where the real-valued functions f, $g_1, ... g_m$ are defined on an open set $D \subset \mathbb{R}^n$ containing X. Let $x^0 \in S$ and let $I(x^0)$ be the set of *effective* (or *active* or *binding*) constraints at x^0, i.e. the set

$$I(x^0) = \{i \mid g_i(x^0) = 0\} \ .$$

The set of non effective constraints is denoted with

$$J(x^0) = I \backslash I(x^0) = \{i \mid i \notin I(x^0)\} \ ,$$

where $I = \{1, ..., m\}$. We shall also denote by g_I the vector-valued function formed by the active constraints at x^0; similarly for g_J.

We shall consider three different approaches for obtaining necessary optimality conditions for (P), where $x^0 \in \text{int}(X)$. The first approach is a linearization theorem, due to Abadie (1967), given here in the generalized version of Mangasarian (1969). First let us introduce the sets

$$I_P(x^0) = \{i \mid g_i(x^0) = 0 \text{ and } g_i \text{ is pseudoconcave at } x^0\} \ ;$$

$$I_{NP}(x^0) = \{i \mid i \in I(x^0),\ i \notin I_P(x^0)\} \ .$$

<u>Theorem 3.5.1</u> (linearization theorem of Abadie-Mangasarian). Let $x^0 \in \text{int}(X)$ be a solution of (P) or simply a point of local minimum of f on S. Let f and g be differentiable at x^0; then the system

$$\begin{cases} z\nabla f(x^0) < 0 & \\ z\nabla g_i(x^0) < 0 \ , & i \in I_{NP}(x^0) \\ z\nabla g_i(x^0) \leqq 0 \ , & i \in I_P(x^0) \end{cases} \tag{2}$$

has no solution z in $\mathbb{R}^n$.

Proof. Let x^0 be a local solution of problem (P) and suppose absurdly that system (2) has a solution z. We shall prove that there exists a $\tilde{\delta} > 0$ such that:

i) $x^0 + \delta z \in X$, $\forall \delta \in (0, \tilde{\delta})$;

ii) $g(x^0 + \delta z) \leqq 0$, $\forall \delta \in (0, \tilde{\delta})$;

iii) $f(x^0 + \delta z) < f(x^0)$, $\forall \delta \in (0, \tilde{\delta})$,

which is in contradiction to the assumptions.

a) As $x^0 \in \text{int}(X)$, there exists a $\hat{\delta} > 0$ such that $x^0 + \delta z \in X$, $\forall \delta \in (0, \hat{\delta})$.

b) As f is differentiable at x^0, we have

$$f(x^0 + \delta z) - f(x^0) = \delta z \nabla f(x^0) + \delta \, \|z\| \, \varepsilon_0(x^0, \delta z) \ ,$$

with $\varepsilon_0(x^0, \delta z) \to 0$ for $\delta \to 0$. Then for δ small enough, say $0 < \delta < \delta_0$, we have

$$z \nabla f(x^0) + \|z\| \, \varepsilon_0(x^0, \delta z) < 0$$

(since $z \nabla f(x^0) < 0$) and hence

$$f(x^0 + \delta z) < f(x^0) \ , \qquad \forall \delta \in (0, \delta_0) \ .$$

c) Similarly, for $i \in I_{NP}(x^0)$ and δ small enough, say $0 < \delta < \delta_i$, we have $g_i(x^0 + \delta z) < g_i(x^0) = 0$, $\forall \delta \in (0, \delta_i)$, $i \in I_{NP}(x^0)$.

d) For $i \in I_P(x^0)$ we have, since g_i is pseudoconcave at x^0 and $\delta z \nabla g_i(x^0) \leqq 0$, $\delta > 0$,

$$g_i(x^0 + \delta z) \leqq g_i(x^0) = 0 \ , \qquad \forall \delta \in (0, \hat{\delta}) \ , \quad \forall i \in I_P(x^0) \ .$$

e) For $i \in J(x^0)$ we have $g_i(x^0) < 0$; hence for δ small enough (say $0 < \delta < \delta_i$), we have

$$g_i(x^0 + \delta z) < 0 \ , \qquad \forall \delta \in (0, \delta_i) \ , \quad \forall i \in J(x^0) \ .$$

Let us now put $\tilde{\delta} = \min\{\hat{\delta}, \delta_0, \delta_i\}$, $i \notin I_P(x^0)$. Then for any δ in the interval $(0, \tilde{\delta})$, the point $x^0 + \delta z$ satisfies conditions i), ii), iii) and the proof is complete. □

We note that in the above theorem the differentiability at x^0 of the constraints g_i, $i \in J(x^0)$, can be substituted by the continuity at x^0.

We are now ready to derive the second approach for obtaining necessary optimality conditions for problem (P), i.e. the approach due to Fritz John (1948), which is the first published result for this kind of problems.

Theorem 3.5.2 (Fritz John necessary optimality conditions). Let $x^0 \in \text{int}(X)$ be a solution of problem (P) or simply a point of local minimum of f on S; let f and g be differentiable at x^0. Then there exist multipliers $u_0 \in \mathbb{R}$ and $u \in \mathbb{R}^m$ such that

$$u_0 \nabla f(x^0) + u \nabla g(x^0) = 0 \tag{3}$$

$$u g(x^0) = 0 \tag{4}$$

$$(u_0, u) \geq 0 \ . \tag{5}$$

Proof. Note that the assumptions of Theorem 3.5.2 are the same as those for Theorem 3.5.1. Hence system (2) has no solution $z \in \mathbb{R}^n$; by Motzkin theorem of the alternative (Theorem 19 of Section 2.4, Chapter II) there exist multipliers u_0, u_{NP}, u_P such that

$$u_0 \nabla f(x^0) + u_{NP} \nabla g_{NP}(x^0) + u_P \nabla g_P(x^0) = 0$$

$$(u_0, u_{NP}) \geq 0 \ , \qquad u_P \geqq 0 \ .$$

Since $g_{NP}(x^0) = 0$ and $g_P(x^0) = 0$, it follows that if we define

$$u_J = 0 \ \text{ and } \ u = (u_{NP}, u_P, u_J)$$

then $u g(x^0) = 0$ and

$$u_0 \nabla f(x^0) + u \nabla g(x^0) = 0$$

with $(u_0, u) \geq 0$, i.e. (u_0, u) a semipositive vector. □

Note that in the previous theorem the vector (u_0, u_{NP}) cannot be the zero vector. The number u_0 is called the *Fritz John multiplier corresponding to the objective function*, while u_i is the *Fritz John multiplier corresponding to constraint* g_i. Note that the Fritz John multipliers corresponding to the noneffective constraints (the vector u_J) are zeros. Conditions (3)–(5) are the *Fritz John conditions* for problem (P) and condition (4) is usually referred to as *complementarity slackness condition.* The Fritz John conditions are necessary optimality conditions, but in general are not sufficient, without further assumptions.

If we choose $u_i = 0, \forall i \in J(x^0)$, we can write the Fritz John conditions in the form:

$$u_o \nabla f(x^0) + \sum_{i \in I(x^0)} u_i \nabla g_i(x^0) = 0 \ ,$$

u_0, u_i $(i \in I(x^0))$ being nonnegative scalars, not all zero.

Now in the above relation (3) two cases can arise: the first case corresponds to the vanishing of u_0, thereby the objective function f disappears from the Fritz John condition (degenerate case); and the second case where $u_0 > 0$. The first case occurs for lack of technical properties of the constraints, called *constraint qualifications*, which assure the positiveness of the first multiplier u_0. In the second case, where $u_0 > 0$, we have the third approach, due to Kuhn and Tucker (1951), for obtaining necessary optimality conditions for problem (P). The Kuhn-Tucker conditions are perhaps the most known necessary optimality conditions for (P), as they are, under some convexity assumptions, also sufficient for optimality. It is therefore important to see which conditions assure the positiveness of the multiplier u_0 in relation (3). As we have already noted, conditions of this nature are usually referred to as *constraint qualifications*, since they involve only the constraints and are independent of the geometric structure of the feasible set S. However, (see Mangasarian (1969)) the degenerate case $u_0 = 0$ may occur, for example, when the optimal point x^0 is at a cusp of S or when S is a singleton made of the point x^0. We now consider some simple constraint qualifications of "algebraic type". In the next section we shall introduce other constraint qualifications, of "topological type", for which the requirement $x^0 \in \text{int}(X)$ is related.

Let $x^0 \in \operatorname{int}(X)$; let us consider the following constraint qualifications for problem (P).

1) *Slater's weak constraint qualification* (Slater (1950), Uzawa (1958), Mangasarian (1969)). The constraints g_i, $i = 1, ..., m$, are said to satisfy Slater's weak constraint qualification at $x^0 \in S$, if g_i, $i \in I(x^0)$, is pseudoconvex at x^0 and there exists a vector $\bar{x} \in S$ such that $g_I(\bar{x}) < 0$.

2) The original *Slater's constraint qualification* requires that D is a convex set in $\mathbb{R}^n$ and that g is convex on D. Slater's constraint qualification is satisfied if there exists $\bar{x} \in S$ such that $g(\bar{x}) < 0$.

3) *Karlin's constraint qualification* (Karlin (1959)). Let D be convex and g be convex on D; Karlin's constraint qualification is satisfied if there exists no vector $p \in \mathbb{R}^m$, $p \geq 0$, such that $p \cdot g(x) \geqq 0$, $\forall x \in D$.

4) *Modified strict constraint qualification* (Mangasarian (1969), Bazaraa, Goode and Shetty (1972b)). It is expressed as: the functions g_i, $i \in I(x^0)$, are pseudoconvex at x^0 and the feasible set S contains at least two distinct points x^1 and x^2 such that g_i, $i \in I(x^0)$, are *strictly quasiconvex at* x^1 *with respect to* x^2, i.e. from $g_i(x^2) \leqq g_i(x^1)$ it follows $g_i(\lambda x^2 + (1-\lambda)\, x^1) < g_i(x^1)$, $\forall \lambda \in (0,1)$.

5) *Weak reverse constraint qualification* (Arrow-Hurwicz-Uzawa (1961), Mangasarian (1969)). It is expressed as: the functions g_i, $i \in I(x^0)$, are pseudoconcave at x^0.

6) *Cottle-Dragomirescu constraint qualification* (Cottle (1963), Dragomirescu (1967)). The vectors $\nabla g_i(x^0)$, $i \in I(x^0)$, are *positively linearly independent*, i.e. the system

$$\sum_{i \in I(x^0)} u_i \nabla g_i(x^0) = 0$$

$$u_i \geqq 0\,, \qquad \forall i \in I(x^0)$$

only admits the zero solution.

7) *Mangasarian-Fromovitz constraint qualification* (Mangasarian and Fromovitz (1967)). The system

$$y\nabla g_i(x^0) < 0 \, , \qquad i \in I(x^0)$$

admits solution.

8) *Nondegeneracy condition* or *rank condition* (Arrow, Hurwicz and Uzawa (1961)). The gradients $\nabla g_i(x^0)$, $i \in I(x^0)$, are linearly independent.

9) *Arrow-Hurwicz-Uzawa first constraint qualification* (Arrow-Hurwicz-Uzawa (1961), Mangasarian (1969)). It is expressed as: there exists a vector $y \in \mathbb{R}^n$ solution of the system

$$y\nabla g_i(x^0) \leqq 0 \, , \qquad i \in I_P(x^0)$$
$$y\nabla g_i(x^0) < 0 \, , \qquad i \in I_{NP}(x^0) \, .$$

We may note that this constraint qualification can also be expressed by the following condition: the system

$$\sum_{i \in I_P(x^0)} u_i \nabla g_i(x^0) + \sum_{i \in I_{NP}(x^0)} u_i \nabla g_i(x^0) = 0$$
$$u_i \geqq 0 \, , \qquad i \in I(x^0)$$

admits solutions only with $u_{NP} = 0$.

Note that conditions 2) and 3) are global conditions, while the remaining ones are local conditions at point $x^0 \in K$.

Theorem 3.5.3. Let $x^0 \in \text{int}(X)$ be feasible and let $C.Q.i$ denote the above i-th constraint qualification ($i = 1, ..., 8$). Then we have the following implications:

$$\begin{array}{ccccc} C.Q.2 & \Leftrightarrow & C.Q.3 & & \\ \Downarrow & & & & \\ C.Q.1 & \Leftarrow & C.Q.4 & & \\ \Downarrow & & & & \\ C.Q.6 & \Leftrightarrow & C.Q.7 & \Leftarrow & C.Q.8 \\ \Downarrow & & & & \\ C.Q.9 & \Leftarrow & C.Q.5 & & \end{array}$$

Proof. $C.Q.2 \Rightarrow C.Q.1$ is trivial; let us prove the equivalence $C.Q.2 \Leftrightarrow C.Q.3$. The implication $C.Q.2 \Rightarrow C.Q.3$ is trivial; for the reverse case let us suppose absurdly that $C.Q.3$ is satisfied but $C.Q.2$ is not satisfied, i.e. the system $x \in D$, $g(x) < 0$ admits no solution. Then, thanks to Theorem 2.9.1 (Fan-Glicksberg-Hoffman theorem), there exists a semipositive vector $u \geq 0$, such that $ug(x) \geqq 0$, $\forall x \in D$ in contradiction with $C.Q.3$ (the equivalence directly follows also from Corollary 2.9.1).
$C.Q.4 \Rightarrow C.Q.1$. Let x^0 satisfy the modified strict $C.Q.$; then by the strict quasiconvexity of g_i at x^1, with respect to x^2, we have

$$g_i(\lambda x^1 + (1-\lambda)\, x^2) < g_i(x^1) \leqq 0 \,,$$
$$\forall \lambda \in (0,1)\,, \quad \forall i \in I(x^0) \,.$$

This implies that there exists some $\bar{x}$ such that $g_i(\bar{x}) < 0$, $i \in I(x^0)$. hence $C.Q.1$ is satisfied.
$C.Q.1 \Rightarrow C.Q.6$. Assume $C.Q.1$ holds and $C.Q.6$ does not hold; then there exists a nonzero solution u to the system

$$\sum_{i \in I(x^0)} u_i \nabla g_i(x^0) = 0$$
$$u_i \geqq 0 \,, \qquad i \in I(x^0) \,.$$

Since every g_i, $i \in I(x^0)$, is pseudoconvex at x^0, then

$$g_i(\bar{x}) - g_i(x^0) < 0 \Rightarrow (\bar{x} - x^0)\nabla g_i(x^0) < 0 \,,$$

i.e.

$$(\bar{x} - x^0)\nabla g_i(x^0) < 0\,, \qquad \forall i \in I(x^0)\,.$$

Since $u_i > 0$ for some $i \in I(x^0)$, it follows that

$$\sum_{i \in I(x^0)} u_i(\bar{x} - x^0)\nabla g_i(x^0) < 0\,,$$

which contradicts the assumption that

$$\sum_{i \in I(x^0)} u_i \nabla g_i(x^0) = 0\,.$$

$C.Q.6 \Leftrightarrow C.Q.7$. Let $C.Q.6$ be verified; by applying Theorem 11 of Section 2.4 (Gordan theorem of the alternative) it follows at once that there exists a vector y such that $y\nabla g_i(x^0) < 0$, $i \in I(x^0)$, admits solution. The implication $C.Q.7 \Rightarrow C.Q.6$ is trivial.
$C.Q.8 \Rightarrow C.Q.7$: trivial.
$C.Q.6 \Rightarrow C.Q.9$: trivial.
$C.Q.5 \Rightarrow C.Q.9$: trivial. □

It is not difficult to find examples where none of the above $C.Q.$s are satisfied. The reader can verify, for example, that at the point $x^0 = (0,0)$ the constraints in $\mathbb{R}^2$:

$$g_1 = -x_2 \leqq 0\,;$$

$$g_2 = x_2 - x_1^3 \leqq 0$$

do not satisfy any of the above $C.Q.$s. Similary for the point $x^0 = (0,0)$ with respect to the constraints in $\mathbb{R}^2$:

$$g_1 = x_1^2 + (x_2 - 1)^2 - 1 \leqq 0\,;$$

$$g_2 = x_1^2 + (x_2 + 1)^2 - 1 \leqq 0\,.$$

From Theorem 3.5.3 it appears that the Arrow-Hurwicz-Uzawa first $C.Q.$ is the weakest among these $C.Q.$s, i.e. it is implied by the remaining ones.

Let us now prove that if the Arrow-Hurwicz-Uzawa $C.Q.$ is satisfied, Theorem 3.5.2 holds, with $u_0 > 0$.

Theorem 3.5.4. Let the assumptions of Theorem 3.5.2 be verified and let $C.Q.9$ be verified. Then in relation (3) we have $u_0 > 0$.

Proof. Assume the contrary, i.e. that in relation (3) $u_0 = 0$; then (3) is given by

$$\sum_{i \in I_P(x^0)} u_i \nabla g_i(x^0) + \sum_{i \in I_{NP}(x^0)} u_i \nabla g_i(x^0) = 0$$

$u_i \geqq 0\,, \quad \forall i \in I(x^0)\,, \quad u_i$ not all zero for $i \in I_{NP}(x^0)$.

By means of Theorem 19 of Section 2.4 (Motzkin transposition theorem) we obtain that the system

$$\begin{cases} y \nabla g_i(x^0) \leqq 0\ , & i \in I_P(x^0) \\ y \nabla g_i(x^0) < 0\ , & i \in I_{NP}(x^0) \end{cases}$$

has no solution, in contrast with the assumed validity of $C.Q.9$. □

Thus if any $C.Q.$ is verified, it is possible to rewrite relations (3)–(5), dividing u_0 and each component of u by u_0. Denoting by λ the m-vector $[u_1/u_0, ..., u_m/u_0]$, the following result is therefore apparent, considered here as the *third approach* for obtaining necessary optimality conditions for problem (P).

Theorem 3.5.5 (Kuhn-Tucker necessary optimality conditions). Let $x^0 \in \text{int}(X)$ be a solution of problem (P) or simply a point of local minimum of f on S; let f and g be differentiable at x^0. Let a constraint qualification be verified; then there exists a vector $\lambda \in I\!R^m$, such that

$$\nabla f(x^0) + \lambda \nabla g(x^0) = 0 \tag{6}$$

$$\lambda \cdot g(x^0) = 0 \tag{7}$$

$$\lambda \geqq 0\ . \tag{8}$$

The nonnegative numbers λ_i are usually called *Kuhn-Tucker multipliers* for problem (P), and again relations (7) are the slack-complementarity conditions. Relations (6)–(7)–(8) can also be written as:

$$\nabla f(x^0) + \sum_{i \in I(x^0)} \lambda_i \nabla g_i(x^0) = 0 \tag{9}$$

$$\lambda_i \geqq 0 \; , \qquad i \in I(x^0) \; . \tag{10}$$

Conditions (6)–(7)–(8) are also called *quasi-saddle-point conditions* in Arrow, Hurwicz and Uzawa (1961); if we denote by $\psi(x, \lambda) = f(x) + \lambda g(x)$ the Lagrangian function for problem (P), then (6)–(7)–(8) can be written in the form:

$$\nabla_x \psi(x, \lambda) = 0 \; ; \qquad \lambda \cdot \nabla_\lambda \psi(x, \lambda) = 0 \; ; \qquad \lambda \geqq 0 \; ,$$

also called *local saddle point conditions.* Note, again, that if the constraints g_i, $i \in I(x^0)$, are all pseudoconcave at x^0, then $C.Q.9$ is trivially satisfied; therefore if, e.g., the constraints g_i, $i = 1, ..., m$, are all *linear,* $C.Q.9$ is automatically satisfied. In other words, programming problems with all linear constraints need no constraint qualification. From a "historic" point of view, we note that this result was previously proved in an early paper on linear programming theory: see Goldmann and Tucker (1956).

Regarding the geometric interpretation of the Kuhn-Tucker conditions, note that any vector of the form $\sum_{i \in I(x^0)} \lambda_i \nabla g_i(x^0)$, where $\lambda_i \geqq 0$ for all $i \in I(x^0)$, belongs to the cone generated by the gradients of the effective constraints. Thus conditions (9)–(10) can be geometrically interpreted as: the vector $-\nabla f(x^0)$ belongs to the cone generated by the gradients of the effective constraints at the optimal point x^0. Finally we note that a point $x^0 \in S$ satisfies the Kuhn-Tucker conditions (6)–(8) *if and only if* the set

$$\Omega(x^0) = \{y \mid y \nabla g_i(x^0) \leqq 0, \; i \in I(x^0) \, ; \, y \nabla f(x^0) < 0\}$$

is empty ($\Omega(x^0) = \emptyset$). This follows easily from applying the Farkas-Minkowski theorem to relations (6)–(8).

3.6. Necessary Optimality Conditions for Problems (P_e) and (P); The Case of a Set Constraint

Till now we have assumed that in problems (P_e) and (P) the optimal point x^0 is interior to $X \subset D \subset I\!R^n$. If this assumption is relaxed we cannot in general obtain the Fritz John conditions; let us consider, with reference to (P_e), the following example, taken from Bazaraa and Shetty (1976).

The problem is to minimize $f(x, y) = -x + y$, subject to $h(x, y) = (x-1)^2 + y^2 - 1 = 0$, where x, y must also belong to the set

$$X = \{(x, y) \mid |x| + |y| \leqq 1\} .$$

It is clear (see also Figure 11) that the optimal point is

$$z^0 = \left(\frac{\sqrt{2}-1}{\sqrt{2}}, \frac{-1}{\sqrt{2}}\right) .$$

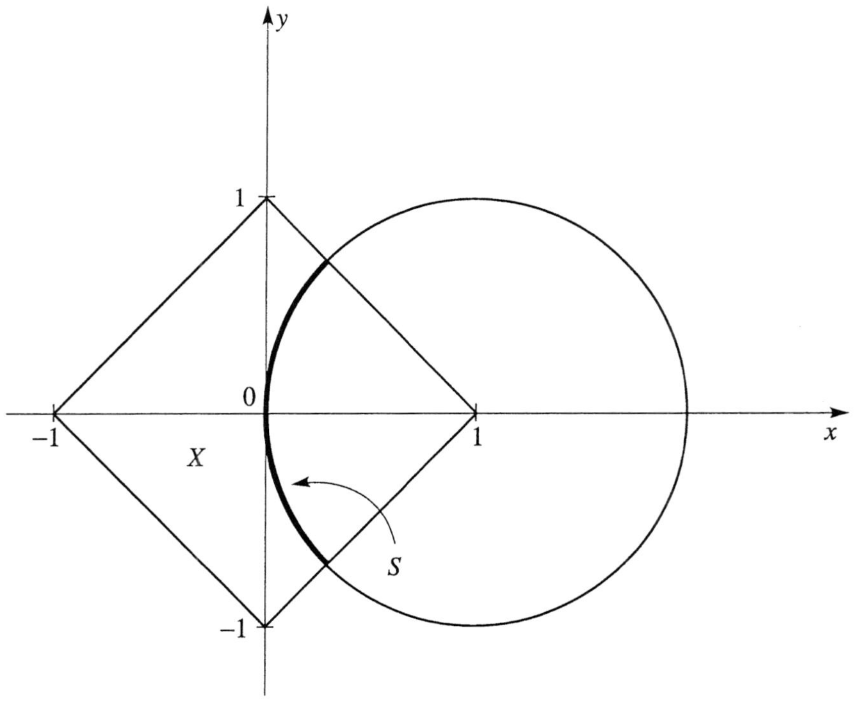

Figure 11.

Note that $z^0 \notin \text{int}(X)$. Now $\nabla f(x_0, y_0) = (-1, 1)$ and $\nabla h(x_0, y_0) = (-\sqrt{2}, -\sqrt{2})$. The related Fritz John conditions are therefore

$$u_0 \nabla f(x_0, y_0) + u \nabla h(x_0, y_0) = 0 ,$$

i.e.

$$u_0(-1,1) + u \cdot (-\sqrt{2}, -\sqrt{2}) = (0,0) \ .$$

These two equations admit only the solution $u_0 = u = 0$ and therefore the Fritz John necessary optimality conditions do not hold at this point.

In this section we shall modify the Fritz John conditions for (P), to also take care of the case where it is not necessarily $x^0 \in \mathrm{int}(X)$, i.e. X is a so-called *set constraint* (not necessarily functionally specified). The same considerations for problem (P_1) will be made in Section 3.8. Moreover, we shall consider a constraint qualification, which assures the validity of the Kuhn-Tucker conditions, also for this case. We recall that if S is a nonempty set in $\mathbb{R}^n$ and $x^0 \in \bar{S}$, then $T(S, x^0)$ is the *Bouligand tangent cone to S at x^0* and $P(S, x^0)$ is the *pseudotangent cone to S at x^0* (see Section 3.4). $T^*(S, x^0)$ and $P^*(S, x^0)$ are the respective *polar cones.*

For problem (P_0) the following result, due to Varaiya (1967), is of fundamental importance, especially for its subsequent developments, and is a sharper version of Theorem 3.2.1.

Theorem 3.6.1. Let x^0 be a solution of problem (P_0) or a point of local minimum of f on X and let f be differentiable at x^0. Then we have $-\nabla f(x^0) \in T^*(X, x^0)$, i.e. $y \nabla f(x^0) \geqq 0$, $\forall y \in T(X, x^0)$.

Proof. Let $y \in T(X, x^0)$; then there exist a sequence $\{x^k\} \in X$ converging to x^0 and a nonnegative sequence $\{\lambda_k\} \in \mathbb{R}$ such that $y = \lim_{k \to +\infty} \lambda_k (x^k - x^0)$. Since f is differentiable at x^0, we have for each k:

$$f(x^k) - f(x^0) = (x^k - x^0)\, \nabla f(x^0) + o(\|x^k - x^0\|) \ ,$$

which implies

$$\lambda_k (f(x^k) - f(x^0)) = \lambda_k (x^k - x^0)\, \nabla f(x^0) + \lambda_k \cdot o(\|x^k - x^0\|) \ .$$

Letting $k \to \infty$ in the second member of the previous equality, we get $y \cdot \nabla f(x^0)$; thus also $\lambda_k (f(x^k) - f(x^0))$ has a finite limit, which must be nonnegative, being $\lambda_k \geqq 0$. Consequently

$$y \nabla f(x^0) \geqq 0 \ , \qquad \forall y \in T(X, x^0) \ ,$$

i.e.

$-\nabla f(x^0) \in T^*(X, x^0)$. □

Theorem 3.6.1 is interesting in and of itself, as it provides a necessary condition for a differentiable function to attain a local minimum over any set $X \subset \mathbb{R}^n$.

Note, moreover, that Theorem 3.6.1 provides a sharper result than Theorem 3.2.1. Guignard (1969) proved the following version of the above theorem (for differentiable functions defined on Banach spaces): A necessary optimality condition for problem (P_0) is $-\nabla f(x^0) \in P^*(X, x^0)$. As, for any set S, we have (see Section 2.3), $S^* = (\overline{\text{conv}(S)})^*$, therefore $T^*(X, x^0) = P^*(X, x^0)$; so the formulation of Guignard is equivalent to the one given in Theorem 3.6.1. Note again that if it is $x^0 \in \text{int}(X)$, then $T(X, x^0) = \mathbb{R}^n$ and $T^*(X, x^0) = \{0\}$: therefore Theorem 3.6.1 recovers also the classical necessary conditions for a free extremum.

Let us now apply Theorem 3.6.1 to the study of necessary optimality conditions for problem (P) with a set constraint $X \subset \mathbb{R}^n$ (similar arguments hold also for problem (P_e); in Section 3.8 we shall study a nonlinear programming problem with both equality and inequality constraints and a set constraint).

Definition 3.6.1. In problem (P) let $x^0 \in S$; the set

$$C(x^0) = \{y \in \mathbb{R}^n \mid y\nabla g_i(x^0) \leqq 0, \ \forall i \in I(x^0)\}$$

is called *linearizing cone* at x^0 for problem (P).

The linearizing cone (called in Arrow-Hurwicz and Uzawa (1961) *cone of locally constrained directions*) is a nonempty closed convex polyhedral cone determined by the active constraints of problem (P). The following lemma was proved by Abadie (1967).

Lemma 3.6.1. In problem (P) let $x^0 \in S$; then it results $P(S, x^0) \subset C(x^0)$, i.e. $C^*(x^0) \subset T^*(S, x^0)$.

Proof. It is sufficient to prove that $T(S,x^0) \subset C(x^0)$, as, being that $C(x^0)$ is a closed convex cone, then $P(S,x^0) = \mathrm{cl}(\mathrm{conv}(T(S,x^0))) \subset C(x^0)$. Let $y \in T(S,x^0)$. Then there exist a sequence $\{x^k\} \in S$ converging to x^0 and a nonnegative sequence $\{\lambda_k\} \in I\!R$, such that $y = \lim_{k\to+\infty} \lambda_k \cdot (x^k - x^0)$. We have, for $i \in I(x^0)$,

$$\lambda_k g_i(x^k) = \lambda_k (x^k - x^0) \cdot \nabla g_i(x^0) + \lambda_k\, o\|x^k - x^0\| \,.$$

If k is large enough and if $y\nabla g_i(x^0) > 0$ for some $i \in I(x^0)$, then the right-hand side of the last expression is positive, so that $g_i(x^k) > 0$, which is in contradiction to $x^k \in S$. □

As a counterexample to the converse of Lemma 3.6.1, consider the following system in $I\!R^2$:

$$g_1(x) = x_2 - x_1^3 \leqq 0$$

$$g_2(x) = -x_2 \leqq 0$$

and take $x^0 = (0,0)$. The cone of tangents is the half-line $x_2 = 0$, $x_1 \geqq 0$, while the linearizing cone is the whole line $x_2 = 0$. It is worthwhile to emphasize that the cone of tangents is a geometrical concept, while the linearizing cone depends only upon the analytical description of the feasible set S. For example, if we add to the above two constraints the third equality

$$g_3(x) = -x_1 - x_2 \leqq 0 \,,$$

the feasible set S remains the same, as well as the cone of tangents at $x^0 = (0,0)$, but now the linearizing cone coincides with the tangent cone.

Definition 3.6.2. The pair (X, g) of problem (P) is said to be *Gould-Tolle regular* at $x^0 \in S$ if and only if for *every* objective function f, differentiable at x^0 and having at x^0 a constrained local minimum, the Kuhn-Tucker conditions (6)–(7)–(8) hold.

The following is a basic result due to Gould and Tolle (1971).

Theorem 3.6.2. In problem (P) the pair (X, g) is Gould-Tolle regular at $x^0 \in S$ if and only if $T^*(S, x^0) \subset C^*(x^0)$.

For the proof of this theorem we need a further lemma.

Lemma 3.6.2. The Kuhn-Tucker conditions (5)–(6)–(7) hold at $x^0 \in S$ if and only if $-\nabla f(x^0) \in C^*(x^0)$.

Proof. The result is a direct consequence of the Farkas-Minkowski theorem of the alternative (Section 2.4, Theorem 2.4.1): the system

$$-\nabla f(x^0) = \sum_{i \in I(x^0)} \lambda_i \nabla g_i(x^0) \; ; \qquad \lambda_i \geqq 0 \, , \quad \forall i \in I(x^0)$$

has solution if and only if $-\nabla f(x^0) \cdot y \leqq 0$ whenever $y \cdot \nabla g_i(x^0) \leqq 0$, for $i \in I(x^0)$, i.e. whenever $y \in C(x^0)$. Therefore the Kuhn-Tucker conditions are equivalent to $-\nabla f(x^0) \in C^*(x^0)$. □

We now turn to proving Theorem 3.6.2. Let $T^*(S, x^0) \subset C^*(x^0)$ and f be any objective function, differentiable at $x^0 \in S$ for problem (P), and having at x^0 a constrained local minimum. By Theorem 3.6.1 $-\nabla f(x^0) \in T^*(S, s^0)$; being that $T^*(S, x^0) \subset C^*(x^0)$, by Lemma 3.6.1 we have $T^*(S, x^0) = C^*(x^0)$ and by Lemma 3.6.2, the Kuhn-Tucker conditions (6)–(7)–(8) of Section 3.5 hold and hence the pair (X, g) is Gould-Tolle regular at x^0. It remains to show that if (X, g) is Gould-Tolle regular at x^0, then $T^*(S, x^0) \subset C^*(x^0)$. In Gould and Tolle (1971) it is in effect proved that if (X, g) is Gould-Tolle regular at x^0, then for every $y \in T^*(S, x^0)$, there exists an objective function f, which is differentiable at x^0, having at x^0 a constrained local minimum and for which $-\nabla f(x^0) = y$. The Gould-Tolle regularity assumption, together with Lemma 3.6.2, then yields $y \in C^*(x^0)$ and hence the result $T^*(S, x^0) \subset C^*(x^0)$. □

See also Bazaraa and Shetty (1976) for another proof of the existence of an objective function f for which $-\nabla f(x^0) = y$, for any $y \in T^*(S, x^0)$ and such that g has a minimum over S at x^0. Both proofs (of Gould and

Tolle and Bazaraa and Shetty) are rather long and intricate, so we prefer to give here only the conceptual aspects in proving Theorem 3.6.2.

The condition $T^*(S,x^0) \subset C^*(x^0)$, i.e. $C(x^0) \subset P(S,x^0)$ or $C(x^0) \subset T^{**}(S,x^0)$, represents a *constraint qualification* (C.Q.) which enjoys the best situation with respect to problem (P), also for the case of a set constraint. This condition was introduced by Guignard (1960), but see also Canon, Cullum and Polak (1970) and Evans (1970). We shall refer to condition $T^*(S,x^0) \subset C^*(x^0)$ as the *Guignard-Gould-Tolle C.Q.* The concept of regularity of the pair (X,g) was also analyzed by Arrow-Hurwicz and Uzawa (1961) but Theorem 3.6.2 was proved by Gould and Tolle (1971); the proof of Arrow-Hurwicz and Uzawa considered only the case of a convex feasible set S; moreover, these authors worked with another constraint qualification. Even if the Guignard-Gould-Tolle C.Q. is the most general C.Q. for the feasible set of problem (P), obviously it is possible to have problems for which $C^*(x^0)$ is a proper subset of $T^*(S,x^0)$ and therefore the Kuhn-Tucker conditions do not hold at x^0: we have already produced a counterexample to the converse of Lemma 3.6.1. For these cases the following considerations and results may be useful (see also Gould and Tolle (1972)): in particular, it will be shown that for *any* problem (P) a nontrivial optimality criterion can be found which is valid for all objective (differentiable) functions with a local constrained minimum at $x^0 \in S$ and *without* imposing the validity of any C.Q.

The following lemma will be utilized.

Lemma 3.6.3. Let us consider the nonempty sets A, H and K, with $0 \in A$, $0 \in H$ and K a convex cone and let $A \cup H = K$. Then $A + H = K$.

Proof. The result follows directly from the following relations:

$$K = A \cup H \subset A + H \subset K + K = K \, . \qquad \square$$

To employ this result, observe that from $C^*(x^0) \subset T^*(S,x^0)$ we have $T^*(S,x^0) \backslash C^*(x^0) \subset T^*(S,x^0)$. The set $T^*(S,x^0) \backslash C^*(x^0)$ does not contain the origin; let us therefore consider the cone $T^*(S,x^0) \backslash C^*(x^0) \cup \{0\}$. Since $T^*(S,x^0)$ is convex and since $T^*(S,x^0) = C^*(x^0) \cup T^*(S,x^0) \backslash C^*(x^0) \cup \{0\}$, it follows then from Lemma 3.6.3 that

$$T^*(S,x^0) = C^*(x^0) + [T^*(S,x^0) \backslash C^*(x^0) \cup \{0\}] \ . \tag{11}$$

If we denote by $B(x^0)$ the *cone of gradients*, i.e. the closed convex cone defined by

$$B(x^0) = \left\{ z \in I\!R^n \mid z = \sum_{i \in I(x^0)} \lambda_i \nabla g_i(x^0), \ \lambda_i \geqq 0 \right\} ,$$

from the previous results, we have

$$B(x^0) = C^*(x^0) \ .$$

Thus relation (11) becomes

$$T^*(S,x^0) = B(x^0) + [T^*(S,x^0) \backslash C^*(x^0) \cup \{0\}] \ .$$

From Theorem 3.6.1 and the above results we have therefore the following criterion, due to Gould and Tolle (1972):

Theorem 3.6.3. If x^0 is a local solution of (P), where f and g_i are differentiable at x^0, then there exist scalars $\lambda_i \geqq 0$, $i \in I(x^0)$, such that

$$-\left[\nabla f(x^0) + \sum_{i \in I(x^0)} \lambda_i \nabla g_i(x^0) \right] \in T^*(S,x^0) \backslash C^*(x^0) \cup \{0\} \ . \tag{12}$$

Let us note that in the special case where the Guignard-Gould-Tolle constraint qualification

$$T^*(S,x^0) = C^*(x^0) \tag{13}$$

is satisfied, then $T^*(S,x^0) \backslash C^*(x^0)$ is empty and (12) reduces to the classical Kuhn-Tucker conditions. However, if (13) is not satisfied for a given problem (P), Theorem 3.6.3 guarantees the existence of nonnegative multipliers such that

$$-\left[\nabla f(x^0) + \sum_{i \in I(x^0)} \lambda_i \nabla g_i(x^0) \right] \in G \ ,$$

where G is a suitable subcone of $T^*(S,x^0)$. In particular G coincides with $T^*(S,x^0) \backslash C^*(x^0) \cup \{0\}$ if no further assumption is made, whereas G may be a proper subcone of this last set, when suitable assumptions are made. Finally G shrinks to the singleton $\{0\}$ when (13) holds.

In general it is not allowed to deduce from the equality

$$T^*(S,x^0)\backslash C^*(x^0) \cup \{0\} = G$$

the other equality

$$T^*(S,x^0) = C^*(x^0) + G$$

even if, in certain cases, numerous cones $G \subset T^*(S,x^0)$ will satisfy the relation

$$\begin{aligned} T^*(S,x^0) &= C^*(x^0) + [T^*(S,x^0)\backslash C^*(x^0) \cup \{0\}] = \\ &= C^*(x^0) + G \ . \end{aligned} \tag{14}$$

Indeed, consider the following example, taken from Gould and Tolle (1972). Let S be given by the constraints

$$g_1(x_1,x_2) = x_2 - x_1^3 \leqq 0$$

$$g_2(x_1,x_2) = -x_2 \leqq 0$$

for which we have already seen that at $x^0 = (0,0)$ the constraint qualification (13) does not hold, and consequently also the Kuhn-Tucker conditions are not satisfied for any objective function f having at x^0 a local minimum. Here (see Figure 12) the cone $C(x^0)$ is the x_1 axis, $C^*(x^0)$ is the x_2 axis; $T(S,x^0)$ is the nonnegative x_1 axis and $T^*(S,x^0)\backslash C^*(x^0) \cup \{0\}$ is the set $\{(x_1,x_2) \mid x_1 < 0\} \cup \{(0,0)\}$.

In this case, if G is taken to be any half ray leading to the left from the origin (i.e. into $T^*(S,x^0)\backslash C^*(x^0)$), then (14) will hold. In particular G can be taken as the nonpositive part of the x_1 axis to obtain the optimality conditions: there exist nonnegative multipliers λ_1^0, λ_2^0 such that

$$-\frac{\partial f}{\partial x_1}(x^0) - \lambda_1 \frac{\partial g_1}{\partial x_1}(x^0) - \lambda_2 \frac{\partial g_2}{\partial x_2}(x^0) \leqq 0$$

$$-\frac{\partial f}{\partial x_1}(x^0) - \lambda_1 \frac{\partial g_1}{\partial x_1}(x^0) - \lambda_2 \frac{\partial g_2}{\partial x_2}(x^0) = 0 \ ,$$

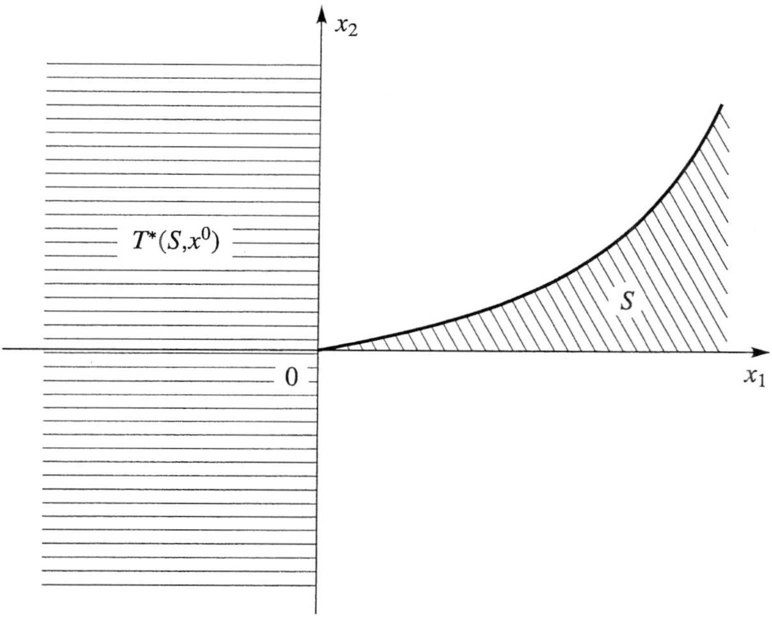

Figure 12.

where f is a (differentiable) objective function having at $x^0 = (0,0)$ a local minimum.

It is important to note again that, when constraint qualification (13) is not satisfied, Theorem 3.6.3 effectively gives a sort of generalization of the Kuhn-Tucker conditions. The fact that in (12) the multiplier of $\nabla f(x^0)$ is just equal to one, is not in contrast with what was specified on the classical Fritz John conditions, as the cone $T^*(S,x^0) \backslash C^*(x^0) \cup \{0\}$ does not necessarily shrink to $\{0\}$ when it is $x^0 \in \text{int}(X)$. This remark induces finding necessary optimality conditions for problem (P), expressed in terms of the Lagrangian function and the cone $T^*(X,x^0)$. A first result of the Kuhn-Tucker type is obtained by Gould and Tolle (1972), under a suitable constraint qualification.

<u>Theorem 3.6.4</u>. If x^0 is a local solution of (P), where f and g_i are differentiable at x^0, and if the following condition holds

$$T^*(S, x^0) = C^*(x^0) + T^*(X, x^0) \ , \tag{15}$$

then there exist scalars $\lambda_i \geqq 0$, $i \in I(x^0)$, such that

$$-\Big[\nabla f(x^0) + \sum_{i \in I(x^0)} \lambda_i \nabla g_i(x^0)\Big] \in T^*(X, x^0) \ . \tag{16}$$

In the same paper the authors prove that condition (15) is both necessary and sufficient for the validity of (16) with respect to *any* objective function f differentiable at x^0 and with a local constrained minimum at x^0. In the same spirit of Theorem 3.6.4 we shall now prove a Fritz John type result for problem (P). We need the following lemma, due to Bazaraa and Goode (1972a).

Lemma 3.6.4. Suppose that Y is a convex set in $\mathbb{R}^n$, K a convex cone in $\mathbb{R}^m$ and A is an (m, n) matrix. If the system $Ax \in K$ has no solution $x \in Y$, then there exists a nonzero vector $p \in K^*$ such that $pAx \geqq 0$, $\forall x \in Y$.

Proof. The set $A(Y) = \{Ax \mid x \in Y\}$ is convex and by hypothesis the set $A(Y) \cap K$ is empty. Then there exists a nonzero vector p such that $pAx \geqq pk$ for each $x \in Y$ and $k \in \bar{K}$. Since $0 \in \bar{K}$, then $pAx \geqq 0$ for each $x \in Y$. It is also clear that $pk \leqq 0$ for every $k \in K$ since otherwise, for any k_0, with $pk_0 > 0$, we would obtain $pAx \geqq \lambda pk_0$ for each $\lambda > 0$, where $pk_0 > 0$, which is impossible. This completes the proof. □

Theorem 3.6.5. Let x^0 be a local solution of problem (P), where f and g_i, $i \in I(x^0)$, are differentiable at x^0. Then for every convex subcone $T_1(X, x^0)$ of $T(X, x^0)$ there exist multipliers $\lambda_0 \geqq 0$, $\lambda_i \geqq 0$, $i \in I(x^0)$, not all zero, such that

$$-\Big[\lambda_0 \nabla f(x^0) + \sum_{i \in I(x^0)} \lambda_i \nabla g_i(x^0)\Big] \in T_1^*(X, x^0) \ .$$

Proof. Consider, for each $x \in X$, the vector-valued function $\alpha(x) = (f(x) - f(x^0), g_i(x))$, $i \in I(x^0)$. As no vector $x \in X$ exists such that

$$\alpha(x) \in \overset{o}{\mathbb{R}}_- \times \overset{o}{\mathbb{R}}_- \times \ldots \times \overset{o}{\mathbb{R}}_- = K \quad \text{and} \quad g_i(x) \leqq 0 \,, \quad \forall i \in J(x^0)$$

(K is obviously an open convex cone), we show that no vector $x \in T(X, x^0)$ exists such that $\nabla\alpha(x^0)\,x \in K$. Ab absurdo, suppose the existence of such a vector x, with $x = \lim_{k\to+\infty} \lambda_k(x^k - x^0)$, with $\{x^k\} \in X$, $x^k \to x^0$, $\{\lambda_k\} \subset I\!R_+$. We have then

$$\lambda_k(\alpha(x^k) - \alpha(x^0)) = \lambda_k \nabla\alpha(x^0)(x^k - x^0) + \lambda_k \cdot o(\|x^k - x^0\|) ,$$

and letting $k \to +\infty$ (being $\alpha(x^0) = 0$),

$$\lim_{k\to+\infty} \lambda_k \alpha(x^k) = \nabla\alpha(x^0)\,x .$$

But being that K is an open cone, we have $\lambda_k\alpha(x^k) \in K$ for k sufficiently large and the absurd conclusion follows: $\alpha(x^k) \in K$, with $x^k \in X$ and $g_i(x^k) < 0$ for $i \in J(x^0)$, thanks to the continuity. Then, no vector $x \in T(X, x^0)$ will satisfy the condition $\nabla\alpha(x^0)\,x \in K$ and the same is true, a fortiori, for $x \in T_1(X, x^0)$.
As $T_1(X, x^0)$ is a convex cone, there exists, thanks to Lemma 3.6.4, a nonzero vector $\lambda \in K^*$, such that $\lambda\nabla\alpha(x^0)\,x \geqq 0$, $\forall x \in T_1(X, x^0)$. Therefore $-\lambda\nabla\alpha(x^0) \in T_1^*(X, x^0)$, from which it follows

$$-\Big[\lambda_0\nabla f(x^0) + \sum_{i\in I(x^0)} \lambda_i \nabla g_i(x^0)\Big] \in T_1^*(X, x^0) ,$$

with $\lambda_0 \geqq 0$, $\lambda_i \geqq 0$ $(i \in I(x^0))$, and $(\lambda_0, \lambda_i) \neq 0$. □

Obviously, in the case where $T(X, x^0)$ is a convex cone, Theorem 3.6.5 gives a sharper result, being $T_1 = T(X, x^0)$. $T(X, x^0)$ is convex if, e.g., X is a convex set (see Section 3.4) or even if X is *star-shaped at* x^0 (equivalently: *convex at* x^0), i.e.

$$tx + (1-t)\,x^0 \in X , \qquad \forall x \in X , \ \ \forall t \in [0, 1] ;$$

see Bazaraa and Shetty (1976).

Penot (1991) has obtained more general conditions on X assuring the convexity of $T(X, x^0)$; following the terminology of this author, in this case the set X is called *tangentially regular at* x^0. If $T(X, x^0)$ is not a convex cone, there exist some convex subcones of the same, that can be chosen to represent $T_1(X, x^0)$ in Theorem 3.6.5. One of these subcones is the Clarke tangent cone, but obviously if we can choose the largest convex subcone of $T(X, x^0)$, Theorem 3.6.5 will be sharper. A convex subcone of $T(X, x^0)$, larger than the Clarke tangent cone, is the *Michel-Penot prototangent cone* (see Section 3.4). More recently Treiman (1991) has shown that there exist infinite convex cones lying between the Clarke tangent cone and the Michel-Penot prototangent cone.

We conclude this section with the following result in which generalized Kuhn-Tucker conditions of the type of (16) are obtained from Theorem 3.6.5, imposing a constraint qualification.

Theorem 3.6.6. Let x^0 be a local solution of problem (P), where f and g_i, $i \in I(x^0)$, are differentiable at x^0 and let the following constraint qualification be fulfilled:

$$\exists \bar{y} \in T_1(X, x^0) : \bar{y} \nabla g_i(x^0) < 0 \ , \qquad \forall i \in I(x^0) \ ,$$

where $T_1(X, x^0)$ is a convex subcone of $T(X, x^0)$. Then there exist multipliers $\lambda_i \geqq 0$, $i \in I(x^0)$, such that

$$-\Big(\nabla f(x^0) + \sum_{i \in I(x^0)} \lambda_i \nabla g_i(x^0)\Big) \in T_1^*(X, x^0) \ .$$

Proof. Assuming in Theorem 3.6.5 that $\lambda_0 = 0$ we should get

$$-\sum_{i \in I(x^0)} \lambda_i \nabla g_i(x^0) \in T_1^*(X, x^0)$$

i.e.

$$\sum_{i \in I(x^0)} \lambda_i y \nabla g_i(x^0) \geqq 0 \ , \qquad \forall y \in T_1(X, x^0)$$

and taking $y = \bar{y}$ we obtain a contradiction with the assumed constraint qualification. □

3.7. Again on Constraint Qualifications

In the previous section we mentioned the Guignard-Gould-Tolle constraint qualification, i.e. $C^*(x^0) = T^*(S, x^0)$ and in Theorem 3.6.4 a generalization of the same was imposed, i.e.

$$T^*(S, x^0) = C^*(x^0) + T^*(X, x^0) .$$

These constraint qualifications are the weakest possible, in the sense that they are necessary and sufficient for the Gould-Tolle regularity of the related necessary optimality condition.

In Section 3.5 we have introduced other constraint qualifications for problem (P), in the case $x^0 \in \mathrm{int}(X)$. All the above constraint qualifications guarantee the positivity of the multiplier of the gradient of the objective function in the various Fritz John conditions, i.e. their validity guarantees that if $x^0 \in S$ is a local solution of (P), then the Kuhn-Tucker conditions hold.

Besides the constraint qualifications already introduced, there are others, with varying degrees of generality. In the following conditions $x^0 \in X$:

1) *Zangwill constraint qualification* (Zangwill 1969)). It is expressed as:

$$C(x^0) \subset \overline{Z(S, x^0)} ,$$

where $Z(S, x^0)$ is the *cone of feasible directions to* S *at* x^0.

2) *Kuhn-Tucker constraint qualification* (Kuhn and Tucker (1951)). It is expressed as:

$$C(x^0) \subset A(S, x^0) ,$$

where $A(S, x^0)$ is the *cone of attainable directions to* S *at* x^0.

3) *Arrow-Hurwicz-Uzawa second constraint qualification* (Arrow, Hurwicz and Uzawa (1961)). It is expressed as

$$C(x^0) \subset \mathrm{cl}(\mathrm{conv}(A(S, x^0)))$$

or equivalently as

$$C(x^0) \subset A^{**}(S, x^0) \ .$$

The cone $A^{**}(S, x^0)$ is also called *cone of weakly attainable directions.*

4) *Abadie constraint qualification* (Abadie (1967)). It is expressed as

$$C(x^0) \subset T(S, x^0) \ .$$

5) We recall the already introduced *Guignard-Gould-Tolle constraint qualification* (Guignard (1969); Gould and Tolle (1971)). It is expressed as

$$C(x^0) \subset P(S, x^0) \ , \qquad \text{i.e.}$$

$$C(x^0) \subset \mathrm{cl}(\mathrm{conv}(T(S, x^0)))$$

or equivalently

$$C(x^0) \subset T^{**}(S, x^0)$$

or equivalently

$$T^*(S, x^0) \subset C^*(x^0) \ .$$

Before going on, we note that in general we find no implications among $A^{**}(S, x^0)$ and $T(S, x^0)$ the following examples, due to Abadie (1967) and Evans (1970), show the non interdependence of the Abadie C.Q. and the Arrow-Hurwicz-Uzawa second C.Q.

Example 3.7.1. Let $X = I\!R^2$, $g_1(x_1, x_2) = x_1 x_2$; $g_2(x_1, x_2) = -x_1$; $g_3(x_1, x_2) = -x_2$. In other words

$$S = \{(x_1, x_2) \mid x_1 = 0, \ x_2 \geqq 0\} \cup \{(x_1, x_2) \mid x_2 = 0, \ x_1 \geqq 0\} \ .$$

If we consider the point $x^0 = (0,0)$, we have $T(S,(0,0)) = S$ and $A(S,(0,0)) = S$. However,

$$A^{**}(S,(0,0)) = \{(x_1, x_2) \mid x_1 \geqq 0,\ x_2 \geqq 0\} .$$

Clearly $T(S,(0,0)) \subset A^{**}(S,(0,0))$, but they are not equal.
Conversely, consider the following functions

$$s(x_1) = \begin{cases} x_1^4 \sin \dfrac{1}{x_1} \ , & x_1 \neq 0 \\ 0 \ , & x_1 = 0 \ ; \end{cases}$$

$$c(x_1) = \begin{cases} x_1^4 \cos \dfrac{1}{x_1} \ , & x_1 \neq 0 \\ 0 \ , & x_1 = 0 \ . \end{cases}$$

The above functions are continuously differentiable; the functions and their derivatives vanish at $x_1 = 0$. Now consider the following constraint functions

$$g_1(x_1, x_2) = x_2 - x_1^2 - s(x_1)$$
$$g_2(x_1, x_2) = -x_2 + x_1^2 + c(x_1)$$
$$g_3(x_1, x_2) = x_1^2 - 1 \ .$$

Again we take $X = \mathbb{R}^2$; it can be verified that

$$A(S,(0,0)) = A^{**}(S,(0,0)) = \{(0,0)\} ,$$

whereas

$$T(S,(0,0)) = \{(x_1, x_2) \mid x_2 = 0\} .$$

So in this case $A^{**}(S,x^0) \subset T(S,x^0)$, but they are unequal.

<u>Theorem 3.7.1</u>. If $x^0 \in \text{int}(X)$, then the Arrow-Hurwicz-Uzawa first C.Q. implies the Zangwill C.Q.

<u>Proof</u>. If we denote by $C_p(x^0)$ the set

$$C_P(x^0) = \{y \in \mathbb{R}^n \mid y\nabla g_i(x^0) \leqq 0,\ \forall\, i \in I_P(x^0);$$
$$y\nabla g_i(x^0) < 0,\ \forall\, i \in I_{NP}(x^0)\}$$

where $I_P(x^0)$ and $I_{NP}(x^0)$ are respectively the sets of pseudoconcave and nonpseudoconcave active constraints at x^0, the Arrow-Hurwicz-Uzawa first constraint qualification, i.e. $C_P(x^0) \neq \emptyset$, can be equivalently expressed as $C(x^0) \subset \overline{C_P(x^0)}$. Indeed, if $C_P(x^0) \neq \emptyset$, then $\overline{C_P(x^0)} = C(x^0)$; on the other hand, since $C(x^0) \neq \emptyset$ (at least it contains the zero vector), if $C(x^0) \subset \overline{C_P(x^0)}$, then $\overline{C_P(x^0)} \neq \emptyset$ and hence $C_P(x^0) \neq \emptyset$. Now we show that $C_P(x^0) \subset Z(S, x^0)$ if $x^0 \in \text{int}(X)$. Let $y \in C_P(x^0)$, i.e. $y \cdot \nabla g_i(x^0) \leqq 0$ for $i \in I_P(x^0)$ and $y \cdot \nabla g_i(x^0) < 0$ for $i \in I_{NP}(x^0)$. Consider $x^0 + \lambda y$ for $\lambda > 0$. Since $x^0 \in \text{int}(X)$, then $x^0 + \lambda y \in X$ for λ sufficiently small. By continuity of g_i, $i \notin I(x^0)$, and since $g_i(x^0) < 0$ it is clear that $g_i(x^0 + \lambda y) < 0$ for all $i \notin I(x^0)$ and for $\lambda > 0$ sufficiently small.

For the pseudoconcave active constraints $\lambda(y \nabla g_i(x^0)) \leqq 0$ implies that $g_i(x^0 + \lambda y) \leqq g_i(x^0) = 0$ for all $i \in I_p(x^0)$ and $\lambda > 0$. Finally for $i \in I_{NP}(x^0)$, $\lambda(y \nabla g_i(x^0)) < 0$ implies that $g_i(x^0 + \lambda y) < 0$ for λ sufficiently small. In other words $x^0 + \lambda y \in S$ for $\lambda > 0$ and sufficiently small, i.e. $y \in Z(S, x^0)$ and the proof is complete. □

On the grounds of Theorem 3.5.3, on the inclusion relationships among cones $Z(S, x^0)$, $A(S, x^0)$, $T(S, x^0)$, $P(S, x^0)$ and Example 3.7.1, under the assumption that $x^0 \in \text{int}(X)$, the implications among the various constraint qualifications can be described by the following diagram (Figure 13).

If $x^0 \notin \text{int}(X)$, the previous diagram is not entirely valid, even if it is easy to verify that it still holds, starting from the Zangwill C.Q. (In Lemma 6.1.1 of Bazaraa and Shetty (1976) the assumption $x^0 \in \text{int}(X)$ is superfluous and actually is never used in the related proof.) In particular, if $x^0 \notin \text{int}(X)$, it is no more true that the A.H.U. first C.Q. implies Zangwill C.Q. In Giorgi and Guerraggio (1994) modifications of the Zangwill C.Q. and of the other constraint qualifications lying in the first half of the diagram are proposed in order to preserve the validity of the whole diagram in the case $x^0 \notin \text{int}(X)$. More precisely, e.g., the Zangwill C.Q. is modified as follows:

$$C(x^0) \subset \overline{Z_1(S, x^0)} ,$$

where

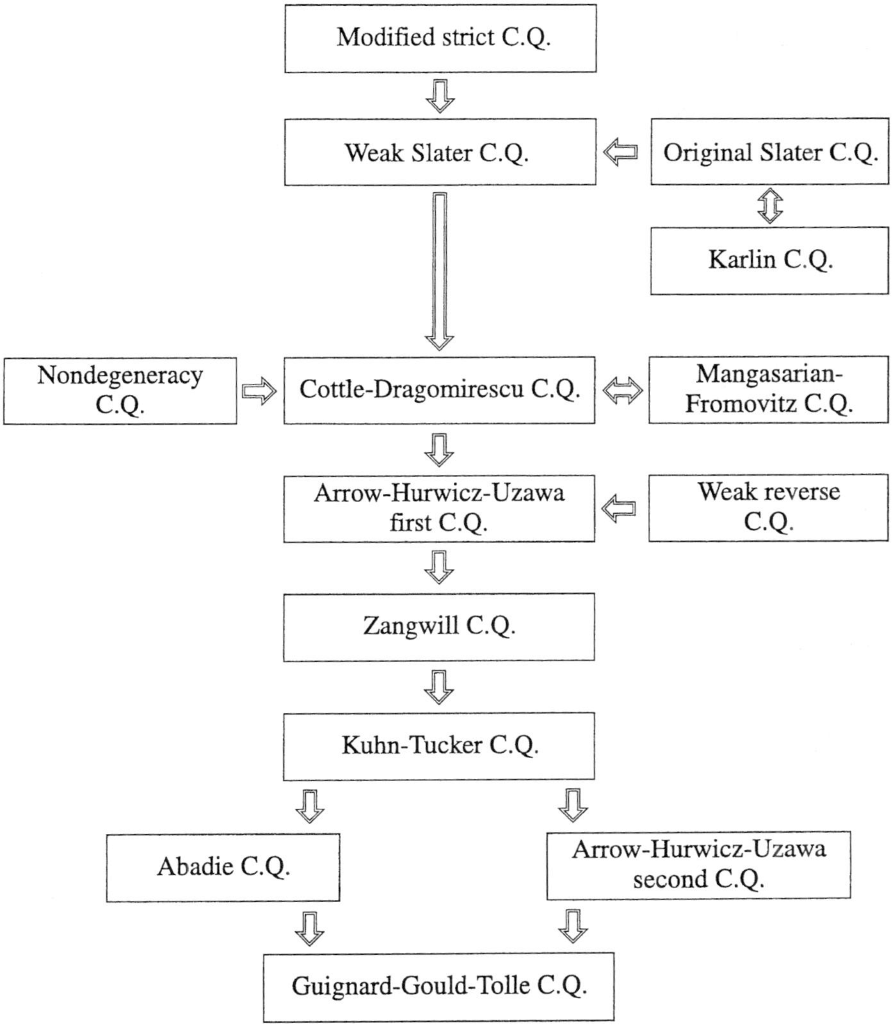

Figure 13.

$$Z_1(S,x^0) = \{x \mid x^0 + \frac{1}{n}x \in X,\ n \in I\!N,$$
$$\exists\, \delta > 0 : g_i(x^0 + \lambda x) \leqq 0,\ \forall i,\ \forall\, \lambda \in [0,\delta]\}\ .$$

In an analogous way the other constraint qualifications which in the diagram "precede" Zangwill C.Q. are modified.

If we define by F_0 the set of objective functions f having a local minimum at $x^0 \in S$ and differentiable at x^0, we already know that there are cases when the classical Kuhn-Tucker conditions (6)–(8) of Section 3.5 do not hold for every $f \in F_0$. We have also mentioned the problem, raised by Gould and Tolle (1972), of regularizing the constraints in every problem in the form of (P), in the sense of forcing the Guignard-Gould-Tolle C.Q. to hold by the addition of redundant constraints. This last problem was studied and solved by Agunwamba (1977).

Definition 3.7.1. For a given integer k, the function q_k is an *x^0-redundant constraint* if and only if $q_k : I\!R^n \to I\!R$, $q_k(x^0) = 0$, $q_k(x) \leqq 0$, $\forall x \in S$ and $\nabla q_k(x^0)$ exists.

Let q be a vector of x^0-redundant constraints for (P) and let us consider the set

$$Q(x^0) = \{y \in I\!R^n \mid y\nabla q_k(x^0) \leqq 0, \ \forall k\} ,$$

i.e. the linearizing cone for $q(x)$ at $x^0 \in S$.

The *Gould-Tolle problem* for (P) is described as follows. Suppose that at $x^0 \in S$ the Guignard-Gould-Tolle C.Q. does not hold. Find a function q such that

$$P(S, x^0) = C(x^0) \cup Q(x^0) .$$

Clearly this problem is one of regularizing the constraints of (P) with respect to the Guignard-Gould-Tolle C.Q. (Actually, in Gould and Tolle (1972) q is required to have a finite number of components.) Agunwamba (1977) not only discusses necessary and sufficient conditions for the existence of a finite set of x^0-redundant constraints which solve the Gould and Tolle problem, but also constructs a general finite solution when certain conditions are satisfied. His main results are given by the following theorem.

Theorem 3.7.2.

i) The Gould-Tolle problem for (P) always has a solution.

ii) A necessary and sufficient condition for the Gould-Tolle problem to have a finite solution is that there exists a subset B of $T^*(S, x^0)$ such that the set $H = B \cap (T^*(S, x^0) \backslash C^*(x^0))$ is a finite set, and every vector $y \in T^*(S, x^0) \backslash C^*(x^0)$ is a nonnegative combination of some vectors in B.

Proof. See Agunwamba (1977). □

The following corollary gives sufficient conditions for the existence of a finite solution for the above Gould-Tolle problem.

Corollary 3.7.1. The Gould-Tolle problem has a finite solution if:

i) $\mathrm{cl}(\mathrm{conv}(T^*(S, x^0) \backslash C^*(x^0)))$ is a finite cone (i.e. convex polyhedral); or if

ii) $T^*(S, x^0)$ is a finite cone; or if

iii) (P, x^0) is a finite cone.

3.8. Necessary Optimality Conditions for (P_1)

Let us now consider a more general nonlinear programming problem which has both equality and inequality constraints, i.e.

$$\min_{x \in S_1} f(x) , \qquad (P_1)$$

$$S_1 = \{x \mid x \in X, \ g_i(x) \leqq 0, \ i = 1, ..., m \, ; \ h_j(x) = 0, \\ j = 1, ..., r \ (< n)\} ,$$

where $X \subset \mathbb{R}^n$, g_i, $i = 1, ..., m$, are differentiable at least at $x^0 \in S_1$ and h_j, $j = 1, ..., r$, are continuously differentiable at least in a neighbourhood of $x^0 \in S_1$. Theorem 3.6.3 can immediately be fitted to (P_1), if this latter is rewritten as:

$$\text{Min} \, f(x) \,, \quad x \in X \,, \quad g(x) \leqq 0 \,, \quad h(x) \leqq 0 \,, \quad h(x) \geqq 0 \,.$$

Let us introduce the following sets:

$$D(x^0) = \{y \in \mathbb{R}^n \mid y \quad \nabla h_j(x^0) = 0, \ \forall j = 1, ..., r\} \,.$$

$D(x^0)$ can thus be viewed as the orthogonal complement of the subspace of $\mathbb{R}^n$ spanned by $\nabla h_j(x^0)$, $j = 1, ..., r$.

$$E(x^0) = C(x^0) \cap D(x^0) \,.$$

Taking Lemma 3.6.1 into account it is immediate to show that $E^*(x^0) \subset T^*(S_1, x^0)$. Indeed, from the quoted lemma, we get

$$T(S_1, x^0) \subset \{y \in \mathbb{R}^n \mid y\nabla g_i(x^0) \leqq 0, \ \forall i \in I(x^0) \,; \, y\nabla h_j(x^0) \leqq 0, \\ \forall j = 1, ..., r \,; \, y\nabla h_j(x^0) \geqq 0, \ \forall j = 1, ..., r\} = E(x^0)$$

and from this result we obtain the above assertion.

From the previous results the following theorem follows.

Theorem 3.8.1. If x^0 is a local solution of (P_1), then there exist scalars $\lambda_i \geqq 0$, $i \in I(x^0)$, $\mu_j \in \mathbb{R}$, $j = 1, ..., r$, such that

$$-\Big[\nabla f(x^0) + \sum_{i \in I(x^0)} \lambda_i \nabla g_i(x^0) + \sum_{j=1}^{r} \mu_j \nabla h_j(x^0)\Big] \in \tag{1}$$

$$\in T^*(S_1, x^0) \backslash E^*(x^0) \cup \{0\} \,.$$

It is obvious that when

$$T^*(S_1, x^0) = E^*(x^0) \,, \tag{2}$$

i.e. the Guignard-Gould-Tolle C.Q. holds for (P_1), then (1) becomes the classical Kuhn-Tucker condition for (P_1), i.e.

$$\nabla f(x^0) + \sum_{i \in I(x^0)} \lambda_i \nabla g_i(x^0) + \sum_{j=1}^{r} \mu_j \nabla h_j(x^0) = 0 ,$$

$$\lambda_i \geqq 0 , \quad i \in I(x^0) , \quad \mu_j \in I\!R , \quad j = 1, ..., r .$$

Or, taking the complementarity slackness conditions into account:

$$\nabla f(x^0) + \sum_{i=1}^{m} \lambda_i \nabla g_i(x^0) + \sum_{j=1}^{r} \mu_j \nabla h_j(x^0) = 0 , \tag{3}$$

$$\lambda_i g_i(x^0) = 0 , \qquad i = 1, ..., m , \tag{4}$$

$$\lambda_i \geqq 0 , \quad i = 1, ..., m ; \qquad \mu_j \in I\!R , \quad j = 1, ..., r . \tag{5}$$

Definition 3.6.2 can be extended to problem (P_1). We say that the triple (g, h, X) is *Gould-Tolle regular* at $x^0 \in S_1$ if for *every* objective function f with a local minimum at $x^0 \in S_1$ there exist vectors $\lambda \in I\!R^m$ and $\mu \in I\!R^n$ such that (3)–(5) hold. It can be proved (see Gould and Tolle (1971)) that the triple (g, h, X) is Gould-Tolle regular at $x^0 \in S_1$ if and only if condition (2) holds. We note that the device to transform an equality constraint into two inequality constraints of opposite sign becomes useless in obtaining the Fritz John conditions for (P_1). Indeed, if we consider the "transformed" problem: $\text{Min}\, f(x)$, subject to $x \in X$, $g(x) \leqq 0$, $h(x) \leqq 0$, $-h(x) \leqq 0$, it is clear that *any* feasible point x^0 solves the relation

$$u_0 \nabla f(x^0) + \sum_{i=1}^{m} u_i \nabla g_i(x^0) + \sum_{j=1}^{r} v_j \nabla h_j(x^0) +$$

$$+ \sum_{j=1}^{r} w_j \nabla(-h_j(x^0)) = 0 ,$$

where $u_0 = u_i = 0$ for each $i = 1, ..., m$; $v_j = w_j = 1$ for each $j = 1, ..., r$. Note that the complementarity slackness conditions hold:

$$\sum_{i=1}^{m} u_i g_i(x^0) + \sum_{j=1}^{r} v_j h_j(x^0) + \sum_{j=1}^{r} w_j h_j(x^0) = 0 .$$

This shows that each feasible point will trivially satisfy the above conditions, which then become useless. In order to fit Theorem 3.6.5 to problem

(P_1), we introduce the following lemmas (see Bazaraa and Goode (1972a) and Bazaraa and Shetty (1976)).

Lemma 3.8.1. Let C be an open convex cone in $\mathbb{R}^m$ and let $b \in \bar{C}$; define the (convex) cone $C_b = \{c - \lambda b : c \in C,\ \lambda \geqq 0\}$. If $a \in C_b$, then $b + \delta a + o(\delta) \in C$ for $\delta \to 0^+$, where $(\|o(\delta)\|/\delta) \to 0$ for $\delta \to 0^+$.

Proof. Let $a = c - \lambda b \in C_b$, where $c \in C$ and $\lambda \geqq 0$. Then

$$b + \delta a + o(\delta) = (1 - \lambda\delta)\, b + \delta\Big(C + \frac{o(\delta)}{\delta}\Big) .$$

Since C is open and $c \in C$ then for δ sufficiently small we have $c + o(\delta)/\delta \in C$. Also for δ sufficiently small $(1 - \lambda\delta) = \mu \geqq 0$. Hence for small enough $\delta > 0$, we have

$$b + \delta a + o(\delta) = (1 - \lambda\delta)\, b + \delta\Big(C + \frac{o(\delta)}{\delta}\Big) \in \bar{C} + C .$$

But $\bar{C} + C = C$ since C is open and hence $b + \delta a + o(\delta) \in C$ and the proof is complete. □

We recall the definition of the *cone of quasi-interior directions to X at $x^0 \in \bar{X}$* (see Section 3.4):

$$Q(X, x^0) = \{x \mid \exists\, N(x) \text{ such that, for each } \delta > 0,$$
$$\exists\, t \in (0, \delta),\ \forall\, y \in N(x) : x^0 + ty \in X\} .$$

$Q(X, x^0)$ is an open cone.

Lemma 3.8.2. Let $X \subset \mathbb{R}^n$ be an arbitrary set and let C be an open convex cone of $\mathbb{R}^m$; let α and h be two vector-valued functions defined on an open set containing X and taking values in $\mathbb{R}^m$ and $\mathbb{R}^r$, respectively. Suppose that α is differentiable at $x^0 \in X$ and h is continuously differentiable in a neighbourhood of $x^0 \in X$ with the Jacobian matrix $\nabla h(x^0)$ of full rank. If the system

$$\begin{cases} \alpha(x) \in C \\ h(x) = 0 \end{cases}$$

has no solution in X, then the system

$$\begin{cases} \nabla\alpha(x^0)\, y \in C_{\alpha(x^0)} \\ \nabla h(x^0)\, y = 0 \ , \end{cases}$$

where $x^0 \in X$ such that $\alpha(x^0) \in \bar{C}$ and $h(x^0) = 0$, has no solution in $Q(X, x^0)$.

Proof. Let us suppose, ab absurdo, that there exists $y \in Q(X, x^0)$ which solves the second system (the first system remaining impossible in X). From the definition of $Q(X, x^0)$, it follows that a neighbourhood $N(0)$ and a sequence $\{\lambda_k\} \to 0^+$ exist such that $x^0 + \lambda_k y + \lambda_k \cdot o(1) \in X$, for every element $o(1) \in N(0)$. Letting $x = (x^1, x^2)$ and $x^0 = (x^{01}, x^{02})$, by the implicit function theorem, from $h(x) = 0$ and $h(x^0) = 0$ it follows that there exists a neighbourhood of x^{01} where one (and only one) function $x^2 = F(x^1)$ is defined such that $h(x^1, F(x^1)) = 0$, $x^{02} = F(x^{01})$. So we have $\nabla_1 h(x^0) + \nabla_2 h(x^0)\, \nabla F(x^{01}) = 0$. By comparing this relation with the assumption $\nabla h(x^0)\, y = 0$, i.e. $\nabla_1 h(x^0)\, y^1 + \nabla_2 \nabla h(x^0)\, y^2 = 0$ (with obvious meaning of the notation) we obtain $y^2 = \nabla F(x^{01})\, y^1$. Now consider the point

$$(x^{01} + \lambda_k y^1, F(x^{01} + \lambda_k y^1)) = (x^{01} + \lambda_k y^1, x^{02} + \lambda_k y^2 + o_1(\lambda_k)) =$$
$$= (x^0 + \lambda_k y + o_2(\lambda_k)) \ ,$$

where $o_2(\lambda_k) = (0, o_1(\lambda_k))$. For k large enough this point belongs to X (being $y \in Q(X, x^0)$), and obviously satisfies the condition $h(x) = 0$; moreover, in force of Lemma 3.8.1, we have

$$\alpha(x^0 + \lambda_k y + o_2(\lambda_k)) =$$
$$= \alpha(x^0) + \lambda_k \cdot \nabla\alpha(x^0)\, y + o_3(\lambda_k) \in C \ .$$

This conclusion is absurd, and in contradiction with the assumed impossibility of the first system in X. □

Theorem 3.8.2. Let x^0 be a local solution of (P_1). Moreover, let $Q(X, x^0)$ be convex; then there exist scalars $\lambda_0 \geqq 0$, $\lambda_i \geqq 0$ $(i \in I(x^0))$, μ_j $(j = 1, 2, ..., r)$, with $(\lambda_0, \lambda, \mu)$ a nonzero vector, such that

$$-\Big[\lambda_0 \nabla f(x^0) + \sum_{i \in I(x^0)} \lambda_i \nabla g_i(x^0) + \sum_{j=1}^{r} \mu_j \nabla h_j(x^0)\Big] \in$$

$$\in Q^*(X, x^0) \ . \tag{6}$$

Proof. Note that if $Q(X, x^0)$ is empty the theorem holds trivially, and it also holds trivially if the rank of $\nabla h(x^0)$ is not full. Without loss of generality, assume therefore that the rank of $\nabla h(x^0)$ is full, suppose that x^0 is a local solution of (P_1) and let $\alpha(x) = (f(x) - f(x^0), g(x))$. On the grounds of what was observed in the proof of Theorem 3.6.5, we can assert that there exists no vector $x \in X$ such that $h(x) = 0$ and $\alpha(x) \in \overset{0}{R}_- \times \overset{0}{R}_- \times ... \times \overset{0}{R}_- = C$, where C is an open convex cone. Therefore, thanks to Lemma 3.8.2, no vector $y \in Q(X, x^0)$ exists such that $\nabla \alpha(x^0)\, y \in C_{\alpha(x^0)}$ and $\nabla h(x^0)\, y = 0$. Then, thanks to Lemma 3.6.4, there exists a nonzero vector $(q, \mu) \in C^*_{\alpha(x^0)} \times I\!R^r$ such that

$$q(\nabla \alpha(x^0)\, y) + \mu(\nabla h(x^0)\, y) \geqq 0 \ , \qquad \forall y \in Q(X, x^0) \ .$$

Hence

$$-[q \nabla \alpha(x^0) + \mu \nabla h(x^0)] \in Q^*(X, x^0) \ ,$$

i.e., with $q = (\lambda_0, \lambda_i)$, $i \in I(x^0)$, we have

$$-\Big[\lambda_0 \nabla f(x^0) + \sum_{i \in I(x^0)} \lambda_i \nabla g_i(x^0) + \sum_{j=1}^{r} \mu_j \nabla h_j(x^0)\Big] \in Q^*(X, x^0) \ .$$

As to the sign of the multipliers, we observe that $C \subset C_{\alpha(x^0)}$ and therefore $C^*_{\alpha(x^0)} \subset C^*$; being $q \in C^*$, it follows that $\lambda_0 \geqq 0$, $\lambda_i \geqq 0$, $\forall i \in I(x^0)$.□

The conditions above may be viewed as generalized Fritz John necessary conditions for (P_1). The following remarks may be useful.

1) Theorem 3.8.2 sharpens a result of Bazaraa and Goode (1972a) who refer to the *cone of interior directions* $I(X, x^0)$; indeed $I(X, x^0) \subset Q(X, x^0)$ and hence $Q^*(X, x^0) \subset I^*(X, x^0)$.

2) Note that Theorem 3.6.5 gives a necessary Fritz-John condition for (P), sharper than the one given by Theorem 3.8.2 for (P_1), as we may have $Q \subset T_1$ and therefore $T_1^*(X, x^0) \subset Q^*(X, x^0)$. This is not surprising, as (P_1) contains also the equality constraints vector $h(x) = 0$.

3) In the assumptions of the theorem we require the convexity of $Q(X, x^0)$; this is a weaker hypothesis than the convexity of X (considered, e.g., in Robinson (1982)).

4) If $x^0 \in \text{int}(X)$, then $Q(X, x^0) = \mathbb{R}^n$ and $Q^*(X, x^0) = \{0\}$. Then Theorem 3.8.2 gives in this case the classical Fritz-John conditions for (P_1):

$$\lambda_0 \nabla f(x^0) + \sum_{i \in I(x^0)} \lambda_i \nabla g_i(x^0) + \sum_{j=1}^{r} \mu_j \nabla h_j(x^0) = 0$$

$$(\lambda_0, \lambda_i) \geqq 0\,, \quad i \in I(x^0)\,, \quad (\lambda_0, \lambda_i, \mu_j) \neq 0\ .$$

Obviously, being that $g_i(x)$, $i \notin I(x^0)$, differentiable at x^0, these conditions may be equivalently written as:

$$\lambda_0 \nabla f(x^0) + \lambda \nabla g(x^0) + \mu \nabla h(x^0) = 0$$

$$\lambda g(x^0) = 0$$

$$(\lambda_0, \lambda) \geqq 0\,, \quad (\lambda_0, \lambda, \mu) \neq 0\ .$$

5) If in Theorem 3.8.2 we could replace $Q(X, x^0)$ by the larger cone $T(X, x^0)$, the theorem would be sharper, as $T^*(X, x^0) \subset Q^*(X, x^0)$. However, this sharper result does not hold in general. Consider the following example, taken from Bazaraa and Goode (1972a). The problem is: $\text{Min}\, f(x)$, subject to: $x \in X$, $h(x) = 0$, where

$$X = \{(x_1, x_2) \mid x_1 \text{ and } x_2 \text{ are rational}\}\ ;$$

$$f(x) = f(x_1, x_2) = x_2\ ; \qquad h(x) = h(x_1, x_2) = x_2 - \sqrt{2}\, x_1\ .$$

It is clear that

$$X \cap \{x \mid h(x) = 0\} = \{(0,0)\} ,$$

so the only admissible point is the origin and hence $x^0 = (0,0)$ solves the above problem. It is clear that $T(X, x^0) = I\!R^2$ and hence $T^*(X, x^0) = \{(0,0)\}$. However, there exists no nonzero vector (λ_0, μ) satisfying relation (6) of Theorem 3.8.2. One should note that in the above example $Q(X, x^0)$ is empty and hence the theorem holds trivially.

6) In Mangasarian (1969, p. 168) the following necessary optimality condition for (P_1), called "minimum principle" condition, is proved.

$$(x - x^0) \cdot \Big[\lambda_0 \nabla f(x^0) + \sum_{i \in I(x^0)} \lambda_i \nabla g_i(x^0) + \sum_{j=1}^{r} \mu_j \nabla h_j(x^0)\Big] \geqq 0 ,$$

$$\forall x \in \bar{X} ,$$

where X is a convex set in $I\!R^n$ with nonempty interior, $(\lambda_0, \lambda_i, \mu_j) \neq 0$, $\lambda_i \geqq 0$, $\forall i \in I(x^0)$. If we put $y = x - x^0$ and $Y = X - x^0$, we have

$$-\Big[\lambda_0 \nabla f(x^0) + \sum_{i \in I(x^0)} \lambda_i \nabla g_i(x^0) + \sum_{j=1}^{r} \mu_j \nabla h_j(x^0)\Big] \in \bar{Y}^* ,$$

and the result of Mangasarian may be viewed as a special case of Theorem 3.8.2. Indeed, under the convexity assumption on X (or on Y) and being $\mathrm{int}(X) \neq \emptyset$, from

$$\bar{Y} \subset \overline{\mathrm{cone}(Y)} = T(Y, 0) ,$$

it follows

$$Q^*(Y, 0) = T^*(Y, 0) \subset \bar{Y}^* .$$

In (6) of Theorem 3.8.2 the multiplier λ_0 may be not positive; if one wants to insure the positivity of λ_0, some constraint qualification for (P_1)

need be imposed. We have already seen that the Guignard-Gould-Tolle condition

$T^*(S_1, x^0) = E^*(x^0)$

insures the validity of the Kuhn-Tucker conditions for (P_1): see Theorem 3.8.1. Keeping the notations already introduced, let us list some other constraint qualifications for (P_1):

a) *Kuhn-Tucker C.Q.*
   It is expressed as: $E(x^0) \subset A(S_1, x^0)$.

b) *Arrow-Hurwicz-Uzawa second C.Q.*
   It is expressed as: $E(x^0) \subset \overline{\text{conv}(A(S_1, x^0))}$.

c) *Abadie C.Q.*
   It is expressed as: $E(x^0) \subset T(S_1, x^0)$.

d) *Mangasarian-Fromovitz C.Q.* (Mangasarian and Fromovitz (1967)).
   It is expressed as: $x^0 \in \text{int}(X)$, $\nabla h_j(x^0)$, $j = 1, ..., r$, are linearly independent and the system

$$\begin{cases} y \nabla g_i(x^0) < 0 \ , & i \in I(x^0) \\ y \nabla h_j(x^0) = 0 \ , & j = 1, ..., r \end{cases}$$

   has solution $y \in \mathbb{R}^n$.

e) *Weak Slater C.Q.*
   It is expressed as: $x^0 \in \text{int}(X)$, g_i, $i \in I(x^0)$, are pseudoconvex at x^0; h_j, $j = 1, ..., r$, are linear and there exists an $\bar{x} \in X$ such that $g_i(\bar{x}) < 0$, $\forall i \in I(x^0)$ and $h_j(\bar{x}) = 0$, $\forall j = 1, ..., r$.

f) *Weak reverse C.Q.*
   It is expressed as: $x^0 \in \text{int}(X)$, g_i, $i \in I(x^0)$, are pseudoconcave at x^0 and h_j, $j = 1, ..., r$, are pseudolinear (i.e. both pseudoconcave and pseudoconvex) at x^0.

g) *Independence C.Q.*
   It is expressed as: $x^0 \in \text{int}(X)$, $\nabla g_i(x^0)$, $i \in I(x^0)$, and $\nabla h_j(x^0)$, $j = 1, ..., r$, are linearly independent.

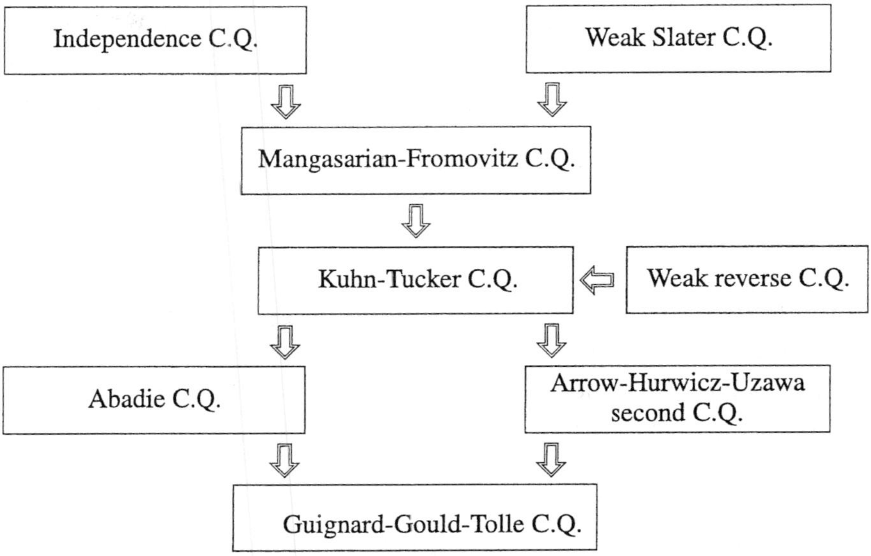

Figure 14.

By means of considerations analogous to the ones made for problem (P), in case $x^0 \in \text{int}(X)$ we have the diagram of Figure 14 showing the various relationships among the constraint qualifications considered for problem (P_1) [see also Bazaraa and Shetty, (1976), Theorem 6.2.3].

It must be noted that the Mangasarian-Fromovitz C.Q. also plays a role in obtaining sensitivity and stability results for a perturbed nonlinear programming problem with inequality and equality constraints. For example, it is shown by Gauvin and Tolle (1977) and Gauvin (1977) that this C.Q. is necessary and sufficient for having the set of Kuhn-Tucker multipliers bounded, an important result in the development of stability analysis for perturbations of (P_1). More precisely, if x^0 is a local minimum for (P_1), let us denote with $K(x^0)$ the set of Kuhn-Tucker multipliers corresponding to x^0, i.e. the set of vectors (λ, μ) such that (3)–(5) hold. Then $K(x^0)$ is a nonempty bounded set if and only if the Mangasarian-Fromovitz C.Q. is satisfied at x^0.

Sensitivity and stability in optimization problems are also studied, e.g., in Evans and Gould (1970), Geoffrion (1971), Greenberg and Pierskalla (1972), Fiacco (1983, 1984), Rockafellar (1993). See also Section 3.10.

3.9. Sufficient First-Order Optimality Conditions for (P) and (P_1)

We present first the following classical theorem of Mangasarian (1969) which gives first order sufficient optimality criteria for problem (P) and which subsumes the sufficiency results of Arrow and Enthoven (1961) and Arrow-Hurwicz and Uzawa (1961).

Theorem 3.9.1. Let $x^0 \in S$ and $f : D \to \mathbb{R}$ be pseudoconvex at x^0 (with respect to D) and $g_i : D \to \mathbb{R}$, $i \in I(x^0)$, be quasiconvex at x^0 (with respect to D). If there exists a vector $\lambda \in \mathbb{R}^m$ such that

$$(x - x^0)\,[\nabla f(x^0) + \lambda \nabla g(x^0)] \geqq 0\ , \qquad \forall x \in S\ ;$$

$$\lambda g(x^0) = 0\ ;$$

$$\lambda \geqq 0\ ,$$

then x^0 solves (P).

Proof. First note that thanks to the complementarity slackness conditions $\lambda g(x^0) = 0$ we have $\lambda_i = 0$, $\forall i \notin I(x^0)$. Since $g_i(x) \leqq 0 = g_i(x^0)$, $\forall i \in I(x^0)$, $\forall x \in S$, it follows by the quasiconvexity of g_i, $i \in I(x^0)$,

$$(x - x^0)\,\nabla g_i(x^0) \leqq 0\ , \qquad \forall i \in I(x^0)\ ,\ \ \forall x \in S\ .$$

Being that $\lambda_i \geqq 0$, $\forall i \in I(x^0)$, and $\lambda_i = 0$, $\forall i \notin I(x^0)$, we get

$$\sum_{i=1}^{m} \lambda_i (x - x^0)\,\nabla g_i(x^0) \leqq 0\ , \qquad \forall x \in S\ .$$

But since

$$(x - x^0)\,[\nabla f(x^0) + \lambda \nabla g(x^0)] \geqq 0\ , \qquad \forall x \in S\ ,$$

then

$$(x - x^0)\,\nabla f(x^0) \geqq 0\,, \qquad \forall x \in S\,,$$

which, by the pseudoconvexity of f at x^0, implies that

$$f(x) \geqq f(x^0)\,, \qquad \forall x \in S\,. \qquad \square$$

From Theorem 3.9.1 the following Kuhn-Tucker sufficient optimality theorem directly follows.

Theorem 3.9.2. Let $x^0 \in S$, f be pseudoconvex at x^0, g_i, $i \in I(x^0)$, be quasiconvex at x^0. If (x^0, λ) satisfies the Kuhn-Tucker conditions (6)–(8) of Section 3.5, then x^0 solves (P).

Note that in Theorems 3.9.1 and 3.9.2 the pseudoconvexity of f at x^0 and the quasiconvexity of g_i, $i \in I(x^0)$, at x^0, are actually required to hold with respect to the feasible set S.

Another first-order sufficient optimality theorem for (P) is obtained by imposing generalized convexity conditions on the Lagrangian function $\psi(x, \lambda) = f(x) + \lambda g(x)$; $\lambda \geqq 0$.

Theorem 3.9.3. Let $x^0 \in S$ and let the pair (x^0, λ) satisfy the Kuhn-Tucker conditions (6)–(8) of Section 3.5. If $\psi(\cdot, \lambda)$ is pseudoconvex at x^0, with respect to S, then x^0 solves (P).

Proof. As $\nabla_x \psi(x^0, \lambda) = 0$, being that $\psi(\cdot, \lambda)$ is pseudoconvex at x^0, it has a minimum at x^0 on S. Therefore for each $x \in S$ we have

$$f(x) + \lambda g(x) \geqq f(x^0) + \lambda g(x^0)\,,$$

i.e.

$$f(x) + \lambda g(x) \geqq f(x^0)\,,$$

and, being that $g(x) \leqq 0$, $\forall x \in S$, and $\lambda \geqq 0$, we obtain

$$f(x) \geqq f(x^0)\,, \qquad \forall x \in S\,. \qquad \square$$

We note that the assumptions made in Theorems 3.9.2 and 3.9.3 are not comparable. Note again that in Theorem 3.9.1 the assumption of the quasiconvexity of g_i, $i \in I(x^0)$, was only used to establish that, for all $x \in S$,

$$\sum_{i \in I(x^0)} \lambda_i (x - x^0) \nabla g_i(x^0) \leqq 0 .$$

But for this purpose it suffices that S is a *convex set* or even a *star-shaped set* at x^0. For then, $(1-\alpha)\, x^0 + \alpha x \in S$, $\forall \alpha \in [0,1]$, $\forall x \in S$, whence

$$\tilde{g}_i(\alpha) = g_i((1-\alpha)\, x^0 + \alpha x) \leqq 0 , \qquad \forall \alpha \in [0,1] , \quad \forall x \in S .$$

As $\tilde{g}_i(0) = g_i(x^0) = 0$, $\forall i \in I(x^0)$, it results ($\tilde{g}_i'(\alpha)$ denoting $d\tilde{g}_i(\alpha)/d\alpha$)

$$\tilde{g}_i'(0) = (x - x^0) \nabla g_i(x^0) \leqq 0 , \qquad \forall i \in I(x^0) , \quad \forall x \in S .$$

So we obtain the following result (see also Arrow and Enthoven (1961) who credit the above remark to Uzawa):

<u>Theorem 3.9.4</u>. Let $x^0 \in S$, f be pseudoconvex at x^0, S convex and conditions (6)–(8) of Section 3.5 be satisfied. Then x^0 solves (P).

Hanson (1981) noted that in proving first-order sufficient optimality conditions for (P), the linear term $(x - x^0)$, appearing in the definitions of pseudoconvex and quasiconvex (differentiable) functions, plays no role. This fact motivated the introduction of invex (and generalized invex) functions (see Section 2.17 of Chapter II). The proof of the following theorem follows the same lines of the proof of Theorem 3.9.1.

<u>Theorem 3.9.5</u>. Let f, g_i, $i \in I(x^0)$, be invex functions at $x^0 \in S$, with respect to the *same* function $\eta(x, x^0) : D \times D \to \mathbb{R}^n$. If the Kuhn-Tucker conditions (6)–(8) of Section 3.5 are satisfied at x^0, then x^0 solves (P).

Theorem 3.9.5 can be further weakened by imposing that f is invex at $x^0 \in S$ and g_i, $i \in I(x^0)$, are quasi-invex at $x^0 \in S$ (invexity and quasi-invexity are assumed with respect to the same function $\eta(x, x^0)$).

Another interesting application of invex functions is made by Hanson and Mond (1987). They consider problem (P), where $X \subset \mathbb{R}^n$ is open, and say that f and g are *Type I objective and constraint functions* for (P), with respect to a vector-valued function η at x^0, if there exists an n-dimensional vector function η, defined for all $x \in S$, such that

$$f(x) - f(x^0) \geqq \eta(x) \cdot \nabla f(x^0)$$

and

$$-g(x^0) \geqq \eta(x) \cdot [\nabla g(x^0)]^T \ .$$

Hanson and Mond prove the following two results.

Theorem 3.9.6. For $x^0 \in S$ to be optimal for (P) it is sufficient that f and g_i, $i \in I(x^0)$, are Type I functions with respect to a common function η at x^0 and the Kuhn-Tucker conditions (6)–(8) of Section 3.5 are satisfied at x^0.

Note that Theorem 3.9.6 is nothing but Theorem 3.9.5, where η depends only on x and not also on x^0. Note, moreover, that if η is identically zero then Theorem 3.9.6 holds trivially and x^0 is a solution of (P) without regard to the Kuhn-Tucker conditions.

Theorem 3.9.7. If $x^0 \in S$ and the number of active constraints at x^0 is k (i.e. $\text{card}(I(x^0)) = k$), with $k < n$, then for x^0 to be solution for (P) it is necessary that f and g are Type I functions, with respect to a common vector η at x^0, *not identically zero* for each $x \in S$.

Therefore, combining together the two last theorems, we can assert that if X is open, $x^0 \in S$, the Kuhn-Tucker conditions (6)–(8) of Section 3.5 hold at x^0 and $\text{card}(I(x^0)) < n$, then for x^0 to be solution for (P) it is necessary and sufficient that f and g are Type I functions with respect to a common vector η at x^0, not identically zero for all $x \in S$.

Other sufficient first-order optimality conditions involving invex and generalized invex functions are given in Craven (1981), Craven and Glover

(1985), Kaul and Kaur (1985), Jeyakumar (1985a), Martin (1985), Rueda and Hanson (1988), Weir and Jeyakumar (1988). See also Kaul and Kaur (1982), Mahajan and Vartak (1977), Giorgi (1984, 1995), Giorgi and Guerraggio (1994), Guignard (1969), Szlobec (1970) for other considerations on first-order sufficiency criteria of the Kuhn-Tucker type for (P).

Mangasarian (1969) proved the following Fritz John-type sufficient criterion for (P):

Theorem 3.9.8. Let $x^0 \in S$, f be convex at x^0 and g_i, $i \in I(x^0)$, strictly convex at x^0. If (x^0, u_0, u) solves the Fritz John conditions (3)–(5) of Section 3.5, then x^0 solves (P).

Note that the above theorem requires a rather stringent assumption on g_i, $i \in I(x^0)$, but also allows for $u_0 = 0$ (i.e. for a zero multiplier of the gradient of the objective function) in the Fritz John conditions. In fact, as shown in Theorem 3.9.14 for (P_1), the thesis of Theorem 3.9.8 also holds if f is pseudoconvex at x^0 and g_i, $i \in I(x^0)$, are strictly pseudoconvex at x^0.

Another Fritz-John sufficient optimality criterion for (P), more general than Theorem 3.9.8, is given by Kaul and Kaur (1985).

Theorem 3.9.9. Let $x^0 \in S$, f invex at x^0 and g_i, $i \in I(x^0)$, strictly invex at x^0, with respect to the same function $\eta(x, x^0) : D \times D \to I\!R^n$. If conditions (3)–(5) of Section 3.5 are satisfied, then x^0 solves (P).

Proof. From

$$u_0 \nabla f(x^0) + \sum_{i \in I(x^0)} u_i \nabla g_i(x^0) = 0$$

$$(u_0, u_i) \geq 0 \,, \qquad i \in I(x^0) \,,$$

it follows, by the Gordan theorem of the alternative, that the system

$$\begin{cases} \eta(x, x^0)\, \nabla f(x^0) & < 0 \\ \eta(x, x^0)\, (\nabla g_I(x^0))^T & < 0 \end{cases} \tag{1}$$

has no solution $x \in D$.
Suppose that x^0 is not a solution of (P); then there exists $x^* \in S$ such that $f(x^*) < f(x^0)$ and $g_I(x^*) \leqq 0 = g_I(x^0)$. But from the invexity of f at x^0 and the strict invexity of g_I at x^0 we obtain the inequalities

$$0 > f(x^*) - f(x^0) \geqq \eta(x^*, x^0)\, \nabla f(x^0)$$

$$0 \geqq g_I(x^*) - g_I(x^0) > \eta(x^*, x^0)\, (\nabla g_I(x^0))^T \,.$$

These inequalities show that x^* is a solution of the system

$$\begin{cases} \eta(x, x^0)\, \nabla f(x^0) & < 0 \\ \eta(x, x^0)\, (\nabla g_I(x^0))^T & < 0 \end{cases}$$

which is a contradiction to the assumption that (1) has no solution. Therefore $f(x) \geqq f(x^0)$, $\forall x \in S$. □

Obviously if $\eta(x, x^0) = x - x^0$, Theorem 3.9.9 becomes Theorem 3.9.8.

Sufficient criteria of the Kuhn-Tucker type for problem (P_1) have been proved by Mangasarian (1969) in the following, rather general, formulation.

Theorem 3.9.10. Let $x^0 \in S_1$, f be pseudoconvex at x^0, g_i, $i \in I(x^0)$, be quasiconvex at x^0 and let h_j, $j = 1, ..., r$, be quasilinear (i.e. quasiconvex and quasiconcave at x^0). If there exist $\lambda \in I\!R^m$, $\mu \in I\!R^r$, such that the Kuhn-Tucker conditions (3)–(5) of Section 3.8 are satisfied at x^0, then x^0 solves (P_1).

Proof. This sufficient optimality criterion follows directly from Theorem 3.9.2 by observing that the equality constraint $h(x) = 0$ can be written as $h(x) \leqq 0$ and $-h(x) \leqq 0$, and that the negative of a quasiconcave function is quasiconvex. □

Singh (1977) gives the following sufficient optimality criteria for (P_1), criteria considered by this author as a generalization of the previous result of Mangasarian (the result of Singh is, however, almost trivial).

Theorem 3.9.11. Let $x^0 \in S_1$, f pseudoconvex at x^0, g and h quasiconvex at x^0 and suppose that there exist $\lambda \in I\!R^m$, $\mu \in I\!R^r$, such that

$$\nabla f(x^0) + \lambda \nabla g(x^0) + \mu \nabla h(x^0) = 0 \ ,$$

$$\lambda g(x^0) = 0 \ ,$$

$$\lambda \geqq 0 \ , \qquad \mu \geqq 0 \ .$$

Then x^0 solves (P_1).

Other first-order sufficient optimality theorems for (P_1) are given by Bector and Gulati (1977), Bhatt and Misra (1975), Mahajan and Vartak (1977). All these results are trivial consequences of already established results for (P). Indeed, consider the following problems:

$$\operatorname{Min} f(x) \ , \qquad x \in S = \{x \mid x \in X, \ g(x) \leqq 0\} \ , \tag{P}$$

$$\operatorname{Min} f(x) \ , \qquad x \in S_2 = \{x \mid x \in X, \ g(x) \leqq 0, \ h(x) \leqq 0\} \ . \tag{P_2}$$

Clearly $S_1 \subset S_2 \subset S$. Now if $\mathcal{C}(x^0)$, $\mathcal{C}_2(x^0)$ are sufficient conditions for a point x^0 to be solution of, respectively, (P), (P_2), then it follows trivially that

$$\mathcal{C}(x^0) \ , \qquad h(x^0) = 0$$

or

$$\mathcal{C}_2(x^0) \ , \qquad h(x^0) = 0$$

are sufficient conditions for x^0 to be solution of (P_1). These conditions are precisely the ones given in Theorem 3.9.11 and in the other papers cited above.

Obviously Theorems 3.9.10 and 3.9.11 can be reformulated under suitable assumptions of invexity and generalized invexity of the functions involved.

We obtain now for problem (P_1) a sufficient optimality condition, under no (generalized) convexity assumption on $h(x)$. Let us consider the *linearizing cone* $C(x^0)$:

$$C(x^0) = \{z \in \mathbb{R}^n \mid z \nabla g_i(x^0) \leqq 0, \ \forall i \in I(x^0)\} \ ;$$

the *cone of gradients* (for the constraints g_i):

$$B(x^0) = \left\{ z \in \mathbb{R}^n \mid z = \sum_{i \in I(x^0)} \lambda_i \nabla g_i(x^0),\ \lambda_i \geqq 0 \right\}$$

and the *cone of gradients* (for the constraints h_j):

$$L(x^0) = \{ z \in \mathbb{R}^n \mid z = \mu \nabla h(x^0),\ \mu \in \mathbb{R}^r \} .$$

In Section 3.6 we have noted that it is

$$B(x^0) = C^*(x^0) .$$

Then we have the following result:

Theorem 3.9.12. Let $x^0 \in S_1$, f pseudoconvex at x^0, g_i, $i \in I(x^0)$, quasiconvex at x^0 and let $L(x^0) \subset C^*(x^0)$. Then if there exist $\vartheta \in \mathbb{R}^m$ and $\mu \in \mathbb{R}^r$ such that

$$\nabla f(x^0) + \vartheta \nabla g(x^0) + \mu \nabla h(x^0) = 0 \tag{2}$$

$$\vartheta g(x^0) = 0$$

$$\vartheta \geqq 0 ,$$

then x^0 solves (P_1).

Proof. As $L(x^0) \subset C^*(x^0) = B(x^0)$, then

$$\mu \nabla h(x^0) = \sum_{i \in I(x^0)} \lambda_i \nabla g_i(x^0) , \qquad \lambda_i \geqq 0 ,\ \ i \in I(x^0) .$$

Let us now define $q \in \mathbb{R}^m$ as:

$$q_i = \begin{cases} \lambda_i , & \forall\, i \in I(x^0) \\ 0 , & \forall\, i \notin I(x^0) . \end{cases}$$

From (2) we obtain

$$\nabla f(x^0) + \vartheta \nabla g(x^0) + q \nabla g(x^0) = 0 .$$

Letting $w = \vartheta + q$ we obtain

$$\nabla f(x^0) + w\nabla g(x^0) = 0$$

$$wg(x^0) = \vartheta g(x^0) + qg(x^0) = 0$$

$$g(x^0) \leqq 0\,, \quad h(x^0) = 0\,, \quad w = \vartheta + q \geqq 0\,.$$

From these relations, using the same arguments used in the proof of Theorem 3.9.2, we obtain

$$f(x^0) \leqq f(x)\,, \qquad \forall x \in S_1\,. \qquad \square$$

Another quite general sufficient first-order optimality condition (of the Kuhn-Tucker type) for problem (P_1) is obtained in Giorgi and Guerraggio (1994) and Giorgi (1995). Let us premise the following definition.

Definition 3.9.1. The set $D \subset I\!R^n$ is said to be *invex at* $x^0 \in D$ with respect to a function $\eta : D \times D \to I\!R^n$, $\eta(x, x^0)$ different from the zero vector, when $\eta(x, x^0) \in P(D, x^0)$, $\forall x \in D$, where $P(D, x^0)$ is the pseudotangent cone to D at x^0.

This definition, which is a generalization of the concept of *pseudoconvex set*, due to Guignard (1969), extends the concept of convexity of a set and also of convexity of a set at a point x^0 ("star-shaped set").

Indeed, if a set $S \subset I\!R^n$ is convex or also star-shaped at $x^0 \in S$, then it is also invex at x^0, with $\eta(x, x^0) = x - x^0$, i.e. it is pseudoconvex at x^0. Since we proved (Theorem 3.4.10) that, if S is convex, we have $P(S, x^0) = T(S, x^0) = \overline{\text{cone}(S - x^0)}$ (these equalities hold also if S is star-shaped at x^0; see Bazaraa and Shetty (1976)), then it is obvious that $x - x^0 = \lim_{k \to +\infty} \lambda_k(x^k - x^0)$; with $\lambda_k = 1$ and $x^k = x$, $k = 1, 2, \ldots$, this means that $x - x^0 \in P(S, x^0)$. An example of an invex set is: $A = \{x \in I\!R \mid x = 0 \text{ or } x = 1/n,\ n \in N_+\}$; this set is invex at $x^0 = 0$ for $\eta(x, x^0) = x - x^0$ and also for, e.g., $\eta(x, x^0) = (x - x^0)^2$.

Theorem 3.9.13. Let $x^0 \in S_1$, let S_1 be invex at x^0 and f be pseudo-invex at x^0, with respect to the same function η. If there exist scalars $\lambda_i \geqq 0$,

$i \in I(x^0)$, and $\mu_j \in I\!R$, $j = 1, ..., r$, such that

$$-\Big[\nabla f(x^0) + \sum_{i \in I(x^0)} \lambda_i \nabla g_i(x^0) + \sum_{j=1}^{r} \mu_j \nabla h_j(x^0)\Big] \in P^*(S_1, x^0) ,$$

then x^0 solves (P_1).

Proof. Let $\varphi = (g_i, h_j)$, $i \in I(x^0)$ and $j = 1, ..., r$, and let $B = \varphi(S_1)$ the image space of the constraints of (P_1). From the invexity of S_1 at x^0, there will exist a vector-valued function $\eta : S_1 \times S_1 \to I\!R^n$ such that $\eta(x, x^0) \in P(S_1, x^0)$, $\forall\, x \in S_1$.

Given an arbitrary $y \in T(S_1, x^0)$, then, by a well-known property of Bouligand tangent cones, we have $\nabla \varphi(x^0) \cdot y \in T(B, \varphi(x^0))$.

Being $\eta(x, x^0) \in P(S_1, x^0)$, we have $\eta(x, x^0) = \lim_{k \to +\infty} y^k$, with $y^k \in \mathrm{conv}(T(S_1, x^0))$, i.e.

$$\eta(x, x^0) = \lim_{k \to +\infty} (t_k y^{1k} + (1 - t_k)\, y^{2k}) ,$$

with $y^{1k}, y^{2k} \in T(S_1, x^0)$, $0 \leqq t_k \leqq 1$. Then it results:

$$\begin{aligned} \nabla \varphi(x^0) \cdot \eta(x, x^0) &= \lim_{k \to +\infty} [t_k \nabla \varphi(x^0) \cdot y^{1k} + (1 - t_k)\, \nabla \varphi(x^0) \cdot y^{2k}] \in \\ &\in \lim_{k \to +\infty} [t_k T(B, \varphi(x^0)) + (1 - t_k) \cdot T(B, \varphi(x^0))] = \\ &= \mathrm{cl}(\mathrm{conv}(T(B, \varphi(x^0)))) = P(B, \varphi(x^0)) . \end{aligned}$$

Now let us prove that $\omega = (\lambda_i, \mu_j)$, $i \in I(x^0)$, $\lambda_i \geqq 0$, belongs to $P^*(B, \varphi(x^0))$. Let us consider the quantity $\omega{\cdot}t$, where $t \in T(\varphi(S_1), \varphi(x^0))$, i.e.

$$t = \lim_{k \to +\infty} \lambda_k^+ (\varphi_k(S_1) - \varphi(x^0)) .$$

Then

$$\begin{aligned} \omega \cdot t &= \lim_{k \to +\infty} \lambda_k^+ (\omega \cdot \varphi_k(S_1) - \omega \cdot \varphi(x^0)) = \\ &= \lim_{k \to +\infty} \lambda_k^+ (\omega \cdot \varphi_k(S_1)) \leqq 0 . \end{aligned}$$

Then if $p \in P(\varphi(S_1), \varphi(x^0))$, i.e.

$$p = \lim_{k \to +\infty} \{t_k \tau^{1,k} + (1 - t_k) \cdot \tau^{2,k}\}$$

with $0 \leqq t_k \leqq 1$, $\tau^{1,k}, \tau^{2,k} \in T(\varphi(S_1), \varphi(x^0))$, then

$$\omega \cdot p = \lim_{k \to +\infty} t_k \omega \tau^{1,k} + (1 - t_k) \cdot \omega \cdot \tau^{2,k} \leqq 0 \ ,$$

i.e. $\omega \in P^*(\varphi(S_1), \varphi(x^0))$.
We have therefore

$$[\omega \cdot (\nabla \varphi(x^0) \cdot \eta(x, x^0))] \leqq 0 \ , \qquad x \in S_1 \ .$$

From

$$-[\nabla f(x^0) + \omega \cdot \nabla \varphi(x^0)] \in P^*(S_1, x^0) \ ,$$

it follows

$$\eta(x, x^0) \, [\nabla f(x^0) + \omega \nabla \varphi(x^0)] \geqq 0 \ , \qquad \forall \, x \in S_1 \ ,$$

i.e.

$$\eta(x, x^0) \, \nabla f(x^0) \geqq 0 \ , \qquad \forall \, x \in S_1$$

and by the pseudo-invexity of f at x^0, $f(x) \geqq f(x^0)$, $\forall \, x \in S_1$. □

Zlobec (1970) presents an asymptotic sufficient optimality condition which is actually more general than the corresponding Guignard's sufficient optimality condition. Also Zlobec's condition can be generalized by means of invexity. More precisely:

i) if S_1 is η-invex at $x^0 \in S_1$;

ii) if f is η-pseudo-invex at x^0 (with respect to the same function η);

iii) if there exists a sequence $\{\omega^k\} \subset P^*(B, \varphi(x^0))$ such that

$$\lim_{k \to +\infty} -[\nabla f(x^0) + \omega^k \nabla \varphi(x^0)] \in P^*(S_1, x^0)$$

then x^0 solves (P_1).

Indeed, from $\eta(x, x^0) \in P(S_1, x^0)$, $\forall\, x \in S_1$, we obtain

$$\nabla \varphi(x^0)\, \eta(x, x^0) \in P(B, \varphi(x^0))$$

and

$$\omega^k \nabla \varphi(x^0)\, \eta(x, x^0) \leqq 0 .$$

Afterwards from

$$\eta(x, x^0)\, [\nabla f(x^0) + \omega^k \nabla \varphi(x^0)] \geqq 0 ,$$

for each $x \in S_1$ and for k large enough, we have, thanks to the η-pseudo-invexity of f at x^0, $f(x) \geqq f(x^0)$.

Also Fritz John-type sufficient optimality conditions have been obtained for (P_1): a rather general criterion is given by Skarpness and Sposito (1980) who amend a previous result of Singh (1977).

We recall first that a function $f : \mathbb{R}^n \to \mathbb{R}$ is said to be *strictly pseudoconvex at* x^0, if for any $x \neq x^0$,

$$(x - x^0)\, \nabla f(x^0) \geqq 0 \Rightarrow f(x) - f(x^0) > 0$$

or, equivalently

$$f(x) - f(x^0) \leqq 0 \Rightarrow (x - x^0)\, \nabla f(x^0) < 0 .$$

(See Chapter II, Section 2.10.)

<u>Theorem 3.9.14</u>. Let $x^0 \in S_1$, f be pseudoconvex at x^0, g_i, $i \in I(x^0)$, and h_j, $j = 1, ..., r$, be strictly pseudoconvex at x^0. If there exist $\lambda_0 \in \mathbb{R}$, $\lambda \in \mathbb{R}^m$, $\mu \in \mathbb{R}^r$ such that

$$\lambda_0 \nabla f(x^0) + \lambda \nabla g(x^0) + \mu \nabla h(x^0) = 0 \tag{3}$$

$$\lambda g(x^0) = 0 \tag{4}$$

$$(\lambda_0, \lambda, \mu) \geq 0 , \tag{5}$$

then x^0 solves (P_1).

<u>Proof</u>. Condition (3) can be written as follows (λ_I and g_I have an obvious significance):

$$\lambda_0 \nabla f(x^0) + \lambda_I \nabla g_I(x^0) + \mu \nabla h(x^0) = 0 \ .$$

Therefore appealing to the Gordan theorem of the alternative (result 11 of Section 2.4), we can affirm that there does not exist any $z \in I\!R^n$ such that

$$\nabla f(x^0)\, z < 0\,, \quad \nabla g_I(x^0)\, z < 0\,, \quad \nabla h(x^0)\, z < 0\ . \tag{6}$$

Then, the system

$$f(x) - f(x^0) < 0$$

$$g_I(x) - g_I(x^0) \leqq 0$$

$$h(x) - h(x^0) = 0$$

has no solution $x \in I\!R^n$. If there did exist a solution $\bar{x}$ to this system ($\bar{x} \neq x^0$), then, thanks to the pseudoconvexity of f and to the strict pseudoconvexity of g_I and h, we would have

$$f(\bar{x}) - f(x^0) < 0 \Rightarrow \nabla f(x^0)\,(\bar{x} - x^0) < 0$$

$$g_I(\bar{x}) - g_I(x^0) \leqq 0 \Rightarrow \nabla g_I(x^0)\,(\bar{x} - x^0) < 0$$

$$h(\bar{x}) - h(x^0) = 0 \Rightarrow \nabla h(x^0)\,(\bar{x} - x^0) < 0\ .$$

But this violates (6), with $z = (\bar{x} - x^0)$. Then $x^0 \in S_1$ is an optimal solution of problem (P_1). □

We obtain now some first-order sufficient *local* optimality conditions for (P_1), under no (generalized) convexity assumptions on the functions involved in the same problem. Let us first consider problem (P_0), i.e. $\text{Min}\, f(x)$, $x \in X \subset I\!R^n$. The following result is due to Hestenes (1975); see also Bazaraa and Goode (1989).

<u>Theorem 3.9.15</u>. Let $x^0 \in X$ and suppose that

$$y\nabla f(x^0) > 0 \;, \qquad \forall y \in T(X, x^0)\,, \;\; y \neq 0 \;; \tag{7}$$

then x^0 is a strict loca minimum for problem (P).

Proof. Suppose by contradiction that x^0 is not a strict local minimum of f on X; then there exists a sequence $\{x^k\}$ converging to x^0 such that $x^k \neq x^0$, $x^k \in X$, and $f(x^k) \leqq f(x^0)$.
Let $y^k = (x^k - x^0)/\|x^k - x^0\|$; then $\|y^k\| = 1$ and, by the Bolzano-Weierstrass theorem, there exist a subsequence $\bar{y}^k$ and a vector y so that $\|y\| = 1$ and $\bar{y}^k \to y$ as $k \to +\infty$. Then $y \in T(X, x^0)$ and by assumption $y\nabla f(x^0) > 0$, being that $\|y\| = 1$.
Now it is

$$f(x^k) - f(x^0)/\|x^k - x^0\| =$$
$$= ((x^k - x^0)/\|x^k - x^0\|)\, \nabla f(x^0) + o(\|x^k - x^0\|)/\|x^k - x^0\| \;,$$

i.e.

$$\frac{f(x^k) - f(x^0)}{\|x^k - x^0\|} = y^k \nabla f(x^0) + \frac{o\,\|x^k - x^0\|}{\|x^k - x^0\|} \;.$$

Since $f(x^k) \leqq f(x^0)$, the left-hand side of the last expression is nonpositive, while the right-hand side converges to a positive number as $k \to +\infty$ in the subsequence $\{\bar{y}^k\}$. This contradiction proves the result. □

It must be remarked that condition (7) cannot be verified if $x^0 \in \text{int}(X)$, as in this case $T(X, x^0) = \mathbb{R}^n$. Moreover, even if X is a closed convex set, it is not allowed to substitute $T(X, x^0)$ with $X - x^0$, being in this case $X - x^0 \subset \overline{\text{cone}(X - x^0)} = T(X, x^0)$. See also Giorgi (1995).

Theorem 3.9.16. Let $x^0 \in S_1$ and let conditions

$$\nabla f(x^0) + \lambda \nabla f(x^0) + \mu \nabla h(x^0) = 0$$

$$\lambda g(x^0) = 0$$

$$\lambda \geqq 0$$

be verified. Suppose further that the system

$$\begin{cases} y\nabla g_i(x^0) \leqq 0 \ , & i \in I(x^0) \\ y\nabla h_j(x^0) = 0 \ , & j = 1, ..., r \end{cases} \tag{8}$$

admits solutions $y \neq 0$ and that for none of these solutions it is $y\nabla f(x^0) = 0$. Then x^0 is a strict local solution of (P_1).

Proof. Under the assumptions of the theorem

$$y\nabla f(x^0) = -y\Big(\sum_{i \in I(x^0)} \lambda_i \nabla g_i(x^0) + \mu \nabla h(x^0)\Big)$$

and for all $y \neq 0$ solution of (8) we have $y\nabla f(x^0) \geqq 0$, however, being that $y\nabla f(x^0) = 0$ is excluded by assumption, then $y\nabla f(x^0) > 0$, $\forall\, y \in E(x^0)$, where $E(x^0)$ is the solution set of (8), i.e. the *linearizing cone* at x^0 for problem (P_1).

Now, as we always have $T(S_1, x^0) \subset E(x^0)$ (see Section 3.8), then $y\nabla f(x^0) > 0$, $\forall y \in T(S_1, x^0)$, $y \neq 0$, and hence the result follows from Theorem 3.9.15. □

Theorem 3.9.17. Let $x^0 \in S_1$ and let the rank of A be n, where

$$A = \begin{bmatrix} \nabla f(x^0) \\ \nabla g_I(x^0) \\ \nabla h(x^0) \end{bmatrix} .$$

If there exist multipliers $\lambda_i > 0$, $i \in I(x^0)$, and $\mu_j \in \mathbb{R}$, $j = 1, ..., r$, such that

$$\nabla f(x^0) + \sum_{i \in I(x^0)} \lambda_i \nabla g_i(x^0) + \mu \nabla h(x^0) = 0 \ , \tag{9}$$

then x^0 is a strict local solution of (P_1).

Proof. Let y be a solution of the system

$$\begin{cases} y\nabla f(x^0) \leqq 0 \\ y\nabla g_i(x^0) \leqq 0 \ , & i \in I(x^0) \\ y\nabla h_j(x^0) = 0 \ , & j = 1, ..., r \ . \end{cases} \tag{10}$$

Obviously the solution set of system (10) is nonempty; then, thanks to (9), we have

$$0 \geqq y\nabla f(x^0) = -\Big(\sum_{i\in I(x^0)} \lambda_i y\nabla g_i(x^0) + \sum_{j=1}^{r} \mu_j y\nabla h_j(x^0)\Big) =$$

$$= - \sum_{i\in I(x^0)} \lambda_i y\nabla g_i(x^0) \geqq 0 \ .$$

This implies ($\lambda_i > 0$, $\forall i \in I(x^0)$):

$$y\nabla f(x^0) = 0 \ ; \qquad y\nabla g_i(x^0) = 0 \, , \quad \forall i \in I(x^0) \ ;$$

$$y\nabla h_j(x^0) = 0 \, , \quad \forall j = 1 \ ..., r \ .$$

Since n among vectors $\nabla f(x^0)$, $\nabla g_i(x^0)$, $i \in I(x^0)$ and $\nabla h_j(x^0)$, $j = 1, ..., r$, are linearly independent, then system (10) admits the unique solution $y = 0$.

Now let us assume by contradiction that x^0 is not a strict local solution of (P_1). Then there exists a sequence $\{x^k\} \subset S_1$, $x^k \neq x^0$, $\lim_{k\to+\infty} x^k = x^0$, such that $f(x^k) \leqq f(x^0)$, $\forall k$.

The vector x^k can then be written as $x^k = x^0 + \delta_k y^k$, where $\|y^k\| = 1$ and δ_k are positive numbers with $\lim_{k\to+\infty} \delta_k = 0$.

By the Bolzano-Weierstrass theorem, the sequence $\{y^k\}$ contains a convergent subsequence we may again denote by $\{y^k\}$. Thus $\lim_{k\to+\infty} y^k = y$, with $\|y\| = 1$. From $f(x^k) \leqq f(x^0)$; $g_i(x^k) - g_i(x^0) = g_i(x^k) \leqq 0$, $i \in I(x^0)$; $h_j(x^k) - h_j(x^0) = 0$, $\forall j = 1, ..., r$, we have

$$\frac{f(x^0 + \delta_k y^k) - f(x^0)}{\delta_k} \leqq 0 \ ;$$

$$\frac{g_i(x^0 + \delta_k y^k) - g_i(x^0)}{\delta_k} \leqq 0 \ , \qquad i \in I(x^0) \ ;$$

$$\frac{h_j(x^0 + \delta_k y^k) - h_j(x^0)}{\delta_k} = 0 \ , \qquad j = 1, ..., r \ .$$

Taking the limits of these quotients, for $k \to +\infty$, namely taking the directional derivatives in the direction y ($\|y\| = 1$), we have

$$y\nabla f(x^0) \leqq 0 \; ; \qquad y\nabla g_i(x^0) \leqq 0 \, , \quad i \in I(x^0) \; ;$$

$$y\nabla h_j(x^0) = 0 \, , \quad j = 1, ..., r \; .$$

Thus the direction y is a solution of system (10) and, under our assumptions, this leads to a contradiction. □

3.10. Second-Order Optimality Conditions

In this section optimality conditions for problem (P_1) which involve second-order derivatives are discussed. The main results are due to McCormick (1967), Fiacco (1968), Fiacco and McCormick (1968), but also see Messerli and Polak (1969), Ben-Tal (1980), Guignard (1969), McCormick (1983) and Robinson (1982) for other approaches.

Before stating the optimality conditions for (P_1) we give the following general result for (P_0), which is sharper than the one found in Theorem 3.2.3.

Theorem 3.10.1. Let the stationary point $x^0 \in X$ be a local solution of (P_0), where f is twice continuously differentiable. Then

$$yHf(x^0)\, y \geqq 0 \, , \qquad \forall\, y \in T(X, x^0) \; .$$

Proof. Let $y \in T(X, x^0)$; then there exist a sequence $\{x^k\} \subset X$ and a sequence $\{\lambda_k\}$, $\lambda_k \geqq 0$, such that $x^k \to x^0$ and $\lambda_k(x^k - x^0) \to y$. Then, being that $\nabla f(x^0) = 0$,

$$\begin{aligned} 0 \; &\leqq \; \lambda_k^2 (f(x^k) - f(x^0) - (x^k - x^0)\, \nabla f(x^0)) = \\ &= \tfrac{1}{2}\, \lambda_k (x^k - x^0)\, Hf(x^0)\, \lambda_k (x^k - x^0) + \\ &\quad + \lambda_k^2\, o(\|x^k - x^0\|^2) \to \tfrac{1}{2}\, yHf(x^0)\, y \; . \end{aligned}$$

□

From now on we assume in this section that the functions involved in (P_1) are twice continuously differentiable at least in a neighbourhood of the point $x^0 \in S_1$. Moreover, in order to simplify the analysis, we suppose $x^0 \in \operatorname{int}(X)$. We state now the following *second-order constraint qualification* (S.O.C.Q.), due to McCormick (1967).

Let $x^0 \in S_1$ and define the set

$$Z(x^0) = \{z \in \mathbb{R}^n \mid z\nabla g_i(x^0) = 0, \ i \in I(x^0),$$
$$z\nabla h_j(x^0) = 0, \ j = 1, ..., r\} .$$

The S.O.C.Q. is said to hold at x^0 if every nonzero $z \in Z(x^0)$ is tangent to a twice differentiable arc incident at x^0; that is, for each $z \in Z(x^0)$ there exists a twice differentiable function $\alpha(\vartheta) : [0, \bar{\vartheta}] \to \mathbb{R}^n$ such that $\alpha(0) = x^0$,

$$g_i[\alpha(\vartheta)] = 0 \, , \quad i \in I(x^0) \, ; \qquad h_j[\alpha(\vartheta)] = 0 \, , \quad j = 1, ..., r$$

for $0 \leqq \vartheta \leqq \bar{\vartheta}$ (1)

and

$$\alpha'(0) = z \, . \tag{2}$$

We may note that, apart from the twice differentiability of α, the previous S.O.C.Q. coincides with the following Kuhn-Tucker C.Q.

$$A(\tilde{S}_1, x^0) = Z(x^0)$$

where

$$\tilde{S}_1 = S_1 \cap \{x \mid g_i(x) = 0, \ i \in I(x^0)\} \, .$$

Theorem 3.10.2 (Second-order necessary conditions). Suppose that x^0 is a local solution of (P_1) and that the Kuhn-Tucker conditions (3)–(5) of Section 3.8 hold at x^0; suppose that at the same point the S.O.C.Q. is satisfied. Then for every $z \in Z(x^0)$, $z \neq 0$, it follows that

$$z\Big[Hf(x^0) + \sum_{i=1}^{m} \lambda_i Hg_i(x^0) + \sum_{j=1}^{r} \mu_j Hh_j(x^0)\Big] z \geqq 0 \, .$$

Proof. Let $z \neq 0$, $z \in Z(x^0)$ and let $\alpha(\vartheta)$ be the twice differentiable vector-valued function guaranteed by the S.O.C.Q. That is, $\alpha(0) = x^0$, $\alpha'(0) = z$. Denote $\alpha''(0)$ by w. From (1) and the chain rule it follows that

$$\frac{d^2 g_i[\alpha(0)]}{d\vartheta^2} = zHg_i(x^0)\,z + w\nabla g_i(x^0) = 0\ , \qquad i \in I(x^0) \tag{3}$$

$$\frac{d^2 h_j[\alpha(0)]}{d\vartheta^2} = zHh_j(x^0)\,z + w\nabla h_j(x^0) = 0\ , \qquad j = 1, \ldots, r\ . \tag{4}$$

From (3) of Section 3.8 and the definition of $Z(x^0)$, we have

$$\frac{df[\alpha(0)]}{d\vartheta} = z\nabla f(x^0) =$$

$$= z\Big[\sum_{i=1}^{m} \lambda_i \nabla g_i(x^0) + \sum_{j=1}^{r} \mu_j \nabla h_j(x^0)\Big] = 0\ .$$

Since x^0 is a local minimum and $df[\alpha(0)]/d\vartheta = 0$, it follows that $d^2 f[\alpha(0)]/d\vartheta^2 \geqq 0$, that is

$$\frac{d^2 f[\alpha(0)]}{d\vartheta^2} = zHf(x^0)\,z + w\nabla f(x^0) \geqq 0\ . \tag{5}$$

Multiplying (3) and (4) by the corresponding multipliers, adding to (5) and taking the Kuhn-Tucker conditions into account, we obtain

$$z\Big[Hf(x^0) + \sum_{i=1}^{m} \lambda_i Hg_i(x^0) + \sum_{j=1}^{r} \mu_j Hh_j(x^0)\Big]\,z \geqq 0\ . \qquad \square$$

Note that the S.O.C.Q. is automatically satisfied if in (P_1) all constraints are linear. The following example, due to McCormick (1967), illustrates that the first-order Kuhn-Tucker C.Q. can be satisfied while the S.O.C.Q. fails to hold.

Example 3.10.1. Let us consider at $x^0 = (0, 0)$ the constraint set given by

$$g_1(x) = x_1^9 - x_2^3 \leqq 0$$

$$g_2(x) = -x_1^9 - x_2^3 \leqq 0$$

$$g_3(x) = -x_1^2 - x_2^2 - 2x_2 \leqq 0\ .$$

The Kuhn-Tucker C.Q. is satisfied at x^0; on the other hand any vector z to be considered for the S.O.C.Q. is of the form $z = (z_1, 0)$, where $z_1 \neq 0$

since z must be nonzero. Since there is no arc along which g_1, g_2 and g_3 remain equal to zero, the S.O.C.Q. fails to hold.

That the S.O.C.Q. does not imply the first-order Kuhn-Tucker C.Q. can be seen in the following example, again due to McCormick (1967).

Example 3.10.2. Let us consider at $x^0 = (0,0)$ the constraint set given by

$$g_1(x) = x_1^2 + x_2^2 - 2x_2 \leqq 0$$

$$g_2(x) = x_1^2 + x_2^2 + 2x_2 \leqq 0$$

$$g_3(x) = -x_1 \leqq 0 \ .$$

The S.O.C.Q. is trivially satisfied because there are no vectors orthogonal to all three gradients $(0\ -2)$, $(0,2)$, $(-1,0)$. On the other hand, the linearizing cone is $C = \{(y_1, 0) \mid y_1 \geqq 0\}$, but there are no arcs pointing into the constraint set, which is a singleton, and hence the first-order Kuhn-Tucker C.Q. fails to hold.

The following theorem gives a sufficient condition for the S.O.C.Q.

Theorem 3.10.3. A sufficient condition for the S.O.C.Q. to be satisfied at a point $x^0 \in S_1$ is that the vectors $\nabla g_i(x^0)$, $i \in I(x^0)$, $\nabla h_j(x^0)$, $j = 1, ..., r$, are linearly independent.

Proof. See McCormick (1967), Fiacco and McCormick (1968). □

We know that the conditions given in Theorem 3.10.3 are also a first-order C.Q. for (P_1). Bazaraa, Sherali and Shetty (1993) and Ben-Tal (1980) claim that if the vectors $\nabla g_i(x^0)$, $i \in I(x^0)$, $\nabla h_j(x^0)$, $j = 1, ..., r$ are linearly independent at x^0, local solution of (P_1), then (obviously) x^0 satisfies the Kuhn-Tucker conditions (3)–(5) of Section 3.8 and, moreover, it is

$$y\Big[Hf(x^0) + \sum_{i=1}^{m} \lambda_i \nabla g_i(x^0) + \sum_{j=1}^{r} \mu_j \nabla h_j(x^0)\Big]\, y \geqq 0 \ ,$$

for all y, such that

$$y \in E(x^0) = \{y \in I\!R^n \mid y\nabla g_i(x^0) \leqq 0, \ i \in I(x^0), \\ y\nabla h_j(x^0) = 0, \ j = 1, ..., r\} .$$

However, this result is false, as proved by the following counterexample. Consider the problem

$$\begin{cases} \operatorname{Min} f(x) = -x^4 \\ g(x) = x^2 - 1 \leqq 0 , \qquad x \in I\!R . \end{cases}$$

Obviously, $x^0 = 1$ is a minimum point, $\nabla g(x^0) \neq 0$ and the Kuhn-Tucker conditions are fulfilled with $\lambda = 2$. Moreover, it is $Hf(x^0) = -12$, $Hg(x^0) = 2$ and hence $H_x\psi(1, \lambda) = -8$. But for $y = -1 \in E(x^0)$ we get $yH_x\psi(1, \lambda)\, y = -8 < 0$.

The following correct result is given by Kyparisis (1985); this author assumes the following constraint qualification, called *strict Mangasarian-Fromovitz C.Q.* (S.M.F.C.Q.):

i) $\nabla g_i(x^0)$, $i \in I^+(x^0)$; $\nabla h_j(x^0) = 0$, $j = 1, ..., r$ are linearly independent, and

ii) there exists $y \in I\!R^n$ such that

$$y\nabla g_i(x^0) < 0 , \qquad i \in I(x^0) \backslash I^+(x^0) ,$$
$$y\nabla g_i(x^0) = 0 , \qquad i \in I^+(x^0) ,$$
$$y\nabla h_j(x^0) = 0 , \qquad j = 1, ..., r ,$$

where

$$I^+(x^0) = \{i \mid i \in I(x^0), \ \lambda_i > 0\}$$

is the set of *strongly active constraints* at x^0 and depends on the multiplier vector λ in the Kuhn-Tucker conditions for (P_1).

Then define the cone

$$\hat{Z}(x^0) = \{z \in \mathbb{R}^n \mid z\nabla g_i(x^0) = 0, \ i \in I^+(x^0) \, ; \, z\nabla g_i(x^0) \leqq 0,$$
$$i \in I(x^0) \backslash I^+(x^0) \, ; \, z\nabla h_j(x^0) = 0, \ j = 1, ..., r\} \, .$$

Then we have the following second-order necessary conditions.

Theorem 3.10.4. Let x^0 be a local minimum point for (P_1) and assume that the S.M.F.C.Q. holds at x^0 with some (λ, μ) satisfying the Kuhn-Tucker conditions (3)–(5) of Section 3.8. Then

$$z\Big[Hf(x^0) + \sum_{i=1}^{m} \lambda_i Hg_i(x^0) + \sum_{j=1}^{r} \mu_j Hh_j(x^0)\Big] \, z \geqq 0$$

for all $z \in \hat{Z}(x^0)$.
Moreover, the S.M.F.C.Q. holds at x^0 if and only if the Kuhn-Tucker multipliers vector (λ, μ) is unique.

Proof. Let

$$\tilde{\tilde{S}}_1 = S_1 \cap \{x \mid g_i(x) = 0, \ \forall i \in I^+(x^0)\} \, .$$

Then

$$\mathcal{L}(x, \lambda, \mu) = f(x) + \sum \lambda_i g_i(x) + \sum \mu_j h_j(x) = f(x) \, ,$$
$$\forall x \in \tilde{\tilde{S}}_1 \, .$$

Thus x^0 is a local minimum for the problem

$$\min_{x \in \tilde{\tilde{S}}_1} \tilde{f}(x) = \mathcal{L}(x, \lambda, \mu) \, .$$

Since

$$\nabla f(x^0) = \nabla_x \mathcal{L}(x^0, \lambda, \mu) = 0$$

we get (recall Theorem 3.10.1)

$$zH\tilde{f}(x^0) \, z = zH_x\mathcal{L}(x^0, \lambda, \mu) \, z \geqq 0 \, ,$$
$$\forall z \in T(\tilde{\tilde{S}}_1, x^0) \, .$$

Obviously the S.M.F.C.Q. is the M.F.C.Q. referred to set $\tilde{\tilde{S}}_1$, which implies the Abadie C.Q. to set $\tilde{\tilde{S}}_1$, i.e.

$$\hat{Z}(x^0) = T(\tilde{\tilde{S}}_1, x^0)$$

and therefore the previous inequality holds for all $z \in \hat{Z}(x^0)$.
The uniqueness of the multipliers is a consequence of the Motzkin alternative theorem (Theorem 19 of Section 2.4).
In fact, let us consider a first set of multipliers

$$\{\lambda_i \geqq 0, \ i = 1, ..., m \, ; \, \mu_j, \ j = 1, ..., r\}$$

with

$$\nabla f(x^0) + \sum_i \lambda_i \nabla g_i(x^0) + \sum_j \mu_j \nabla h_j(x^0) = 0$$

and $\lambda_i g_i(x^0) = 0, \, i = 1, ..., m.$
Consider then another set of multipliers

$$\{\bar{\lambda}_i \geqq 0, \ i = 1, ..., m \, ; \, \bar{\mu}_j, \ j = 1, ..., r\}$$

with

$$\nabla f(x^0) + \sum_i \bar{\lambda}_i \nabla g_i(x^0) + \sum_j \bar{\mu}_j \nabla h_j(x^0) = 0$$

and $\bar{\lambda}_i g_i(x^0) = 0, \, i = 1, ..., m.$
We get

$$\sum_i (\bar{\lambda}_i - \lambda_i) \nabla g_i(x^0) + \sum_j (\bar{\mu}_j - \mu_j) \nabla h_j(x^0) = 0$$

and therefore

a) $\bar{\lambda}_i = \lambda_i = 0, \, \forall i \notin I(x^0).$

b) $\bar{\lambda}_i - \lambda_i = \bar{\lambda}_i \geqq 0, \, \forall i \in I(x^0) \backslash I^+(x^0)$; supposing by contradiction that $\bar{\lambda}_i > 0$, for an index $i \in I(x^0) \backslash I^+(x^0)$, by means of the Motzkin transposition theorem applied to the last equality, we obtain that the S.M.F.C.Q. does not hold. Therefore $\bar{\lambda}_i - \lambda_i = 0, \, \forall i \in I(x^0) \backslash I^+(x^0).$

c) Finally, because of the S.M.F.C.Q., vectors $\nabla g_i(x^0)$, $i \in I^+(x^0)$, $\nabla h_j(x^0)$, $j = 1, ..., r$, are linearly independent and therefore $\bar{\lambda}_i - \lambda_i = 0$, $\forall\, i \in I^+(x^0)$ and $\bar{\mu}_j - \mu_j = 0$, $\forall\, j = 1, ..., r$.

Conversely assume that the S.M.F.C.Q. does not hold because ii) does not hold. Then, thanks to the Motzkin alternative theorem, it is possible to find scalars s_i and t_j such that

$$\sum_{i \in I(x^0) \setminus I^+(x^0)} s_i \nabla g_i(x^0) + \sum_{i \in I^+(x^0)} s_i \nabla g_i(x^0) + \sum_{j=1}^{r} t_j \nabla h_j(x^0) = 0$$

with $s_i \geqq 0$, not all zero, for each $i \in I(x^0) \setminus I^+(x^0)$ and such that $\max_{i \in I^+(x^0)} |s_i| < \min_{i \in I^+(x^0)} \lambda_i$.
As the Kuhn-Tucker conditions hold at x^0 we get

$$\nabla f(x^0) + \sum_{i \in I(x^0) \setminus I^+(x^0)} (s_i + \lambda_i) \nabla g_i(x^0) +$$

$$+ \sum_{i \in I^+(x^0)} (s_i + \lambda_i) \nabla g_i(x^0) + \sum_{j=1}^{r} (t_j + \mu_j) \nabla h_j(x^0) = 0 .$$

Thus we should obtain two different sets of multipliers. The same conclusion holds if the S.M.F.C.Q. does not hold because i) does not hold. □

We should remark that the S.M.F.C.Q. could not properly be considered a constraint qualification, since the set $I^+(x^0)$ is not known before the validation of the Kuhn-Tucker conditions.

Another remark, concerning the S.O.C.Q. of McCormick, is that it requires that the arc $\alpha(\vartheta)$ must be contained in the manifold

$$\{x \mid g_i(x) = 0, \; i \in I(x^0), \; h_j(x) = 0, \; j = 1, ..., r\} ,$$

i.e.

$$g_i[\alpha(\vartheta)] = 0 , \quad i \in I(x^0) , \quad h_j[\alpha(\vartheta)] = 0 , \quad j = 1, ..., r .$$

However, Theorem 3.10.2 can be obtained under a weaker S.O.C.Q.: (S.O.C.Q.II) For each $z \neq 0$, $z \in Z(x^0)$, there exists a twice differentiable feasible arc $\alpha(\vartheta) : [0, \bar{\vartheta}] \to I\!R^n$ such that:

$$\alpha(0) = x^0 , \qquad \alpha'(0) = z$$

and

$$\frac{d^2}{d\vartheta^2} g_i[\alpha(\vartheta)]|_{\vartheta=0} = 0 , \qquad i \in I(x^0) .$$

The proof of Theorem 3.10.2, under S.O.C.Q.II is left to the reader. For second-order optimality conditions in topological vector spaces, the reader is referred to Ben-Tal and Zowe (1982) and to Maurer and Zowe (1979). See also Hettich and Jongen (1977) for another S.O.C.Q.

Let us now consider sufficient second-order conditions for optimality in problem (P_1). Such conditions were essentially derived by Pennisi (1953), Hestenes (1966, 1975), McCormick (1967); see also Pallu de la Barrière (1963) and the unpublished Master's Thesis of Karush (1939).

We recall first the definition of the cone $\hat{Z}(x^0)$, previously introduced:

$$\hat{Z}(x^0) = \{z \in I\!R^n \mid z\nabla g_i(x^0) = 0, \ i \in I^+(x^0) ; \ z\nabla g_i(x^0) \leqq 0, \\ i \in I(x^0) \backslash I^+(x^0) ; \ z\nabla h_j(x^0) = 0, \ j = 1, ..., r\} .$$

We have the following theorem, the proof of which is taken from Fiacco and McCormick (1968).

<u>Theorem 3.10.5</u>. Let $x^0 \in S_1$; if there exist vectors λ, μ such that the Kuhn-Tucker conditions (3)–(5) of Section 3.8 are satisfied at x^0 and if, for every $z \in \hat{Z}(x^0)$, $z \neq 0$, it follows that

$$z\Big[Hf(x^0) + \sum_{i=1}^{m} \lambda_i H g_i(x^0) + \sum_{j=1}^{r} \mu_j H h_j(x^0)\Big]\, z > 0 , \tag{6}$$

then x^0 is a strict local minimum for problem (P_1).

<u>Proof</u>. Assume the Kuhn-Tucker conditions (3)–(5) of Section 3.8 hold and that x^0 is not a strict local minimum. Then there exists a sequence $\{z^k\}$ of feasible points $z^k \neq x^0$ converging to x^0 and such that for each z^k

$$f(x^0) \geqq f(z^k) \ . \tag{7}$$

Let $z^k = x^0 + \vartheta_k y^k$, where $\vartheta_k > 0$ for each k and $\|y^k\| = 1$. Without loss of generality, assume that the sequence $\{\vartheta_k, y^k\}$ converges to $(0, \bar{y})$, where $\|\bar{y}\| = 1$. Since the points z^k are feasible,

$$g_i(z^k) - g_i(x^0) = \vartheta_k y^k \nabla g_i(x^0 + \eta_{i,k}\vartheta_k y^k) \leqq 0 \ , \quad i \in I(x^0) \ , \tag{8}$$

$$h_j(z^k) - h_j(x^0) = \vartheta_k y^k \nabla h_j(x^0 + \bar{\eta}_{j,k}\vartheta_k y^k) = 0 \ , \quad j = 1, ..., r \tag{9}$$

and from (7)

$$f(z^k) - f(x^0) = \vartheta_k y^k \nabla f(x^0 + \eta_k \vartheta_k y^k) \leqq 0 \ , \tag{10}$$

where η_k, $\eta_{i,k}$, $\bar{\eta}_{j,k}$ are numbers between 0 and 1.
Dividing (8), (9) and (10) by ϑ_k and taking limits, we get

$$\bar{y}\nabla g_i(x^0) \leqq 0 \ , \qquad i \in I(x^0) \tag{11}$$

$$\bar{y}\nabla h_j(x^0) = 0 \ , \qquad j = 1, ..., r \tag{12}$$

$$\bar{y}\nabla f(x^0) \leqq 0 \ . \tag{13}$$

Suppose that (11) holds with a strict inequality for some $i \in I^+(x^0)$. Then combining (3)–(5) of Section 3.8, (11) and (12), we obtain

$$0 \geqq \bar{y}\nabla f(x^0) = -\sum_{i=1}^{m} \lambda_i \bar{y}\nabla g_i(x^0) - \sum_{j=1}^{r} \mu_j \bar{y}\nabla h_j(x^0) > 0 \ , \tag{14}$$

which is a contradiction.
Therefore $\bar{y}\nabla g_i(x^0) = 0$, $\forall i \in I^+(x^0)$ (or $I^+(x^0)$ is empty) and so $\bar{y} \in \hat{Z}(x^0)$. From Taylor's expansion formula we obtain

$$g_i(z^k) = g_i(x^0) + \vartheta_k y^k \nabla g_i(x^0) + \tfrac{1}{2}(\vartheta_k)^2 \cdot \cdot y^k [H g_i(x^0 + \xi_{i,k}\vartheta_k y^k)]\, y^k \leqq 0 \ , \qquad i = 1, ..., m \tag{15}$$

$$h_j(z^k) = h_j(x^0) + \vartheta_k y^k \nabla h_j(x^0) + \tfrac{1}{2}(\vartheta_k)^2 \cdot \cdot y^k [H h_j(x^0 + \bar{\xi}_{j,k}\vartheta_k y^k)]\, y^k = 0 \ , \qquad j = 1, ..., r \tag{16}$$

and

$$f(z^k) - f(x^0) = \vartheta_k y^k \nabla f(x^0) + \tfrac{1}{2}(\vartheta_k)^2 \cdot \\ \cdot y^k [Hf(x^0 + \xi_k \vartheta_k y^k)]\, y^k \leqq 0 \,, \tag{17}$$

where ξ_k, $\xi_{i,k}$, $\bar{\xi}_{j,k}$ are again numbers between 0 and 1.
Multiplying (15) and (16) by the corresponding λ_i and μ_j and adding to (17) yields

$$\vartheta_k y^k \left\{\nabla f(x^0) + \sum_{i=1}^{m} \lambda_i \nabla g_i(x^0) + \sum_{j=1}^{r} \mu_j \nabla h_j(x^0)\right\} +$$

$$+ \tfrac{1}{2}(\vartheta_k)^2 y^k \Big[Hf(x^0 + \xi_k \vartheta_k y^k) + \sum_{i=1}^{m} \lambda_i H g_i(x^0 + \xi_{i,k}\vartheta_k y^k) +$$

$$+ \sum_{j=1}^{r} \mu_j H h_j(x^0 + \bar{\xi}_{j,k}\vartheta_k y^k)\Big]\, y^k \leqq 0 \,.$$

The expression in brackets vanishes by the Kuhn-Tucker conditions; dividing the remaining portion by $\frac{1}{2}(\vartheta_k)^2$ and taking limits we obtain

$$\bar{y}\Big[Hf(x^0) + \sum_{i=1}^{m} \lambda_i H g_i(x^0) + \sum_{j=1}^{r} \mu_j H h_j(x^0)\Big]\, \bar{y} \leqq 0 \,. \tag{17}$$

Since $\bar{y}$ is nonzero and belongs to $\hat{Z}(x^0)$, it follows that (17) contradicts (6). □

Note that Theorem 3.10.5 contains the classical sufficient conditions for a point x^0 to be a strict local unconstrained minimum of the twice continuously differentiable function $f : X \to I\!R$ and for x^0 to be a strict local minimum for problem (P_e), i.e. Theorems 3.2.5 and 3.3.2 are cases of Theorem 3.10.5. Hettich and Jongen (1977) include in the definition of $\hat{Z}(x^0)$ also the equation $z \nabla f(x^0) = 0$. Although this is valid, it does not further restrict $\hat{Z}(x^0)$, since when x^0 is a point satisfying the Kuhn-Tucker conditions (3)–(5) of Section 3.8 and $z \in \hat{Z}(x^0)$, we have automatically $z \nabla f(x^0) = 0$.

Fiacco (1968) extends Theorem 3.10.5 to sufficient conditions for a not necessarily strict minimum for problem (P_1).

Let us again consider the set $\hat{Z}(x^0)$ and let $z \in \hat{Z}(x^0)$, $z \neq 0$. Define the Lagrangian function for (P_1):

$$\mathcal{L}(x, \lambda, \mu) = f(x) + \lambda g(x) + \mu h(x)$$

and define the set

$$Y(\varepsilon, \delta) = \{y \mid \|y - z\| \leqq \varepsilon \text{ for some } z \in \hat{Z}(x^0),\ x^0 + \delta_y y \text{ is feasible for some } \delta_y,\ 0 < \delta_y < \delta,\ \varepsilon > 0 \text{ and } \|y\| = 1\} .$$

Theorem 3.10.6 (Neighbourhood sufficiency theorem). Let $x^0 \in S_1$; if there exist vectors λ, μ satisfying the Kuhn-Tucker conditions (3)–(5) of Section 3.8 and if there exists $\varepsilon' > 0$, $\delta' > 0$ such that for every $y \in Y(\varepsilon', \delta')$ it follows that

$$y[H_x\mathcal{L}(x^0 + t\delta_y y, \lambda, \mu)]\, y \geqq 0 , \tag{18}$$

$\forall t \in (0, 1)$, then x^0 is a local minimum for (P_1).

Proof. See Fiacco (1968) and Fiacco and McCormick (1968). □

If in (18) we have

$$y[H\mathcal{L}_x(x^0 + t\delta_y y, \lambda, \mu)]\, y > 0 ,$$

for all $t \in (0, 1)$, then x^0 is a strict local minimum for problem (P_1). Fiacco (1968) proves that Theorem 3.10.5 of McCormick can be obtained, as a corollary, from Theorem 3.10.6.

Hestenes (1975) and Robinson (1982) (the second author considers a more general problem than (P_1)) prove that the sufficient optimality conditions of Theorem 3.10.5 actually provide a sharper result. Indeed the following theorem holds.

Theorem 3.10.7. Let $x^0 \in S_1$; if there exist vectors λ, μ such that the Kuhn-Tucker conditions (3)–(5) of Section 3.8 are satisfied at x^0 and if for every $z \in \hat{Z}(x^0)$, $z \neq 0$, it follows that

$$z\Big[Hf(x^0) + \sum_{i=1}^{m} \lambda_i Hg_i(x^0) + \sum_{j=1}^{r} \mu_j Hh_j(x^0)\Big]\, z > 0 ,$$

then there are a neighbourhood $N(x^0)$ and a constant $m > 0$ such that

$$f(x) \geqq f(x^0) + m \|x - x^0\|^2 , \qquad \forall x \in S_1 \cap N(x^0) .$$

Proof. See Hestenes (1975). □

Robinson (1982) asks whether the strict local minimizer obtained by means of Theorems 3.10.5 or 3.10.7 is also an *isolated* local minimum, that is, whether there is some neighbourhood of x^0 containing no other local minimizer for problem (P_1). Robinson considers the following example:

minimize $\frac{1}{2} x^2$

subject to: $x^6 \sin(1/x) = 0$

$$[\sin(1/0) := 0] .$$

The feasible region is $\{0\} \cup \{(n\pi)^{-1}, n = \pm 1, \pm 2, ...\}$. The second-order sufficient conditions of Theorem 3.10.5 are satisfied at the origin. However, the origin is a cluster point of the feasible region and every feasible point is a local minimizer. It is true that this is a "bad" problem; but this anomaly can be excluded by means of some "regularity" conditions on the constraints.

More precisely, Robinson shows that if at the feasible point x^0 the Kuhn-Tucker conditions (3)–(5) of Section 8 hold for (P_1), the Mangasarian-Fromovitz constraint qualification is satisfied at x^0 and, moreover, the following General Second-Order Sufficient Conditions hold at x^0:

G.S.O.S.C.: Relation (6) of Theorem 3.10.5 holds for every $z \in \hat{Z}(x^0)$, $z \neq 0$, and for every λ, μ such that (x^0, λ, μ) satisfies the Kuhn-Tucker conditions.

Then x^0 is actually an *isolated* local minimum point for (P_1). Note that if the Independence C.Q., or the Strict Mangasarian-Fromovitz C.Q. hold, then in the Kuhn-Tucker conditions the multipliers λ, μ are unique and so Theorem 3.10.5 assures that x^0 is an isolated local minimum point for (P_1), being that the General Second-Order Sufficient Condition is automatically verified.

Another area of application of the second-order sufficient optimality conditions is in the study of the sensitivity analysis in nonlinear programming. General results, depending on Theorem 3.10.5, are given by Fiacco and McCormick (1968) and Fiacco (1983). We present only the following result.

Theorem 3.10.8. Let f, g_i, $i = 1, ..., m$; h_j, $j = 1, ..., r$, be twice continuously differentiable on the open set $X \subset \mathbb{R}^n$; consider the problem $P(b, c)$:

$$\text{Min } f(x)$$

subject to:

$$g_i(x) \leqq b_i \ , \qquad i = 1, ..., m$$

$$h_j(x) = c_j \ , \qquad j = 1, ..., r \ .$$

Let x^0 be a local solution of this problem, where $b_i = 0$, $i = 1, ..., m$, and $c_j = 0$, $j = 1, ..., r$, and let x^0 satisfy the following conditions:

i) The gradients $\nabla g_i(x^0)$, $i \in I(x^0)$, $\nabla h_j(x^0)$, $j = 1, ..., r$, are linearly independent.

ii) The second-order sufficient conditions of Theorem 3.10.5 are satisfied at x^0.

iii) In the Kuhn-Tucker conditions $\lambda_i > 0$, $\forall i \in I(x^0)$ (i.e. the strict complementary slackness conditions hold).

Then there is a continuously differentiable vector-valued function $x^0(b, c)$, defined on a neighbourhood of $(0, 0)$ in $\mathbb{R}^m \times \mathbb{R}^r$, such that $x^0(0, 0) = x^0$ and such that for every (b, c) in a neighbourhood of $(0, 0)$, $x^0(b, c)$ is a strict local solution of problem $P(b, c)$. Moreover,

$$\nabla_b f(x^0(b, c))|_{(0,0)} = -\lambda$$

$$\nabla_c f(x^0(b, c))|_{(0,0)} = -\mu \ .$$

Proof. See Fiacco (1983), McCormick (1983), Luenberger (1984). □

The reader may note that the above theorem extends to a problem with both inequality and equality constraints, the results on the interpretation of Lagrange multipliers, given in Theorem 3.3.5.

3.11. Linearization Properties of a Nonlinear Programming Problem

In this section we study some interesting characterizations of the solutions of a smooth nonlinear programming problem, with respect to various linearized forms of the same problem. For simplicity we shall consider problem (P) and suppose that $X \subset \mathbb{R}^n$ is open and that all functions are differentiable:

$$\operatorname*{Min}_{x \in S} f(x) \tag{P}$$

$$S = \{x \mid x \in X,\ g_i(x) \leqq 0,\ i = 1, ..., m\} .$$

Consider then the following problems, where $x^0 \in S$:

$$(L_1)\quad \operatorname{Min}\{(x - x^0)\nabla f(x^0) \mid x \in X,\ g_i(x) \leqq 0,\ i = 1, ..., m\} ;$$

$$(L_2)\quad \operatorname{Min}\{(x - x^0)\nabla f(x^0) \mid x \in X,\ (x - x^0)\nabla g_i(x^0) \leqq 0,\ i \in I(x^0)\} ;$$

$$(L_3)\quad \operatorname{Min}\{f(x) \mid x \in X,\ (x - x^0)\nabla g_i(x^0) \leqq 0,\ i \in I(x^0)\} .$$

(L_1), (L_2) and (L_3) represent various "degrees of linearization" of problem (P). In this section we shall establish necessary and sufficient conditions such that a solution x^0 of one of the four problems above is also a solution of the other three remaining problems. This points out a kind of invariance property of the solution of (P) with respect to the three linearizations considered (see also Kortanek and Evans (1967)). If x^0 is a solution, e.g., of (P) and also solution, e.g., of (L_1), we shall write for sake of brevity $(P) \Rightarrow (L_1)$.

Theorem 3.11.1. Let $x^0 \in S$.

i) If x^0 is a solution of (P), at x^0 the constraints of (P) satisfy a C.Q. and every g_i, $i \in I(x^0)$, is quasiconvex at x^0, then x^0 is also a solution of (L_1); if x^0 is a solution of (P) and at x^0 the constraints of (P) satisfy a C.Q., then x^0 is also a solution of (L_2); if x^0 is a solution of (P), at x^0 the constraints of (P) satisfy a C.Q. and f is pseudoconvex at x^0, then x^0 is also a solution of (L_3).

ii) If x^0 is a solution of (L_1) and f is pseudoconvex at x^0, then x^0 is also a solution of (P); if x^0 is a solution of (L_2), f is pseudoconvex at x^0 and every g_i, $i \in I(x^0)$, is quasiconvex at x^0, then x^0 is also a solution of (P); if x^0 is a solution of (L_3) and every g_i, $i \in I(x^0)$, is quasiconvex at x^0, then x^0 is also a solution of (P).

Proof. We prove the following implications:

1) $(P) \Rightarrow (L_1)$. Note first that S is the feasible set of both (P) and (L_1). Suppose that x^0 is optimal for (P); then, thanks to the quasiconvexity of g_i, $i \in I(x^0)$, we have the following implication, for each $x \in S$:

$$g_i(x) \leqq g_i(x^0) = 0\,, \quad i \in I(x^0) \Rightarrow (x - x^0)\,\nabla g_i(x^0) \leqq 0\,. \tag{1}$$

Being a constraint qualification satisfied at x^0, the Kuhn-Tucker conditions will hold and, taking (1) into account, we have

$$(x - x^0)\,\nabla f(x^0) \geqq 0\,, \qquad \forall\, x \in S\,,$$

i.e. x^0 is also optimal for (L_1).

2) $(P) \Rightarrow (L_2)$. Denote with D_2 the feasible set of (L_2); as at x^0, solution of (P), a constraint qualification is satisfied by g_i, $i = 1, 2, ..., m$, the Kuhn-Tucker conditions will hold, so that we can write

$$(x - x^0)\,\nabla f(x^0) = -\sum_{i \in I(x^0)} \lambda_i (x - x^0)\,\nabla g_i(x^0) \geqq 0\,,$$

$$\forall\, x \in D_2\,. \tag{2}$$

Hence x^0, being in D_2, is also an optimal solution for (L_2).

3) $(L_1) \Rightarrow (P)$. Suppose that x^0 is optimal for (L_1), i.e.

$$(x - x^0)\, \nabla f(x^0) \geqq 0\,, \qquad \forall x \in S\,. \tag{3}$$

   Then, being that f is pseudoconvex at x^0, from (3) we can draw the inequality $f(x) \geqq f(x^0)$, $\forall x \in S$. Hence x^0 is also optimal for (P).

4) $(L_3) \Rightarrow (P)$. The objective functions of the two problems are equal; suppose then that x^0 is optimal for (L_3). Then we have, because of the quasiconvexity of g_i, $i \in I(x^0)$,

$$S \subset \{x \mid x \in X,\ g_i(x) \leqq g_i(x^0) = 0,\ \forall i \in I(x^0)\} \subset$$
$$\subset \{x \mid x \in X,\ (x - x^0)\, \nabla g_i(x^0) \leqq 0,\ \forall i \in I(x^0)\}\,.$$

   So the assertion is trivial.

5) Under the pseudoconvexity of f we have $(L_2) \Rightarrow (L_3)$. The proof parallels the one given for proving the implication $(L_1) \Rightarrow (P)$.

6) For completeness let us prove also that, under the quasiconvexity of g_i, $i \in I(x^0)$, we have $(L_2) \Rightarrow (L_1)$. The proof parallels the one given for proving the implication $(L_3) \Rightarrow (P)$. □

From the previous theorem it appears that given the validity of a C.Q. at $x^0 \in S$, the pseudoconvexity of f and the quasiconvexity of every g_i, $i \in I(x^0)$, the four problems (P), (L_1), (L_2) and (L_3) are equivalent, in the sense that they have the same solutions. In particular, we note that (L_2) is a linear program, both in the objective function and in the constraints and therefore it may be viewed, under the said assumptions of equivalence, as a "linear test" for the optimality in the nonlinear problem (P).

The inclusions proved in Theorem 3.11.1 are depicted in the following diagram (Figure 15).

<u>Corollary 3.11.1</u>. Let f be pseudoconvex at $x^0 \in S$ and every g_i, $i \in I(x^0)$, quasiconvex at x^0. If at x^0 a constraint qualification is satisfied, then a

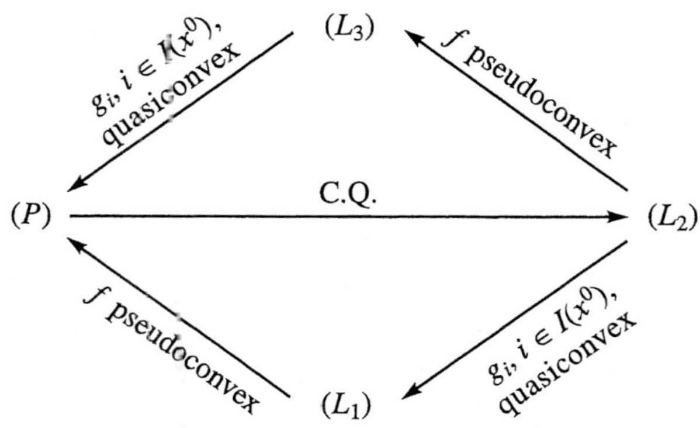

Figure 15.

necessary and sufficient condition for x^0 to be a solution of (P) is that the system of inequalities

$$\begin{cases} y\nabla g_i(x^0) \leqq 0 \ , & i \in I(x^0) \\ y\nabla f(x^0) < 0 \end{cases} \tag{4}$$

admits no solution $y \in \mathbb{R}^n$.

Proof. Under our assumptions, thanks to Theorem 3.11.1, x^0 is a solution of (P) if and only if it is a solution of (L_2), i.e.

$$(x - x^0)\,\nabla g_i(x^0) \leqq 0\,, \quad i \in I(x^0) \ \Rightarrow$$

$$\Rightarrow \ x\nabla f(x^0) \geqq x^0\nabla f(x^0)\,. \tag{5}$$

Write $y = x - x^0$; then (5) becomes

$$y\nabla g_i(x^0) \leqq 0\,, \quad i \in I(x^0) \Rightarrow y\nabla f(x^0) \geqq 0\,.$$

The sufficiency part springs directly from the Farkas-Minkowski theorem of the alternative, since from the inconsistency of (4) we can establish the existence of multipliers $\lambda_i \geqq 0$, $i \in I(x^0)$, such that

$$-\nabla f(x^0) = \sum_{i \in I(x^0)} \lambda_i \nabla g_i(x^0)\,.$$

However, being that f is pseudoconvex at x^0 and every g_i, $i \in I(x^0)$, quasiconvex at x^0, this last relation qualifies x^0 as a solution of problem (P).□

Historically every constraint qualification for problem (P) or (P_1) affects only the constraints of the said problems. There are in the literature, especially for the nondifferentiable convex case, also conditions affecting the constraints together with the objective function, conditions assuring the validity of Kuhn-Tucker type relations. It is more appropriate to call these last conditions "regularity conditions". These are often investigated with reference to duality or saddle points theorems for a convex program. (See Sections 3.14 and Chapter 5. See also Cambini (1986), Geoffrion (1971), Giannessi (1984), Martein (1985), Rockafellar (1970, 1974).)

A necessary and sufficient regularity condition for the differentiable problem (P) is simply obtained from the previous results. In fact, we have the following:

Corollary 3.11.2. The point $x^0 \in S$ is a solution of (L_2) if and only if there exists a vector λ such that the pair (x^0, λ) fulfills the Kuhn-Tucker conditions for (P):

$$\nabla f(x^0) + \lambda \nabla g(x^0) = 0$$

$$\lambda g(x^0) = 0$$

$$\lambda \geqq 0 \ .$$

Corollary 3.11.2 has also been obtained by Martein (1985) in a wider context of regularity conditions.

3.12. Some Specific Cases

In this section we briefly consider some particular, or, more generally, some quite important cases of problems (P) or (P_1).

a) In many mathematical programming problems arising from economic-financial models, the independent variables are requested to be non-negative:

$$\text{Min}\,\{f(x) \mid x \in X \in I\!R^n,\ g(x) \leqq 0,\ x \geqq 0\}\ . \tag{1}$$

Let $x^0 \in \text{int}(X)$ be a local solution of problem (1) and let at x^0 some C.Q. be satisfied. It is easy to see that the Kuhn-Tucker conditions for (1) are expressed by

$$\nabla f(x^0) + \sum_{i=1}^{m} \lambda_i \nabla g_i(x^0) \geqq 0$$

$$x^0\Big[\nabla f(x^0) + \sum_{i=1}^{m} \lambda_i \nabla g_i(x^0)\Big] = 0$$

$$\lambda g(x^0) = 0$$

$$\lambda \geqq 0\ .$$

If we write also the feasibility conditions for x^0, i.e.

$$x^0 \geqq 0\ , \qquad g(x^0) \leqq 0$$

and set $\psi(x,\lambda) = f(x) + \lambda g(x)$, all the previous conditions may be equivalently written in a "symmetric" manner:

$$x^0 \geqq 0\ , \qquad \nabla_x \psi(x^0,\lambda) \geqq 0\ , \qquad x^0 \nabla_x \psi(x^0,\lambda) = 0$$

$$\lambda \geqq 0\ , \qquad \nabla_\lambda \psi(x^0,\lambda) \leqq 0\ , \qquad \lambda \nabla_\lambda \psi(x^0,\lambda) = 0\ .$$

b) A more general problem than (1) has been considered by Mangasarian (1969): let $X^0 \subset I\!R^{n_1} \times I\!R^{n_2}$, X^0 open, and consider the problem

$$\text{Min}\,\{f(x,y) \mid (x,y) \in X^0,\ g(x,y) \leqq 0,\ h(x,y) = 0,\ y \geqq 0\}\ . \tag{2}$$

If (x^0, y^0) is a local solution of (2) and if the constraints of (2) satisfy at (x^0, y^0) some C.Q., it is easy to see that the Kuhn-Tucker conditions for this problem are expressed by

$$\nabla_x \mathcal{L}(x^0, y^0, \lambda, \mu) = 0 \ ,$$
$$\nabla_y \mathcal{L}(x^0, y^0, \lambda, \mu) \geqq 0 , \quad y^0(\nabla_y \mathcal{L}(x^0, y^0, \lambda, \mu) = 0 , \quad y^0 \geqq 0 \ ,$$
$$\lambda \geqq 0 , \quad \nabla_\lambda \mathcal{L}(x^0, y^0, \lambda, \mu) \leqq 0 , \quad \lambda(\nabla_\lambda \mathcal{L}(x^0, y^0, \lambda, \mu) = 0 \ ,$$
$$\mu \in \mathbb{R}^r \ , \quad \nabla_\mu \mathcal{L}(x^0, y^0, \lambda, \mu) = 0 \ ,$$

where

$$\mathcal{L}(x, y, \lambda, \mu) = f(x, y) + \lambda g(x, y) + \mu h(x, y) \ .$$

These conditions are also sufficient if at (x^0, y^0), $f(x, y)$ is pseudoconvex, $g_i(x, y)$, $i \in \{i \mid g_i(x^0, y^0) = 0\}$ is quasiconvex and $h_j(x, y)$, $j = 1, ..., r$, is quasilinear (i.e. both quasiconvex and quasiconcave).

c) Let $X \subset \mathbb{R}^n$ open, $f : X \to \mathbb{R}$, $S = X \cap \mathbb{R}^n_+$; a necessary condition for $x^0 \in S$ to be optimal for f over S is:

$$\nabla f(x^0) \geqq 0 \ , \qquad x^0 \nabla f(x^0) = 0 \ .$$

These conditions are also sufficient for $x^0 \in S$ to be optimal for f over S if f is pseudoconvex at x^0 (with respect to S).

d) A very important case of (P) and (P_1) is when all the constraints are linear.
Let $\Omega \in \mathbb{R}^{n_1+n_2+n_3}$ be the set defined by the solution of the system

$$A_{11}x^1 + A_{12}x^2 + A_{13}x^3 \geqq b^1$$
$$A_{21}x^1 + A_{22}x^2 + A_{23}x^3 = b^2$$
$$A_{31}x^1 + A_{32}x^2 + A_{33}x^3 \leqq b^3$$
$$x^1 \geqq 0 \ , \qquad x^2 \in \mathbb{R}^{n_2} \ , \qquad x^3 \leqq 0 \ ,$$

with $x^j \in \mathbb{R}^{n_j}$, $b^i \in \mathbb{R}^{m_i}$ and A_{ij} $(i, j = 1, 2, 3)$ is an $m_i \cdot n_j$ (real) matrix.
Then a necessary condition, such that $\bar{x} = (\bar{x}^1, \bar{x}^2, \bar{x}^3) \in \Omega$ is a local minimum of the differentiable function $f(x^1, x^2, x^3) : \mathbb{R}^{n_1+n_2+n_3} \to \mathbb{R}$, is that there exist multipliers $\lambda_i \in \mathbb{R}^{m_i}$ $(i = 1, 2, 3)$, such that

$$\nabla_1 = \nabla_{x^1} f(\bar{x}) - (\lambda^1 A_{11} + \lambda^2 A_{21} + \lambda^3 A_{31}) \geqq 0 \,,$$
$$\bar{x}^1 \nabla_1 = 0 \,;$$

$$\nabla_{x^2} f(\bar{x}) - (\lambda^1 A_{12} + \lambda^2 A_{22} + \lambda^3 A_{32}) = 0$$

$$\nabla_3 = \nabla_{x^3} f(\bar{x}) - (\lambda^1 A_{13} + \lambda^2 A_{23} + \lambda^3 A_{33}) \leqq 0 \,,$$
$$\bar{x}^3 \nabla_3 = 0 \,;$$

$$\lambda^1 \geqq 0 \,, \qquad \lambda^1(A_{11}\bar{x}^1 + A_{12}\bar{x}^2 + A_{13}\bar{x}^3 - b^1) = 0$$

$$\lambda^3 \leqq 0 \,, \qquad \lambda^3(A_{31}\bar{x}_1 + A_{32}\bar{x}^2 + A_{33}\bar{x}^3 - b^3) = 0 \,.$$

If $f(x)$ is pseudoconvex at $\bar{x}$, the above conditions are also sufficient for a (global) minimum of f on Ω.

e) A special case of d) is when the objective function f is linear; we have in this case the *linear programming problem*. The literature for this important class of mathematical programming problems is enormous; we quote only the book of Bazaraa, Jarvis and Sherali (1990) and, for a historical account, the paper of Dantzig (1983).
Consider the following standard formulation:

$$\operatorname*{Min}_{x} \; \{cx \mid Ax \geqq b, \; x \geqq 0\} \,,$$

where $b \in \mathbb{R}^m$, $c \in \mathbb{R}^n$, A is a real matrix of order m, n. By means of the Kuhn-Tucker conditions it is quite immediate to prove that necessary and sufficient conditions, such that a feasible $\bar{x}$ solves the linear program above, are that there exists a vector $\lambda \in \mathbb{R}^m$ such that

$$\lambda A \leqq c \,, \qquad \lambda(A\bar{x} - b) = 0 \,,$$

$$\lambda \geqq 0 \,, \qquad (\lambda A - c)\, \bar{x} = 0 \,.$$

f) Another important specific case of mathematical programming problems, rather close to linear programming problems from an analytical and computational point of view is the *quadratic programming problem*, i.e.

$$\operatorname*{Min}_{x} \; \{cx + \tfrac{1}{2}\, xCx \mid Ax \geqq b, \; x \geqq 0\} \; ,$$

where $b \in I\!R^m$, $c \in I\!R^n$, A is a matrix of order m, n and C a symmetric matrix of order n. A necessary condition such that a feasible $\bar{x} \in I\!R^n$ is optimal for this problem is that there exists a vector $\lambda \in I\!R^m$ such that

$$c + C\bar{x} - \lambda A \geqq 0 \; , \qquad \bar{x}(c + C\bar{x} - \lambda A) = 0 \; ,$$

$$\lambda \geqq 0 \; , \qquad \lambda(-b + A\bar{x}) = 0 \; .$$

If C is positive semidefinite, then the objective function is convex (strictly convex if C is positive definite) and in this case the above conditions become also sufficient for the optimality of the feasible point x.

A classical book on quadratic programming is Boot (1964). See also Avriel (1976), Bazaraa, Sherali and Shetty (1993) and the references quoted therein.

g) Another problem that has recently received much attention is the *fractional* (or *hyperbolic*) *programming problem* (see, e.g., Craven (1988), Schaible (1978), Singh and Dass (1989), Schaible (1981), Cambini, Castagnoli, Martein, Mazzoleni and Schaible (1990)).
A programming problem, e.g. of the form

$$\operatorname{Min}\{f(x) \mid x \in X, \; g_i(x) \leqq 0, \; i = 1, ..., m\}$$

is called a fractional programming problem, when the objective function f is given by

$$f(x) = \frac{u(x)}{v(x)} \; , \qquad v(x) \neq 0 \; .$$

We have seen, in Section 2.15 of Chapter II, many criteria for establishing the generalized convexity (concavity) of a ratio between two functions. These criteria may be applied for obtaining further information on the optimality conditions for a fractional programming problem. E.g. if

$$f(x) = \frac{cx + \alpha}{dx + \beta} , \qquad dx + \beta \neq 0 ,$$

i.e. we have a linear fractional objective function, $f(x)$ will be pseudolinear (i.e. both pseudoconvex and pseudoconcave).

h) Another widely studied class of programming problems is *geometric programming* (see, e.g. Duffin, Peterson and Zener (1967) and Avriel, Rijckaert and Wilde (1973).
First define a *polynomial* g as a real function consisting of a finite sum of terms

$$g(x) = \sum_i c_i \prod_{j=1}^{n} (x_j)^{a_{ij}} ,$$

where $c_i > 0$ and $a_{ij} \in \mathbb{R}$ are given constants and $x_j > 0$, $\forall j = 1, ..., n$. (Note that, in general, polynomials are neither convex nor concave; however, they have some properties of generalized convexity.) A nonlinear program given by

$$\text{Min} \{g_0(x) \mid g_k(x) \leqq 1, \ k = 1, ..., m, \ x > 0\} ,$$

where $g_0, g_1, ..., g_m$ are polynomials, is called a *geometric programming problem*. These problems have a variety of applications, especially in the engineering sciences.

i) The following problem, called discrete *min-max problem*, *Maximum component minimum problem*, goes back to the classical paper of Kuhn and Tucker (1951) and has been subsequently studied by many authors. See, e.g. Bram (1966), Dem'yanov and Malozemov (1974), Danskin (1966, 1967), Kapur (1973), Angrisani (1982) and Pallu de la Barrière (1963) who calls it the "Kantorovich problem":

$$\min_{x \in S_1} \ \max_{1 \leqq i \leqq q} \ f_i(x) , \qquad (P_K)$$

$$S_1 = \{x \mid x \in X, \ g(x) \leqq 0, \ h(x) = 0\} ,$$

where $X \subseteq I\!R^n$ is an open set, $g: I\!R^n \to I\!R^m$, $h: I\!R^n \to I\!R^r$. If all the functions involved in this problem are twice continuously differentiable at the point $x^0 \in S_1$, it is possible to prove the following necessary and sufficient optimality conditions for (P_K) (see, e.g., Pallu de la Barrière (1963)). Let

$$\tilde{f}(x) = \max_{1 \leqq i \leqq q} f_i(x) ,$$

$$I(x^0) = \{i \mid g_i(x^0) = 0\} ,$$

$$H(x^0) = \{i \mid \tilde{f}(x^0) = f_i(x^0)\} .$$

a) Necessary conditions of optimality.
If x^0 is a point of local minimum for $\tilde{f}$ on S_1 and if the vectors $\nabla g_i(x^0)$, $i \in I(x^0)$ and $\nabla h_j(x^0)$, $j = 1, ..., r$, are linearly independent, then there exist multipliers $t_i \geqq 0$, $i \in H(x^0)$, with

$$\sum_{i \in H(x^0)} t_i = 1 ;$$

$$\lambda_i \geqq 0 , \quad i \in I(x^0) ; \qquad v_j \in I\!R, \quad j = 1, ..., r$$

such that

$$\sum_{i \in H(x^0)} t_i \nabla f_i(x^0) + \sum_{i \in I(x^0)} \lambda_i \nabla g_i(x^0) + \\ + \sum_{j=1}^{r} v_j \nabla h_j(x^0) = 0 . \tag{1}$$

b) Sufficient conditions for local optimality.
Let $\tilde{\mathcal{L}}(x, t, \lambda, v) = tf(x) + \lambda g(x) + vh(x)$; if there exist multipliers t_i $(i \in H(x^0))$, λ_i $(i \in I(x^0))$, v_j $(j = 1, ..., r)$ satisfying the properties sub a), if relation (1) is satisfied at $x^0 \in S_1$ and if

$$y H_x \tilde{\mathcal{L}}(x^0, t, \lambda, v)\, y > 0$$

for every $y \neq 0$ of the subspace defined by

$$y\nabla f_i(x^0) = 0 \;, \qquad \forall i \in H(x^0)\,, \;\; \text{with } t_i > 0$$

$$y\nabla g_i(x^0) = 0 \;, \qquad \forall i \in I(x^0)\,, \;\; \text{with } \lambda_i > 0$$

$$y\nabla h_j(x^0) = 0 \;, \qquad \forall k = 1,...,r \;,$$

then x^0 is a point of strict local minimum of $\tilde{f}$ on S_1.

c) An extention of problem (P) (or (P_1)) may be obtained by requiring that the constraints belong to some convex cone (not necessarily the nonpositive orthant). This extension is studied, e.g., by Craven (1978) who considers the problem

$$\text{Min}\,\{f(x) \mid x \in X,\; g(x) \in V\} \;, \tag{2}$$

where V is a closed convex cone of $I\!R^m$. The Kuhn-Tucker conditions necessary for a minimum of (2) at the feasible point $x^0 \in \text{int}(X)$ are, assuming a suitable constraint qualification:
there exists a vector of multipliers $\lambda \in V^*$ (the polar of V) such that

$$\nabla f(x^0) + \lambda \nabla g(x^0) = 0$$

$$\lambda g(x^0) = 0 \;.$$

See also Bazaraa and Goode (1972), Glover (1983), Guignard (1969), Massam and Zlobec (1974, 1978), Varaiya (1967), Nagahisa and Sakawa (1969).

3.13. Extensions to Topological Spaces

(These hints require some basic notions of topology.)
Since the early years of the development of the studies concerning mathematical programming problems, many articles considered such problems defined on topological spaces, especially Banach spaces, and with various differentiability assumptions. See, e.g., Bazaraa and Goode (1973), Bender (1978), Borwein (1978), Das (1975), Girsanov (1972), Holmes (1972), Hurwicz (1958), Massam and Zlobec (1974, 1978), Nagahisa and Sakawa

(1969), Neustadt (1976), Ritter (1969, 1970), Zaffaroni (1993), Ben-Tal and Zowe (1982), Maurer and Zowe (1979), Russel (1966), Luenberger (1969). An extension to Banach space of the approach given in the previous section for problems (P) and (P_1) is presented in Guignard (1969) and Gould and Tolle (1972, 1975), under Fréchet differentiability assumption of the functions involved. We note, however, that if the problems are not defined in a normed linear space, even the assumpion of Fréchet differentiability may not be possible; in this case one must make use of more general notions of differentiability (see Chapter IV, in addition to the above cited papers).

Here we briefly describe the results of Gould and Tolle (1972, 1975). Let X and Y be Banach spaces and let $g : X \to Y$ be (Fréchet) differentiable; suppose $A_x \subset X$, $A_y \subset Y$ and define the constraint set

$$S = A_x \cap g^{-1}(A_y) ,$$

i.e.

$$S = \{x \mid x \in A_x,\ g(x) \in A_y\} .$$

Suppose $f : X \to \mathbb{R}$, f differentiable; then the optimization problem of interest is

$$\min_{x \in S} f(x) .$$

The set of all objective functions f which have a local constrained minimum $x^0 \in S$ will be denoted by F_0 and the set of all derivatives at x^0 of elements in F_0 will be denoted by DF_0. Denote the topological duals of X, Y as X^*, Y^*, respectively, and for any set $N^* \subset X^*$ let $\bar{N}^*$ denote the closure of N^* in the weak * topology. For B a nonempty subset of X, the (*negative*) *polar cone* of B, B^-, is the subset of X^* given by

$$B^- = \{x^* \in X^* \mid x^*(b) \leqq 0,\ \forall b \in B\} .$$

The following properties of polar cones, already given for Euclidean spaces, are also true in Banach spaces:

i) If $B_1 \subset B_2$, then $B_2^- \subset B_1^-$;

ii) $B^- = (\overline{\text{conv}(B)})^-$;

iii) B^- is a closed convex cone;

iv) $B \subset (B^-)^-$, with equality if and only if B is a closed convex cone.

The Bouligand tangent cone to a subset B of the Banach space X at $x^0 \in \bar{B}$ is defined as in Section 3.4, but here we also define the *weak tangent cone* as follows.

Definition 3.13.1. The *weak tangent cone* to B at $x^0 \in \bar{B}$ is the set

$$T_w(B, x^0) = \{x \in X \mid \exists \{\lambda_n\} \in I\!R,\ \lambda_n \geqq 0,\ \forall n\,;$$
$$\exists \{x^n\} \in B : x^n \to x^0,\ \lambda_n(x^n - x^0) \to x \text{ weakly,}$$
$$\text{i.e. } \lambda_n x^*(x^n - x^0) \to x^*(x),\ \forall x^* \in X^*\}\ .$$

The *weak pseudotangent cone* to B at $x^0 \in \bar{B}$, denoted by $P_w(B, x^0)$, is the closure of the convex hull of $T_w(B, x^0)$.

Definition 3.13.2. Let $x^0 \in S$, the *pseudolinearizing cone at* x^0, $K(x^0)$, and the *weak pseudolinearizing cone at* x^0, $K_w(x^0)$, are defined by

$$K(x^0) = \{x \in X \mid Dg(x^0)(x) \in P(A_y, g(x^0))\}\ ,$$

where $Dg(x^0)$ denotes the derivative of g at x^0;

$$K_w(x^0) = \{x \in X \mid Dg(x^0)(x) \in P_w(A_y, g(x^0))\}\ .$$

By using the properties of the tangent cones, it can be easily verified that $K(x^0)$ and $K_w(x^0)$ are closed convex cones in X and that $K(x^0) \subset K_w(x^0)$, with equality holding if Y is finite dimensional or if A_y is convex.

Definition 3.13.3. The *cone of gradients at* $x^0 \in S$, $B^*(x^0)$, and the *weak cone of gradients at* $x^0 \in S$, $B^*_w(x^0)$, are defined by

$$B^*(x^0) = \{x^* \in X^* \mid x^* = y^* \cdot Dg(x^0),$$
$$\text{for some } y^* \in T^-(A_y, g(x^0))\}\ ;$$

$$B_w^*(x^0) = \{x^* \in X^* \mid x^* = y^* \cdot Dg(x^0),$$
$$\text{for some } y^* \in T_w^-(A_y, Dg(x^0)\} \; .$$

Then $B_w^*(x^0) \subset B^*(x^0)$, with equality holding if A_y is convex or Y is finite dimensional.

Varaiya (1967), in the case where A_y is convex, and Guignard (1969), more generally, have shown that the following relations hold:

$$\overline{B^*(x^0)} = K^-(x^0) \subset T^-(S, x^0)$$
$$-DF_0 \subset T^-(S, x^0) \; . \tag{1}$$

Thus, if the constraint qualification

$$T^-(S, x^0) \subset K^-(x^0)$$

holds, it follows that the optimality condition

$$-DF_0 \subset \overline{B^*(x^0)} \tag{2}$$

is true. It should be noted that, in the case when $B^*(x^0)$ is closed and (1) is satisfied, it follows from (2) that, for any $f \in F_0$, there is a $y^* \in T^-(A_y, g(x^0))$ such that $-Df(x^0) = y^* \cdot Dg(x^0)$. This is a direct extension of the Kuhn-Tucker conditions previously given for Euclidean spaces. Gould and Tolle (1972, 1975) prove the following, more general results.

Theorem 3.13.1. The following relations hold

$$\overline{B_w^*(x^0)} = K_w^-(x^0) \subset T_w^-(S, x^0) \; ;$$
$$-DF_0 \subset T_w^-(S, x^0) \; .$$

Corollary 3.13.1. The optimality condition

$$-DF_0 \subset \overline{B_w^*(x^0)} \tag{3}$$

holds, if the weak constraint qualification

$$T_w^-(S, x^0) \subset K_w^-(x^0) \tag{4}$$

is satisfied.

Corollary 3.13.2. If A_y is convex and the constraint qualification

$$T_w^-(S, x^0) \subset K^-(x^0)$$

is satisfied, then the optimality condition (2) holds.

It is natural to ask whether the constraint qualification (4) is the weakest which will ensure the validity of the optimality condition (3). The following theorem answers this question in the affirmative, under an additional assumption.

Theorem 3.13.2. If X is reflexive, then

$$-DF_0 = T_w^-(S, x^0) \ .$$

Therefore it follows from the previous relations, that for X reflexive, the weak constraint qualification (4) and the optimality condition (3) are equivalent.

For further considerations we refer the reader to the cited papers of Gould and Tolle (1972, 1975).

3.14. Optimality Criteria of the Saddle Point Type

In the last few years the words "nonsmooth optimization" generally refer to nonlinear programming problems (or also to problems of calculus of variations or optimal control), where the functions involved are not differentiable (in the sense of Fréchet), but rather satisfy weaker assumptions concerning various kinds of limits of various kinds of differential quotients, in order to obtain generalized gradients or generalized directional derivatives.

After the classical work on Convex Analysis of R.T. Rockafellar, the theory of *subgradients* of convex functions is by now widely known, together with its numerous applications in mathematical programming. Another important step in this direction has been achieved by F.H. Clarke, who extended the theory of subgradients to nonconvex (locally Lipschitz) functions. Other contributions for building an axiomatic theory of generalized directional derivatives are more recent and rely on topological and algebraic properties of various local conical approximations of sets (Section 3.4).

All these approaches will be treated in the next chapter; in the present section we are concerned with the *saddle point* characterization of optimality conditions. Indeed, this was the first approach used for treating a nonlinear (convex) programming problem in absence of differentiability of the functions involved. We take first into consideration the problem

$$\operatorname*{Min}_{x \in S} f(x) , \tag{P}$$

$$S = \{x \mid x \in X,\ g(x) \leqq 0\} ,$$

where $X \subset \mathbb{R}^n$, $f : X \to \mathbb{R}$, $g : X \to \mathbb{R}^m$.

With regard to (P) let us consider the *Lagrangian function*

$$\psi(x, \lambda) = f(x) + \lambda g(x) \tag{1}$$

defined for $x \in X$, $\lambda \geqq 0$. The vector λ in (1) is called "Lagrange multipliers vector" or also "dual variables vector". We say that (x^0, λ^0), with $x^0 \in X$, $\lambda^0 \geqq 0$, is a *saddle point* (more precisely: a *Kuhn-Tucker saddle point*) for ψ (with respect to X) and for (P) if:

$$\psi(x^0, \lambda) \leqq \psi(x^0, \lambda^0) \leqq \psi(x, \lambda^0) , \qquad \forall\, x \in X , \ \forall\, \lambda \geqq 0 , \tag{2}$$

i.e. if $\psi(x, \lambda^0)$ admits minimum at x^0 over X and $\psi(x^0, \lambda)$ admits maximum at λ^0, for each $\lambda \geqq 0$:

$$\psi(x^0, \lambda^0) = \min_{x \in X} \psi(x, \lambda^0) = \max_{\lambda \geqq 0} \psi(x^0, \lambda) .$$

In most cases the surface generated by a Lagrangian function $\psi : \mathbb{R}^2 \to \mathbb{R}$, which admits a saddle point, looks like a horse saddle (see Figure 1 of Section 3.2).

Lemma 3.14.1. A point (x^0, λ^0), $x^0 \in X$, $\lambda^0 \geqq 0$, is a saddle point of $\psi(x, \lambda)$ if and only if

a) x^0 minimizes $\psi(x, \lambda^0)$ over X;

b) $g(x^0) \leqq 0$;

c) $\lambda^0 g(x^0) = 0$;

i.e. (x^0, λ^0), $x^0 \in S$, $\lambda^0 \geqq 0$, is a saddle point of ψ if and only if

$$f(x^0) \leqq f(x) + \lambda^0 g(x) , \qquad \forall x \in X .$$

Proof. Suppose that (x^0, λ^0) is a saddle point for ψ. By the first inequality of (2) we get $(\lambda - \lambda^0)\, g(x^0) \leqq 0$, $\forall \lambda \geqq 0$, which is possible only for $g(x^0) \leqq 0$; thus b) is fulfilled. Especially, with $\lambda = 0$ we get $\lambda^0 g(x^0) \geqq 0$ and, being $\lambda^0 \geqq 0$ and $g(x^0) \leqq 0$, also $\lambda^0 g(x^0) \leqq 0$. Hence it is $\lambda^0 g(x^0) = 0$, so c) holds. The second inequality of (2) just means a). Conversely, from b) and c) we get immediately, with $\lambda \geqq 0$,

$$\lambda g(x^0) \leqq 0 = \lambda^0 g(x^0) .$$

By adding $f(x^0)$ we have the first inequality of (2). The second inequality of (2) is assured by a). □

Relation sub c) is known as the *complementarity slackness condition.* A first result concerning sufficient optimality criteria in terms of saddle point of the Lagrangian function is that no convexity assumption on (P) is required (and obviously also no differentiability assumption).

Theorem 3.14.1. If (x^0, λ^0), $x^0 \in X$, $\lambda^0 \geqq 0$, is a saddle point for ψ with respect to X, then x^0 solves (P).

Proof. If (x^0, λ^0) is a saddle point for ψ, then thanks to Lemma 3.14.1, we have

$$g(x^0) \leqq 0 , \quad \lambda^0 \geqq 0 , \quad \lambda^0 g(x^0) = 0$$

and

$$f(x^0) \leqq f(x) + \lambda^0 g(x) , \qquad \forall x \in X$$

i.e.

$$f(x^0) \leqq f(x) \ , \qquad \forall x \in S \ . \qquad \square$$

It should be remarked that an analogous theorem holds also for a constrained minimization problem with both inequality and equality constraints. Indeed, as no convexity assumption was made in Theorem 3.14.1, equality constraints can be handled by replacing them with two inequality constraints of opposite sign. So if we define the problem

$$\min_{x \in S_1} f(x) \qquad (P_1)$$

$$S_1 = \{x \mid x \in X, \ g(x) \leqq 0, \ h(x) = 0\} \ ,$$

where $f : X \to \mathbb{R}$, $g : X \to \mathbb{R}^m$ and $h : X \to \mathbb{R}^p$, and define the Lagrangian function for (P_1)

$$\mathcal{L}(x, \lambda, \mu) = f(x) + \lambda g(x) + \mu h(x) \ ,$$

defined for $x \in X$, $\lambda \geqq 0$, $\mu \in \mathbb{R}^p$, the triplet (x^0, λ^0, μ^0) is a saddle point for $\mathcal{L}$ (with respect to X) and for (P_1) if

$$\mathcal{L}(x^0, \lambda, \mu) \leqq \mathcal{L}(x^0, \lambda^0, \mu^0) \leqq \mathcal{L}(x, \lambda^0, \mu^0) \ ,$$
$$\forall x \in X \ , \quad \forall \lambda \geqq 0 \ , \quad \forall \mu \in \mathbb{R}^p \ .$$

It is then quite immediate to prove the following result.

Theorem 3.14.2. If (x^0, λ^0, μ^0) is a saddle point for $\mathcal{L}(x, \lambda, \mu)$, with respect to X, then x^0 solves (P_1).

The necessary saddle point optimality conditions require first of all the convexity assumption on the functions involved in (P). We take therefore into consideration the *convex* nonlinear programming problem:

$$\min_{x \in S} f(x) \ , \qquad (P_c)$$

$$S = \{x \mid x \in X, \ g(x) \leqq 0\} \ ,$$

where $X \subset \mathbb{R}^n$ is a convex set, $f : X \to \mathbb{R}$ and $g : X \to \mathbb{R}^m$ are convex functions on X.

The first theorem we give is a Fritz John saddle point necessary optimality theorem. Define first the *Fritz John-Lagrange function* $\psi(x, \vartheta, u) = \vartheta f(x) + ug(x)$, where $x \in X$, $\vartheta \in \mathbb{R}$, $u \in \mathbb{R}^m$, $(\vartheta, u) \geq 0$. The following result is due to Uzawa (1958), Karlin (1959), Berge (1963).

Theorem 3.14.3. If x^0 solves (P_c), then there exists $\vartheta_0 \in \mathbb{R}$, $u^0 \in \mathbb{R}^m$, with $(\vartheta_0, u^0) \geq 0$, such that (x^0, ϑ_0, u^0) is a saddle point for ψ, i.e.

$$\psi(x^0, \vartheta_0, u) \leqq \psi(x^0, \vartheta_0, u^0) \leqq \psi(x, \vartheta_0, u^0) , \quad \forall x \in X , \quad \forall u \geqq 0 . \tag{3}$$

Moreover, $u^0 g(x^0) = 0$.

Proof. As x^0 is a solution of (P_c), the system

$$\begin{cases} f(x) - f(x^0) < 0 \\ g(x) \leqq 0 \\ x \in X \end{cases}$$

has no solution. Hence a fortiori the system

$$\begin{cases} f(x) - f(x^0) < 0 \\ g(x) < 0 \\ x \in X \end{cases}$$

has no solution. Thus, by the Fan-Glicksberg-Hoffman theorem (Theorem 2.9.1), there exists $\vartheta_0 \geqq 0$, $u^0 \geqq 0$, with $(\vartheta_0, u^0) \neq 0$ (i.e. $(\vartheta_0 . u^0)$ *semipositive vector*) such that

$$\vartheta_0[f(x) - f(x^0)] + u^0 g(x) \geqq 0 , \qquad \forall x \in X . \tag{4}$$

By letting $x = x^0$ in the above relation, we get $u^0 g(x^0) \geqq 0$. But since $u^0 \geqq 0$ and $g(x^0) \leqq 0$, we also have $u^0 g(x^0) \leqq 0$ and hence we obtain the complementarity condition $u^0 g(x^0) = 0$. From (4) we obtain also

$$\psi(x^0, \vartheta_0, u^0) = \vartheta_0 f(x^0) \leqq \vartheta_0 f(x) + u^0 g(x) = \\ = \psi(x, \vartheta_0, u^0) .$$

The first inequality of the Fritz John-Lagrange saddle point characterization (3) is trivially obtained. □

Theorem 3.14.3 obviously can be extended to problem (P_1), where f and g are convex functions on the convex set $X \subset \mathbb{R}^n$ and $h(x)$ is a linear affine function on $\mathbb{R}^n$.

Moreover, it should be remarked that if in the same theorem we have $\vartheta_0 = 0$, then the objective function f does not appear in the saddle point necessary optimality conditions and Theorem 3.14.3 is not, in this case, very expressive. In order to exclude $\vartheta_0 = 0$, i.e. to assure in (3) $\vartheta_0 > 0$ (and obviously in this case it is possible to choose $\vartheta_0 = 1$), we have to introduce a suitable *constraint qualification* (C.Q.) for problem (P_c). Here we consider the following constraint qualifications:

I) *Slater's C.Q.* (Slater (1950)).
(P_c) satisfies Slater's C.Q. if there exists $\bar{x} \in X$ such that $g(\bar{x}) < 0$.

II) *Karlin's C.Q.* (Karlin (1959)).
(P_c) satisfies Karlin's C.Q. if there exists no semipositive vector $p \geq 0$ such that $pg(x) \geqq 0$, $\forall\, x \in X$.

III) *Strict constraint qualification* (Mangasarian (1969)).
(P_c) satisfies the strict C.Q. if there exist two vectors $x^1, x^2 \in S$, $x^1 \neq x^2$, and g is strictly convex at x^1.

Theorem 3.14.4.

Slater's C.Q. $\Leftrightarrow$ Karlin's C.Q. $\Leftarrow$ Strict C.Q.

Proof. By the first corollary to the Fan-Glicksberg-Hoffman theorem (Corollary 2.9.1), Slater's C.Q. and Karlin's C.Q. are equivalent. Let us now prove that strict C.Q. implies Slater's C.Q.
Since X is convex we have

$$(1-\lambda)\,x^1 + \lambda x^2 \in X\,, \qquad \forall\,\lambda \in [0,1]$$

and since g is strictly convex at x^1 it follows that

$$g[(1-\lambda)\,x^1 + \lambda x^2] < (1-\lambda)\,g(x^1) + \lambda g(x^2) \leqq 0 \;,$$
$$\forall \lambda \in (0,1) \;,$$

where the last inequality follows from the fact that $g(x^1) \leqq 0$ and $g(x^2) \leqq 0$. Thus g satisfies at $\bar{x} = (1-\lambda)\,x^1 + \lambda x^2$ the Slater's C.Q. □

Regarding Slater's C.Q. we note that if $g : I\!R^n \to I\!R^m$ is continuous and $S = \{x \mid g(x) \leqq 0\}$, then the validity of Slater's C.Q. implies that S has a nonempty interior. The converse is, however, not true; consider e.g. the following example:

$$n = m = 1 \,, \quad g(x) = \begin{cases} 0 \,, & \text{if } |x| \leqq 1 \\ (|x|-1)^2 \,, & \text{if } |x| > 1 \,. \end{cases}$$

For other necessary and/or sufficient conditions assuring that in (2) we have $\vartheta_0 > 0$, the reader is referred to the paper of Martein (1985), where *regularity conditions* are taken into account, i.e. conditions involving both the objective function and the constraints.

Theorem 3.14.5 (Kuhn and Tucker (1951), Uzawa (1958), Karlin (1959)). Let x^0 be a solution of (P_c) and let one of the constraint qualifications I), II), III) be satisfied; then there exists $\lambda^0 \geqq 0$ such that (x^0, λ^0) is a saddle point of the Lagrangian function $\psi(x,\lambda) = f(x) + \lambda g(x)$ (and consequently $\lambda^0 g(x^0) = 0$).

Proof. Let us assume that Karlin's C.Q. holds (i.e. that Slater's C.Q. holds); by Theorem 3.14.3 there exists a triplet (x^0, ϑ_0, u^0) such that (3) holds. If $\vartheta_0 > 0$, Theorem 3.14.5 is proved as, in any case, $\lambda^0 g(x^0) = 0$. If $\vartheta_0 = 0$, then $u^0 \geq 0$ and from the second inequality of (3) it follows $0 \leqq u^0 g(x)$, $\forall x \in X$ (since $\vartheta_0 = 0$ and $u^0 g(x^0) = 0$). But this contradicts Karlin's C.Q. and therefore $\vartheta_0 > 0$. □

It is easy to construct examples where the constraint qualifications I) or II) or III) do not hold and the Lagrangian function $\psi(x,\lambda)$ admits no saddle point, though x^0 is a solution of (P).

Consider, e.g., the problem

$\text{Min}\, f(x) = x$

subject to: $x^2 \leqq 0, \quad x \in \mathbb{R}$.

The optimal solution is at $x_0 = 0$; the corresponding saddle point problem of the Lagrangian is to find $\lambda_0 \geqq 0$ such that

$$x_0 + \lambda x_0^2 \leqq x_0 + \lambda_0 x_0^2 \leqq x + \lambda_0 x^2$$

for every $x \in \mathbb{R}$ and for every $\lambda \geqq 0$, or equivalently

$$0 \leqq x + \lambda_0 x^2 \ .$$

Now, for $\lambda_0 \geqq 0$, the above inequality does not hold for any $x \in \mathbb{R}$. Thus there exists no λ_0 such that (x_0, λ_0) is a saddle point for our problem. Let us now consider the saddle point necessary optimality conditions for a constrained minimization problem with both convex inequality constraints and *linear affine* equality constraints:

$$\underset{x \in S_{1L}}{\text{Min}} \ f(x) \qquad (P_{1L})$$

$$S_{1L} = \{x \mid x \in \mathbb{R}^n, \ g(x) \leqq 0, \ h(x) = 0\} \ ,$$

where $f : \mathbb{R}^n \to \mathbb{R}$, $g : \mathbb{R}^n \to \mathbb{R}^m$ are convex functions and $h(x) = Ax - b$, with A a real matrix of order p, n and $b \in \mathbb{R}^p$.

We note that at present no necessary saddle point optimality condition is known for a programming problem with *nonlinear* equality constraints and where the usual Lagrangian function $\mathcal{L}(x, \lambda, \mu) = f(x) + \lambda g(x) + \mu h(x)$ is utilized.

Theorem 3.14.6. Let x^0 be a solution of problem (P_{1L}) and let g and h satisfy any of the following constraint qualifications:

i) (Generalized Slater C.Q.). There exists $\bar{x} \in \mathbb{R}^n$ such that $g(\bar{x}) < 0$, $A\bar{x} = b$.

ii) (Generalized Karlin C.Q.). There exist no $p \geq 0$, $p \in \mathbb{R}^m$, and no $v \in \mathbb{R}^p$ such that $pg(x) + v(Ax - b) \geqq 0$, $\forall \, x \in \mathbb{R}^n$.

iii) (Generalized strict C.Q.). There exist $x^1, x^2 \in S_{1L}$, $x^1 \neq x^2$, such that g is strictly convex at x^1.

Then there exist vectors $\lambda^0 \in \mathbb{R}^m$, $\lambda^0 \geqq 0$, and $\mu^0 \in \mathbb{R}^p$ such that

$$\mathcal{L}(x^0, \lambda, \mu) \leqq \mathcal{L}(x^0, \lambda^0, \mu^0) \leqq \mathcal{L}(x, \lambda^0, \mu^0) ,$$

$$\forall \lambda \in \mathbb{R}^m , \quad \lambda \geqq 0 , \quad \forall \mu \in \mathbb{R}^p , \quad \forall x \in \mathbb{R}^n$$

where $\mathcal{L}(x, \lambda, \mu) = f(x) + \lambda g(x) + \mu(Ax - b)$. Moreover, $\lambda^0 g(x^0) = 0$.

Proof. See Mangasarian (1969), Uzawa (1958). We note that the following implications hold for the constraint qualifications introduced here: iii) $\Rightarrow$ i) $\Rightarrow$ ii). Moreover, we point out that under assumption i) (Generalized Slater C.Q.), Theorem 3.14.6 is an immediate consequence of the Berge's extension of the Farkas-Minkowski theorem (Theorem 2.9.2). □

For other more general saddle point optimality conditions, implying more general Lagrangian functions and/or convexity assumptions, the reader is referred, e.g., to Arrow, Gould and Howe (1973), Bazaraa (1973), Cambini (1986), Elster and Nehse (1980), Giannessi (1984), Rockafellar (1974), Jeyakumar (1985b, 1988).

The optimality conditions contained in the above theorem are also known as *global* Kuhn-Tucker optimality conditions, in contraposition to the *local* Kuhn-Tucker optimality conditions, given in the previous sections for the differentiable case. It is interesting to relate then, for the differentiable case, the saddle point characterizations with the local Kuhn-Tucker optimality conditions. We have the following results.

Theorem 3.14.7. Let problem (P) be given, where f and g are differentiable on the open set $X \subset \mathbb{R}^n$. Then if (x^0, λ^0) is a saddle point for $\psi(x, \lambda)$, with respect to X, then (x^0, λ^0) satisfies the local Kuhn-Tucker conditions:

$$\nabla_x \psi(x^0, \lambda^0) = 0 \tag{5}$$

$$\nabla_\lambda \psi(x^0, \lambda^0) \leqq 0 \tag{6}$$

$$\lambda^0 \nabla_\lambda \psi(x^0, \lambda^0) = 0 \tag{7}$$

$$\lambda^0 \geqq 0 . \tag{8}$$

Proof. By definition of saddle point we have that the function $\psi(x, \lambda^0)$ has at x^0 a minimum; being that X is open we have then relation (5). Thanks to Lemma 3.14.1 we have $\nabla_\lambda \psi(x^0, \lambda^0) = g(x^0) \leqq 0$, i.e. relation (6) and $\lambda^0 \nabla_\lambda \psi(x^0, \lambda^0) = \lambda^0 g(x^0) = 0$, i.e. relation (7). □

Theorem 3.14.8. Let in problem (P_c) the functions f and g be differentiable on the open convex set $X \subset \mathbb{R}^n$. Then if there exists a vector (x^0, λ^0), satisfying the Kuhn-Tucker conditions (5)–(8), then (x^0, λ^0) is a saddle point for the Lagrangian function $\psi(x, \lambda) = f(x) + \lambda g(x)$.

Proof. As ψ is convex with respect to x, we have

$$\psi(x, \lambda^0) - \psi(x^0, \lambda^0) \geqq (x - x^0)\, \nabla_x \psi(x^0, \lambda^0) = 0 \;, \quad \forall\, x \in X$$

and therefore

$$\psi(x^0, \lambda^0) \leqq \psi(x, \lambda^0) \;, \qquad \forall\, x \in X \;.$$

Being that $\lambda^0 g(x^0) = 0$, $\lambda^0 \geqq 0$ and $g(x^0) \leqq 0$, then

$$\psi(x^0, \lambda) \leqq \psi(x^0, \lambda^0) \;, \qquad \forall\, \lambda \geqq 0 \;.$$ □

The assertions of Theorems 3.14.7 and 3.14.8 remain true also for problem (P_1). In Theorem 3.14.8 we can also assume the pseudoconvexity of the Lagrangian function $\mathcal{L}(x, \lambda)$, with respect to x.

Combining together Theorems 3.14.7 and 3.14.8 we obtain at once the following classical result of Kuhn and Tucker (1951).

Theorem 3.14.9. Let us consider problem (P_c), where $f : X \to \mathbb{R}$ and $g : X \to \mathbb{R}^m$ are convex differentiable functions on the open convex set $X \subset \mathbb{R}^n$ and let any of the constraint qualifications I), II), III) be satisfied. Then

$$\{x^0 \text{ solution of } (P_c)\} \Leftrightarrow$$

$$\Leftrightarrow \{\exists\, \lambda^0 \geqq 0 \text{ such that } (x^0, \lambda^0) \text{ is a saddle point for } \psi(x, \lambda)\} \Leftrightarrow$$

$\Leftrightarrow \{\exists\, \lambda^0$ such that (x^0, λ^0) satisfies the

Kuhn-Tucker conditions (5)–(8)$\}$.

The following diagram shows the relationships between the local solutions of the minimization problem (P), the global solutions, the saddle point characterization for $\psi(x, \lambda)$ and the Kuhn-Tucker conditions (5)–(8).

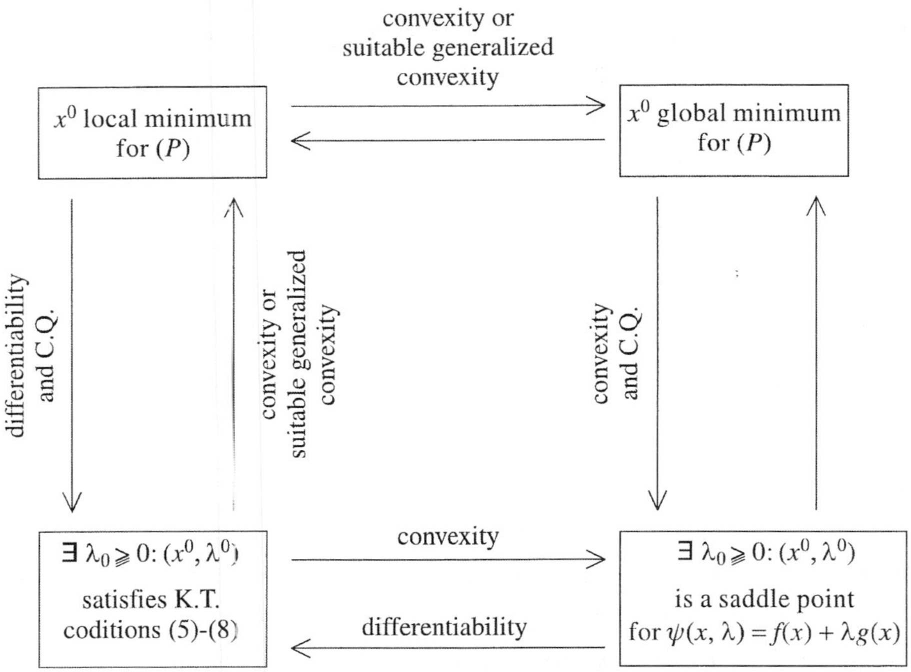

Figure 16.

The above relationships may obviously be reformulated with regard to problem (P_{1L}).

References to Chapter III

J. ABADIE (1967), On the Kuhn-Tucker theorem; in J. Abadie (Ed.), Nonlinear Programming, North Holland, Amsterdam, 19–36.

S.N. AFRIAT (1971), Theory of maxima and the method of Lagrange, SIAM J. Applied Math., 20, 343–357.

C.C. AGUNWAMBA (1977), Optimality condition: constraint regularization, Math. Programming, 13, 38–48.

M. ANGRISANI (1982), Problemi di minimax di tipo discreto: un approccio unificato mediante il teorema dell'alternativa di Motzkin, Rivista di Matematics per le Scienze Economiche e Sociali, 5, 97–113.

K.J. ARROW and A.C. ENTHOVEN (1961), Quasiconcave programming, Econometrica, 29, 779–800.

K.J. ARROW, F.J. GOULD and S.M. HOWE (1973), A general saddle point result for constrained minimization, Math. Programming, 5, 225–234.

K.J. ARROW, L. HURWICZ and H. UZAWA (1961), Constraint qualifications in maximization problems, Naval Res. Logistics Quart., 8, 175–191.

J.P. AUBIN and I. EKELAND (1984), Applied Nonlinear Analysis, J. Wiley & Sons, New York.

J.P. AUBIN and H. FRANKOWSKA (1990), Set-valued Analysis, Birkhäuser Verlag, Basel, 1990.

M. AVRIEL (1976), Nonlinear Programming: Analysis and Methods, Prentice Hall, Englewood Cliffs, N.J.

M. AVRIEL, M.J. RIJCKAERT and D.J. WILDE (Eds.) (1973), Optimization and Design, Prentice Hall, Englewood Cliffs, N.J.

M. AVRIEL, W.E. DIEWERT, S. SCHAIBLE and I. ZANG (1987), Generalized Concavity, Plenum Press, New York.

M.S. BAZARAA (1973), A theorem of the alternative with application to convex programming, optimality, duality and stability, J. Math. Anal. Appl. 41, 701–715.

M.S. BAZARAA and J.J. GOODE (1972), Necessary optimality criteria in mathematical programming in the presence of differentiability, J. Math. Anal. Appl. 40, 609–621.

M.S. BAZARAA and J.J. GOODE (1973), Necessary optimality criteria in mathematical programming in normed linear space, J.O.T.A., 11, 235–244.

M.S. BAZARAA and J.J. GOODE (1982), Sufficient conditions for a globally exact penalty function without convexity, Math. Programming Study, 19, 1–15.

M.S. BAZARAA, J.J. GOODE and M.S. NASHED (1974), On the cone of tangents with applications to mathematical programming, J.O.T.A., 13, 389–426.

M.S. BAZARAA, J.J. GOODE and C.M. SHETTY (1972), Constraint qualifications revisited, Management Science, 18, 567–573.

M.S. BAZARAA, J.J. JARVIS and H.D. SHERALI (1990), Linear Programming and Network Flows, J. Wiley, New York.

M.S. BAZARAA and C.M. SHETTY (1976), Foundations of Optimization, Springer Verlag, Berlin.

M.S. BAZARAA, H.D. SHERALI and C.M. SHETTY (1993), Nonlinear Programming – Theory and Algorithms, J. Wiley, New York.

C.R. BECTOR and T.R. GULATI (1977), Sufficient optimality conditions in presence of quasiconvex equality constraints, Math. Operationsforsch. und Statistik, Ser. Optimization, 8, 181–184.

E.J. BELTRAMI (1967), A constructive proof of the Kuhn-Tucker multiplier rule, J. Math. Anal. Appl., 26, 297–306.

P.J. BENDER (1978), Nonlinear programming in normed linear spaces, J.O.T.A., 24, 263–285.

A. BEN-ISRAEL, A. BEN-TAL and S. ZLOBEC (1981), Optimality in Nonlinear Programming. A Feasible Direction Approach, J. Wiley, New York.

A. BEN-ISRAEL and B. MOND (1985), On sufficient optimality conditions for mathematical programs with equality constraints, J.O.T.A., 47, 253–255.

A. BEN-TAL (1980), Second-order and related extremality conditions in nonlinear programming, J.O.T.A., 31, 143–165.

A. BEN-TAL and J. ZOWE (1982), A unified theory of first and second-order conditions for extremum problems in topological vector spaces, Math. Programming Study, 19, 39–76.

S.K. BHATT and S.K. MISRA (1975), Sufficient optimality criteria in nonlinear programming in the presence of convex equality and inequality constraints, Zeitschrift für Op. Res., 19, 101–105.

J.M. BORWEIN (1978), Weak tangent cones and optimization in Banach spaces, SIAM J. Control, 16, 512–522.

J.C.C. BOOT (1964), Quadratic Programming, North Holland, Amsterdam.

G. BOULIGAND (1932), Introduction à la Géométrie Infinitésimale Directe, Gauthiers-Villars, Paris.

J. BRAM (1966), The Lagrange multiplier theorem for max-min with several constraints, SIAM J. Applied Math., 14, 665–667.

E. BURGER (1955), On extrema with side conditions, Econometrica, 23, 451–452.

A. CAMBINI (1986), Non-linear separation theorems, duality and optimality conditions; in: R. Conti, E. De Giorgi and F. Giannessi (Eds.), Optimization and Related Fields, Springer Verlag, Berlin, 57–93.

A. CAMBINI, E. CASTAGNOLI, L. MARTEIN, P. MAZZOLENI and S. SCHAIBLE (Eds.) (1990), Generalized Convexity and Fractional Programming with Economic Applications, Springer Verlag, Berlin.

M.D. CANON, C.D. CULLUM and E. POLAK (1966), Constrained minimization problems in finite dimensional spaces, SIAM J. Control, 4, 528–547.

M.D. CANON, C.D. CULLUM and E. POLAK (1970), Theory of Optimal Control and Mathematical Programming, McGraw-Hill, New York

Y. CHABRILLAC and J.-P. CROUZEIX (1984), Definiteness and semidefiniteness of quadratic forms revisited, Linear Algebra and Its Appl.,

63, 283–292.

F.H. CLARKE (1975), Generalized gradients and applications, Trans. Amer. Math. Soc., 205, 247–262.

F.H. CLARKE (1976), A new approach to Lagrange multipliers, Math. Oper. Res., 1, 165–174.

F.H. CLARKE (1983), Optimization and Nonsmooth Analysis, J. Wiley & Sons, New York.

L. COLLATZ and W. WETTERLING (1975), Optimization Problems, Springer Verlag, Berlin.

B. CORNET (1987), Regularity properties of open tangent cones, Math. Programming Study, 30, 17–33.

R.W. COTTLE (1963), A theorem of Fritz John in mathematical programming, RAND Corporation Memo, RM-3858-PR.

B.D. CRAVEN (1978), Mathematical Programming and Control Theory, Chapman & Hall, London.

B.D. CRAVEN (1981), Invex functions and constrained local minima, Bull. Austral. Math. Soc., 24, 357–366.

B.D. CRAVEN (1988), Fractional Programming, Heldermann Verlag, Berlin.

B.D. CRAVEN and B.M. GLOVER (1985), Invex functions and duality, J. Austral. Math. Soc., Ser. A, 39, 1–20.

J.M. DANSKIN (1966), The theory of max-min with applications, SIAM J. Appl. Math., 14, 641–664.

J.M. DANSKIN (1967), The Theory of Max-Min and Its Application to Weapons Allocation Problems, Springer Verlag, Berlin.

G.B. DANTZIG (1983), Reminescences about the origin of linear programming; in: A. Bachem, M. Grötschel and B. Korte (Eds.), Mathematical Programming. The State of the Art, Springer Verlag, Berlin.

P.C. DAS (1975), Constrained optimization problems in Banach space, J.O.T.A., 17, 279–291.

G. DEBREU (1952), Definite and semidefinite quadratic forms, Econometrica, 20, 295–300.

V.F. DEM'YANOV and V.N. MALOZEMOV (1974), Introduction to Minimax, J. Wiley, New York.

S. DOLECKI (1982), Tangency and differentiation: some applications of convergence theory, Annali Mat. Pura ed Applicata, 130, 223–255.

M. DRAGOMIRESCU (1967), Sur la nécessité des conditions de Kuhn-Tucker, C.R. Acad. Sci. Paris, 264, Série A, 1073–1075.

A.J. DUBOVITSKIJ and A.A. MILJUTIN (1963), Extremum problems with constraints, Soviet Math. Dokl., 4, 452–455.

A.J. DUBOVITSKIJ and A.A. MILJUTIN (1965), Extremum problems in the presence of restrictions, USSR Comput. Math. and Math. Phys., 5, 1–80.

R.J. DUFFIN, E.L. PETERSON and L. ZENER (1967), Geometric Programming. Theory and Applications, J. Wiley & Sons, New York.

K.-H. ELSTER and R. NEHSE (1980), Optimality conditions for some nonconvex problems; in K. Iracki, K. Malanowski and S. Walukiewicz (Eds.), Optimization Techniques, Springer Verlag, Berlin, Part II, 1–9.

K.-H. ELSTER and J. THIERFELDER (1985a), On cone approximations of sets; in: P. Kenderov (Ed.), Mathematical Methods in Operations Research (Summer School on Operations Research – Primorsko, 1984), Sofia, 33–59.

K.-H. ELSTER and J. THIERFELDER (1985b), A general concept of cone approximations in nondifferentiable optimization; in: V.F. Demyanov and D. Pallaschke (Eds.), Nondifferentiable Optimization: Motivations and Applications, Springer Verlag, Berlin, 170–189.

K.-H. ELSTER and J. THIERFELDER (1988a), On cone approximations and generalized directional derivatives; in: F.H. Clarke, V.F. Demyanov and F. Giannessi (Eds.), Nonsmooth Optimization and Related Topics, Plenum Press, new York, 133–154.

K.-H. ELSTER and J. THIERFELDER (1988b), Generalized notions of directional derivatives, Publ. N. 155, Group of Optimization and Operations Research, Univ. of Pisa (Italy).

K.-H. ELSTER and J. THIERFELDER (1988c), Abstract cone approximations and generalized differentiability in nonsmooth optimization, Optimization, 19, 315–341.

J.P. EVANS (1970), On constraint qualifications in nonlinear programming, Naval Res. Logistics Quart., 17, 281–286.

J.P. EVANS and F.J. GOULD (1970), Stability in nonlinear programming, Operations Res., 18, 107–118.

A.V. FIACCO (1968), Second order sufficient conditions for weak and strict constrained minima, SIAM J. Appl. Math., 16, 105–108.

A.V. FIACCO (1983), Introduction to Sensitivity and Stability Analysis in Nonlinear Prgoramming, Academic Press, New York.

A.V. FIACCO (Ed.) (1984), Sensitivity, Stability and Parametric Analysis, Math. Programming Study, 21, North Holland, Amsterdam.

A.V. FIACCO and G.P. McCORMICK (1968), Nonlinear Programming: Sequential Unconstrained Minimization Techniques, J. Wiley & Sons, New York.

R.V. GAMKRELIDZE (1967), Extremal problems in finite-dimensional spaces, J.O.T.A., 1, 173–193.

F.R. GANTMACHER (1959), The Theory of Matrices (2 volumes), Chelsea Publishing Co., New York.

J. GAUVIN (1977), A necessary and sufficient regularity condition to have bounded multipliers in nonconvex programming, Math. Programming, 12, 136–138.

J. GAUVIN and J.W. TOLLE (1977), Differential stability in nonlinear programming, SIAM J. Control and Optimization, 15, 294–311.

A.M. GEOFFRION (1971), Duality in nonlinear programming: a simplified applications-oriented development, SIAM Review, 13, 1–36.

F. GIANNESSI (1984), Theorems of the alternative and optimality conditions, J.O.T.A., 42, 331–365. Errata corrige, J.O.T.A., 44, 363–364.

G. GIORGI (1983), Inclusions among constraint qualifications in nonlinear programming, Proceedings VII Convegno A.M.A.S.E.S., ITEC Editrice, Milan, 215–246.

G. GIORGI (1984), Quasiconvex programming revisited, Calcolo, 21, 307–316.

G. GIORGI (1985), Programmi tangenti di programmi non lineari, Proceedings IX Convegno A.M.A.S.E.S., ITEC Editrice, Milan, 171–185.

G. GIORGI (1995), On first-order sufficient conditions for constrained optima; in: T. Maruyama and W. Takahashi (Eds.), Nonlinear and Convex Analysis in Economic Theory, Springer Verlag, Berlin, 53–66.

G. GIORGI and A. GUERRAGGIO (1992a), On the notion of tangent cone in mathematical programming, Optimization, 25, 11–23.

G. GIORGI and A. GUERRAGGIO (1992b), Approssimazioni coniche locali: proprietà algebriche e topologiche, Studi Matematici, N. 14, Univ. L. Bocconi, Milan.

G. GIORGI and A. GUERRAGGIO (1992c), On a characterization of Clarke's tangent cone, J.O.T.A., 74, 369–372.

G. GIORGI and A. GUERRAGGIO (1993), Constraint qualifications: the invex case, the set constraint case and other remarks, Proceedings XVII Convegno A.M.A.S.E.S., Ischia, 8–11 September 1993, Tipografia G. Giglio, Napoli, 485–509.

G. GIORGI and A. GUERRAGGIO (1994), First order generalized optimality conditions for programming problems with a set constraint; in: S. Komlosi, T. Rapcsàk, S. Schaible (Eds.), Generalized Convexity, Proceedings, Pe'cs, Hungary, 1992, Springer Verlag, Berlin, 171–185.

G. GIORGI and E. MOLHO (1992), Generalized invexity: relationships with generalized convexity and applications to optimality and duality conditions; in: P. Mazzoleni (Ed.), Generalized Concavity for Economic Applications, Proceedings of a Workshop held in Pisa, April 2, 1992, Tecnoprint, Bologna, 53–70.

I.V. GIRSANOV (1972), Lectures on Mathematical Theory of Extremum Problems, Lecture Notes in Economics and Math. Systems N. 67, Springer Verlag, Berlin.

B.M. GLOVER (1983), Differentiable programming in Banach spaces, Math. Operationsforsch. und Statistik, Ser. Optimization, 14, 499–508.

A.J. GOLDMAN and A.W. TUCKER (1956), Theory of linear programming; in: H.W. Kuhn and A.W. Tucker (Eds.), Linear Inequalities and Related Systems, Princeton Univ. Press, Princeton, 19–40.

F.J. GOULD and J.W. TOLLE (1971), A necessary and sufficient qualification for constrained optimization, SIAM J. Applied Math., 20, 164–172.

F.J. GOULD and J.W. TOLLE (1972), Geometry of optimality conditions and constraint qualifications, Math. Programming, 2, 1–18.

F.J. GOULD and J.W. TOLLE (1975), Optimality conditions and constraint qualifications in Banach spaces, J.O.T.A., 15, 667–684.

H.J. GREENBERG and W. PIERSKALLA (1972), Extensions of the Evans-Gould stability theorem for mathematical programs, Operations Res., 20, 143–153.

M. GUIGNARD (1969), Generalized Kuhn-Tucker conditions for mathematical programming problems in a Banach space, SIAM J. Control, 7, 232–141.

G. HADLEY (1964), Nonlinear and Dynamic Programming, Addison-Wesley, Reading, Mass., 1964.

H. HALKIN and L.W. NEUSTADT (1966), General necessary conditions for optimization problems, Proc. Nat. Acad. Sci. U.S.A., 56, 1066–1071.

H. HANCOCK (1960), Theory of Maxima and Minima, Dover Publications, New York (original publication: 1917).

M.A. HANSON (1981), On sufficiency of the Kuhn-Tucker conditions, J. Math. Anal. Appl., 80, 545–550.

M.A. HANSON and B. MOND (1987), Necessary and sufficient conditions in constrained optimization, Math. Programming, 37, 51–58.

M.R. HESTENES (1966), Calculus of Variations and Optimal Control Theory, J. Wiley, New York.

M.R. HESTENES (1975), Optimization Theory. The Finite-Dimensional Case, J. Wiley, New York.

R. HETTICH and H.T. JONGEN (1977), On first and second-order conditions for local optima for optimization problems in finite dimensions, Methods of Operations Research, XXIII, 83–97.

J.B. HIRIART-URRUTY (1979), Tangent cones, generalized gradients and mathematical programming in Banach spaces, Math. Operations Res., 4, 79–97.

R.B. HOLMES (1975), A Course in Optimization and Best Approximation, Springer Verlag, Berlin.

L. HURWICZ (1958), Programming in linear spaces; in: K.J. Arrow, L. Hurwicz and H. Uzawa (Eds.), Studies in Linear and Nonlinear Programming, Stanford Univ. Press, Stanford, 38–102.

V. JEYAKUMAR (1985a), Strong and weak invexity in mathematical programming, Methods of Operations Research, 55, Verlag Anton Hain, Meisenheim, 109–125.

V. JEYAKUMAR (1985b), Convexlike alternative theorems and mathematical programming, Optimization, 16, 643–652.

V. JEYAKUMAR (1988), Equivalence of saddle-points and optima, and duality for a class of non-smooth non-convex problems, J. Math. Anal. Appl., 130, 334–343.

F. JOHN (1948), Extremum problems with inequalities as side conditions; in: K.O. Friedrichs, O.E. Neugebauer and J.J. Stoker (Eds.), Studies and Essays – Courant Anniversary Volume, Wiley-Interscience, New York, 187–204.

K.C. KAPUR (1973), On max-min problems, Naval Res. Logistics Quarterly, 20, 639–644.

S. KARLIN (1959), Mathematical Methods and Theory in Games, Programming and Economics, Addison-Wesley, Reading, Mass.

W. KARUSH (1939), Minima of Functions of Several Variables with Inequalities as Side Conditions, M.S. Thesis, Department of Mathematics, Univ. of Chicago.

R.N. KAUL and S. KAUR (1982), Sufficient optimality conditions using generalized convex functions, Opsearch, 19, 212–224.

R.N. KAUL and S. KAUR (1985), Optimality criteria in nonlinear programming involving nonconvex functions, J. Math. Anal. Appl., 105, 104–112.

R.P. KING (1966), Necessary and sufficient conditions for inequality constrained extreme values, Ind. & Eng. Chem. Fund., 5, 484–489.

K.O. KORTANEK and J.P. EVANS (1967), Pseudoconcave programming and Lagrange regularity, Op. Research, 15, 882–891.

V. KRIVAN (1991), On the intersection of contingent cones, J.O.T.A., 70, 397–404.

H.W. KUHN and A.W. TUCKER (1951), Nonlinear programming; in: J. Neyman (Ed.), Proceedings of the Second Berkely Symposium on Mathematical Statistics and Probability, Univ. of California Press, Berkeley, 481–492.

S. KURCYUSZ (1976) On existence and nonexistence of Lagrange multipliers in Banach space, J.O.T.A., 20, 81–110.

J. KYPARISIS (1985), On the uniqueness of Kuhn-Tucker multipliers in nonlinear programming, Math. Programming, 32, 242–246.

P.J. LAURENT (1972), Approximation et optimisation, Hermann, Paris.

J.K. LENSTRA, A.H.G. RINNOOY KAN and A. SCHRIJVER (Eds.) (1991), History of Mathematical Programming, C.W.I. – North Holland, Amsterdam.

D.G. LUENBERGER (1969), Optimization by Vector Space Methods, John Wiley, New York, 1969.

D.G. LUENBERGER (1984), Linear and Nonlinear Programming, Addison-Wesley, Reading, Mass.

D.G. MAHAJAN and M.N. VARTAK (1977), Generalizations of some duality theorems in nonlinear programming, Math. Programming, 12, 293–317.

O.L. MANGASARIAN (1969), Nonlinear Programming, McGraw-Hill, New York.

O.L. MANGASARIAN and S. FROMOVITZ (1967), The Fritz John necessary optimality conditions in the presence of equality and inequality constraints, J. Math. Anal. Appl., 17, 33–47.

W.H. MARLOW (1978), Mathematics for Operations Research, J. Wiley & Sons, New York.

D.H. MARTIN (1985), The essence of invexity, J.O.T.A., 47, 65–76.

D.H. MARTIN and G.G. WATKINS (1985), Cores of tangent cones and Clarke's tangent cone, Mathematis of Op. Research, 10, 565–575.

L. MARTEIN (1985), Regularity conditions for constrained extremum problems, J.O.T.A., 47, 217–233.

B. MARTOS (1975), Nonlinear Programming. Theory and Methods, North Holland, Amsterdam.

H. MASSAM and S. ZLOBEC (1974), Various definitions of the derivative in mathematical programming, Math. Programming, 7, 144–161.

H. MASSAM and S. ZLOBEC (1978), Various definitions of the derivative in mathematical programming – An addendum, Math. Programming, 14, 108–111.

H. MAURER and J. ZOWE (1979), First and second-order necessary and sufficient optimality conditions for infinite-dimensional programming problems, Math. Programming Study, 16, 98–110.

G.P. McCORMICK (1967), Second order conditions for constrained minima, SIAM J. Applied Math., 15, 641–652.

G.P. McCORMICK (1976), Optimality criteria in nonlinear programming; in: R.W. Cottle and C.E. Lemke (Eds.), Nonlinear Programming, SIAM-AMS Proceedings, Vol. IX, American Mathematical Society, Providence, 27–38.

G.P. McCORMICK (1983), Nonlinear Programming – Theory, Algorithms and Applications, John Wiley & Sons, New York.

E.J. MESSERLI and E. POLAK (1969), On second order necessary conditions of optimality, SIAM J. Control, 7, 272–291.

P. MICHEL and J.P. PENOT (1984), Calcul sous-différentiel pour des fonctions Lipschitziennes et non Lipschitziennes, C.R. Acad. Sci. Paris, t. 298, 269–272.

Y. NAGAHISA and Y. SAKAWA (1969), Nonlinear programming in Banach spaces, J.O.T.A., 4, 182–190.

L. NEUSTADT (1976), Optimization, Princeton University Press, Princeton.

J. PALATA (1989), A survey of conical approximations used in the optimization, Optimization, 20, 147–161.

R. PALLU DE LA BARRIERE (1963), Compléments à la théorie des multipicateurs en programmation non linéaire, Rev. Française Recherche Opérat., 7, 163–180.

L. PENNISI (1953), An indirect proof for the problem of Lagrange with differential inequalities as added side conditions, Trans. Amer. Math. Soc., 74, 177–198.

J.P. PENOT (1979), The use of generalized subdifferential calculus in optimization theory, Methods of Oper. Research, 31, Verlag Anton Hain, Meisenheim, Königstein, 495–511.

J.P. PENOT (1981), A characterization of tangential regularity, Nonlinear Analysis, Theory, Methods and Appl., 5, 625–643.

J.P. PENOT (1982), On regularity conditions in mathematical programming, Math. Programming Study, 19, 167–199.

D.W. PETERSON (1973), A review of constraint qualifications in finite-dimensional spaces, SIAM Review, 15, 639–654.

C.G. PHIPPS (1952), Maxima and minima under restraint, Amer. Math. Monthly, 59, 230–235.

J. PONSTEIN (1980), Approaches to the Theory of Optimization, Cambridge Univ. Press, Cambridge.

B.H. POURCIAU (1980), Modern multiplier rules, The American Math. Monthly, 87, 433–452.

V. PREDA (1994), On sufficiency and duality for generalized quasiconvex programs, J. Math. Anal. Appl., 181, 77–88.

A. PREKOPA (1980), On the development of optimization theory, The American Math. Monthly, 87, 527–542.

K. RITTER (1969, 1970), Optimization theory in linear spaces I, II, III, Mathematische Annalen, 182, 189–206; 183, 169–180; 184, 133–154.

R.T. ROCKAFELLAR (1970), Convex Analysis, Princeton Univ. Press, Princeton.

R.T. ROCKAFELLAR (1974), Augmented Lagrange multiplier functions and duality in nonconvex programming, SIAM J. Control, 12, 268–285.

R.T. ROCKAFELLAR (1976), Lagrange multipliers in optimization; in: R.W. Cottle and C.E. Lemke (Eds.), Nonlinear Programming, SIAM-AMS Proceedings, Vol. IX, American Math. Society, Providence, 145–168.

R.T. ROCKAFELLAR (1980), Generalized directional derivatives and subgradients of nonconvex functions, Canadian J. of Math., 32, 251–280.

R.T. ROCKAFELLAR (1981), The Theory of Subgradients and Its Applications to Problems of Optimization: Convex and Nonconvex Functions, Heldermann Verlag, Berlin.

R.T. ROCKAFELLAR (1993), Lagrange multipliers and optimality, SIAM Review, 35, 183–238.

S.M. ROBINSON (1976), Stability theory for systems of inequalities, Part II: Differentiable nonlinear systems, SIAM J. Numer. Anal., 13, 497–513.

S.M. ROBINSON (1982), Generalized equations and their solutions, Part II: Applications to nonlinear programming, Math. Programming Study, 19, 200–221.

E.D. ROGAK and J.F. SCOTT-THOMAS (1973), On necessary conditions for optimality in a Banach space, J. Math. Anal. Appl., 41, 44–53.

N.G. RUEDA and M.A. HANSON (1988), Optimality criteria in mathematical programming involving generalized invexity, J. Math. Anal. Appl., 130, 375–385.

D.L. RUSSELL (1966), The Kuhn-Tucker conditions in Banach spaces with an application to control theory, J. Math. Anal. Appl., 15, 200–212.

E. SACHS (1978), Differentiability in optimization theory, Math. Operationsforsch. und Statistik, Ser. Optimization, 9, 497–513.

S. SAKS (1937), Theory of the Integral, Hafner Publ. Co., New York.

S. SCHAIBLE (1978), Analyse und Anwendungen von Quotientenprogrammen, Hain-Verlag, Meisenheim am Glan.

S. SCHAIBLE (1981), A survey on fractional programming; in: S. Schaible and W.T. Ziemba (Eds.), Generalized Concavity in Optimization and Economics, Academic Press, New York, 417–440.

S. SCHAIBLE and T. IBARAKI (1983), Fractional programming, European J. of Operations Res., 12, 325–338.

F. SEVERI (1930), Su alcune questioni di topologia infinitesimale, Ann. Soc. Polon. Math., 9, 97–108.

F. SEVERI (1935), Sulla differenziabilità totale delle funzioni di più variabili reali, Annali Mat. Pura e Appl., Serie 13, 1–35.

C. SINGH (1977), Sufficient optimality criteria in nonlinear programming for generalized equality-inequality constraints, J.O.T.A., 22, 631–635.

C. SINGH and B.K. DASS (Eds.) (1989), Continuous-time Programming, Fractional Programming and Multi-objective Programming, Analytic Publishing Co., Delhi.

B. SKARPNESS and V.A. SPOSITO (1980), A modified Fritz John optimality criterion, J.O.T.A., 31, 113–115.

M. SLATER (1950), Lagrange multipliers revisited: a contribution to nonlinear programming, Cowles Commission Discussion Paper, Mathematics, 403.

A. TAKAYAMA (1985), Mathematical Economics, Cambridge Univ. Press, Cambridge.

J.S. TREIMAN (1991), An infinite class of convex tangent cones, J.O.T.A., 68, 563–581.

C. URSESCU (1982), Tangent sets' calculus and necessary conditions for extremality, SIAM J. Control Optim., 20, 563–574.

H. UZAWA (1958), The Kuhn-Tucker theorem in concave programming; in: K.J. Arrow, L. Hurwicz and H. Uzawa (Eds.), Studies in Linear and Nonlinear Programming, Stanford Univ. Press, Stanford, 32–37.

P.P. VARAIYA (1967), Nonlinear programming in Banach space, SIAM J. Appl. Math., 15, 284–293.

M. VLACH (1970), On necessary conditions of optimality in linear spaces, Comm. Math. Univ. Carolinae, 11, 501–513.

M. VLACH (1981), Approximation operators in optimization theory, Zeitschrift für Operations Research (Theory), 25, 15–24.

D.E. WARD (1987), Convex subcones of the contingent cone in nonsmooth calculus and optimization, Trans. Amer. Math. Soc., 302, 661–682. (Corrigendum in Trans. Amer. Math. Soc., 311, 429–431, 1989.)

D.E. WARD (1988), The quantificational tangent cones, Canadian J. of Math., 40, 666–694.

T. WEIR and V. JEYAKUMAR (1988), A class of nonconvex functions and mathematical programming, Bull. Austral. Math. Soc., 38, 177–189.

J. WERNER (1984), Optimization – Theory and Applications, F. Vieweg & Sohn, Braunschweig/Wiesbaden.

P. WHITTLE (1971), Optimization under Constraints, Wiley-Interscience, New York.

A. ZAFFARONI (1993), Condizioni necessarie di ottimo per funzioni direzionalmente derivabili, Atti XVII Convegno A.M.A.S.E.S., Ischia, 8–11 Settembre 1993, Tip. G. Giglio, Naples, 899–917.

W.I. ZANGWILL (1969), Nonlinear Programming: A Unified Approach, Prentice Hall, Englewood Cliffs, N.J.

S. Zlobec (1970), Asymptotic Kuhn-Tucker conditions for mathematical programming problems in a Banach space, SIAM J. Control, 8, 505–512.

G. ZOUTENDIJK (1960), Methods of Feasible Directions, Elsevier, Amsterdam.

J. ZOWE and S. KURCYUSZ (1979), Regularity and stability for the mathematical programming problems in Banach spaces, Appl. Math. Optimiz., 5, 49–62

CHAPTER IV. NONSMOOTH OPTIMIZATION PROBLEMS

4.1. Preliminary remarks

In the previous chapters we have discussed smooth optimization problems. All the functions involved were assumed to be differentiable or continuously differentiable, i.e. they could be replaced by "simple" linear approximations. By means of suitable constraint qualifications, imposed in order to keep small the failure of these linear approximations, we were able to construct necessary and sufficient optimality conditions for various types of optimization problems.

Frequently, however, the assumption of classical differentiability is not too realistic since in many modern applications one can find functions which are not differentiable. Moreover, there are many functions, which are theoretically differentiable but the structure of the calculation of their derivatives is so intricate that we can detect a nonsmooth behavior in a numerical sense.

For this reason in the last 20 years there have been many efforts to provide suitable tools for the study of more general problems in mathematical optimization but also in many areas of analysis in absence of differentiability. In this connection many approaches have been presented in view of generalizing the classical notion of differentiability. These approaches are related with

- generalizations of the space structure,

- generalizations of the directional derivatives (by modified difference quotients and modified convergence notions),

- generalizations of the gradient (by replacing the gradient mapping by a set-valued mapping).

In this chapter we want to give a short survey about the most important generalized differentiability notions introduced in connection with the discussion with mathematical optimization problems.

We shall try to compare these notions and shall point out the common properties. Later in Section 4.6 we present a general principle for the construction of generalized directional derivatives by means of abstract local cone approximations of sets. In such a way a duality between the approximation of functions and the approximation of sets is pointed out and moreover, assuming suitable properties for the approximating cones we can give a general approach for obtaining necessary optimality conditions for nonsmooth optimization problems.

We restrict our treatment to first-order differentiability notions. Regarding generalized differentiability notions of higher order, the problem is more complicated and cannot be discussed in this framework. For this subject we refer especially to the papers of Ben-Tal/Zowe (1982), Hiriart-Urruty (1982) and Maurer/Zowe (1979).

First of all we shall give some remarks about (classical) differentiability.

4.2. Differentiability

Let $f : \mathbb{R}^n \to \mathbb{R}$ and $x^0 \in \text{int}(D)$. For the classical differentiability concepts it is required that the difference $f(x^0 + y) - f(x^0)$ for $y \in \mathbb{R}^n$, with $\|y\|$ sufficiently small, can be approximated in a certain sense by the linear term $y\nabla f(x^0)$. In any case the vector $\nabla f(x^0) \in \mathbb{R}^n$ is called the *gradient* of f at the point x^0 and the scalar product $y\nabla f(x^0)$ is called the *directional derivative* of f at x^0 in the direction y.

In the past, several possibilities of such linear approximations were developed by many authors. We shall discuss in this section only some of the most important concepts which are used in mathematical optimization and which are fundamental for the comparison and the classification of the later-introduced generalized differentiability notions. For further details we refer to the basic works of Averbuch/Smoljyanov (1967), Bastiani (1964), Fröhlicher/Bucher (1966), Gähler (1978) and Marinescu (1957).

All the classical differentiability concepts have in common that the directional derivatives are described as limits of a special difference quotient. In this manner the simplest way is given if for each direction $y \in \mathbb{R}^n$ the relation

$$\lim_{t \downarrow 0} \frac{f(x^0 + ty) - f(x^0)}{t} = y \nabla f(x^0) \tag{1}$$

holds which can be expressed equivalently by

$$f(x^0 + ty) = f(x^0) + ty\nabla f(x^0) + o(|t|) \tag{2}$$

for small $t > 0$.

A function f with this property is called *Gâteaux differentiable* or *weakly differentiable* at x^0. Naturally, each Gâteaux differentiable function at x^0 is also partially differentiable at x^0 and therefore the components of the gradient $\nabla f(x^0)$ are again the partial derivatives $\frac{\partial}{\partial x_i} f(x^0)$, $i = 1, ..., n$.

We should remark that a Gâteaux differentiable function is not necessarily continuous. This is demontrated in the following example.

Example 4.2.1. Let $f : \mathbb{R}^2 \to \mathbb{R}$ be given by

$$f(x_1, x_2) = \begin{cases} ((x_1)^2 + (x_2)^2)\,\dfrac{x_2}{x_1} & \text{if } x_1 \neq 0 \\ 0 & \text{if } x_1 = 0 \ . \end{cases}$$

Obviously, f is not continuous at $x^0 = (0,0)$ since for the arc

$$x(t) = \left(\sqrt{\frac{\sin^3 t}{\cos t}}, \sqrt{\sin t \cos t}\right), \qquad t \in (0, \pi/2) ,$$

which tends to x^0 with decreasing t, we get $f(x(t)) \equiv 1 \neq f(x^0)$. But f is Gâteaux differentiable at this point. Really, for each direction $y = (y_1, y_2)$ with $y_1 \neq 0$ we have

$$\lim_{t \downarrow 0} \frac{f(x^0 + ty) - f(x^0)}{t} = \lim_{t \downarrow 0} t((y_1)^2 + (y_2)^2)\,\frac{y_2}{y_1} = 0 \ .$$

For $y = (y_1, y_2)$ with $y_1 = 0$ this relation holds trivially. Hence, it is $\nabla f(x^0) = 0$.

The concept of Gâteaux differentiability can be used for the discussion of free optimization problems, which is a simple consequence of the defintion: Assuming f to be Gâteaux differentiable at the point $x^0 \in \operatorname{int}(D)$ and x^0 a local minimum point of f, then of course for each direction $y \in \mathbb{R}^n$ it is $y\nabla f(x^0) \geqq 0$, i.e. it holds the classical optimality condition $\nabla f(x^0) = 0$ mentioned in Theorem 3.2.2.

For the study of constrained optimization problems, however, the notion of Gâteaux differentiability cannot be fitted. For this reason we regard

Example 4.2.2. Let us consider the functions

$$f(x_1, x_2) = \begin{cases} x_2 & \text{if } x_2 = (x_1)^2 \\ x_1 & \text{else} \end{cases}$$

$$h(x_1, x_2) = x_2 - (x_1)^2 \ .$$

Then of course $x^0 = (0,0)$ is a local minimum point for the constrained optimization problem $\operatorname{Min}_{x \in S} f(x)$ where the feasible set is described by $S = \{x \in \mathbb{R}^2 \mid h(x) = 0\}$. Both functions f and h are Gâteaux differentiable at x^0 with $\nabla f(x^0) = (1,0)$ and $\nabla h(x^0) = (0,1)$ but the Lagrange multiplier conditions in Theorem 3.3.1 and Theorem 3.3.2 are not fulfilled since both vectors $\nabla f(x^0)$ and $\nabla h(x^0)$ are linearly independent.

Naturally, the objective function f, discussed in Example 4.2.2, is not continuous at the points $(x_1, x_2) \neq (0,0)$ with $x_2 = (x_1)^2$. So one could guess that by use of continuous functions the applications of the Gâteaux differentiability notion to constrained optimization problems becomes possible. However, this is not true. If we replace the function f in Example 4.2.2 by the continuous function f_1 according to

$$f_1(x_1, x_2) = \begin{cases} \dfrac{2(x_1)^2 x_2}{(x_1)^4 + (x_2)^2} \sqrt{(x_1)^2 + (x_2)^2} + x_1 & \text{if } (x_1, x_2) \neq (0,0) \\ 0 & \text{else} \end{cases}$$

then the reader should convince himself that the same previous troubles arise. Also here the function is Gâteaux differentiable at the minimum point $x^0 = (0,0)$ with $\nabla f(x^0) = (1,0)$ such that the Lagrange multiplier conditions fail at this point.

Therefore, for the discussion of constrained optimization problems it is necessary to sharpen the notion of Gâteaux differentiability. For this reason with respect to the topological structure of the space one claims the uniformity of the limit in (1) for all directions y of the unit ball $B_1(0)$ and thus the uniformity on all bounded sets of $\mathbb{R}^n$. This means

$$\lim_{y\to 0, y\neq 0} \frac{f(x^0+y) - f(x^0) - y\nabla f(x^0)}{\|y\|} = 0 \tag{3}$$

or equivalently

$$f(x^0+y) = f(x^0) + y\nabla f(x^0) + o(\|y\|) \tag{4}$$

for small $\|y\|$. Functions with this property are called *Fréchet differentiable* or *completely differentiable* (or for short: *differentiable*) at x^0.

Naturally each function which is Fréchet differentiable at x^0 is also Gâteaux differentiable at x^0 with the same gradient. Moreover, Fréchet differentiability is sufficient for continuity, which can be derived directly from relation (4).

For the comparison of the generalized differentiability notions discussed in the next sections it is useful to formulate the following assertion.

Theorem 4.2.1. The function f is Fréchet differentiable at x^0 if and only if for all $y \in \mathbb{R}^n$

$$\lim_{\substack{t\downarrow 0\\ z\to y}} \frac{f(x^0+tz) - f(x^0)}{t} = y\nabla f(x^0) \ . \tag{5}$$

Proof. Let f be Fréchet differentiable at x^0. Then by (3) we have for all $y \in \mathbb{R}^n$

$$\lim_{\substack{t\downarrow 0\\ z\to y, z\neq 0}} \left\{\frac{f(x^0+tz) - f(x^0)}{t\,\|z\|} - \frac{z}{\|z\|}\nabla f(x^0)\right\} = 0$$

which is sufficient for (5).

For the inverse implication let (5) be fulfilled for all $y \in \mathbb{R}^n$. Then of course f is Gâteaux differentiable at x^0. We show that the limit in (1) is uniform for all $y \in B_1(0)$ which is equivalent to the Fréchet differentiability at x^0. Assuming the opposite, we could find a number $\varepsilon > 0$, a sequence $\{y^k\} \subseteq B_1(0)$ and a sequence $\{t_k\} \subseteq \mathbb{R}_+$ tending to zero such that

$$\left| \frac{f(x^0 + t_k y^k) - f(x^0)}{t_k} - y^k \nabla f(x^0) \right| > \varepsilon \; .$$

Since the unit ball $B_1(0)$ is compact, without loss of generality we can assume that the sequence $\{y^k\}$ converges to a vector $y \in B_1(0)$. But now we would have a contradiction to (5). □

We should remark that Theorem 4.2.1 is not true in general topological vector spaces (e.g. in Banach spaces) since the unit ball $B_1(0)$ is no more a compact set. While in all vector spaces the Gâteaux differentiability is equivalent to the uniform convergence of (1) on finite sets, for the stronger differentiability notion in general topological vector spaces we have to distinguish between the uniform convergence of (1) on all bounded sets and the uniform convergence of (1) on all compact sets. In the latter case the function f is called *Hadamard differentiable* at x^0.

We have seen that a Fréchet differentiable function is Gâteaux differentiable but the converse implication does not hold generally (see e.g. the function f in Example 4.2.1 and Example 4.2.2). Naturally, for $n = 1$, both concepts are the same. For $n > 1$ we have

Theorem 4.2.2. A Lipschitzian function around x^0 is Fréchet differentiable at x^0 if and only if it is Gâteaux differentiable at x^0.

Proof. If f is Lipschitzian around x^0 with Lipschitz constant $L > 0$ then for all $y \in \mathbb{R}^n$ we have

$$\left| \frac{f(x^0 + tz) - f(x^0)}{t} - \frac{f(x^0 + ty) - f(x^0)}{t} \right| =$$

$$= \left| \frac{f(x^0 + tz) - f(x^0 + ty)}{t} \right| \leqq L \, \|z - y\|$$

for z contained in a small neighborhood of y and $t > 0$ sufficiently small. Therefore, (1) and (5) coincide. □

We can apply this assertion to convex and concave functions defined on a convex set $D \in I\!R^n$. Obviously such functions are Lipschitzian around all points $x \in \operatorname{int}(D)$ and thus Theorem 4.2.2 is applicable. Moreover, in Section 2.6 we have seen that in this case the classical directional derivative of f at x^0 according to

$$f'(x^0, y) = \inf_{t>0} \frac{f(x^0 + ty) - f(x^0)}{t} = \lim_{t \downarrow 0} \frac{f(x^0 + ty) - f(x^0)}{t}$$

exists for each direction $y \in I\!R^n$ and provides a sublinear (i.e. positively homogeneous and convex) function of the directional vector y. If we assume that the partial derivatives at x^0 exist then of course for the orthogonal unit vectors $e^1, ..., e^n$ we have the equality

$$f'(x^0, \pm e^i) = \pm \frac{\partial}{\partial x_i} f(x^0) = \pm\, e^i \nabla f(x^0)$$

where $\nabla f(x^0) = \Big(\frac{\partial}{\partial x_1} f(x^0), ..., \frac{\partial}{\partial x_n} f(x^0)\Big)$. Regarding the sublinearity of the directional derivative we can conclude that the equality

$$f'(x^0, y) = y \nabla f(x^0)$$

holds even for all $y \in I\!R^n$. That means that f is Gâteaux differentiable at x^0 and – using Theorem 4.2.2 – that f is Fréchet differentiable at x^0. Thus we can state that for convex (likewise concave) functions the notions of partial differentiability, Gâteaux differentiability and Fréchet differentiability are the same.

As we have seen in the previous chapter the concept of Fréchet differentiability can be used to describe necessary optimality conditions for constrained nonlinear optimization problems. We were able to derive the classical Lagrange multiplier rules and the Fritz-John and Kuhn/Tucker theory by means of this type of differentiability.

For certain discussions, however (e.g. in connection with parametric optimization problems and with stability assertions), the Fréchet differentiability concept is not sharp enough. So the notion of *strict differentiability* is introduced: a function f is called *strictly differentiable* at x^0 if

$$\lim_{\substack{y \to 0, y \neq 0 \\ x \to x^0}} \frac{f(x+y) - f(x) - y\nabla f(x^0)}{\|y\|} = 0 \ , \tag{6}$$

or equivalently, if for each $\varepsilon > 0$ there exists a number $\delta > 0$ such that

$$|f(x^1) - f(x^2) - (x^1 - x^2)\,\nabla f(x^0)| < \varepsilon(\|x^1 - x^2\|) \tag{7}$$

for all $\|x^1 - x^0\| < \delta$ and $\|x^2 - x^0\| < \delta$.

Naturally, by definition strict differentiability is sufficient for Fréchet differentiability. Moreover, if f is strictly differentiable at x^0 then it is also Lipschitzian around x^0 which can be derived from (7) by

$$|f(x^1) - f(x^2)| < (\|\nabla f(x^0)\| + \varepsilon)\,\|x^1 - x^2\|$$

for $x^1, x^2 \in B_\delta(x^0)$. Analogous to Theorem 4.2.1 we can formulate

Theorem 4.2.3. The function f is strictly differentiable at x^0 if and only if for all $y \in I\!R^n$

$$\lim_{\substack{t \downarrow 0 \\ z \to y \\ x \to x^0}} \frac{f(x+tz) - f(x)}{t} = y\nabla f(x^0) \ . \tag{8}$$

The proof of this assertion is analogous to the proof of Theorem 4.2.1.

In Theorem 4.2.2 we have stated that a Lipschitzian function around x^0 is Fréchet differentiable at x^0 if it is Gâteaux differentiable at this point which means that by this assumption we can neglect the variation of the directional vector y in (5). Taking this into account also for strict differentiability a similar assertion can be formulated using relation (8). The following theorem shows that this assertion holds even without Lipschitz assumption.

Theorem 4.2.4. The function f is strictly differentiable at x^0 if and only if for all $y \in I\!R^n$

$$\lim_{\substack{t \downarrow 0 \\ x \to x^0}} \frac{f(x+ty) - f(x)}{t} = y \nabla f(x^0) \; . \tag{9}$$

Proof. We have to show that (9) is sufficient for (8). This shall be demonstrated for the case $n = 2$.
For this let (9) be fulfilled. With the representation

$$z = z_1 e^1 + z_2 e^2$$

where e^1, e^2 are the orthogonal unit vectors, we have

$$\frac{f(x+tz) - f(x)}{t} = \frac{f(x + tz_2 e^2 + tz_1 e^1) - f(x + tz_2 e^2)}{t} + \\ + \frac{f(x + tz_2 e^2) - f(x)}{t} \tag{10}$$

which converges for $t \downarrow 0$, $x \to x^0$ and $z \to y$ (since $x + tz_2 e^2 \to x^0$) to

$$y_1 e^1 \nabla f(x^0) + y_2 e^2 \nabla f(x^0) = y \nabla f(x^0)$$

and we get (8).
For higher dimensional spaces, i.e. in case $n > 2$, the proof is the same. Here we have to decompose (10) as a sum of n special difference quotients (for each direction $e^1, ..., e^n$) which converges to

$$y_1 e^1 \nabla f(x^0) + ... + y_n e^n \nabla f(x^0) = y \nabla f(x^0) \; . \qquad \square$$

In view of a comparison between Fréchet differentiability and strict differentiability in the following example we shall see that both notions are not the same even by Lipschitz assumption.

Example 4.2.3. Let the function $f : I\!R \to I\!R$ be given by

$$f(x) = \begin{cases} x^2 \sin \dfrac{1}{x} & \text{if } x \neq 0 \\ 0 & \text{if } x = 0 \; . \end{cases}$$

Obviously, f is (Gâteaux and Fréchet) differentiable at $x^0 = 0$ with the classical derivative $\nabla f(x^0) = f'(x^0) = 0$. Also f is Lipschitzian around

$x^0 = 0$ which can be stated by the mean value theorem (since the derivative according to $2x \sin \frac{1}{x} - \cos \frac{1}{x}$ is bounded near zero). But f is not strictly differentiable at $x^0 = 0$. For this, we regard the sequences $\{x^k\}$ and $\{t_k\}$ according to

$$x^k = \frac{1}{2k\pi + \pi/2}\,, \qquad t_k = \frac{1}{2k\pi - \pi/2} - \frac{1}{2k\pi + \pi/2}\,.$$

Both sequences tend to zero and we get with $y = 1$

$$\lim_{k\to\infty} \frac{f(x^k + t_k y) - f(x^k)}{t_k} = - \lim_{k\to\infty} \frac{\left(\frac{1}{2k\pi - \pi/2}\right)^2 + \left(\frac{1}{2k\pi + \pi/2}\right)^2}{\frac{1}{2k\pi - \pi/2} - \frac{1}{2k\pi + \pi/2}} =$$

$$= - \lim_{k\to\infty} \frac{8k^2\pi^2 + \pi^2/2}{(4k^2\pi^2 - \pi^2/4)\,\pi} = -2/\pi$$

which does not coincide with $y\nabla f(x^0) = 0$.

In practice it is not easy to check the strict differentiability of a function. Therefore it is useful to find a sufficient condition. For this purpose assume f to be *continuously Gâteaux differentiable* at x^0. That means that f is Gâteaux differentiable on a neighbourhood of x^0 and the gradient mapping $\nabla f(x)$ is continuous at x^0. It holds

Theorem 4.2.5. If f is continuously Gâteaux differentiable at x^0, then it is strictly differentiable at x^0.

Proof. Let $x \in \mathbb{R}^n$ with $\|x - x^0\|$ sufficiently small and $z \in \mathbb{R}^n$ be given. Then for small $t > 0$ the function $\varphi_{x,z}(t) = f(x + tz)$ is continuous and differentiable. Using the classical mean value theorem we get

$$\frac{f(x+tz) - f(x)}{t} = \frac{\varphi_{x,z}(t) - \varphi_{x,z}(0)}{t} = \varphi'_{x,z}(\xi) = z\nabla f(x + \xi z)$$

with $\xi \in (0, t)$. Using the continuity of the gradient mapping at x^0 we obtain

$$\lim_{\substack{t\downarrow 0\\ z\to y\\ x\to x^0}} \frac{f(x+tz) - f(x)}{t} = y\nabla f(x^0)\ .$$

Since this equality is true for all $y \in \mathbb{R}^n$, by Theorem 4.2.3 the function f is strictly differentiable at x^0. □

It is useful to remark that one can weaken the hypothesis in Theorem 4.2.5 by assuming only the continuous partial differentiability of f at x^0 (which is a well-known result of analysis). Here the result can be obtained by the repeated use of the mean value theorem for each direction $e^1, ..., e^n$ of orthogonal unit vectors. Especially for $n = 2$ we get from (10)

$$\frac{f(x+tz)-f(x)}{t} = z_1 \frac{\partial}{\partial x_1} f(x + \xi_1 z_1 e^1 + t z_2 e^2) + \\ + z_2 \frac{\partial}{\partial x_2} f(x + \xi_2 z_2 e^2)$$

with $\xi_1, \xi_2 \in (0,t)$ and this sum tends to $y \nabla f(x^0)$ for $x \longrightarrow x^0$, $z \longrightarrow y$ and $t \downarrow 0$ because of the continuity of the partial derivatives. Thus the function f is strictly differentiable at x^0.

Naturally, if f is continuously Gâteaux differentiable not only at the point x^0 but even on a neighborhood of x^0, then f is also continuously Fréchet differentiable and even continuously strictly differentiable at this neighborhood and all these notions coincide (frequently denoted by $f \in C^1$).

Now we are able to give the following table in which the connections between the several kinds of differentiability, the Lipschitz property and the continuity of a function are demonstrated:

Continuous differentiability	
↓	
Strict differentiability	↓
↓	Lipschitz property
Fréchet differentiability	↓
↓	Continuity
Gâteaux differentiability	

In our previous remarks we pointed out that the converse implications do not hold in general. This shall be confirmed by the following example in which the function f of Example 4.2.3 is modified.

Example 4.2.4. Let $\alpha \geqq 0$ and $f : I\!R \to I\!R$ be given by

$$f(x) = \begin{cases} x^\alpha \sin \dfrac{1}{x} & \text{if } x \neq 0 \\ 0 & \text{if } x = 0\ . \end{cases}$$

We discuss the properties of the function at $x^0 = 0$. Obviously f is continuous at $x^0 = 0$ for $\alpha > 0$ and Lipschitzian around $x^0 = 0$ for $\alpha \geqq 2$ (as in the former example it can be derived directly from the mean value theorem). For $\alpha < 2$ the function is not Lipschitzian around $x^0 = 0$. This can be demonstrated by the sequences $\{x^k\}$ and $\{z^k\}$ according to

$$x^k = \frac{1}{2k\pi + \pi/2}\ , \qquad z^k = \frac{1}{2k\pi - \pi/2}\ .$$

Here for the difference quotients we get

$$\frac{f(z^k) - f(x^k)}{\|z^k - x^k\|} = -\frac{\Big(\dfrac{1}{2k\pi - \pi/2}\Big)^\alpha + \Big(\dfrac{1}{2k\pi + \pi/2}\Big)^\alpha}{\dfrac{1}{2k\pi - \pi/2} - \dfrac{1}{2k\pi + \pi/2}} =$$

$$= -\frac{1}{\pi}\Big(\frac{4k^2\pi^2 - \pi^2/4}{(2k\pi - \pi/2)^\alpha} + \frac{4k^2\pi^2 - \pi^2/4}{(2k\pi + \pi/2)^\alpha}\Big)$$

which tends to $-\infty$ for $k \to \infty$ (i.e. $x^k \to 0$ and $z^k \to 0$). Therefore, no Lipschitz constant can be found. Regarding the differentiability we see that f is (Gâteaux and Fréchet) differentiable at $x^0 = 0$ for $\alpha > 1$ and continuously differentiable (and hence strictly differentiable) at this point for $\alpha > 2$.

In Example 4.2.3 we have seen that for $\alpha = 2$ f is not strictly differentiable at $x^0 = 0$. We summarize the results in the following table

	Continuity	Differen-tiability	Lipschitz property	Strict differ-entiability
$\alpha = 0$	$-$	$-$	$-$	$-$
$0 < \alpha \leq 1$	$+$	$-$	$-$	$-$
$1 < \alpha < 2$	$+$	$+$	$-$	$-$
$\alpha = 2$	$+$	$+$	$+$	$-$
$\alpha > 2$	$+$	$+$	$+$	$+$

4.3. Directional Derivatives and Subdifferentials for Convex Functions

If we want to extend the classical differentiability concepts mentioned in Section 4.2, then it is useful to start with the discussion of convex functions. Here with respect to the epigraphs of such functions, the well-known assertions about the algebraic and topological properties of convex sets (especially separation theorems) provide an approach for generalizing, in an excellent manner, the notion of the directional derivative and the notion of the gradient of differentiable functions. Today these generalizations are basic tools in convex analysis and permit a good application for the discussion of mathematical optimization problems.

Convex functions and their properties are discussed earlier in Sections 2.5 and 2.6. In this section we shall summarize once more the most important results regarding the directional differentiability and the structure of the subdifferential of such functions – also in connection with their application to convex optimization problems. These results allow further generalizations for not-necessary convex functions in the next sections.

Let $f : \mathbb{R}^n \to \mathbb{R} \cup \{\pm\infty\}$ be an extended real-valued convex function and $x^0 \in \mathbb{R}^n$ be a point where f is finite. As we have seen in Sections 2.5 and 2.6, the convexity of f is equivalent to the monotonic increase of the difference quotient

$$\frac{f(x^0 + ty) - f(x^0)}{t}$$

with respect to the parameter $t > 0$ for all fixed vectors $y \in \mathbb{R}^n$. Thus the *directional derivative* of f at x^0 according to

$$f'(x^0, y) = \inf_{t>0} \frac{f(x^0 + ty) - f(x^0)}{t} = \lim_{t \downarrow 0} \frac{f(x^0 + ty) - f(x^0)}{t} \tag{1}$$

exists for each direction y (possibly with values $\pm\infty$) and provides an extended real-valued function $f'(x^0, .)$. From the convexity of f we could derive directly that $f'(x^0, .)$ is also convex (even sublinear, since it is positively homogeneous).

Obviously the value $f'(x^0, y)$ is finite if $x^0 \pm ty \in \operatorname{dom}(f)$ for any $t > 0$. This is a simple consequence of the inequalities

$$-\infty < -\frac{f(x^0 - ty) - f(x^0)}{t} \leqq -f'(x^0, -y) \leqq$$

$$\leqq f'(x^0, y) \leqq \frac{f(x^0 + ty) - f(x^0)}{t} < +\infty .$$

Clearly, if $x^0 \in \operatorname{int}(\operatorname{dom}(f))$ then $f'(x^0, y)$ is finite for all y and the directional derivative $f'(x^0, .)$ presents a continuous function on $\mathbb{R}^n$. Since in case of $x^0 \in \operatorname{int}(\operatorname{dom}(f))$ the convex function f is Lipschitzian around x^0, then we can estimate

$$\left| \frac{f(x^0 + tz) - f(x^0)}{t} - \frac{f(x^0 + ty) - f(x^0)}{t} \right| =$$

$$= \left| \frac{f(x^0 + tz) - f(x^0 + ty)}{t} \right| \leqq L \, \|z - y\|$$

for $\|z - y\|$ and $t > 0$ sufficiently small (here L is the Lipschitz constant) and in analogy to Theorem 4.2.2 we even get

$$f'(x^0, y) = \lim_{\substack{t \downarrow 0 \\ z \to y}} \frac{f(x^0 + tz) - f(x^0)}{t} \tag{2}$$

for any $y \in \mathbb{R}^n$, i.e. f is *uniformly directionally differentiable* at x^0.

Comparing (1) and (2) with (1) and (5) of Section 4.2 we can recognize that in case of $x^0 \in \operatorname{int}(\operatorname{dom}(f))$ the function f is (Gâteaux, even Fréchet) differentiable at x^0 if and only if the directional derivative $f'(x^0,.)$ is linear with respect to the directional vector y (see also our remarks in Section 4.2 regarding the differentiability of convex functions).

In Section 2.5 we have seen that for convex functions which are differentiable at x^0, it holds the well-known inequality

$$(x - x^0)\nabla f(x^0) \leqq f(x) - f(x^0) \qquad \forall x \in \mathbb{R}^n \ .$$

Taking this into account, as a generalization of the gradient $\nabla f(x^0)$ in Section 2.6 we have introduced the *subdifferential* of f at the point $x^0 \in \mathbb{R}^n$ as a set of vectors by

$$\partial f(x^0) = \{u \in \mathbb{R}^n \mid u(x - x^0) \leqq f(x) - f(x^0)\ \forall x \in \mathbb{R}^n\} \ .$$

Obviously the subdifferential is the intersection of closed halfspaces and therefore it is a closed convex set. If $x^0 \in \operatorname{rel.int}(\operatorname{dom}(f))$, we have shown that $\partial f(x^0) \neq \emptyset$. Moreover, if $x^0 \in \operatorname{int}(\operatorname{dom}(f))$ then the subdifferential is compact and it is single-valued iff f is differentiable at x^0.

Both concepts of directional derivative and subdifferential of a convex function are closely related. So we have

$$\partial f(x^0) = \{u \in \mathbb{R}^n \mid uy \leqq f'(x^0, y)\ \forall y \in \mathbb{R}^n\} \ , \tag{3}$$

which means (since $f'(x^0, 0) = 0$) that $\partial f(x^0)$ is the subdifferential of $f'(x^0,.)$ at $y^0 = 0$, i.e. we have

$$\partial f(x^0) = \partial(f'(x^0,.))(0) \ .$$

Conversely, in case of $x^0 \in \operatorname{rel.int}(\operatorname{dom}(f))$ it holds

$$f'(x^0, y) = \sup\{uy \mid u \in \partial f(x^0)\} \qquad \forall y \in \mathbb{R}^n \tag{4}$$

i.e. $f'(x^0,.)$ is the supporting function of the set $\partial f(x^0)$.

Especially from (3) and (4) we can derive the above-mentioned equivalencies to the differentiability of f in the form

f is differentiable at x^0

$\Leftrightarrow f'(x^0, .)$ is linear
(and $f'(x^0, y) = y\nabla f(x^0)\ \forall\, y \in \mathbb{R}^n$)

$\Leftrightarrow \partial f(x^0)$ contains exactly one element
(namely $\partial f(x^0) = \{\nabla f(x^0)\}$).

Regarding all these properties we can state that the directional derivative $f'(x^0, y)$ and the subdifferential $\partial f(x^0)$ are authentic generalizations of the directional derivative $y\nabla f(x^0)$ and the gradient $\nabla f(x^0)$ of a differentiable function. This shall be reflected in the following optimality conditions for convex optimality problems.

The first assertion which concerns free optimization problems is already given partly in Theorem 2.8.4.

Theorem 4.3.1. Let f be a convex function and $x^0 \in \mathbb{R}^n$ be a point where f is finite. Then x^0 is a (global) minimum point of f if and only if the following equivalent conditions hold:

i) $f'(x^0, y) \geqq 0,\ \forall\, y \in \mathbb{R}^n$,

ii) $0 \in \partial f(x^0)$.

Proof. If x^0 is a minimum point of f then

$$f'(x^0, y) = \inf_{t>0} \frac{f(x^0 + ty) - f(x^0)}{t} \geqq 0 \qquad \forall\, y \in \mathbb{R}^n .$$

The implication i) $\Rightarrow$ ii) is a consequence of relation (3). Finally, from the definition of the subdifferential we get the optimality of x^0 if ii) is fulfilled.□

Now we shall consider the convex constrained optimization problem

$$\operatorname*{Min}_{x \in S} f(x) \tag{P_c}$$

where the feasible set is given by

$$S = \{x \in \mathbb{R}^n \mid g_i(x) \leqq 0,\ i = 1, ..., m\} .$$

All functions are assumed to be convex.

First we give a geometrical optimality assertion in which only the convexity of the objective function and the convexity of the feasible set are considered. Here $T(S, x^0)$ denotes the contingent cone to the set S at x^0 introduced in Section 3.4.

Theorem 4.3.2. Let f be a convex function, S be a convex set and $x^0 \in \text{int}(\text{dom}(f))$ be a feasible point. Then x^0 is a (global) minimum point of the problem (P_c) if and only if the following equivalent conditions hold:

i) $f'(x^0, y) \geq 0, \forall y \in T(S, x^0)$,

ii) $0 \in \partial f(x^0) + T^*(S.x^0)$.

Proof. Let x^0 be a minimum point of (P_c). The proof that i) holds is analogous to the proof of Theorem 3.6.1 using the definition of the contingent cone and the representation of the directional derivative according to relation (2) which is guaranteed by the assumption $x^0 \in \text{int}(\text{dom}(f))$. Now let i) be fulfilled. Then the nonempty convex cones

$$A = \{(y, \alpha) \in \mathbb{R}^n \times \mathbb{R} \mid \alpha \geqq f'(x^0, y)\}$$

$$B = \{(y, \alpha) \in \mathbb{R}^n \times \mathbb{R} \mid y \in T(S, x^0), \ \alpha < 0\}$$

are disjoint. By the separation theorem (Theorem 2.2.5) we can find a vector $(u, \xi) \in \mathbb{R}^n \times \mathbb{R}$, $(u, \xi) \neq (0, 0)$, such that

$$(u, \xi)(y, \alpha) \leqq 0 \qquad \forall (y, \alpha) \in A ,$$

$$(u, \xi)(y, \alpha) \geqq 0 \qquad \forall (y, \alpha) \in B .$$

Obviously, $\xi \leqq 0$ and $\xi = 0$ provides a contradiction since by assumption $f'(x^0, .)$ is finite and $uy \leqq 0$ for all y would imply $u = 0$. Therefore we have $\xi < 0$ and without loss of generality we can set $\xi = -1$. Now we get

$$uy \leqq f'(x^0, y) \qquad \forall y \in \mathbb{R}^n$$

$$uy \geqq 0 \qquad \forall y \in T(S, x^0)$$

i.e. by relation (3) we have

$u \in \partial f(x^0) \cap (-T^*(S, x^0))$

which is equivalent to ii).
Finally, let ii) be fulfilled, i.e. there exists a vector $u \in \partial f(x^0)$ with $u \in -T^*(S, x^0)$. Since S is convex, for an arbitrary feasible point $x \in S$ we have

$x - x^0 \in \overline{\text{cone}}(S - x^0) = T(S, x^0)$

(see Theorem 3.4.10) and regarding the definition of the subdifferential we get

$f(x) - f(x^0) \geqq u(x - x^0) \geqq 0$.

Hence, x^0 is a (global) minimal point of the optimization problem. □

Now we will also formulate optimality conditons for the problem (P_c) in terms of multiplier rules by means of directional derivatives and subdifferentials. For this we have to include in our considerations the analytic structure of the feasible set. First we give an assertion of Fritz John type.

Theorem 4.3.3. Let f and g_i, $i = 1, ..., m$, be convex functions and let x^0 be a point in which all functions are finite and continuous. If x^0 is a minimum point of the problem (P_c) then there exist multipliers $u_i \geqq 0$, $i = 0, ..., m$, not all vanishing, such that $u_i g_i(x^0) = 0$ for $i = 1, ..., m$, and such that the following equivalent conditions hold:

i) $u_0 f'(x^0, y) + \sum_{i=1}^{m} u_i g_i'(x^0, y) \geqq 0 \qquad \forall y \in I\!R^n$,

ii) $0 \in u_0 \partial f(x^0) + \sum_{i=1}^{m} u_1 \partial g_i(x^0)$.

Proof. Let x^0 be a minimum point of (P_c). Analogous to the proof of Theorem 3.5.1 we can state that no vector $y \in I\!R^n$ exists which is a solution of the system

$$\begin{cases} f'(x^0, y) < 0 \\ g_i'(x^0, y) < 0 \qquad i \in I(x^0) \ . \end{cases}$$

Here $I(x^0)$ is the index set of active constraints.
By The Fan-Glicksberg-Hoffmann theorem (Theorem 2.9.1) we can find multipliers $u_i \geq 0$, $i \in \{0\} \cup I(x^0)$, not all vanishing, such that

$$u_0 f'(x^0, y) + \sum_{i \in I(x^0)} u_i g_i'(x^0, y) \geqq 0 \qquad \forall y \in \mathbb{R}^n .$$

Setting $u_i = 0$ for $i \notin I(x^0)$ we get i).
Now by definition of the subdifferential, i) can be described equivalently by

$$0 \in \partial\Big(u_0 f'(x^0, .) + \sum_{i=1}^{m} u_i g_i'(x^0, .)\Big)\,(0) .$$

Since all the functions are continuous ar x^0 also the directional derivatives are continuous at $y^0 = 0$. Using Theorem 2.6.8 about the subdifferential of the sum of convex functions and regarding relation (3) we can state that i) is equivalent to

$$0 \in u_0 \partial(f'(x^0, .))\,(0) + \sum_{i=1}^{m} u_i \partial(g_i'(x^0, .))\,(0) =$$

$$= u_0 \partial f(x^0) + \sum_{i=1}^{m} u_i \partial g_i(x^0)$$

which means ii). □

Naturally, for the formulation of a Kuhn/Tucker assertion we have to assume a suitable constraint qualification. Here we use the Slater constraint qualification: there must exist a point $\bar{x} \in S$ with $g_i(\bar{x}) < 0$ for $i = 1, ..., m$.

Theorem 4.3.4. Let f and g_i, $i = 1, ..., m$, be convex functions and let $x^0 \in S$ be a point in which all functions are finite and continuous. Further let the Slater constraint qualification be fulfilled. Then x^0 is a (global) minimum point of the problem (P_c) if and only if there exist multipliers $u_i \geqq 0$, $i = 1, ..., m$, such that $u_i g_i(x^0) = 0$ for $i = 1, ..., m$, and such that the following equivalent conditions hold:

i) $f'(x^0, y) + \sum_{i=1}^{m} u_i g_i'(x^0, y) \geqq 0 \qquad \forall y \in \mathbb{R}^n$,

ii) $0 \in \partial f(x^0) + \sum_{i=1}^{m} u_i \partial g_i(x^0)$.

Proof. Let x^0 be a minimum of (P_c). According to Theorem 4.3.3 there are multipliers $u_i \geqq 0$, $i = 1, ..., m$, not all vanishing, such that

$$u_0 f'(x^0, y) + \sum_{i \in I(x^0)} u_i g_i'(x^0, y) \geqq 0 \qquad \forall y \in \mathbb{R}^n .$$

Assuming $u_0 = 0$ we would get

$$\sum_{i \in I(x^0)} u_i g_i'(x^0, y) \geqq 0 \qquad \forall y \in \mathbb{R}^n .$$

By the Fan-Glucksberg-Hoffmann theorem (Theorem 2.9.1) again we see that the system

$$g_i'(x^0, y) < 0 , \qquad i \in I(x^0)$$

is not solvable. But assuming the Slater condition the solvability of this system is ensured because of

$$g_i'(x^0, \bar{x} - x^0) \leqq g_i(\bar{x}) - g_i(x^0) < 0 \qquad \forall i \in I(x^0)$$

and we have a contradiction. Thus $u_0 > 0$. Without loss of generality we can set $u_0 = 1$ and i) is fulfilled.
The implication i) $\Rightarrow$ ii) can be shown analogously to Theorem 4.3.3. Finally let ii) and the complementary slackness condition be fulfilled. Then there exists subgradients $v^0 \in \partial f(x^0)$ and $v^i \in \partial g_i(x^0)$, $i \in I(x^0)$, with

$$v^0 + \sum_{i \in I(x^0)} u_i v^i = 0 .$$

Since for arbitrary $x \in S$ the inequalities

$$v^i(x - x^0) \leq g_i(x) - g_i(x^0) \leqq 0 \qquad \forall i \in I(x^0)$$

hold, we get

$$f(x) - f(x^0) \geqq v^0(x - x^0) = \Big(- \sum_{i \in I(x^0)} u_i v^i\Big)(x - x^0) =$$

$$= - \sum_{i \in I(x^0)} u_i v^i (x - x^0) \geqq 0 \ ,$$

which means the minimality of x^0. □

We shall remark that for the proofs of the last assertions the convexity of the functions is not required directly but the convexity of the directional derivatives, which is a weaker assumption, is. In the next sections we present some approaches about the possibility for the construction of convex directional derivatives for nonconvex functions.

4.4. Generalized Directional Derivatives

For nonconvex functions we cannot expect that in general the limits of the difference quotients discussed in Sections 4.2 and 4.3 respectively exist, not to mention the linearity or convexity of these derivatives. Therefore other approaches were developed to find suitable generalized directional derivatives useful for applications in the nonconvex case. The simplest way for this purpose is the replacement of the limit operation by the upper and lower limits.

So, given a function $f : \mathbb{R}^n \to \mathbb{R} \cup \{\pm\infty\}$ and a point $x^0 \in \mathbb{R}^n$ where f is finite, the *upper Dini directional derivative* and the *lower Dini directional derivative* at the point x^0 in the direction $y \in \mathbb{R}^n$ are defined by

$$f'_U(x^0, y) = \limsup_{t \downarrow 0} \frac{f(x^0 + ty) - f(x^0)}{t} \ , \tag{1}$$

$$f'_L(x^0, y) = \liminf_{t \downarrow 0} \frac{f(x^0 + ty) - f(x^0)}{t} \ . \tag{2}$$

It is quite obvious that

$$f'_L(x^0, \cdot) \leqq f'_U(x^0, \cdot)$$

and that in the case of the equality of the above limits we get the classical directional derivative $f'(x^0, y)$ discussed in Sections 2.6 and 4.3. Naturally, in general this equality does not ensure the convexity of the directional derivative which would be very important for the application.

In this connection Pshenichnyi (1972) and Ioffe/Tichomirov (1979) have proposed a concept in which the convexity of the directional derivative is assumed. To be exact, a function f is called *quasidifferentiable* (according to Pshenichnyi) or *locally convex* (according to Ioffe/Tichomirov) at the point x^0 if

$$f'_L(x^0, \cdot) = f'_U(x^0, \cdot)$$

(i.e. that $f'(x^0, y)$ exists for each $y \in \mathbb{R}^n$) and if this directional derivative is finite and sublinear in y. In the mentioned books of Pshenichnyi and Ioffe/Tichomirov it is pointed out that the class of such functions is relatively extensive. Also one can find calculus rules for this concept which extend the calculus rules for differentiable and convex functions and which make possible the application of this notion for the handling of (special) nonconvex optimization problems.

Of course, in case of the equality and linearity of the above-introduced directional derivatives the function f is Gâteaux differentiable according to relation (1) in Section 4.2.

The meaning of the directional derivative for convex functions is not only connected with the sublinearity but also with the uniform convergence with respect to the directional vector y according to relation (2) of Section 4.3 in the case of $x^0 \in \operatorname{int}(\operatorname{dom}(f))$. This property is important for the discussion of constrained optimization problems.

Taking this into account, the *uniform upper Dini directional derivative* and the *uniform lower Dini directional derivative* (also *upper* and *lower Dini-Hadamard directional derivative*) of the function f at the point x^0 in the direction $y \in \mathbb{R}^n$ are introduced by

$$f_U^+(x^0, y) = \limsup_{\substack{t \downarrow 0 \\ z \longrightarrow y}} \frac{f(x^0 + tz) - f(x^0)}{t}, \tag{3}$$

$$f_L^+(x^0,y) = \liminf_{\substack{t\downarrow 0\\ z\to y}} \frac{f(x^0+tz)-f(x^0)}{t} \,. \tag{4}$$

Here we have used the topological structure of the space such that we can expect topological properties of these directional derivatives. So by definition, $f_U^+(x^0,.)$ is upper semi continuous and $f_L^+(x^0,.)$ is lower semi continuous with respect to the directional vector.

In comparison with (1) and (2) we have

$$f_L^+(x^0,.) \leqq f_L'(x^0,.) \leqq f_U'(x^0,.) \leq f_U^+(x^0,.) \,.$$

We can state that the function f is uniformly directional differentiable at the point x^0 similar to (2) in Section 4.3 if all these directional derivatives coincide, especially if

$$f_L^+(x^0,.) = f_U^+(x^0,.) \,.$$

Then of course, $f_L^+(x^0,0) = f_U^+(x^0,0) = 0$ and this directional derivative is finite and continuous on $I\!R^n$. In analogy to our former remarks the function f is called *regular locally convex* (see Ioffe/Tichomirov (1979)) if both directional derivatives are equal and present a sublinear function in y. Finally, using Theorem 4.2.1 we see that the function f is Fréchet differentiable at x^0 iff this directional derivative is also linear.

Regarding the limit (8) in Section 4.2 we can also define the following generalized directional derivatives at x^0 by

$$f_U^*(x^0,y) = \limsup_{\substack{t\downarrow 0\\ z\to y\\ x\to x^0}} \frac{f(x+tz)-f(x)}{t} \,, \tag{5}$$

$$f_L^*(x^0,y) = \liminf_{\substack{t\downarrow 0\\ z\to y\\ x\to x^0}} \frac{f(x+tz)-f(x)}{t} \,. \tag{6}$$

Also here, $f_U^*(x^0,.)$ and $f_L^*(x^0,.)$ are upper semi continuous and lower semi continuous respectively which can be derived directly from the definitions.

Both directional derivatives, however, are very interesting because of their convexity properties. In Theorem 4.4.4 we will prove that $f_U^*(x^0,.)$ is convex. Now because of

$$f_L^*(x^0,.) = -(-f)_U^*(x^0,.)$$

we can state immediately that $f_L^*(x^0,.)$ is concave.

Comparing all the directional derivatives introduced in this section we have the following inequalities

$$\begin{aligned} f_L^*(x^0,.) \ &\leqq \ f_L^+(x^0,.) \leqq f_L'(x^0,.) \ \leqq \\ &\leqq \ f_U'(x^0,.) \leqq f_U^+(x^0,.) \leqq f_U^*(x^0,.) \ . \end{aligned}$$

In case of

$$f_L^*(x^0,.) = f_U^*(x^0,.)$$

all directional derivatives coincide, are finite and convex as well as concave, i.e. linear. This is equivalent with the strict differentiability of f at x^0 discussed in Section 4.2 (see Theorem 4.2.3).

We shall remark that the description of some directional derivatives can be simplified if f is assumed to be Lipschitzian around x^0. Then analogous to our remarks in Section 4.2 in this case the variation of the directional vector y can be omitted since

$$\begin{aligned} &\left| \frac{f(x+tz)-f(x)}{t} - \frac{f(x+ty)-f(x)}{t} \right| = \\ &= \left| \frac{f(x+tz)-f(x+ty)}{t} \right| \leqq L\, \|z-y\| \to 0 \end{aligned}$$

for $x \to x^0$, $z \to y$ and $t \downarrow 0$ (here L is the Lipschitz constant). Therefore we get in this case

$$f_L^+(x^0,y) = f_L'(x^0,y) \qquad , \qquad \forall\, y \in \mathbb{R}^n \ ,$$

$$f_U^+(x^0,y) = f_U'(x^0,y) \qquad , \qquad \forall\, y \in \mathbb{R}^n \ ,$$

$$f_L^*(x^0,y) = \liminf_{\substack{t \downarrow 0 \\ x \to x^0}} \frac{f(x+ty)-f(x)}{t} \qquad , \qquad \forall\, y \in \mathbb{R}^n \ ,$$

$$f_U^*(x^0, y) = \limsup_{\substack{t \downarrow 0 \\ x \to x^0}} \frac{f(x+ty) - f(x)}{t}, \qquad \forall\, y \in \mathbb{R}^n .$$

The last limit, generally denoted by

$$f^0(x^0, y) = \limsup_{\substack{t \downarrow 0 \\ x \to x^0}} \frac{f(x+ty) - f(x)}{t} \tag{7}$$

is known as the *Clarke generalized derivative*. It is introduced in the famous paper of Clarke (1975). In Theorem 4.4.4 we shall state that this directional derivative is convex too, also if f is not assumed to be Lipschitzian around x^0. Analogous, because of

$$\liminf_{\substack{t \downarrow 0 \\ x \to x^0}} \frac{f(x+ty) - f(x)}{t} = -(-f)_U^0(x^0, y)$$

this directional derivative is concave.

Before we introduce and discuss further directional derivatives by more complicated convergence concepts, we shall analyze which of the above mentioned notions are suitable for the discussion of optimality conditions. For unconstrained optimization problems we have the following result.

Theorem 4.4.1. If $x^0 \in \mathbb{R}^n$ is a local minimum point of the function f then

$$f_L^+(x^0, y) \geqq 0 \qquad \forall\, y \in \mathbb{R}^n .$$

Proof. If $x^0 \in S$ is a local minimum point of f then for each $y \in \mathbb{R}^n$, each z in a small neighborhood of y and each $t > 0$ sufficiently small, we have

$$\frac{f(x^0 + tz) - f(x^0)}{t} \geqq 0$$

such that

$$f_L^+ = \liminf_{\substack{t \downarrow 0 \\ z \to y}} \frac{f(x^0 + tz) - f(x^0)}{t} \geqq 0 . \qquad \square$$

Of course Theorem 4.4.1 remains true if $f_L^+(x^0,.)$ is substituted with all other directional derivatives which are larger. Concerning the directional derivative $f_L^*(x^0,.)$, however, the assertion would be false in general. So e.g. for the function $f(x) = \|x\|$ the point $x^0 = 0$ is the minimum of f but because of $f_L^*(0,y) = -\|y\|$ the mentioned condition is not verified.

Now we regard constrained optimization problems of the form

$$\operatorname*{Min}_{x \in S} f(x) . \qquad (P_0)$$

In analogy to the assertions stated in Theorem 3.6.1 and Theorem 4.3.2 we can formulate two geometrical optimality conditions using the directional derivatives $f_L^+(x^0,.)$, $f_U^+(x^0,.)$, the contingent cone $T(S,x^0)$ and the cone of interior directions $I(S,x^0)$.

Theorem 4.4.2. If $x^0 \in S$ is a local minimum point of the problem (P_0) then

i) $f_U^+(x^0,y) \geqq 0, \ \forall y \in T(S,x^0)$.

ii) $f_L^+(x^0,y) \geqq 0, \ \forall y \in I(S,x^0)$.

Proof. Let $y \in T(S,x^0)$. Then there exist a sequence $\{y^k\} \subseteq I\!R^n$ converging to $y \in I\!R^n$ and a positive sequence $\{t_k\} \subseteq I\!R$ converging to zero such that

$$x^0 + t_k y^k \in S$$

for each $k \in I\!N$. Since x^0 is a local minimum of f on S we get

$$f_U^+(x^0,y) \geqq \limsup_{k \to \infty} \frac{f(x^0 + t_k y^k) - f(x^0)}{t_k} \geqq 0 .$$

For the second condition let $y \in I(S,x^0)$. Then for each sequence $\{y^k\} \subseteq I\!R^n$ converging to $y \in I\!R^n$ and for each positive sequence $\{t_k\} \subseteq I\!R$ converging to zero we have

$$x^0 + t_k y^k \in S$$

for large $k \in \mathbb{N}$. Since x^0 is a local minimum of f on S we get for all these sequences

$$\liminf_{k\to\infty} \frac{f(x^0 + t_k y^k) - f(x^0)}{t_k} \geqq 0 \ ,$$

i.e. $f_L^+(x^0, y) \geqq 0$. □

That the stronger condition

$$f_L^+(x^0, y) \geqq 0 \qquad \forall\, y \in T(S, x^0)$$

in general does not hold can be demonstrated by our former Example 4.2.2. For the objective function

$$f(x_1, x_2) = \begin{cases} x_2 & \text{if } x_2 = (x_1)^2 \\ x_1 & \text{else} \end{cases}$$

we get with $x^0 = (0,0)^T$ and $y = (y^1, y^2)$

$$f_L^+(x^0, y) = \begin{cases} 0 & \text{if } y_2 = 0,\ y_1 > 0 \\ y_1 & \text{else} \end{cases}$$

$$f_U^+(x^0, y) = \begin{cases} 0 & \text{if } y_2 = 0,\ y_1 < 0 \\ y_1 & \text{else .} \end{cases}$$

The contingent cone to the feasible set $S = \{x \in \mathbb{R}^2 \mid x_2 = (x_1)^2\}$ at the minimum point $x^0 = (0,0)^T$ has the form

$$T(S, x^0) = \{y \mid y_2 = 0\} \ .$$

Thus we can state the condition of Theorem 4.4.2

$$f_U^+(x^0, y) \geqq 0 \qquad \forall\, y \in T(S, x^0)$$

but with $y^0 = (-1, 0)^T \in T(S, x^0)$ we get $f_L^+(x^0, y^0) = -1 < 0$.

However, it holds

Theorem 4.4.3. If $x^0 \in S$ and

$$f_L^+(x^0, y) > 0 \qquad \forall\, y \in T(S, x^0)$$

then x^0 is a local minimum point of the problem (P_0).

Proof. Assuming that x^0 is not a local minimum point of (P_0), then we can find a sequence $\{x^k\} \subseteq S$ converging to x^0 with $f(x^k) < f(x^0)$ for each k. We set

$$y^k = \frac{x^k - x^0}{\|x^k - x^0\|} , \qquad t_k = \|x^k - x^0\| .$$

Then the sequence $\{y^k\}$ is contained in the compact unit sphere and without loss of generality we can assume that it tends to the vector y of this sphere. Since $x^0 + t_k y^k = x^k \in S$ for all k, by definition $y \in T(S, x^0)$, but

$$f_L^+(x^0, y) \leqq \liminf_{k \to \infty} \frac{f(x^0 + t_k y^k) - f(x^0)}{t_k} \leqq 0$$

which is a contradiction. □

If we want to construct multiplier rules for the optimization problems of the type of (P) and (P_1), i.e. when the feasible set is described by equality and inequality constraints, then the assertions of Theorem 4.4.2 provide a suitable approach. For this purpose, however, we have to use the separation theorems of convex analysis and therefore convexity of the directional derivatives must be ensured.

Essentially, there are three approaches for providing convex (sublinear) directional derivatives.

First of all (this is the simplest possibility), one can impose the convexity of the directional derivative introduced above. Thus is e.g. the case of the quasidifferentiable functions according to Pshenichnyi and the (regular) locally convex functions according to Ioffe and Tichomirov. We know that for convex functions this property is automatically fulfilled.

Second, we can regard upper sublinear approximations of the above introduced directional derivatives. Following Ioffe (1979) and Pshenichnyi (1980) a sublinear function $h : \mathbb{R}^n \to \mathbb{R}$ is called

* *upper convex approximation* of f at the point x^0 if

$$f'_U(x^0, y) \leqq h(y) \qquad \forall\, y \in \mathbb{R}^n \; ; \tag{8}$$

* *regular upper convex approximation* of f at the point x^0 if

$$f^{+}_U(x^0, y) \leqq h(y) \qquad \forall\, y \in \mathbb{R}^n \; . \tag{9}$$

Clearly, each regular upper convex approximation is also an upper convex approximation of f at x^0 but not conversely (in general). However, if f is Lipschitzian around x^0 then both notions are equivalent because of

$$f'_U(x^0, .) = f^{+}_U(x^0, .) \; .$$

Moreover, in this case we have (with Lipschitz constant $L > 0$)

$$f^{*}_U(x^0, y) = \limsup_{\substack{t \downarrow 0 \\ z \to y \\ x \to x^0}} \frac{f(x + tz) - f(x)}{t} \leqq L\,\|y\| \; .$$

Both functions $h_1(.) = f^{*}_U(x^0, .)$ and $h_2(.) = L\,\|.\|$ are sublinear, continuous and – thanks to the relation $f^{+}_U(x^0, .) \leq f^{*}_U(x^0, .)$ – regular upper approximations of f at x^0.

We see that (8) and (9) are equivalent to

$$\limsup_{t \downarrow 0} \frac{f(x^0 + ty) - f(x^0) - h(ty)}{t} \leqq 0 \qquad \forall\, y \in \mathbb{R}^n \; ,$$

$$\limsup_{\substack{t \downarrow 0 \\ z \to y}} \frac{f(x^0 + tz) - f(x^0) - h(ty)}{t} \leqq 0 \qquad \forall\, y \in \mathbb{R}^n \; ,$$

respectively.

Using the uniform convergence in sense of (3) in Section 4.2, Giannessi (1985) has introduced the notion of *upper semi-differentiability*. The function f is called *upper semi-differentiable* at x^0 if there exists a sublinear function $h : \mathbb{R}^n \to \mathbb{R}$ such that

$$\limsup_{y \to 0,\; y \neq 0} \frac{f(x^0 + y) - f(x^0) - h(y)}{\|y\|} \leqq 0 \; .$$

We see that in this case h is a regular upper convex approximation.

A third possibility is the construction of special directional derivatives which are convex in any case. This can be done by use of more complicated convergence concepts. So e.g. for the directional derivatives $f_U^*(x^0,.)$ and $f^0(x^0,.)$ introduced in (5) and (7) according to

$$f_U^*(x^0,y) = \limsup_{\substack{t\downarrow 0\\ z\to y\\ x\to x^0}} \frac{f(x+tz)-f(x)}{t} =$$

$$= \inf_{\varepsilon>0} \limsup_{\substack{t\downarrow 0\\ x\to x^0}} \sup_{\substack{z:\\ \|z-y\|<\varepsilon}} \frac{f(x+tz)-f(x)}{t} ,$$

$$f^0(x^0,y) = \limsup_{\substack{t\downarrow 0\\ x\to x^0}} \frac{f(x+ty)-f(x)}{t} ,$$

we have stated the convexity which we shall prove in Theorem 4.4.4. Beside this, Rockafellar (1979), (1980), (1981) has analyzed the following sublinear directional derivatives, today denoted as *Rockafellar upper subderivative*:

$$f^{\nearrow}(x^0,y^0) = \lim \sup_{\substack{t\downarrow 0\\ x\to x^0}} \inf_{z\to y} \frac{f(x+tz)-f(x)}{t} =$$

$$= \sup_{\varepsilon>0} \limsup_{\substack{t\downarrow 0\\ x\to x^0}} \inf_{\substack{z:\\ \|z-y\|<\varepsilon}} \frac{f(x+tz)-f(x)}{t} . \tag{10}$$

It holds

<u>Theorem 4.4.4</u>. The directional derivatives $f_U^*(x^0,.)$, $f^0(x^0,.)$ and $f^{\nearrow}(x^0,.)$ are convex.

<u>Proof</u>. For the proof of the convexity of $f_U^*(x^0,.)$ let $y^1, y^2 \in \mathbb{R}^n$ and $\{x^k\} \subseteq \mathbb{R}^n$, $\{t_k\} \in \mathbb{R}_+$, $\{z^k\} \subseteq \mathbb{R}^n$ be sequences with $x^k \to x^0$, $t_k \downarrow 0$,

$z^k \to y^1 + y^2$ such that

$$\lim_{k\to\infty} \frac{f(x^k + t_k z^k) - f(x^k)}{t_k} = f_U^*(x^0, y^1 + y^2) .$$

Then of course $\{z^k\}$ can be decomposed in two sequences $\{u^k\}$ and $\{v^k\}$ such that $u^k + v^k = z^k$ for all k and $u^k \to y^1$, $v^k \to y^2$. We get

$$\begin{aligned} f_U^*(x^0, y^1 + y^2) = &\lim_{p\to\infty} \frac{f(x^k + t_k(u^k + v^k)) - f(x^k)}{t_k} \leqq \\ &\leqq \limsup_{k\to\infty} \frac{f(x^k + t_k v^k + t_k u^k) - f(x^k + t_k v^k)}{t_k} + \\ &\quad + \limsup_{k\to\infty} \frac{f(x^k + t_k v^k) - f(x^k)}{t_k} = \\ &\leqq f_U^*(x^0, y^1) + f_U^*(x^0, y^2) . \end{aligned}$$

That means that $f_U^*(x^0, .)$ is subadditive and (because it is positively homogeneous) convex.

The convexity of $f^0(x^0, .)$ can be shown analogously.

Now we will show the convexity of $f^{\nearrow}(x^0, .)$. For this let $y^1, y^2 \in \mathbb{R}^n$, $\varepsilon > 0$ and $\{x^k\} \subseteq \mathbb{R}^n$, $\{t_k\} \in \mathbb{R}_+$ be sequences with $x^k \to x^0$ and $t_k \downarrow 0$. By definition of $f^{\nearrow}(x^0, y^2)$ there exists a sequence $\{v^k\} \subseteq B_{\varepsilon/2}(y^2)$ such that

$$\limsup_{k\to\infty} \frac{f(x^k + t_k v^k) - f(x^k)}{t_k} \leqq f^{\nearrow}(x^0, y^2) .$$

Since $x^k + t_k v^k \to x^0$, by the definition of $f^{\nearrow}(x^0, y^1)$ again we can find a sequence $\{u^k\} \subseteq B_{\varepsilon/2}(y^1)$ such that

$$\limsup_{k\to\infty} \frac{f(x^k + t_k v^k + t_k u^k) - f(x^k + t_k v^k)}{t_k} \leqq f^{\nearrow}(x^0, y^1) .$$

Thus with $y^k = u^k + v^k$, $k \in \mathbb{N}$, we have found a sequence $\{y^k\} \in B_\varepsilon(y^1 + y^2)$ with

$$\limsup_{k\to\infty} \frac{f(x^k + t_k y^k) - f(x^k)}{t_k} \leqq$$

$$\leqq \limsup_{k\to\infty} \frac{f(x^k + t_k v^k + t_k u^k) - f(x^k + t_k v^k)}{t_k} +$$

$$+ \limsup_{k\to\infty} \frac{f(x^k + t_k v^k) - f(x^k)}{t_k} \leqq$$

$$\leqq f^{\nearrow}(x^0, y^1) + f^{\nearrow}(x^0, y^2) .$$

Since this equality is true for each $\varepsilon > 0$, each sequence $\{x^k\} \to x^0$ and each sequence $\{t_k\} \downarrow 0$, we get

$$f^{\nearrow}(x^0, y^1 + y^2) = \sup_{\varepsilon>0} \limsup_{\substack{t\downarrow 0 \\ z\to y \\ z\to x^0}} \inf_{\substack{z: \\ \|z-y^1-y^2\|<\varepsilon}} \frac{f(x+tz) - f(x)}{t} \leqq$$

$$\leqq f^{\nearrow}(x^0, y^1) + f^{\nearrow}(x^0, y^2)$$

which is sufficient for the convexity. □

With respect to the comparability of the last introduced directional derivatives we see that

$$f^{\nearrow}(x^0, .) \leqq f^0(x^0, .) \leqq f_U^*(x^0, .) .$$

Clearly, if f is Lipschitzian around x^0 then the variation in the directional vector can be omitted according to our former remarks. This means that in this case we have equality. A more detailed description of these directional derivatives is given in Section 4.7.

Finally, we should state that by definition it holds

$$f_L^+(x^0, .) \leqq f^{\nearrow}(x^0, .) \quad \text{and} \quad f_U^+(x^0, .) \leqq f_U^*(x^0, .) .$$

Thus, by Theorems 4.4.1 and 4.4.2 these directional derivatives can be used to describe multiplier rules for nonconvex optimization problems (see Section 4.8 for a general approach).

Nevertheless we shall remark that frequently these directional derivatives are regarded as too large for the construction of sharp optimality conditions. Therefore, in the last few years some other notions have been proposed. We mention only the *Michel-Penot directional derivatives* introduced by Michel and Penot (1984) in which the variation in x is replaced by $x^0 + tu \to x^0$ for $t \downarrow 0$ with fixed direction $u \in \mathbb{R}^n$ according to

$$f^{\oplus}(x^0, y) = \sup_{u \in \mathbb{R}^n} \limsup_{\substack{t \downarrow 0 \\ z \to y}} \frac{f(x^0 + tu + tz) - f(x^0 + tu)}{t},$$

$$f^{\square}(x^0, y) = \sup_{u \in \mathbb{R}^n} \limsup_{t \downarrow 0} \frac{f(x^0 + tu + ty) - f(x^0 + tu)}{t},$$

$$f^{\oslash}(x^0, y) = \sup_{u \in \mathbb{R}^n} \lim \sup_{t \downarrow 0} \inf_{z \to y} \frac{f(x^0 + tu + tz) - f(x^0 + tu)}{t}.$$

Also these directional derivatives are convex (the proof is analogous to that of Theorem 4.4.4) and we have

$$f^{\oslash}(x^0, .) \leqq f^{\square}(x^0, .) \leqq f^{\oplus}(x^0, .) .$$

By definition these directional derivatives are smaller than the Clarke/Rockafellar directional derivatives $f^*_U(x^0, .)$, $f^0(x^0, .)$ and $f^{\nearrow}(x^0, .)$. It holds

$$\begin{array}{ccccc} f^+_U(x^0, .) & \leq & f^{\oplus}_U(x^0, .) & \leq & f^*_U(x^0, .) \\ \vee\!| & & \vee\!| & & \vee\!| \\ f'_U(x^0, .) & \leq & f^{\square}(x^0, .) & \leq & f^0(x^0, .) \\ \vee\!| & & \vee\!| & & \vee\!| \\ f^+_L(x^0, .) & \leq & f^{\oslash}(x^0, .) & \leq & f^{\nearrow}(x^0, .) . \end{array}$$

We shall not extend these investigations. For further results we refer e.g. to the papers of Michel/Penot (1984) and Jofré/Penot (1989).

Most of the directional derivatives mentioned in this section will be discussed once more in Section 4.7. We shall see that these notions are closely connected with the local cone approximations introduced in Section 3.4 and in this manner we have an interesting tool for the classification of the generalized differentiability notions. Moreover, by means of an abstract definition of local cone approximations and the associated abstract directional derivatives, in Section 4.8 we will provide a general approach for the derivation of general necessary optimality conditions for nonsmooth optimization problems.

4.5 Generalized Gradient Mappings

In the last section we have discussed differentiability notions as generalizations of the classical directional derivative $y\nabla f(x^0)$ of a differentiable function $f : \mathbb{R}^n \to \mathbb{R}$ at the point x^0 of the domain. Now we shall give another kind of generalization in which the classical gradient $\nabla f(x^0)$ and the gradient mapping $\nabla f : \mathbb{R}^n \to \mathbb{R}^n$ is replaced by a set and a set-valued mapping respectively.

An excellent example for this purpose is the subdifferential for convex functions

$$\partial f(x^0) = \{u \in \mathbb{R}^n \mid u(x - x^0) \leqq f(x) - f(x^0),\ \forall x \in \mathbb{R}^n\}$$

which is introduced in Section 2.6 and which is mentioned in Section 4.3. We have shown that many properties of the classical calculus could be extended for this notion.

As a generalization of the equivalent description of this subdifferential according to

$$\partial f(x^0) = \{u \in \mathbb{R}^n \mid uy \leqq f'(x^0, y),\ \forall y \in \mathbb{R}^n\}$$

(see Theorem 2.6.5 and relation (3) of Secton 4.3), one can introduce other (generalized) subdifferentials, if the directional derivative $f'(x^0, .)$ is replaced by some other generalized directional derivatives, discussed in the previous section. Then regarding such directional derivatives which are larger than $f_L^+(x^0, .)$, we get necessary conditions for x^0 to be a local minimum of the function f in the form $0 \in \partial f(x^0)$ – see Theorems 4.3.1 and 4.4.1.

One cannot expect that in general these subdifferential notions can be used for the discussion of constrained optimization problems since many important properties known for the subdifferential of convex functions (e.g. (4) of Section 4.3 and the calculus rules) may fail. Since most of these properties are derived by the separation theorems, convexity assumptions for the applied directional derivatives are necessary. A general approach to construct necessary optimality conditions using sublinear directional derivatives and the associated subdifferentials is given in Section 4.8.

Another proposal for the generalization of the gradient is given by Clarke (1975). Regarding locally Lipschitzian functions this author has reminded that such functions are differentiable (in the Fréchet sense) almost everywhere. Moreover, Clarke has introduced a subdifferential (the so-called *generalized gradients* or the *Clarke subdifferential*) for locally Lipschitzian functions by

$$\begin{aligned}\partial_{Cl} f(x^0) = \operatorname{conv}\{u \in \mathbb{R}^n \mid \exists \{x^k\} \subseteq \mathbb{R}^n : x^k \to x^0, \\ f \text{ is differentiable at } x^k\ \forall k \in \mathbb{N} : \nabla f(x^k) \to u\} .\end{aligned}$$

Clarke has shown that by this assumption the set $\partial_{Cl} f(x^0)$ is a nonempty, compact convex set. Moreover, if f is convex and $x^0 \in \operatorname{int}(\operatorname{dom}(f))$ (obviously f is Lipschitzian around x^0 in this case) then it holds $\partial_{Cl} f(x^0) = \partial f(x^0)$. For strictly differentiable functions in x^0 we have $\partial_{Cl} f(x^0) = \{\nabla f(x^0)\}$. (See the paper of Clarke (1975).)

It is very interesting that also the Clarke subdifferential is closely connected with special directional derivatives. We have seen in the former section that for a function f which is Lipschitzian around x^0 the equality

$$f_U^*(x^0, y) = f^0(x^0, y) = f^{\nearrow}(x^0, y)$$

holds. Clarke has shown the following relations

$$\partial_{Cl} f(x^0) = \{u \in \mathbb{R}^n \mid uy \leqq f^0(x^0, y) \quad \forall y \in \mathbb{R}^n\} ,$$

$$f^0(x^0, y) = \max\{uy \mid u \in \partial_{Cl} f(x^0)\} \qquad \forall y \in \mathbb{R}^n ,$$

i.e. $f^0(x^0, .)$ is the support function to $\partial_{Cl} f(x^0)$.

Today the Clarke subdifferential is one of the most important generalized differentiability notions, not only from the theoretical point of view but also in connection with the development of numerical methods. In

Sections 4.7 and 4.8 we shall extend the concept to not necessarily Lipschitzian functions and we will derive optimality assertions as a special case of a more general setting.

Regarding more results we refer to the papers and books of Clarke (1975), (1976), (1983) and Rockafellar (1979), (1980), (1981).

In Section 4.4 we mentioned quasidifferentiable functions according to Pshenichnyi, i.e. functions such that the quantity

$$f'(x^0, y) = \lim_{t \downarrow 0} \frac{f(x^0 + ty) - f(x^0)}{t}$$

exists for all $y \in I\!R^n$ and the directional derivative $f'(x^0, .)$ is a finite and sublinear function in the directional vector y. Obviously, by Theorem 2.6.6 this is equivalent to the existence of a nonempty convex compact set $M(x^0) \subseteq I\!R^n$ (take $M(x^0) = \partial(f'(x^0, .))(0)$ using the classical subdifferential for convex functions) such that

$$f'(x^0, y) = \max\{uy \mid u \in M(x^0)\} \qquad \forall y \in I\!R^n$$

for all $y \in I\!R^n$.

Naturally, each Gâteaux differentiable function at x^0 and each convex function which is continuous at x^0 is quasidifferentiable at x^0 in this sense with $M(x^0) = \{\nabla f(x^0)\}$ and $M(x^0) = \partial f(x^0)$ respectively. Nevertheless, the class of these functions is restricted and it is useful to extend this notion in a natural manner.

Therefore, Dem'yanov and Rubinov (1980) have suggested composing the directional derivative $f'(x^0, .)$ as a difference of two sublinear functions. Doing so, they call a function *quasidifferentiable* (to avoid confusion we shall call it *D-R-quasidifferentiable*) at x^0 if there exist two nonempty convex compact sets $\underline{\partial} f(x^0), \bar{\partial} f(x^0) \in I\!R^n$ such that

$$f'(x^0, y) = \max\{uy \mid u \in \underline{\partial} f(x^0)\} - \max\{uy \mid u \in \bar{\partial} f(x^0)\}$$

for all $y \in I\!R^n$. The pair $(\underline{\partial} f(x^0), \bar{\partial} f(x^0)$ is called the *quasidifferential* of f at the x^0.

Obviously the pair is not unique. Really, with $(\underline{\partial} f(x^0), \bar{\partial} f(x^0))$ also the pair $(\underline{\partial} f(x^0) + A, \bar{\partial} f(x^0) + A)$ is a quasidifferential, if $A \subseteq I\!R^n$ is an arbitrary compact convex set.

It is quite obvious that for Gâteaux differentiable functions at x^0 the pairs $(\{\nabla f(x^0)\}, \{0\})$ and $(\{0\}, \{-\nabla f(x^0)\})$ are suitable quasidifferentials of f at x^0. If f is a convex function with $x^0 \in \mathrm{int}(\mathrm{dom}(f))$ or f is quasidifferentiable according to Pshenichnyi, then of course quasidifferentials of f at x^0 can be described by $(\partial f(x^0), \{0\})$ and $(M(x^0), \{0\})$. Also for concave functions we can present a quasidifferential by $(\{0\}, \partial(-f)(x^0))$ and many calculus rules permit the discussion of a wide class of functions which is closed for many algebraic operations (e.g. linear combination, product, quotient, minimum, maximum).

A necessary optimality condition for x^0 to be a local minimum of a D-R-quasidifferentiable function f can be given in the following manner.

Theorem 4.5.1. If x^0 is a local minimum point of f of the D-R-quasidifferentiable function f then

$$\bar{\partial} f(x^0) \subseteq \underline{\partial} f(x^0) .$$

Proof. We assume the opposite. Then there exists a vector $v \in \bar{\partial} f(x^0)$ with $v \notin \underline{\partial} f(x^0)$. Using the strict separation theorem (Theorem 2.2.6) we can find a vector $y^0 \in I\!R^n$, $y^0 \neq 0$, such that

$$\max\{uy^0 \mid u \in \underline{\partial} f(x^0)\} < vy^0 \leqq$$
$$\leqq \max\{uy^0 \mid u \in \bar{\partial} f(x^0)\} .$$

Thus we have

$$f'(x^0, y^0) = \max\{uy^0 \mid u \in \underline{\partial} f(x^0)\} - \max\{uy^0 \mid u \in \bar{\partial} f(x^0)\} < 0$$

i.e. y^0 is a direction of descent for the function f. But this is a contradiction to the optimality of x^0. □

The quasidifferentiability concept of Dem'yanov/Rubinov can also be used for the discussion of constrained optimization problems. We shall not follow this concept. Here the reader is referred to the papers and books of Dem'yanov, cited in the bibliographical references.

4.6. Abstract Cone Approximations of Sets and Relating Differentiability Notions

In this section we shall give an axiomatic approach for constructing generalized directional derivatives and generalized subdifferentials of extended real-valued functions $f : \mathbb{R}^n \to \mathbb{R} \cup \{\pm\infty\}$. The basic idea is the fact that each directional derivative discussed in the former sections is positively homogeneous (as a function of the directional vector y) and therefore its epigraph is a cone which can be interpreted geometrically as a local cone approximation of the epigraph of f. Thus, conversely we can develop new directional derivatives in such a way, that the epigraph of f is locally replaced by certain local cone approximation. Using the special approximations of Section 3.4, in the next section we state many close connections to known differentiability notions.

In this section, however, we give a more general approach. Here abstract directional derivatives and subdifferentials are introduced regarding an axiomatic notion of local cone approximations. In this definition the essential properties of known cone approximations are included (see also Elster/Thierfelder (1985), (1988)).

Definition 4.6.1. A set-valued mapping $K : 2^{\mathbb{R}^n} \times \mathbb{R}^n \to 2^{\mathbb{R}^n}$ is called a *local cone approximation,* if to each set $S \subseteq \mathbb{R}^n$ and to each point $x \in \mathbb{R}^n$ a cone $K(S, x)$ is associated, with the following properties:

$$K(S - x, 0) = K(S, x) \; , \tag{1}$$

$$K(\varphi(S), \varphi(x)) = \varphi(K(S, x)) \; , \quad \text{where } \varphi : \mathbb{R}^n \to \mathbb{R}^n \text{ is a non-singular linear mapping} \; , \tag{2}$$

$$K(S \cap N(x), x) = K(S, x) \; , \quad \text{for each neighborhood } N(x) \text{ of } x \; , \tag{3}$$

$$K(S, x) = \emptyset \qquad \forall x \notin \bar{S} \; , \tag{4}$$

$$K(S, x) = \mathbb{R}^n \qquad \forall x \in \operatorname{int}(S) \; , \tag{5}$$

$$K(S,x) + C \subseteq K(S,x) \ , \qquad \text{for each cone } C \subseteq \mathbb{R}^n \text{ with } S + C \subseteq S \, . \tag{6}$$

The special cone approximations introduced in Section 3.4 are defined especially for the case $x \in \bar{S}$. If we extend these definitions by the agreement $K(S,x) = \emptyset$ for $x \notin \bar{S}$, then it is easy to show that all properties of Definition 4.6.1 are fulfilled by those special cones.

In Definition 4.6.1 only very general properties are included. So conditions (1) and (2) claim a compatibility of the cone approximation with the linear structure of the space, i.e. an invariance regarding translations and non singular linear transformations (e.g. rotations and reflections).

Conditions (3), (4) and (5) reflect a compatibility of the approximation with the topological structure. Here the local character of the cone approximation is pointed out. In fact, the cone approximation is completely determined by the geometry of the set S near the point x (i.e. in any small neighborhood of the point x).

Condition (6) can be expressed by the Rockafellar recession cone

$$0^+S = \{y \in \mathbb{R}^n \mid S + ty \subseteq S \quad \forall t > 0\}$$

(see Definition 3.4.9). By the agreement $0^+\emptyset = \mathbb{R}^n$ (6) is equivalent to

$$0^+S \subseteq 0^+K(S,x) \, .$$

The meaning of this condition will become clear later, when we use the notion of local cone approximation for the development of directional derivatives.

We shall notice that the conditions (1),...,(6) are independent. In fact, if $K(.,.)$ is a local cone approximation according to Definition 4.6.1 then the set-valued mappings $K_i(.,.)$, $i = 1, ..., 6$, according to

$$K_1(S,x) = K(S,x) + x$$

$$K_2(S,x) = K(S,x) + C \qquad \text{(where } C \subseteq \mathbb{R}^n \text{ is an arbitrary cone)}$$

$$K_3(S,x) = \begin{cases} \operatorname{cone}(S - x) & \text{if } x \in \bar{S} \\ \emptyset & \text{else} \end{cases}$$

$$K_4(S,x) = \mathbb{R}^n$$

$$K_5(S,x) = \emptyset$$

$$K_6(S,x) = \begin{cases} \mathbb{R}^n & \text{if } x \in \operatorname{int}(S) \\ \{0\} & \text{if } x \in \bar{S} \backslash \operatorname{int}(S) \\ \emptyset & \text{if } x \notin \bar{S} \end{cases}$$

fulfill all conditions except (i).

Having any local cone approximation $K(.,.), K_1(.,.), ..., K_p(.,.)$ then it is easy to show that also

$$\operatorname{int}(K(.,.)), \quad \overline{K(.,.)}, \quad \operatorname{conv} K(.,.), \quad \operatorname{cone} K(.,.),$$

$$\bigcup_{i=1}^{p} K_i(.,.), \quad \bigcap_{i=1}^{p} K_i(.,.), \quad \sum_{i=1}^{p} K_i(.,.)$$

are local cone approximations according to Definition 4.6.1.

Regarding the algebraic complementation we get the following result.

Theorem 4.6.1. If $K(.,.)$ is a local cone approximation, then also the mapping $L(.,.)$ according to

$$L(S,x) = \mathbb{R}^n \backslash K(\mathbb{R}^n \backslash S, x)$$

is a local cone approximation.

Proof. Using the relations (1),...,(6) for $K(.,.)$ we get:

$$\begin{aligned} 1)\ \ L(S-x,0) &= \mathbb{R}^n \backslash K(\mathbb{R}^n \backslash (S-x), 0) = \\ &= \mathbb{R}^n \backslash K(\mathbb{R}^n \backslash S) - x, 0) = \\ &= \mathbb{R}^n \backslash K(\mathbb{R}^n \backslash S, x) = \\ &= L(S,x) \,. \end{aligned}$$

$$\begin{aligned} 2)\ \ L(\varphi(S),\varphi(x)) &= \mathbb{R}^n \backslash K(\mathbb{R}^n \backslash \varphi(S), \varphi(x)) = \\ &= \mathbb{R}^n \backslash K(\varphi(\mathbb{R}^n \backslash S), \varphi(x)) = \\ &= \mathbb{R}^n \backslash \varphi(K(\mathbb{R}^n \backslash S, x)) = \\ &= \varphi(\mathbb{R}^n \backslash K(\mathbb{R}^n \backslash S, x)) = \\ &= \varphi(L(S,x)) \end{aligned}$$

for each non singular linear mapping.

3) $L(S \cap N(x), x) = \mathbb{R}^n \backslash K(\mathbb{R}^n \backslash (S \cap N(x)), x) =$
$$= \mathbb{R}^n \backslash K((\mathbb{R}^n \backslash S) \cap N(x), x) =$$
$$= \mathbb{R}^n \backslash K(\mathbb{R}^n \backslash S, x) =$$
$$= L(S, x)$$
for any neighborhood $N(x)$ of x.

4) If $x \notin \bar{S}$, then $x \in \operatorname{int}(\mathbb{R}^n \backslash S)$ and we get
$$L(S, x) = \mathbb{R}^n \backslash K(\mathbb{R}^n \backslash S, x) = \mathbb{R}^n \backslash \mathbb{R}^n = \emptyset \ .$$

5) If $x \in \operatorname{int}(S)$, then $x \notin \overline{\mathbb{R}^n \backslash S}$ and we get
$$L(S, x) = \mathbb{R}^n \backslash K(\mathbb{R}^n \backslash S, x) = \mathbb{R}^n \backslash \emptyset = \mathbb{R}^n \ .$$

6) Since
$$\begin{aligned} 0^+(\mathbb{R}^n \backslash S) &= \{y \in \mathbb{R}^n \mid (\mathbb{R}^n \backslash S) + ty \subseteq \mathbb{R}^n \backslash S \ \ \forall t > 0\} = \\ &= \{y \in \mathbb{R}^n \mid \mathbb{R}^n \backslash S \subseteq (\mathbb{R}^n \backslash S) - ty \ \ \forall t > 0\} = \\ &= \{y \in \mathbb{R}^n \mid \mathbb{R}^n \backslash S \subseteq \mathbb{R}^n \backslash (S - ty) \ \ \forall t > 0\} = \\ &= \{y \in \mathbb{R}^n \mid S - ty \subseteq S \ \ \forall t > 0\} = \\ &= -0^+ S \end{aligned}$$
we obtain
$$\begin{aligned} 0^+ S &= -0^+(\mathbb{R}^n \backslash S) \subseteq \\ &\subseteq -0^+(K(\mathbb{R}^n \backslash S, x)) = \\ &= 0^+(\mathbb{R}^n \backslash K(\mathbb{R}^n \backslash S, x)) = \\ &= 0^+ L(S, x) \end{aligned}$$
which is equivalent to property (6) in Definition 4.6.1. □

Thus we are able to give

Definition 4.6.2. Let $K(.,.)$ be a local cone approximation. The local cone approximation $L(.,.)$, defined by

$$L(S,x) = I\!R^n \setminus K(I\!R^n \setminus S, x) \ ,$$

is called the *dual local cone approximation* of $K(.,.)$.

If $L(.,.)$ is dual to $K(.,.)$ then naturally, by twice complementation we can state that also $K(.,.)$ is dual to $L(.,.)$. This means that $\{K(.,.), L(.,.)\}$ forms a dual pair of local cone approximations.

Regarding the special cone approximations introduced in Section 3.4 we can recognize the following dual pairs:

$$\{Z(.,.), F(.,.)\} \ , \qquad \{Z_m(.,.), F_m(.,.)\} \ ,$$

$$\{I(.,.), T(.,.)\} \ , \qquad \{I_m(.,.), T_m(.,.)\} \ ,$$

$$\{A(.,.), Q(.,.)\} \ , \qquad \{A_m(.,.), Q_m(.,.)\} \ .$$

This duality relation is very interesting. So each (algebraic or topological) property of the one local cone approximation is associated with any dual property of the other. In this manner the discussion and the comparison of special local cone approximation can be simplified considerably as we have seen in Section 3.4.

We shall agree that in the following the notation "$K(.,.)$ has any property" (convexity, closeness or so on) means that the cones $K(S,x)$ admit this property for all sets S and all points x.

Now the concept of local cone approximations shall be used for the construction of generalized differentiability notions. Let $f : I\!R^n \to I\!R \cup \{\pm\infty\}$ be an extended real-valued function and $x^0 \in I\!R^n$ be a point in which f is finite. As we have mentioned above we will introduce generalized directional derivatives of f at x^0 by approximating the epigraph of f locally at the point $(x^0, f(x^0))$.

Definition 4.6.3. Let $K(.,.)$ be a local cone approximation. Then the extended real-valued function $f^K(x^0,.) : I\!R^n \to I\!R \cup \{\pm\infty\}$, defined by

$$f^K(x^0,y) := \inf \{\xi \in R \mid (y,\xi) \in K(\text{epi}\, f, (x^0, f(x)))\} \ , \qquad y \in I\!R^n \ ,$$

is called the K-*directional derivative* of f at x^0.

To avoid confusions we agree $\inf \emptyset = +\infty$ (analogous $\sup \emptyset = -\infty$).

Obviously $f^K(x^0,.)$ is a positively homogeneous function. Moreover, we have the following relations regarding the epigraph and the strict epigraph of $f^K(x^0,.)$:

$$\mathrm{epi}^0\, f^K(x^0,.) \subseteq K(\mathrm{epi}\, f, (x^0, f(x^0))) \subseteq \mathrm{epi}\, f^K(x^0,.) \; . \tag{7}$$

The second inclusion is obvious. The first inclusion is a consequence of property (6) in Definition 4.6.1. In fact, since

$$\{0\} \times I\!R_+ \subseteq 0^+(\mathrm{epi}\, f)$$

we also get

$$\{0\} \times I\!R_+ \subseteq 0^+(K(\mathrm{epi}\, f, (x^0, f(x^0))))$$

which means that

$$(y, f^K(x^0, y) + \xi) \in K(\mathrm{epi}\, f, (x^0, f(x^0)))$$

for all $y \in \mathrm{dom}(f^K(x^0,.))$ and all $\xi > 0$.

Regarding (7), the set $K(\mathrm{epi}\, f, (x^0, f(x^0)))$ is nearly the epigraph or the strict epigraph of a new function, namely $f^K(x^0,.)$. For the exact comparison we have to consider only the "vertical closure" and the "vertical interior" of the cone $K(\mathrm{epi}\, f, (x^0, f(x^0)))$ in the sense of

<u>Theorem 4.6.1</u>.

$$\begin{aligned} \mathrm{epi}\, f^K(x^0,.) \; &= \{(y,\xi) \mid \forall \varepsilon > 0 : (y, \xi + \varepsilon) \in K(\mathrm{epi}\, f, (x^0, f(x^0)))\} \\ \mathrm{epi}^0\, f^K(x^0,.) &= \{(y,\xi) \mid \exists \varepsilon > 0 : (y, \xi - \varepsilon) \in K(\mathrm{epi}\, f, (x^0, f(x^0)))\} \; . \end{aligned}$$

The assertion is a simple consequence of the relations between the epigraph and the strict epigraph of a function.

Naturally, if $K(\mathrm{epi}\, f, (x^0, f(x^0)))$ is closed, then

$$\mathrm{epi}\, f^K(x^0,.) = K(\mathrm{epi}\, f, (x^0, f(x^0)))$$

and $f^K(x^0,.)$ is lower semi-continuous. In the same manner, if $K(\text{epi} f, (x^0, f(x^0)))$ is an open set, then

$$\text{epi}^0 f^K(x^0,.) = K(\text{epi} f, (x^0, f(x^0)))$$

and $f^K(x^0,.)$ is upper semi-continuous.

Regarding dual cone approximations $K(.,.)$ and $L(.,.)$, an interesting relation between the directional derivatives $f^K(x^0,.)$ and $f^L(x^0,.)$ can be stated. By Definition 4.6.2 we get immediately

$$\begin{aligned} f^L(x^0,y) &= \inf\{\xi \in R \mid (y,\xi) \in L(\text{epi} f, (x^0, f(x^0)))\} = \\ &= \inf\{\xi \in R \mid (y,\xi) \notin K(\text{hypo}^0 f, (x^0, f(x^0)))\} = \\ &= \sup\{\xi \in R \mid (y,\xi) \in K(\text{hypo}^0 f, (x^0, f(x^0)))\} = \\ &= -\inf\{\xi \in R \mid (y,\xi) \in K(\text{epi}^0(-f), (x^0, -f(x^0)))\} . \qquad (8) \end{aligned}$$

The last equality can be deduced from property (2) of Definition 4.6.1, using the reflection $\varphi : \mathbb{R}^{n+1} \to \mathbb{R}^{n+1}$ defined by

$$\varphi(y,\xi) = (y, -\xi) \qquad \forall\, (y,\xi) \in \mathbb{R}^{n+1} .$$

Here φ is a non-singular linear mapping and

$$\varphi(\text{epi} f) = \text{hypo}(-f) , \qquad \varphi(\text{epi}^0 f) = \text{hypo}^0(-f) .$$

Analogously we have

$$\begin{aligned} -(-f)^K(x^0,y) &= -\inf\{\xi \in R \mid (y,\xi) \in K(\text{epi}(-f), (x^0, -f(x^0)))\} = \\ &= -\inf\{\xi \in R \mid (y,\xi) \notin L(\text{hypo}^0(-f), (x^0, -f(x^0)))\} = \\ &= -\sup\{\xi \in R \mid (y,\xi) \in L(\text{hypo}^0(-f), (x^0, -f(x^0)))\} = \\ &= \inf\{\xi \in R \mid (y,\xi) \in L(\text{epi}^0 f, (x^0, f(x^0)))\} . \qquad (9) \end{aligned}$$

In order to obtain the equality of (8) and (9) we have to regard connections between the cones $K(\text{epi}(-f), (x^0, -f(x^0)))$ and $K(\text{epi}^0(-f), (x^0, -f(x^0)))$ or (and this is equivalent) between the cones $L(\text{epi} f, (x^0, f(x^0)))$ and $L(\text{epi}^0 f, (x^0, f(x^0)))$ in the form

$$\inf \{\xi \in R \mid (y, \xi) \in K(\text{epi}(-f), (x^0, -f(x^0)))\} =$$
$$= \inf \{\xi \in R \mid (y, \xi) \in K(\text{epi}^0(-f), (x^0, -f(x^0)))\}$$

and

$$\inf \{\xi \in R \mid (y, \xi) \in L(\text{epi}\, f, (x^0, f(x^0)))\} =$$
$$= \inf \{\xi \in R \mid (y, \xi) \in L(\text{epi}^0\, f, (x^0, f(x^0)))\}$$

respectively. This means that in this case for the description of the directional derivatives in Definition 4.6.3 the epigraph of the function can be replaced by the strict epigraph. Such a relation cannot be derived directly from the general definition of a local cone approximation. But the reader can convince himself that this is fulfilled for the special local cone approximations discussed in Section 3.4.

We will summarize the result in

<u>Theorem 4.6.3</u>. For any $y \in \mathbb{R}^n$ the following assertions are equivalent:

i) $(-f)^K(x^0, y) = \inf \{\xi \in R \mid (y, \xi) \in K(\text{epi}^0(-f), (x^0, -f(x)))\}$;

ii) $f^L(x^0, y) = \inf \{\xi \in R \mid (y, \xi) \in L(\text{epi}^0\, f, (x^0, f(x)))\}$;

iii) $f^L(x^0, y) = -(-f)^K(x^0, y)$.

It is interesting to analyze the directional derivatives of a function f which are associated with $\text{int}(K(.,.))$ and $\overline{K(.,.)}$ where $K(.,.)$ is an arbitrary local cone approximation. Clearly, we have

$$f^{\bar{K}}(x^0, .) \leqq f^K(x^0, .) \leqq f^{\text{int}(K)}(x^0, .) . \tag{10}$$

Obviously $f^{\bar{K}}(x^0, .)$ is lower semi-continuous, $f^{\text{int}(K)}(x^0, .)$ is upper continuous and it holds

$$f^{\bar{K}}(x^0, y) = \liminf_{z \to y} f^K(x^0, z) \tag{11}$$

$$f^{\text{int}(K)}(x^0, y) = \limsup_{z \to y} f^K(x^0, z) \tag{12}$$

for any $y \in \mathbb{R}^n$.

To get more details we shall assume the local cone approximation $K(.,.)$ to be convex (which means that the cones $K(S,x)$ are convex for all sets S and all points x). Then naturally $f^K(x^0,.)$ but also $f^{\bar{K}}(x^0,.)$ and $f^{\text{int}(K)}(x^0,.)$ are sublinear functions. Moreover, in case of $\text{int}(K(\text{epi}\, f, (x^0, f(x^0)))) \neq \emptyset$ it holds

$$\overline{K(\text{epi}\, f, (x^0, f(x^0)))} = \overline{\text{int}(K(\text{epi}\, f, (x^0, f(x^0))))}\ , \tag{13}$$

$$\text{int}(K(\text{epi}\, f, (x^0, f(x^0)))) = \text{int}(\overline{K(\text{epi}\, f, (x^0, f(x^0)))}) \tag{14}$$

and we even get

$$f^{\bar{K}}(x^0, y) = \liminf_{z \to y}\ f^{\text{int}(K)}(x^0, z) \tag{15}$$

$$f^{\text{int}(K)}(x^0, y) = \limsup_{z \to y}\ f^{\bar{K}}(x^0, z)\ . \tag{16}$$

Now we can formulate

<u>Theorem 4.6.4</u>. Let $K(.,.)$ be a convex local cone approximation. For $y \in \text{dom}(f^{\text{int}(K)}(x^0,.)$ we have

$$f^{\bar{K}}(x^0, y) = f^K(x^0, y) = f^{\text{int}(K)}(x^0, y)\ .$$

<u>Proof</u>. If $f^{\text{int}(K)}(x^0, y) = -\infty$ then the equality is trivial.
Now let $f^{\text{int}(K)}(x^0, y)$ be finite. Because of the upper semi continuity we have even $y \in \text{int}(\text{dom}(f^{\text{int}(K)}(x^0,.)))$ and therefore $f^{\text{int}(K)}(x^0,.)$ is continuous at y since it is convex (see Theorem 2.5.15). Hence, from (15) we get

$$f^{\bar{K}}(x^0, y = \liminf_{z \to y}\ f^{\text{int}(K)}(x^0, z) = f^{\text{int}(K)}(x^0, y)$$

which is equivalent to the assertion. □

As a conclusion we formulate the following result:

Theorem 4.6.5. Let $K(.,.)$ be a convex local cone approximation. If $0 \in \mathrm{dom}(f^{\mathrm{int}(K)}(x^0,.))$ (i.e. $\mathrm{dom}(f^{\mathrm{int}(K)}(x^0,.)) = \mathbb{R}^n$), then

$$f^{\bar{K}}(x^0,y) = f^K(x^0,y) = f^{\mathrm{int}(K)}(x^0,y)$$

for all $y \in \mathbb{R}^n$.

Proof. If $f^{\mathrm{int}(K)}(x^0,0) = -\infty$, then we have

$$f^{\bar{K}}(x^0,.) = f^K(x^0,.) = f^{\mathrm{int}(K)}(x^0,.) \equiv -\infty \ .$$

In case of $f^{\mathrm{int}(K)}(x^0,0) = 0$, analogous to the former proof we can state that $f^{\mathrm{int}(K)}(x^0,.)$ is continuous at $y = 0$ and therefore (since it is sublinear) finite and continuous on $\mathbb{R}^n$.
Hence the assertion holds according to Theorem 4.6.4. □

Before we give further results let us introduce the K-subdifferential which is associated with the K-directional derivative in the manner described in the previous section.

Definition 4.6.4. Let $K(.,.)$ be a local cone approximation. Then the set

$$\partial_K f(x^0) = \{u \in \mathbb{R}^n \mid uy \leqq f^K(x^0,y) \ \ \forall y \in \mathbb{R}^n\}$$

is called the K-*subdifferential* of f at x^0.
The elements of $\partial_K f(x^0)$ are called the K-*subgradients* of f at x^0.

Obviously we have

$$\begin{aligned} \partial_K f(x^0) &= \{u \in \mathbb{R}^n \mid (u,-1)(y, f^K(x^0,y)) \leqq 0 \ \ \forall y \in \mathbb{R}^n\} = \\ &= \{u \in \mathbb{R}^n \mid (u,-1)(y,\xi) \leqq 0 \ \ \forall (y,\xi) \in \mathrm{epi}\, f^K(x^0,.)\} = \\ &= \{u \in \mathbb{R}^n \mid (u,-1)(y,\xi) \leqq 0 \ \forall (y,\xi) \in K(\mathrm{epi}\, f, (x^0, f(x^0)))\} = \\ &= \{u \in \mathbb{R}^n \mid (u,-1) \in (K(\mathrm{epi}\, f, (x^0, f(x^0))))^*\} \qquad (17) \end{aligned}$$

where $(.)^*$ denotes the polar cone introduced in Section 2.3. Using the closedness and the convexity of the polar cone we can formulate

<u>Theorem 4.6.6</u>. The K-subdifferential is a closed convex set and it holds

$$\partial_{\bar{K}} f(x^0) = \partial_K f(x^0) \; .$$

If $K(.,.)$ is a convex local cone approximation and $\mathrm{dom}(f^{\mathrm{int}(K)}(x^0,.)) \neq \emptyset$ (i.e. $\mathrm{int}(K(\mathrm{epi}\, f, (x^0, f(x^0)))) \neq \emptyset$), then it holds even

$$\partial_{\mathrm{int}(K)} f(x^0) = \partial_K f(x^0) \; .$$

<u>Proof</u>. The first assertion is obvious.
The second assertion is a consequence of (13) and (17). □

In the following theorem we give an assertion about the existence of K-subgradients and some connections between the K-subdifferential and the K-directional derivative which are similar to those found in Theorem 2.6.6.

<u>Theorem 4.6.7</u>. Let $K(.,.)$ be a convex local cone approximation. Then it holds

i) $\partial_K f(x^0) \neq \emptyset$ if and only if $f^{\bar{K}}(x^0, 0) \geqq 0$.

ii) If $f^{\bar{K}}(x^0, 0) = 0$, then

$$f^{\bar{K}}(x^0, y) = \sup \{uy \mid u \in \partial_K f(x^0)\} \qquad \forall\, y \in \mathbb{R}^n \; .$$

iii) If $f^{\bar{K}}(x^0, 0) = 0$ and $f^{\mathrm{int}(K)}(x^0, y) < \infty$, then

$$f^K(x^0, y) = \max \{uy \mid u \in \partial_K f(x^0)\} \; .$$

iv) If $f^{\mathrm{int}(K)}(x^0, 0) = 0$ then $\partial_K f(x^0)$ is a bounded set and

$$f^K(x^0, y) = \max \{uy \mid u \in \partial_K f(x^0)\} \qquad \forall\, y \in \mathbb{R}^n \; .$$

<u>Proof</u>.
1) If $f^{\bar{K}}(x^0, 0) = \infty$ then $\overline{K(\mathrm{epi}\, f, (x^0, f(x^0)))} = \emptyset$ and hence we have $\partial_K f(x^0) = \mathbb{R}^n$. Let therefore $f^{\bar{K}}(x^0, 0) = 0$. Then the closed convex cone $\overline{K(\mathrm{epi}\, f, (x^0, f(x^0)))}$ and the point $(0, -1)$ can be separated strictly by Theorem 2.2.6, i.e. there exists a vector $(u, \alpha) \in \mathbb{R}^n \times \mathbb{R}$, $(u, \alpha) \neq (0, 0)$, such that

$$\sup\{(u,\alpha)(y,\xi) \mid (y,\xi) \in \overline{K(\operatorname{epi} f, (x^0, f(x^0)))}\} <$$
$$< (u,\alpha)(0,-1) = -\alpha \ .$$

From this we get $\alpha < 0$ and

$$\sup\{(u,\alpha)(y,\xi) \mid (y,\xi) \in \overline{K(\operatorname{epi} f, (x^0, f(x^0)))}\} =$$
$$= \sup\{(u,\alpha)(y,\xi) \mid (y,\xi) \in \operatorname{epi} f^{\bar{K}}(x^0, .)\} = 0 \ .$$

Without loss of generality we can settle $\alpha = -1$ and we especially get

$$(u,-1)(y, f^{\bar{K}}(x^0,y)) = uy - f^{\bar{K}}(x^0,y) \leqq 0 \qquad \forall\, y \in I\!R^n$$

i.e.

$$u \in \partial_{\bar{K}} f(x^0) = \partial_K f(x^0) \ .$$

Conversely, if $\partial_K f(x^0) = \partial_{\bar{K}} f(x^0) \neq \emptyset$, then by definition we have

$$f^{\bar{K}}(x^0, 0) \geqq u0 = 0 \qquad \forall\, u \in \partial_K f(x^0) \ .$$

2) By the definition of the K-subdifferentials we get

$$f^{\bar{K}}(x^0,y) \geqq \sup\{uy \mid u \in \partial_{\bar{K}} f(x^0)\} = \sup\{uy \mid u \in \partial_K f(x^0)\} \ .$$

For the converse relation let $y^0 \in I\!R^n$. Since by assumption $f^{\bar{K}}(x^0,y^0) > -\infty$ (otherwise we would get $f^{\bar{K}}(x^0,0) = -\infty$ because of the lower semi continuity) we can choose a number $\xi_0 < f^{\bar{K}}(x^0,y^0)$, i.e. we have

$$(y^0,\xi_0) \notin \overline{K(\operatorname{epi} f, (x^0, f(x^0)))} \ .$$

Then

$$\operatorname{conv}\{(0,-1),(y^0,\xi_0)\} \cap \overline{K(\operatorname{epi} f, (x, f(x)))} = \emptyset$$

and both sets can be separated strictly. So there exists a vector $(v,\alpha) \in I\!R^n \times I\!R$, $(v,\alpha) \neq (0,0)$ such that

$$\sup\{(v,\alpha)(y,\xi) \mid (y,\xi) \in \overline{K(\operatorname{epi} f,(x^0,f(x^0)))}\} <$$
$$< \inf\{(v,\alpha)(y,\xi) \mid (y,\xi) \in \operatorname{conv}\{(0,-1),(y^0,\xi_0)\}\} .$$

Analogous to the first part of the proof we can state that $\alpha < 0$ and that

$$\sup\{(v,\alpha)(y,\xi) \mid (y,\xi) \in \overline{K(\operatorname{epi} f,(x^0,f(x^0)))}\} =$$
$$= \sup\{(v,\alpha)(y,\xi) \mid (y,\xi) \in \operatorname{epi} f^{\bar{K}}(x^0,.)\} = 0 .$$

Setting $\alpha = -1$ we get

$$vy - f^{\bar{K}}(x^0,y) \leqq 0 \qquad \forall\, y \in \mathbb{R}^n$$

and

$$vy^0 - \xi_0 > 0 .$$

That means $v \in \partial_{\bar{K}} f(x^0) = \partial_K f(x^0)$ and

$$\xi_0 < vy^0 \leqq \sup\{uy^0 \mid u \in \partial_K f(x^0)\} .$$

This relation is true for all $\xi_0 < f^{\bar{K}}(x^0,y^0)$, hence

$$f^{\bar{K}}(x^0,y^0) \leqq \sup\{uy^0 \mid u \in \partial_K f(x^0)\} .$$

3) Let $y^0 \in \operatorname{dom}(f^{\operatorname{int}(K)}(x^0,.))$. Then according to Theorem 4.6.4 we have

$$f^{\bar{K}}(x^0,y^0) = f^K(x^0,y^0) = f^{\operatorname{int}(K)}(x^0,y^0)$$

and this value is finite because of $f^{\bar{K}}(x^0,0) = 0$. Using the first separation theorem (Theorem 2.2.5) to the sets $K(\operatorname{epi} f,(x^0,f(x^0)))$ and $\{\lambda(y^0,f^K(x^0,y^0)) \mid \lambda \geqq 0\}$ we can find a vector $(u,\alpha) \in \mathbb{R}^n \times \mathbb{R}$, $(u,\alpha) \neq (0,0)$, such that

$$\sup\{(u,\alpha)(y,\xi) \mid (y,\xi) \in K(\operatorname{epi} f,(x^0,f(x^0)))\} \leqq$$
$$\leqq \inf\{\lambda(u,\alpha)(y^0,f^K(x^0,y^0)) \mid \lambda \geq 0\} .$$

Obviously, we have $\alpha \leqq 0$ and both values of the inequality are equal to zero. Assuming $\alpha = 0$ we would get

$$uy \leqq uy^0 \qquad \forall\, y \in \mathrm{dom}(f^K(x^0,.)) \ ,$$

but this means $u = 0$ since

$$y^0 \in \mathrm{dom}(f^{\mathrm{int}(K)}(x^0,.)) \subseteq \mathrm{int}(\mathrm{dom}(f^K(x^0,.))) \ .$$

Thus, $\alpha < 0$ and without loss of generality we can set $\alpha = -1$. We get

$$uy - f^K(x^0, y) \leqq 0 \qquad \forall\, y \in \mathbb{R}^n$$

and

$$uy^0 - f^K(x^0, y^0) \geqq 0 \ ,$$

i.e. $u \in \partial_K f(x^0)$ and regarding ii)

$$f^K(x^0, y^0) = uy^0 = \max\{uy^0 \mid u \in \partial_K f(x^0)\} \ .$$

4) If $f^{\mathrm{int}(K)}(x^0, 0) = 0$, then according to Theorem 4.6.5

$$f^{\bar{K}}(x^0,.) = f^K(x^0,.) = f^{\mathrm{int}(K)}(x^0,.)$$

and this function is continuous for all y. From iii) we get

$$f^K(x^0, y) = \max\{uy \mid u \in \partial_K f(x^0)\}$$

for all $y \in \mathbb{R}^n$. Assuming $\partial_K f(x^0)$ is not bounded, we would find a sequence $\{u^k\} \in \partial_K f(x^0)$ with $\|u^k\| \to \infty$. Then with $y^k = \dfrac{u^k}{\|u^k\|}$ we get

$$f^K(x^0, y^k) \geqq u^k y^k = u^k \frac{u^k}{\|u^k\|} = \|u^k\| \ ,$$

i.e. the directional derivative $f^K(x^0,.)$ is unbounded on the unit sphere, which is a contradiction to the continuity. □

Finally, we shall give some calculus rules for the K-directional derivative and the K-subdifferential. Regarding the multiplication with a positive number we have

Theorem 4.6.8. For any real number $\lambda > 0$ it holds

i) $(\lambda f)^K(x^0, y) = \lambda f^K(x^0, y)$, $\forall\, y \in \mathbb{R}^n$;

ii) $\partial_K(\lambda f)(x^0) = \lambda \partial_K f(x^0)$.

Proof. For a fixed number $\lambda > 0$ the mapping $\varphi_\lambda : \mathbb{R}^{n+1} \to \mathbb{R}^{n+1}$, according to

$$\varphi_\lambda(y, \xi) = (y, \lambda\xi) \qquad \forall\, (y, \xi) \in \mathbb{R}^{n+1} ,$$

is linear and non singular. Using property (2) of Definition 4.6.1 we have

$$\begin{aligned} K(\mathrm{epi}(\lambda f), (x^0, \lambda f(x^0))) &= K(\varphi_\lambda(\mathrm{epi}\, f), \varphi_\lambda(x^0, f(x^0))) = \\ &= \varphi_\lambda(K(\mathrm{epi}\, f, (x^0, f(x^0)))) = \\ &= \{(y, \lambda\xi) \mid (y, \xi) \in K(\mathrm{epi}\, f, (x^0, f(x^0)))\} . \end{aligned}$$

From this we get

$$(\lambda f)^K(x^0, y) = \lambda f^K(x^0, y) \qquad \forall\, y \in \mathbb{R}^n$$

and

$$\begin{aligned} \partial_K(\lambda f)(x^0) &= \{u \in \mathbb{R}^n \mid uy \leqq (\lambda f)^K(x^0, y) \;\; \forall\, y \in \mathbb{R}^n\} = \\ &= \{u \in \mathbb{R}^n \mid uy \leqq \lambda f^K(x^0, y) \;\; \forall\, y \in \mathbb{R}^n\} = \\ &= \lambda \partial_K f(x^0) . \end{aligned}$$

□

We cannot expect that the local cone approximation in this general structure permits rules regarding the representation of the K-directional derivatives and K-subdifferentials for the sum of functions, similar to Theorem 2.6.8. We can give, however, an assertion which compares conditions regarding the sum of K-directional derivatives and the sum of K-subdifferentials for different functions. Such assertions are very important for the construction of optimality conditions for constrained optimization problems. Naturally, from the definition we get

$$0 \in \partial_K f(x^0) \Leftrightarrow f^K(x, y) \geqq 0 \qquad \forall\, y \in \mathbb{R}^n$$

and this equivalence can be used for the discussion of unconstrained optimization problems. As an extension of this relation we have

Theorem 4.6.9. Let $f_1, ..., f_p : \mathbb{R}^n \to \mathbb{R} \cup \{\pm\infty\}$ be real-valued functions, which are finite in the point $x^0 \in \mathbb{R}^n$, $K_1(.,.), ..., K_p(.,.)$ be local cone approximations and $\lambda_1, ..., \lambda_p$ be positive numbers, Then it holds

i) If $0 \in \sum_{i=1}^{p} \lambda_i \partial_{K_i} f_i(x^0)$ then $\sum_{i=1}^{p} \lambda_i f_i^{K_i}(x^0, y) \geqq 0, \ \forall y \in \mathbb{R}^n$.

ii) Let $K_i(.,.)$ be convex, $f_i^{\bar{K}_i}(x^0, 0) = 0$, $i = 1, ..., p$ and $\bigcap_{i=1}^{p} \operatorname{dom}(f_i^{\operatorname{int}(K_1)}(x^0, .)) \neq \emptyset$. Then the converse implication is true, i.e.

$$\text{if } \sum_{i=1}^{p} \lambda_i f_i^{K_i}(x^0, y) \geqq 0 \quad \forall y \in \mathbb{R}^n \quad \text{then} \quad 0 \in \sum_{i=1}^{p} \lambda_i \partial_{K_i} f_i(x^0) .$$

Before proving the assertion, it is useful to remark that in Theorem 4.6.7 we have pointed out that the condition $f_i^{\bar{K}_i}(x^0, 0) = 0$ implies $\partial_{K_i} f_i(x^0) \neq \emptyset$. This means especially that $f_i^{K_i}(x^0, y) > -\infty$ for all $y \in \mathbb{R}^n$, and thus the terms $\sum_{i=1}^{p} \lambda_i \partial_{K_i} f_i(x^0)$ and $\sum_{i=1}^{p} \lambda_i f_i^{K_i}(x^0, y)$ are determined correctly.

Proof of Theorem 4.6.9.

1) If $0 \in \sum_{i=1}^{p} \lambda_i \partial_{K_i} f_i(x^0)$ we get from the definition

$$\begin{aligned}\sum_{i=1}^{p} \lambda_i f_i^{K_i}(x^0, y) &\geqq \sum_{i=1}^{p} \lambda_i \sup \{u_i y \mid u_i \in \partial_{K_i} f_i(x^0)\} = \\ &= \sup \Big\{ \sum_{i=1}^{p} \lambda_i u_i y \mid u_i \in \partial_{K_i} f_i(x^0), \ i = 1, ..., p \Big\} = \\ &= \sup \Big\{ uy \mid u \in \sum_{i=1}^{p} \lambda_i \partial_{K_i} f_i(x^0) \Big\} \geqq 0\end{aligned}$$

for all $y \in \mathbb{R}^n$.

2) Let $\sum_{i=1}^{p} \lambda_i f_i^{K_i}(x^0, y) \geqq 0$, $\forall\, y \in \mathbb{R}^n$. Then

$$0 \in \left\{ u \in \mathbb{R}^n \mid uy \leqq \sum_{i=1}^{p} \lambda_i f_i^{K_i}(x^0, y) \quad \forall\, y \in \mathbb{R}^n \right\}$$

and we have to show that this set is contained in $\sum_{i=1}^{p} \lambda_i \partial_{K_i} f_i(x^0)$. Because of Theorem 4.6.8, for simplification we can assume that $\lambda_i = 1$ for all $i = 1, ..., p$. Moreover, it is sufficient to show this implication only for $p = 2$. Via complete induction one can extend the assertion for any other finite set of functions.
Now let $uy \leqq f_1^{K_1}(x^0, y) + f_2^{K_2}(x^0, y)$. With

$$g(y) = f_1^{K_1}(x^0, y)$$

$$h(y) = uy - f_2^{K_2}(x^0, y)$$

we get

$$g(y) \geqq h(y) \qquad \forall\, y \in \mathbb{R}^n \ ,$$

i.e.

$$\operatorname{int}(\operatorname{epi} g) \cap (\operatorname{hypo} h) = \emptyset \ .$$

By assumption both sets are nonempty convex cones. Using the separation theorem (Theorem 2.2.5) we can find a vector $(v, \alpha) \in \mathbb{R}^n \times \mathbb{R}$, $(v, \alpha) \neq (0, 0)$, with

$$(v, \alpha)(y, \xi) \leqq 0 \qquad \forall\, (y, \xi) \in \operatorname{epi} g \ ,$$

$$(v, \alpha)(y, \xi) \geqq 0 \qquad \forall\, (y, \xi) \in \operatorname{hypo} h \ .$$

Obviously, we have $\alpha \leqq 0$. Assuming $\alpha = 0$ we would get

$$vy \leqq 0 \qquad \forall\, y \in \operatorname{dom}(f_1^{K_1}(x^0, .)) \ ,$$

$$vy \geqq 0 \qquad \forall\, y \in \operatorname{dom}(f_2^{K_2}(x^0, .)) \ .$$

Since by assumption

$$\emptyset \neq \text{dom}(f_1^{\text{int}(K_1)}(x^0,.)) \cap \text{dom}(f_2^{\text{int}(K_2)}(x^0,.)) \subseteq$$
$$\subseteq \text{int}(\text{dom}(f_1^{K_1}(x^0,.))) \cap \text{int}(\text{dom}(f_2^{K_2}(x^0,.))) \ , \tag{18}$$

this is only possible for $v = 0$ which would be a contradiction.
Hence, $\alpha < 0$ and without loss of generality we can set $\alpha = -1$. That means especially

$$vy - f_1^{K_1}(x^0, y) \leqq 0 \qquad \forall\, y \in \text{dom}(f_1^{K_1}(x^0,.))$$
$$vy - (uy - f_2^{K_2}(x^0, y)) \geqq 0 \qquad \forall\, y \in \text{dom}(f_2^{K_2}(x^0,.))$$

i.e.

$$v \in \partial_{K_1} f_1(x^0) \ , \qquad u - v \in \partial_{K_2} f_2(x^0)$$

and

$$u = v + (u - v) \in \partial_{K_1} f_1(x^0) + \partial_{K_2} f_2(x^0) \ . \qquad \square$$

Remark. For the proof it was sufficient to claim only

$$\emptyset \neq \text{dom}(f_1^{\text{int}(K_1)}(x^0,.)) \cap \text{dom}(f_2^{K_2}(x^0,.)) \subseteq$$
$$\subseteq \text{int}(\text{dom}(f_1^{K_1}(x^0,.))) \cap \text{dom}(f_2^{K_2}(x^0,.))$$

instead of (18).
Therefore in the general case the assumed interior point condition of Theorem 4.6.9 can be weakened by requiring

$$\bigcap_{\substack{i=1 \\ i \neq i_0}}^{p} \text{dom}(f_i^{\text{int}(K_i)}(x^0,.)) \cap \text{dom}(f_{i_0}^{K_{i_0}}(x^0,.)) \neq \emptyset$$

for any $i_0 \in \{1, ..., p\}$.

4.7. Special K-Directional Derivative

In this section we analyze the special K-directional derivatives which are associated with the special cone approximations discussed in Section 3.4. Analogous to the other sections let $f : I\!R^n \to I\!R \cup \{\pm\infty\}$ be an extended real-valued function and $x^0 \in I\!R^n$ be a point in which f is finite.

We start with the cone of feasible direction $Z(.,.)$ and the radial tangent cone $F(.,.)$ which are dual in the sense of Definition 4.6.2. Here we get the following result:

<u>Theorem 4.7.1</u>. For any $y \in I\!R^n$ it holds

i) $f^Z(x^0, y) = \limsup\limits_{t \downarrow 0} \dfrac{f(x^0 + ty) - f(x^0)}{t}$,

ii) $f^F(x^0, y) = \liminf\limits_{t \downarrow 0} \dfrac{f(x^0 + ty) - f(x^0)}{t}$,

iii) $f^Z(x^0, y) = -(-f)^F(x^0, y)$.

<u>Proof</u>.

1) $$\begin{aligned} \operatorname{epi} f^Z(x^0, .) &= \{(y,\xi) \in I\!R^n \times I\!R \mid \forall \varepsilon > 0 : \\ &\qquad (y, \xi + \varepsilon) \in Z(\operatorname{epi} f, (x^0, f(x^0)))\} = \\ &= \{(y,\xi) \in I\!R^n \times I\!R \mid \forall \varepsilon > 0 \;\; \exists \lambda > 0 \;\; \forall t \in (0, \lambda) : \\ &\qquad (x^0 + ty, f(x^0) + t(\xi + \varepsilon)) \in \operatorname{epi} f\} = \\ &= \Big\{(y,\xi) \in I\!R^n \times I\!R \mid \forall \varepsilon > 0 \;\; \exists \lambda > 0 \;\; \forall t \in (0, \lambda) : \\ &\qquad \frac{f(x^0 + ty) - f(x^0)}{t} \leqq \xi + \varepsilon\Big\} \\ &= \Big\{(y, \xi) \in I\!R^n \times I\!R \mid \limsup_{t \downarrow 0} \frac{f(x^0 + ty) - f(x^0)}{t} \leqq \xi\Big\} . \end{aligned}$$

2) $\operatorname{epi} f^F(x^0, .) = \{(y,\xi) \in I\!R^n \times I\!R \mid \forall \varepsilon > 0 :$

$$
\begin{aligned}
&(y,\xi+\varepsilon)\in F(\operatorname{epi} f,(x^0,f(x^0)))\} = \\
&= \{(y,\xi)\in \mathbb{R}^n\times\mathbb{R} \mid \forall\varepsilon>0 \ \ \forall\lambda>0 \ \ \exists t\in(0,\lambda): \\
&\qquad (x^0+ty, f(x^0)+t(\xi+\varepsilon))\in \operatorname{epi} f\} = \\
&= \Big\{(y,\xi)\in \mathbb{R}^n\times\mathbb{R} \mid \forall\varepsilon>0 \ \ \forall\lambda>0 \ \ \exists t\in(0,\lambda): \\
&\qquad \frac{f(x^0+ty)-f(x^0)}{t} \leqq \xi+\varepsilon\Big\} \\
&= \Big\{(y,\xi)\in \mathbb{R}^n\times\mathbb{R} \mid \liminf_{t\downarrow 0} \frac{f(x^0+ty)-f(x^0)}{t} \leqq \xi\Big\} .
\end{aligned}
$$

3) $$
\begin{aligned}
f^Z(x^0,y) &= \limsup_{t\downarrow 0} \frac{f(x^0+ty)-f(x^0)}{t} = \\
&= -\liminf_{t\downarrow 0} \frac{(-f)(x^0+ty)-(-f)(x^0)}{t} = \\
&= -(-f)^F(x^0,y) .
\end{aligned}
$$
□

Thus we recognize the special directional derivative (upper and lower Dini directional derivative) $f'_U(x^0,.)$ and $f'_L(x^0,.)$ already discussed in Section 4.4. Also we see the dual character of both directional derivatives in the third relation.

In conclusion we obtain the following simple assertion which we mentioned in Section 4.4.

Theorem 4.7.2. f is Gâteaux differentiable at $x^0 \in \mathbb{R}^n$ if and only if $f^F(x^0,.) = f^Z(x^0,.)$ and this directional derivative is finite and linear.

If f is a convex function, then $\operatorname{epi} f$ is a convex set and because of Theorem 3.4.10 we get

$$
\begin{aligned}
Z(\operatorname{epi} f,(x^0,f(x^0))) &= F(\operatorname{epi} f,(x^0,f(x^0))) = \\
&= \operatorname{cone}(\operatorname{epi} f - (x^0,f(x^0))) .
\end{aligned}
$$

Therefore we can state once more the well-known relation

$$
f^Z(x^0,y) = f^F(x^0,y) = \lim_{t\downarrow 0} \frac{f(x^0+ty)-f(x^0)}{t}
$$

for all $y \in \mathbb{R}^n$, which we have derived in Sections 2.6 and 4.3.

For the cone approximations $I(.,.)$ and $T(.,.)$ (i.e. the cone of interior directions and the contingent cone respectively) we get the following result:

Theorem 4.7.3. For any $y \in \mathbb{R}^n$ it holds

i) $$f^I(x^0, y) = \limsup_{\substack{t \downarrow 0 \\ z \to y}} \frac{f(x^0 + tz) - f(x^0)}{t} \,,$$

ii) $$f^T(x^0, y) = \liminf_{\substack{t \downarrow 0 \\ z \to y}} \frac{f(x^0 + tz) - f(x^0)}{t} \,,$$

iii) $f^I(x^0, y) = -(-f)^T(x^0, y)$.

Proof.

1) Since $I(\text{epi}\, f, (x^0, f(x^0)))$ is open, we have

$$\begin{aligned}
\text{epi}^0 f^I(x^0, .) &= I(\text{epi}\, f, (x^0, f(x^0))) = \\
&= \{(y,\xi) \in \mathbb{R}^n \times \mathbb{R} \mid \exists \varepsilon > 0 \;\; \exists N(y) \;\; \exists \lambda > 0 \\
&\qquad \forall t \in (0, \lambda) \;\; \forall z \in N(y) \;\; \forall \zeta \text{ with } |\zeta - \xi| < \varepsilon : \\
&\qquad (x^0 + tz, f(x^0) + t\zeta) \in \text{epi}\, f\} = \\
&= \Big\{(y,\xi) \in \mathbb{R}^n \times \mathbb{R} \mid \exists \varepsilon > 0 \;\; \exists N(y) \;\; \exists \lambda > 0 \\
&\qquad \forall t \in (0, \lambda) \;\; \forall z \in N(y) \;\; \forall \zeta \text{ with } |\zeta - \xi| < \varepsilon : \\
&\qquad \frac{f(x^0 + tz) - f(x^0)}{t} \leqq \zeta\Big\} = \\
&= \Big\{(y,\xi) \in \mathbb{R}^n \times \mathbb{R} \mid \exists \varepsilon > 0 \;\; \exists N(y) \;\; \exists \lambda > 0 \\
&\qquad \forall t \in (0, \lambda) \;\; \forall z \in N(y) : \\
&\qquad \frac{f(x^0 + tz) - f(x^0)}{t} \leqq \xi - \varepsilon\Big\} =
\end{aligned}$$

$$= \left\{ (y, \xi) \in I\!R^n \times I\!R \mid \limsup_{\substack{t \downarrow 0 \\ z \to y}} \frac{f(x^0 + tz) - f(x^0)}{t} < \xi \right\} .$$

2) Since $T(\operatorname{epi} f, (x^0, f(x^0)))$ is closed, we have

$$\begin{aligned}
\operatorname{epi} f^T(x^0, .) = T(\operatorname{epi} f, (x^0, f(x^0))) = \\
&= \{(y,\xi) \in I\!R^n \times I\!R \mid \forall \varepsilon > 0 \;\; \forall N(y) \;\; \forall \lambda > 0 \\
&\qquad \exists t \in (0, \lambda) \;\; \exists z \in N(y) \;\; \exists \zeta \text{ with } |\zeta - \xi| < \varepsilon : \\
&\qquad (x^0 + tz, f(x^0) + t\zeta) \in \operatorname{epi} f\} = \\
&= \Big\{(y,\xi) \in I\!R^n \times I\!R \mid \forall \varepsilon > 0 \;\; \forall N(y) \;\; \forall \lambda > 0 \\
&\qquad \exists t \in (0, \lambda) \;\; \exists z \in N(y) \;\; \exists \zeta \text{ with } |\zeta - \xi| < \varepsilon : \\
&\qquad \frac{f(x^0 + tz) - f(x^0)}{t} \leqq \zeta \Big\} = \\
&= \Big\{(y,\xi) \in I\!R^n \times I\!R \mid \forall \varepsilon > 0 \;\; \forall N(y) \;\; \forall \lambda > 0 \\
&\qquad \exists t \in (0, \lambda) \;\; \exists z \in N(y) : \\
&\qquad \frac{f(x^0 + tz) - f(x^0)}{t} < \xi + \varepsilon \Big\} = \\
&= \left\{(y, \xi) \in I\!R^n \times I\!R \mid \liminf_{\substack{t \downarrow 0 \\ z \to y}} \frac{f(x^0 + tz) - f(x^0)}{t} \leqq \xi \right\} .
\end{aligned}$$

3) $$\begin{aligned}
f^I(x^0, y) &= \limsup_{\substack{t \downarrow 0 \\ z \to y}} \frac{f(x^0 + tz) - f(x^0)}{t} = \\
&= -\liminf_{\substack{t \downarrow 0 \\ z \to y}} \frac{(-f)(x^0 + tz) - (-f)(x^0)}{t} = \\
&= -(-f)^T(x^0, y) .
\end{aligned}$$

□

Also here we see that these special K-directional derivatives are dual and coincide with the directional derivatives $f_U^+(x^0,.)$ and $f_L^+(x^0,.)$ (the upper and lower Dini-Hadamard derivative) discussed in Section 4.4. In this representation, the upper and the lower semi-continuity respectively of $f^I(x^0,.)$ and $f^T(x^0,.)$ are consequences of the topological properties of the cone approximations $I(.,.)$ and $T(.,.)$. Moreover, because of $0 \in T(\operatorname{epi} f,(x^0,f(x^0)))$ and $0 \notin I(\operatorname{epi} f,(x^0,f(x^0)))$ we get $f^T(x^0,0) \leqq 0$ and $f^I(x^0,0) \geqq 0$.

Naturally in case of

$$f^T(x^0,.) = f^I(x^0,.)$$

the function f is uniformly directionally differentiable at x^0. Then $f^T(x^0,0) = f^I(x^0,0) = 0$ and this directional derivative is finite on $\mathbb{R}^n$.

Regarding the Fréchet differentiability of a function we can give the following assertion mentioned already in Section 4.4.

Theorem 4.7.4. f is Fréchet differentiable at $x^0 \in \mathbb{R}^n$ if and only if $f^T(x^0,.) = f^I(x^0,.)$ and this directional derivative is linear.

Proof. The assertion is a consequence of Theorem 4.2.1. □

The directional derivatives $f^I(x^0,.)$ and $f^T(x^0,.)$ can be used to describe (semi-) continuity properties of function f. It holds

Theorem 4.7.5.

i) If $f^I(x^0,0) = 0$ (this is equivalent to $f^I(x^0,.) < \infty$), then f is upper semi-continuous at x^0.

ii) If $f^T(x^0,0) = 0$ (this is equivalent to $f^T(x^0,.) > -\infty$), then f is lower semi-continuous at x^0.

iii) If $f^I(x^0,0) = f^T(x^0,0) = 0$ (this is equivalent to the finiteness of $f^I(x^0,.)$ and $f^T(x^0,.)$), then f is continuous at x^0.

Proof.

1) Assuming f is not upper semi-continuous at x^0, we could find a positive number $\varepsilon > 0$ and a sequence $\{x^k\} \in \mathbb{R}^n$, $x^k \to x^0$ such that $f(x^k) -$

$f(x^0) > \varepsilon$. Setting $y^k = \dfrac{x^k - x^0}{\|x^k - x^0\|}$ and $t_k = \|x^k - x^0\|$, without loss of generality we can assume that $\{y^k\}$ is convergent (since it is contained in the compact unit sphere) and that $y^k \to y^0$. Then

$$\begin{aligned} f^I(x^0, y^0) &\geqq \limsup_{k\to\infty} \frac{f(x^0 + t_k y^k) - f(x^0)}{t_k} = \\ &= \limsup_{k\to\infty} \frac{f(x^k) - f(x^0)}{\|x^k - x^0\|} = \infty \end{aligned}$$

which is a contradiction.

2) Because of $(-f)^I(x^0, 0) = -f^T(x^0, 0) = 0$ we can conclude that $-f$ is upper continuous at x^0, hence f is lower semi-continuous at x^0.

3) The assertion is a consequence of i) and ii) and the fact that $f(x^0)$ is finite. □

Now let f be a convex function and $x^0 \in \text{int}(\text{dom}(f))$ (in this case f is continuous at x^0). Obviously $(0,1) \in \text{int}(\text{epi}\, f)$ and according to Theorem 3.4.10 we have

$$I(\text{epi}\, f, (x^0, f(x^0))) = \text{cone}(\text{int}(\text{epi}\, f) - (x^0, f(x^0))) \neq \emptyset$$

and

$$T(\text{epi}\, f, (x^0, f(x^0))) = \overline{I(\text{epi}\, f, (x^0, f(x^0)))}\ .$$

Because of $f^I(x^0, 0) \leqq 0$ (we have even $f^I(x^0, 0) = 0$), by Theorem 4.6.5 we see that

$$f^T(x^0, y) = f^I(x^0, y) = \lim_{\substack{t\downarrow 0 \\ z\to y}} \frac{f(x^0 + tz) - f(x^0)}{t}$$

for all $y \in I\!R^n$. Thus we get once more the well-known result (discussed in Section 4.3) that a continuous convex function is uniformly directional differentiable.

We investigate now the K-directional derivative related to the cone approximations $A(.,.)$ and $Q(.,.)$ (i.e. the cone of attainable directions and the cone of quasi-interior directions respectively).

Theorem 4.7.6. For any $y \in \mathbb{R}^n$ it holds

i) $\displaystyle f^A(x^0, y) = \lim\sup_{t \downarrow 0} \inf_{z \to y} \frac{f(x^0 + tz) - f(x^0)}{t}$,

ii) $\displaystyle f^Q(x^0, y) = \lim\inf_{t \downarrow 0} \sup_{z \to y} \frac{f(x^0 + tz) - f(x^0)}{t}$,

iii) $f^A(x^0, y) = -(-f)^Q(x^0, y)$.

Proof.

1) Since $A(\text{epi}\, f, (x^0, f(x^0)))$ is closed, we have

$$
\begin{aligned}
\text{epi}\, f^A(x^0, .) &= A(\text{epi}\, f, (x^0, f(x^0))) = \\
&= \{(y,\xi) \in \mathbb{R}^n \times \mathbb{R} \mid \forall \varepsilon > 0 \ \ \forall N(y) \ \ \exists \lambda > 0 \\
&\qquad \forall t \in (0, \lambda) \ \ \exists z \in N(y) \ \ \exists \zeta \text{ with } |\zeta - \xi| < \varepsilon : \\
&\qquad (x^0 + tz, f(x^0) + t\zeta) \in \text{epi}\, f\} = \\
&= \Big\{(y,\xi) \in \mathbb{R}^n \times \mathbb{R} \mid \forall \varepsilon > 0 \ \ \forall N(y) \ \ \exists \lambda > 0 \\
&\qquad \forall t \in (0, \lambda) \ \ \exists z \in N(y) \ \ \exists \zeta \text{ with } |\zeta - \xi| < \varepsilon : \\
&\qquad \tfrac{f(x^0+tz)-f(x^0)}{t} \leqq \zeta\Big\} = \\
&= \Big\{(y,\xi) \in \mathbb{R}^n \times \mathbb{R} \mid \forall \varepsilon > 0 \ \ \forall N(y) \ \ \exists \lambda > 0 \\
&\qquad \forall t \in (0, \lambda) \ \ \exists z \in N(y) : \\
&\qquad \frac{f(x^0 + tz) - f(x^0)}{t} < \xi + \varepsilon\Big\} = \\
&= \Big\{(y,\xi) \in \mathbb{R}^n \times \mathbb{R} \mid \\
&\qquad \mid \lim\sup_{t \downarrow 0} \inf_{z \to y} \frac{f(x^0 + tz) - f(x^0)}{t} \leqq \xi\Big\} .
\end{aligned}
$$

2) Since $Q(\operatorname{epi} f, (x^0, f(x^0)))$ is open, we have

$$
\begin{aligned}
\operatorname{epi}^0 f^Q(x^0,.) &= Q(\operatorname{epi} f, (x^0, f(x^0))) = \\
&= \{(y,\xi) \in I\!R^n \times I\!R \mid \exists \varepsilon > 0 \;\; \exists N(y) \;\; \forall \lambda > 0 \\
&\qquad \exists t \in (0,\lambda) \;\; \forall z \in N(y) \;\; \forall \zeta \text{ with } |\zeta - \xi| < \varepsilon : \\
&\qquad (x^0 + tz, f(x^0) + t\zeta) \in \operatorname{epi} f\} = \\
&= \Big\{(y,\xi) \in I\!R^n \times I\!R \mid \exists \varepsilon > 0 \;\; \exists N(y) \;\; \forall \lambda > 0 \\
&\qquad \exists t \in (0,\lambda) \;\; \forall z \in N(y) \;\; \forall \zeta \text{ with } |\zeta - \xi| < \varepsilon : \\
&\qquad \frac{f(x^0 + tz) - f(x^0)}{t} \leqq \zeta\Big\} = \\
&= \Big\{(y,\xi) \in I\!R^n \times I\!R \mid \exists \varepsilon > 0 \;\; \exists N(y) \;\; \forall \lambda > 0 \\
&\qquad \exists t \in (0,\lambda) \;\; \forall z \in N(y) : \\
&\qquad \frac{f(x^0 + tz) - f(x^0)}{t} \leqq \xi - \varepsilon\Big\} = \\
&= \Big\{(y,\xi) \in I\!R^n \times I\!R \mid \\
&\qquad \mid \liminf_{t \downarrow 0} \sup_{z \to y} \frac{f(x^0 + tz) - f(x^0)}{t} < \xi\Big\} .
\end{aligned}
$$

3)
$$
\begin{aligned}
f^A(x^0, y) &= \limsup_{t \downarrow 0} \inf_{z \to y} \frac{f(x^0 + tz) - f(x^0)}{t} = \\
&= -\liminf_{t \downarrow 0} \sup_{z \to y} \frac{(-f)(x^0 + tz) - (-f)(x^0)}{t} = \\
&= -(-f)^Q(x^0, y) .
\end{aligned}
$$
□

Because of $0 \in A(\operatorname{epi} f, (x^0, f(x^0)))$ and the closedness of this cone, the directional derivative $f^A(x^0,.)$ is lower semi-continuous with $f^A(x^0, 0) \leqq 0$. In the same manner, since $0 \notin Q(\operatorname{epi} f, (x^0, f(x^0)))$ and this cone is open, the directional derivative $f^Q(x^0,.)$ is upper semi-continuous with $f^Q(x^0, 0) \geqq 0$.

If f is convex, i.e. if $\operatorname{epi} f$ is a convex set, then from Theorem 3.4.10 we know that

$$Q(\operatorname{epi} f, (x^0, f(x^0))) = I(\operatorname{epi} f, (x^0, f(x^0))) ,$$
$$A(\operatorname{epi} f, (x^0, f(x^0))) = T(\operatorname{epi} f, (x^0, f(x^0))) .$$

Hence, it holds

$$f^Q(x^0, .) = f^I(x^0, .) \quad \text{and} \quad f^A(x^0, .) = f^T(x^0, .) .$$

If the convex function f is assumed even to be continuous at x^0 (i.e. if $x^0 \in \operatorname{int}(\operatorname{dom}(f))$, then because of our previous remarks we have $f^I(x^0, .) = f^T(x^0, .)$ and of course all directional derivatives above are equal.

So by the definition of the K-directional derivatives, the inclusion

$$K_1(\operatorname{epi} f, (x^0, f(x^0))) \subseteq K_2(\operatorname{epi} f, (x^0, f(x^0)))$$

implies the relation

$$f^{K_1}(x^0, .) \geqq f^{K_2}(x^0, .) .$$

Regarding the inclusion graph for the classical cone approximations, discussed in Section 3.4, we can give a corresponding graph with respect to the associated directional derivatives.

$$\begin{array}{ccccc} f^I(x^0,.) & \geq & f^Q(x^0,.) & & \\ \text{\rotatebox{90}{$\leq$}} & & \text{\rotatebox{90}{$\leq$}} & & \\ f^Z(x^0,.) & \geq & f^F(x^0,.) & & \\ \text{\rotatebox{90}{$\leq$}} & & \text{\rotatebox{90}{$\leq$}} & & \\ f^A(x^0,.) & \geq & f^T(x^0,.) & \geq & f^P(x^0,.) . \end{array}$$

Here the directional derivative $f^P(x^0, .)$ is associated with the pseudo tangent cone which is the convex hull of the contingent cone. Therefore, $f^P(x^0, .)$ is the convex envelope of $f^T(x^0, .)$.

Before investigating further properties regarding these special directional derivatives, we shall analyze the K-directional derivatives associated with the modified cone approximations, introduced in Section 3.4 (second part). In doing this, we first declare the following notations for limits (see Rockafellar (1980), (1981)):

$$(x,\mu) \downarrow_f x^0 \quad \text{means} \quad (x,\mu) \to (x^0, f(x^0)) \text{ with } \mu \geqq f(x) , \tag{1}$$

$$(x,\mu) \uparrow_f x^0 \quad \text{means} \quad (x,\mu) \to (x^0, f(x^0)) \text{ with } \mu \leqq f(x) , \tag{2}$$

$$x \to_f x^0 \quad \text{means} \quad (x, f(x)) \to (x^0, f(x^0)) . \tag{3}$$

Obviously, the following simple assertions are true.

Lemma 4.7.1.
If f is lower semi-continuous at x^0, then (1) is sufficient for (3).
If f is upper semi-continuous at x^0, then (2) is sufficient for (3).
If f is continuous at x^0, then (3) is equivalent to $x \to x^0$.

Now we get the following results regarding the dual local cone approximations $Z_m(.,.) = H(.,.)$ and $F_m(.,.)$, $I_m(.,.) = E(.,.)$ and $T_m(.,.)$, $A_m(.,.) = TC(.,.)$ and $Q_m(.,.)$ respectively.

Theorem 4.7.7. For any $y \in I\!R^n$ it holds

i) $$f^{Z_m}(x^0,y) = f^H(x^0,y) = \limsup_{\substack{t\downarrow 0 \\ (x,\mu)\downarrow_f x^0}} \frac{f(x+ty)-\mu}{t} ,$$

ii) $$f^{F_m}(x^0,y) = \liminf_{\substack{t\downarrow 0 \\ (x,\mu)\uparrow_f x^0}} \frac{f(x+ty)-\mu}{t} ,$$

iii) $f^{Z_m}(x^0,y) = -(-f)^{F_m}(x^0,y)$.

Theorem 4.7.8. For any $y \in I\!R^n$ it holds

i) $$f^{I_m}(x^0,y) = f^E(x^0,y) = \limsup_{\substack{t\downarrow 0 \\ (x,\mu)\downarrow_f x^0 \\ z\to y}} \frac{f(x+tz)-\mu}{t} ,$$

ii) $$f^{T_m}(x^0,y) = \liminf_{\substack{t\downarrow 0 \\ (x,\mu)\uparrow_f x^0 \\ z\to y}} \frac{f(x+tz)-\mu}{t} ,$$

iii) $f^{I_m}(x^0, y) = -(-f)^{T_m}(x^0, y)$.

Theorem 4.7.9. For any $y \in \mathbb{R}^n$ it holds

i) $f^{A_m}(x^0, y) = f^{TC}(x^0, y) = \lim \sup_{\substack{t \downarrow 0 \\ (x,\mu) \downarrow_f x^0}} \inf_{z \to y} \dfrac{f(x + tz) - \mu}{t}$,

ii) $f^{Q_m}(x^0, y) = \lim \inf_{\substack{t \downarrow 0 \\ (x,\mu) \uparrow_f x^0}} \sup_{z \to y} \dfrac{f(x + tz) - \mu}{t}$,

iii) $f^{A_m}(x^0, y) = -(-f)^{Q_m}(x^0, y)$.

Since the proofs of all assertions are similar, we prove only the last theorem.

Proof of Theorem 4.7.9.
1) Since $A_m(\operatorname{epi} f, (x^0, f(x^0)))$ is closed, we have

$$\operatorname{epi} f^{A_m}(x^0, .) = A_m(\operatorname{epi} f, (x^0, f(x^0))) =$$

$$= \{(y,\xi) \in \mathbb{R}^n \times \mathbb{R} \mid \forall \varepsilon > 0 \;\; \forall N(y) \;\; \exists \lambda > 0 \;\; \exists \delta > 0 \;\; \exists V(x^0) :$$

$$\forall (x, \mu) \in \operatorname{epi} f \text{ with } x \in V(x^0) \text{ and } |\mu - f(x^0)| < \varepsilon,$$

$$\forall t \in (0, \lambda) \;\; \exists z \in N(y) \;\; \exists \zeta \text{ with } |\zeta - \xi| < \varepsilon :$$

$$(x + tz, \mu + t\zeta) \in \operatorname{epi} f\} =$$

$$= \Big\{(y, \xi) \in \mathbb{R}^n \times \mathbb{R} \mid \forall \varepsilon > 0 \;\; \forall N(y) \dots$$

$$\dots \exists z \in N(y) \;\; \exists \zeta \text{ with } |\zeta - \xi| < \varepsilon :$$

$$\frac{f(x + tz) - \mu}{t} \leqq \zeta\Big\} =$$

$$= \Big\{(y, \xi) \in \mathbb{R}^n \times \mathbb{R} \mid \forall \varepsilon > 0 \;\; \forall N(y) \dots \exists z \in N(y) :$$

$$\frac{f(x + tz) - \mu}{t} < \xi + \varepsilon\Big\} =$$

$$= \Big\{(y, \xi) \in \mathbb{R}^n \times \mathbb{R} \mid \lim \sup_{\substack{t \downarrow 0 \\ (x,\mu) \downarrow_f x^0}} \inf_{z \to y} \frac{f(x + tz) - \mu}{t} \leqq \xi\Big\} .$$

2) Since $Q_m(\operatorname{epi} f, (x^0, f(x^0)))$ is open, we have

$$\begin{aligned}
&\operatorname{epi}^0 f^{Q_m}(x^0,.) = Q_m(\operatorname{epi} f, (x^0, f(x^0))) = \\
&= \{(y,\xi) \in \mathbb{R}^n \times \mathbb{R} \mid \exists \varepsilon > 0 \;\; \exists N(y) \;\; \forall \lambda > 0 \;\; \forall \delta > 0 \;\; \forall V(x^0) : \\
&\qquad \exists (x,\mu) \in \operatorname{hypo}^0 f \text{ with } x \in V(x^0) \text{ and } |\mu - f(x^0)| < \varepsilon, \\
&\qquad \exists t \in (0,\lambda) \;\; \forall z \in N(y) \;\; \forall \zeta \text{ with } |\zeta - \xi| < \varepsilon : \\
&\qquad (x + tz, \mu + t\zeta) \in \operatorname{epi} f\} = \\
&= \Big\{(y,\xi) \in \mathbb{R}^n \times \mathbb{R} \mid \exists \varepsilon > 0 \;\; \exists N(y) \ldots \\
&\qquad\qquad \ldots \forall z \in N(y) \;\; \forall \zeta \text{ with } |\zeta - \xi| < \varepsilon : \\
&\qquad\qquad \frac{f(x+tz) - \mu}{t} \leqq \zeta \Big\} = \\
&= \Big\{(y,\xi) \in \mathbb{R}^n \times \mathbb{R} \mid \exists \varepsilon > 0 \;\; \exists N(y) \ldots \forall z \in N(y) \; : \\
&\qquad\qquad \frac{f(x+tz) - \mu}{t} \leqq \xi - \varepsilon \Big\} = \\
&= \Big\{(y,\xi) \in \mathbb{R}^n \times \mathbb{R} \mid \lim \inf_{\substack{t \downarrow 0 \\ (x,\mu) \uparrow_f x^0}} \sup_{z \to y} \frac{f(x+tz) - \mu}{t} < \xi \Big\} .
\end{aligned}$$

3)
$$\begin{aligned}
f^{A_m}(x^0,y) &= \lim \sup_{\substack{t \downarrow 0 \\ (x,\mu) \downarrow_f x^0}} \inf_{z \to y} \frac{f(x+tz) - \mu}{t} = \\
&= -\lim \inf_{\substack{t \downarrow 0 \\ (x,-\mu) \uparrow_{-f} x^0}} \sup_{z \to y} \frac{(-f)(x+tz) - (-\mu)}{t} = \\
&= -(-f)^{Q_m}(x^0,y) .
\end{aligned}$$
□

The representation of the directional derivatives can be simplified if semi-continuity is assumed. Indeed, if f is lower semi-continuous at x^0, then by Lemma 4.7.1 the relation $(x,\mu) \downarrow_f x^0$ is equivalent to

$$x \to_f x^0 , \qquad \mu \to f(x^0) , \qquad \mu \geqq f(x) .$$

On the other hand, if f is upper semi-continuous at x^0, then the relation $(x, \mu) \uparrow_f x^0$ is equivalent to

$$x \to_f x^0 , \qquad \mu \to f(x^0) , \qquad \mu \leqq f(x) .$$

Thus, in Theorem 4.7.7, in Theorem 4.7.8 and in Theorem 4.7.9 we can replace the number μ by $f(x)$. We get

Theorem 4.7.10. If f is lower semi-continuous at x^0, then for any $y \in \mathbb{R}^n$ it holds

i) $$f^{Z_m}(x^0, y) = f^H(x^0, y) = \limsup_{\substack{t \downarrow 0 \\ x \to_f x^0}} \frac{f(x+ty) - f(x)}{t} ,$$

ii) $$f^{I_m}(x^0, y) = f^E(x^0, y) = \limsup_{\substack{t \downarrow 0 \\ x \to_f x^0 \\ z \to y}} \frac{f(x+tz) - f(x)}{t} ,$$

iii) $$f^{A_m}(x^0, y) = f^{TC}(x^0, y) = \lim \sup_{\substack{t \downarrow 0 \\ x \to_f x^0}} \inf_{z \to y} \frac{f(x+tz) - f(x)}{t} .$$

Theorem 4.7.11. If f is upper semi-continuous at x^0, then for any $y \in \mathbb{R}^n$ it holds

i) $$f^{F_m}(x^0, y) = \liminf_{\substack{t \downarrow 0 \\ x \to_f x^0}} \frac{f(x+ty) - f(x)}{t} ,$$

ii) $$f^{T_m}(x^0, y) = \liminf_{\substack{t \downarrow 0 \\ x \to_f x^0 \\ z \to y}} \frac{f(x+tz) - f(x)}{t} ,$$

iii) $$f^{Q_m}(x^0, y) = \lim \sup_{\substack{t \downarrow 0 \\ x \to_f x^0}} \inf_{z \to y} \frac{f(x+tz) - f(x)}{t} .$$

Naturally, if f is continuous at x^0 then the requirement $x \to_f x^0$ can be replaced by the simple requirement $x \to x^0$. In this case we recognize some of the directional derivatives discussed in Section 4.4, e.g. it holds

$$\begin{aligned}
f^{Z_m}(x^0,.) &= f^H(x^0,.) &&= f^0(x^0,.) \,, \\
f^{A_m}(x^0,.) &= f^{TC}(x^0,.) &&= f^{\nearrow}(x^0,.) \,, \\
f^{I_m}(x^0,.) &= f^E(x^0,.) &&= f^*_U(x^0,.) \,, \\
f^{T_m}(x^0,.) & &&= f^*_L(x^0,.) \,,
\end{aligned}$$

where in particular, $f^H(x^0,.)$ is the Clarke generalized derivative and $f^{TC}(x^0,.)$ is the Rockafellar subderivative at x^0, which are convex functions of the directional vector y.

We have stated in Theorems 3.4.13, 3.4.14 and 3.4.15 that the cone approximations $E(.,.)$, $H(.,.)$ and $TC(.,.)$ provide convex cones for each argument and that the cone approximations $T_m(.,.)$, $F_m(.,.)$ and $Q_m(.,.)$ provide complements of convex cones for each argument. Therefore, convexity and concavity behaviour of the associated directional derivatives can be derived (also without continuity assumptions) which shall be formulated in the following assertion.

Theorem 4.7.12.

i) The directional derivatives $f^E(x^0,.)$, $f^H(x^0,.)$ and $f^{TC}(x^0,.)$ are convex.

ii) The directional derivatives $f^{T_m}(x^0,.)$, $f^{F_m}(x^0,.)$ and $f^{Q_m}(x^0,.)$ are concave.

In Theorems 3.4.16 and 3.4.18 some further interesting properties of the cone approximations $E(.,.)$ and $TC(.,.)$ were given. For the associated directional derivatives we get

Theorem 4.7.13.

i) If $f^E(x^0,y) < \infty$, then

$$f^{TC}(x^0,y) = f^H(x^0,y) = f^E(x^0,y) \,.$$

ii) If $f^E(x^0, 0) < \infty$ (i.e. $f^E(x^0, .)$ is finite), then

$$f^{TC}(x^0, y) = f^H(x^0, y) = f^E(x^0, y)$$

for all $y \in \mathbb{R}^n$.

Proof. By assumption we have $E(\operatorname{epi} f, (x^0, f(x^0))) \neq \emptyset$ such that we can use Theorem 3.4.16. We get

$$E(\operatorname{epi} f, (x^0, f(x^0))) = \operatorname{int}(TC(\operatorname{epi} f, (x^0, f(x^0))))$$

and

$$TC(\operatorname{epi} f, (x^0, f(x^0))) = \overline{E(\operatorname{epi} f, (x^0, f(x^0)))} \ .$$

Now the assertion is a consequence of Theorem 4.6.4 and Theorem 4.6.5 respectively. □

Theorem 4.7.14. For each $y \in \mathbb{R}^n$ it holds

i) $f^E(x^0, y) = (-f)^E(x^0, -y)$,

ii) $f^{TC}(x^0, y) = (-f)^{TC}(x^0, -y)$.

Proof. Using Theorem 3.4.18, the topological properties and the duality of the cone approximations $E(.,.)$ and $T_m(.,.)$ in the sense of Definition 4.6.2, we get

$$\begin{aligned} \operatorname{epi}^0 f^E(x^0, .) &= E(\operatorname{epi} f, (x^0, f(x^0))) = -E(\mathbb{R}^n \backslash \operatorname{epi} f, (x^0, f(x^0))) = \\ &= -\{\mathbb{R}^n \backslash T_m(\operatorname{epi} f, (x^0, f(x^0)))\} = -\operatorname{hypo}^0 f^{T_m}(x^0, .) \ . \end{aligned}$$

Hence (using the third part of Theorem 4.7.8) for all $y \in \mathbb{R}^n$ we have

$$f^E(x^0, y) = -f^{T_m}(x^0, -y) = (-f)^E(x^0, -y) \ .$$

The second assertion can be proved analogously using the duality of the cone approximations $TC(.,.)$ and $Q_m(.,.)$. □

Now we are able to give an assertion for the equivalent description of the strict differentiability of a function using the above-mentioned directional derivatives.

Theorem 4.7.15. The following assertions are equivalent:

i) f is strictly differentiable at x^0,

ii) $f^E(x^0,.) = f^{T_m}(x^0,.)$,

iii) $f^E(x^0,.)$ is linear,

iv) $f^{T_m}(x^0,.)$ is linear.

Proof.
i) $\Rightarrow$ ii): If f is strictly differentiable at x^0 then f is continuous at x^0 and we get

$$f^{T_m}(x^0,.) = f_L^*(x^0,.) = f_U^*(x^0,.) = f^E(x^0,.) \ .$$

ii) $\Rightarrow$ iii) and iv): Let $f^E(x^0,.) = f^{T_m}(x^0,.)$. As in our previous remarks, from the topological properties of $E(.,.)$ and $T_m(.,.)$ we get $f^{T_m}(x^0,0) = f^E(x^0,0) = 0$ and this directional derivative is finite on $\mathbb{R}^n$. Thus, $f^{T_m}(x^0,.)$ and $f^E(x^0,.)$ are linear (convex and concave).
iii) $\Leftrightarrow$ iv): By Theorem 4.7.14 and the duality of the directional derivatives we get

$$f^{T_m}(x^0,y) = -(-f)^E(x^0,y) = -f^E(x^0,-y) \ .$$

Thus the linearity of $f^E(x^0,.)$ is equivalent to the linearity of $f^{T_m}(x^0,.)$.
iii) and iv) $\Rightarrow$ i): From the relations between the associated local cone approximations we get

$$f^{T_m}(x^0,.) \leqq f^T(x^0,.) \leqq f^I(x^0,.) \leqq f^E(x^0,.) \ .$$

Since $f^{T_m}(x^0,.)$ and $f^E(x^0,.)$ are linear, all directional derivatives are equal and especially it holds $f^T(x^0,0) = f^I(x^0,0) = 0$. By Theorem 4.7.5 we see that f is continuous at x^0 and therefore

$$f^{T_m}(x^0,y) = f^E(x^0,y) = \lim_{\substack{t\downarrow 0 \\ z\rightarrow y \\ x\rightarrow x^0}} \frac{f(x+tz)-f(x)}{t}$$

for all $y \in I\!R^n$. By Theorem 4.2.3 this is equivalent to the strict differentiability of f at x^0 since the directional derivative is finite, convex and concave, i.e. linear. □

We shall remark that we can also describe the local Lipschitz property of a function by means of the directional derivatives $f^E(x^0, .)$ and $f^{T_m}(x^0, .)$. It holds

Theorem 4.7.16. The following assertions are equivalent:

i) $f^E(x^0, 0) = 0$ (i.e. $f^E(x^0, .)$ is finite),

ii) $f^{T_m}(x^0, 0) = 0$ (i.e. $f^{T_m}(x^0, .)$ is finite),

iii) f is Lipschitzian around x^0.

Proof.
i) $\Leftrightarrow$ ii): The equivalence can be derived from Theorem 4.7.14 and the duality of the directional derivatives, according to

$$f^E(x^0, 0) = (-f)^E(x^0, 0) = -f^{T_m}(x^0, 0) \ .$$

i) and ii) $\Rightarrow$ iii): From the relations between the associated cone approximations we get

$$0 = f^{T_m}(x^0, 0) \leqq f^T(x^0, 0) \leqq f^I(x^0, 0) \leqq f^E(x^0, 0) = 0$$

such that by Theorem 4.7.5 f is continuous at x^0. Assuming f not to be Lipschitzian around x^0, we would find sequences $\{x^k\}$ and $\{z^k\} \subseteq I\!R^n$ tending to x^0 such that

$$\frac{|f(z^k) - f(x^k)|}{\|z^k - x^k\|} \to \infty \ .$$

We set $y^k = \dfrac{z^k - x^k}{\|z^k - x^k\|}$ and $t_k = \|z^k - x^k\|$. Without loss of generality we can assume that $\{y^k\}$ is convergent (since it is contained in the compact unit sphere) and that $y^k \to y^0$. Then

$$\limsup_{k \to \infty} \frac{|f(x^k + t_k y^k) - f(x^k)|}{t_k} = \infty \ .$$

But this is a contradiction since (we have shown that f is continuous at x^0)

$$\limsup_{\substack{t\downarrow 0 \\ x\to x^0 \\ z\to y}} \frac{f(x+tz)-f(x)}{t} = f^E(x^0,y) < \infty$$

and

$$\liminf_{\substack{t\downarrow 0 \\ x\to x^0 \\ z\to y}} \frac{f(x+tz)-f(x)}{t} = f^{T_m}(x^0,y) > -\infty$$

for all $y \in I\!R^n$.
iii) $\Rightarrow$ i): Let f be Lipschitzian around x^0 with the Lipschitz constant $L > 0$. Then

$$f^E(x^0,0) = \limsup_{\substack{t\downarrow 0 \\ x\to x^0 \\ z\to 0}} \frac{f(x+tz)-f(x)}{t} \leqq \limsup_{z\to 0} L\,\|z\| = 0$$

and

$$f^E(x^0,0) = -\liminf_{\substack{t\downarrow 0 \\ x\to x^0 \\ z\to 0}} \frac{-f(x+tz)+f(x)}{t} \geqq -\liminf_{z\to 0} L\,\|z\| = 0\,,$$

hence $f^E(x^0,0)=0$. □

Finally, we can use Theorem 3.4.17 to get a representation of the directional derivatives for convex functions.

Theorem 4.7.17. If f is convex, then for each $y \in I\!R^n$ it holds

i) $f^E(x^0,y) = f^I(x^0,y)$,

ii) $f^{TC}(x^0,y) = f^T(x^0,y)$.

Regarding the inclusion diagram for the cone approximations in Section 3.4 we can give the following table which extends the table on p. 422, but also the relations given in Section 4.4. It holds

$$\begin{array}{ccccccc}
f^E(x^0,.) & \geq & f^I(x^0,.) & \geq & f^Q(x^0,.) & \geq & f^{Q_m}(x^0,.) \\
\vee| & & \vee| & & \vee| & & \vee| \\
f^H(x^0,.) & \geq & f^Z(x^0,.) & \geq & f^F(x^0,.) & \geq & f^{F_m}(x^0,.) \\
\vee| & & \vee| & & \vee| & & \vee| \\
f^{TC}(x^0,.) & \geq & f^A(x^0,.) & \geq & f^T(x^0,.) & \geq & f^{T_m}(x^0,.)
\end{array}$$

The following assertions can be summarized:

1) The directional derivatives lying opposite to each other with respect to the fictitious mid-point of the scheme are dual according to the relation

$$f^{K_1}(x^0,.) = -(-f)^{K_2}(x^0,.) \ .$$

2) The directional derivatives in the first row of the scheme are upper semi-continuous and it holds

$$f^K(x^0, 0) \geqq 0 \ .$$

The directional derivatives in the third row of the scheme are lower semi-continuous and it holds

$$f^K(x^0, 0) \leqq 0 \ .$$

Concerning the directional derivatives in the second row of the scheme it holds

$$f^K(x^0, 0) = 0 \ .$$

3) The directional derivatives in the first column of the scheme are convex. The directional derivatives in the last column of the scheme are concave. The directional derivatives lying in the second and in the third column of the scheme have the property:

$$f_1 \leqq f_2 \ , \quad f_1(x^0) = f_2(x^0) \quad \Rightarrow \quad f_1^K(x^0,.) \leqq f_2^K(x^0,.) \ .$$

4) The function f is directional differentiable in classical sense at x^0 iff

$$f^Z(x^0,.) = f^F(x^0,.) \ .$$

The function f is Gâteaux differentiable at x^0 iff

$$f^Z(x^0,.) = f^F(x^0,.)$$

and this directional derivative is finite and linear.
The function f is uniformly directional differentiable at x^0 iff

$$f^I(x^0,.) = f^T(x^0,.) \ .$$

The function f is Fréchet differentiable at x^0 iff

$$f^I(x^0,.) = f^T(x^0,.)$$

and this directional derivative is linear.
The function f is strictly directional differentiable at x^0 iff

$$f^E(x^0,.) = f^{T_m}(x^0,.) \ .$$

We conclude this section by remarking that there are several other types of local cone approximations and therefore several other associated K-directional derivatives. For example the directional derivatives $f^{\oplus}(x^0,.)$, $f^{\square}(x^0,.)$ and $f^{\oslash}(x^0,.)$, mentioned in Section 4.4, can be generated by the *prototangent cone* of Michel/Penot and related cones. We shall not pursue these special approaches, but we refer e.g. to the papers of Michel/Penot (1984), Jofré/Penot (1989), Ioffe (1986) and Ward (1988).

4.8. Generalized Optimality Conditions

In this section we shall use the concept of abstract local cone approximations given in Definition 4.6.1 and the associated abstract differentiability notions introduced in Definition 4.6.3 and Definition 4.6.4 respectively in order to construct necessary optimality conditions for general optimization problems with inequality constraints, i.e. for the problem

$$\underset{x \in S}{\text{Min}}\ f(x) \qquad (P)$$

where the feasible set is described by

$$S = \{x \in I\!R^n \mid g_i(x) \leqq 0,\ i = 1, ..., m\} .$$

All the functions $f : I\!R^n \to I\!R \cup \{\pm\infty\}$ and $g_i : I\!R^n \to I\!R \cup \{\pm\infty\}$, $i = 1, ..., m$, are assumed to be finite and continuous at a certain point $x^0 \in S$.

If $x^0 \in \text{int}(S)$, then of course for the discussion, if the point x^0 is a local minimum point, the problem can be regarded as unconstrained. In this case we can give the following simple optimality condition.

Theorem 4.8.1. If $x^0 \in \text{int}(S)$ is a local minimum point of the function f, then for all local cone approximations $K(.,.)$ with $K(.,.) \subseteq T(.,.)$ (here $T(.,.)$ is the contingent cone approximation) it holds

i) $f^K(x^0, y) \geqq 0, \quad \forall y \in I\!R^n$,

ii) $0 \in \partial_K f(x^0)$.

Proof. Let $x^0 \in I\!R^n$ be a local minimum point of f. Then by Theorems 4.7.3 and 4.4.1 we get

$$f^K(x^0, y) \geqq f^T(x^0, y) = f_L^+(x^0, y) \geqq 0 \qquad \forall y \in I\!R^n$$

which is equivalent to

$$0 \in \partial_K f(x^0)$$

by the definition of the K-subdifferential. □

An approach for the discussion of the general constraint optimization problem is given in Theorem 4.4.2. Using the tangent cone and the cone of interior directions we can write

$$f^T(x^0, y) = f_L^+(x^0, y) , \qquad f^I(x^0, y) = f_U^+(x^0, y)$$

(see Theorem 4.7.3) and we can formulate immediately the following necessary optimality criterion:

Theorem 4.8.2. If $x^0 \in S$ is a local minimum point of (P), then for all local cone approximations $K_1(.,.)$ and $K_2(.,.)$ with $K_1(.,.) \subseteq T(.,.)$ and $K_2(.,.) \subseteq I(.,.)$ it holds

i) $f^{K_2}(x^0, y) \geqq 0 \quad \forall y \in K_1(S, x^0)$,

ii) $f^{K_1}(x^0, y) \geqq 0 \quad \forall y \in K_2(S, x^0)$.

The assertion is interesting if we use only one local cone approximation $K(.,.)$ with $K(.,.) \subseteq T(.,.)$ and $\mathrm{int}(K(.,.)) \subseteq I(.,.)$. Then of course for the local minimum point we have the necessary optimality conditions

$$f^{\mathrm{int}(K)}(x^0, y) \geqq 0 \qquad \forall y \in \overline{K(S, x^0)}\ , \tag{1}$$

$$f^{\bar{K}}(x^0, y) \geqq 0 \qquad \forall y \in \mathrm{int}(K(S, x^0))\ . \tag{2}$$

In general the conditions in Theorem 4.8.2, especially conditions (1) and (2) respectively, cannot be expressed in terms of the K-subdifferentials. For this – but also in order to construct sharper optimality conditions – we should assume the local cone approximation $K(.,.)$ to be convex. With this assumption we are able to formulate the following result.

Theorem 4.8.3. Let $K(.,.)$ be a convex local cone approximation with $K(.,.) \subseteq T(.,.)$ and $\mathrm{int}(K(.,.)) \subseteq I(.,.)$. If $x^0 \in S$ is a local minimum point of (P) and if one of the following conditions

$(B1)$ $\mathrm{dom}(f^{\mathrm{int}(K)}(x^0, .)) \cap \overline{K(S, x^0)} \neq \emptyset$

$(B2)$ $\mathrm{dom}(f^{\bar{K}}(x^0, .)) \cap \mathrm{int}(K(S, x^0)) \neq \emptyset$

be fulfilled, then

i) $f^K(x^0, y) \geqq 0 \quad \forall y \in K(S, x^0)$,

ii) $0 \in \partial_K f(x^0) + (K(S, x^0))^*$.

Proof. Let x^0 be a local minimum point of (P).
If $(B1)$ is fulfilled, then using the above-mentioned optimality condition (1) according to

$$f^{\mathrm{int}(K)}(x^0, y) \geqq 0 \qquad \forall y \in \overline{K(S, x^0)}\ ,$$

the nonempty convex cones

$$A = \operatorname{epi}^0 f^{\operatorname{int}(K)}(x^0, .) = \operatorname{int}(K(\operatorname{epi} f, (x^0, f(x^0))))$$
$$B = \overline{K(S, x^0)} \times \mathbb{R}_-$$

are disjoint. By the separation theorem (Theorem 2.2.5) we can find a vector $(u, \alpha) \in \mathbb{R}^{n+1}$, $(u, \alpha) \neq (0, 0)$ with

$$(u, \alpha)(y, \xi) \leqq 0 \qquad \forall (y, \xi) \in A ,$$
$$(u, \alpha)(y, \xi) \geqq 0 \qquad \forall (y, \xi) \in B .$$

Obviously, $\alpha \leqq 0$, and for $\alpha = 0$, because of the assumption $(B1)$, we would get $u = 0$ which is a contradiction.
Without loss of generality we set $\alpha = -1$ and we can conclude

$$uy \leqq f^{\operatorname{int}(K)}(x^0, y) \qquad \forall y \in \operatorname{dom}(f^{\operatorname{int}(K)}(x^0, .)) ,$$
$$uy \geqq 0 \qquad \forall y \in \overline{K(S, x^0)} ,$$

which means (see Theorem 4.6.6) since $\operatorname{dom}(f^{\operatorname{int}(K)}(x^0, .)) \neq \emptyset$

$$u \in \partial_{\operatorname{int}(K)} f(x^0) = \partial_K f(x^0)$$

and

$$-u \in (\overline{K(S, x^0)})^* = (K(S, x^0))^* .$$

Hence $0 = u + (-u) \in \partial_K f(x^0) + (K(S, x^0))^*$.
If $(B2)$ is fulfilled, then because of the above-mentioned condition (2) according to

$$f^{\bar{K}}(x^0, y) \geqq 0 \qquad \forall y \in \operatorname{int}(K(S, x^0))$$

the nonempty convex cones

$$C = \operatorname{epi} f^{\bar{K}}(x^0, .) = \overline{K(\operatorname{epi} f, (x^0, f(x^0)))}$$
$$D = \operatorname{int}(K(S, x^0)) \times \operatorname{int}(\mathbb{R}_-)$$

are disjoint. In the same manner as in the first part of the proof, by use of the separation theorem we can find a vector $u \in \mathbb{R}^n$ with

$$uy \leqq f^{\bar{K}}(x^0, y) \qquad \forall\, y \in \mathrm{dom}(f^{\bar{K}}(x^0, .)) ,$$

$$uy \geqq 0 \qquad \forall\, y \in \mathrm{int}(K(S, x^0)) .$$

Therefore (since $\mathrm{int}(K(S,x^0)) \neq \emptyset$)

$$u \in \partial_{\bar{K}} f(x^0) = \partial_K f(x^0)$$

and

$$-u \in (\mathrm{int}(K(S, x^0)))^* = (K(S, x^0))^* .$$

Hence $0 = u + (-u) \in \partial_K f(x^0) + (K(S, x^0))^*$.
Thus, ii) is fulfilled.
Finally, for the special vector $u \in \partial_K f(x^0) \cap (-K(S, x^0))^*$ it holds

$$uy \geqq 0 \qquad \forall\, y \in K(S, x^0)$$

which is sufficient to i) according to

$$f^K(x^0, y) \geq \sup\{uy \mid u \in \partial_K f(x^0)\} \geqq 0 \qquad \forall\, y \in K(S, x^0) . \qquad \square$$

The assertions formulated above have a geometrical structure, since the feasible set S is replaced, locally, by a cone which is determined only by the geometrical structure of set S.

Since the feasible set is described by level sets of several functions, one should discuss other approximations for S using the increasing behaviour of the active constraints. Therefore as an extension to the definitions of Chapter 3 we introduce the *linearizing cone* and the *cone of decreasing directions* by means of the K-directional derivatives. Let

$$S = \{x \in \mathbb{R}^n \mid g_i(x) \leqq 0,\ i = 1, ..., m\}$$

and $x^0 \in S$ be a feasible point. All functions are assumed to be continuous as x^0. Thus, for local considerations we have to regard only the active constraints. The index set is denoted by $I(x^0)$, i.e.

$$I(x^0) = \{i \mid g_i(x^0) = 0\} .$$

<u>Definition 4.8.1</u>. Let $K(.,.)$ be a local cone approximation.

1) The set

$$C^K(x^0) = \{y \in \mathbb{R}^n \mid g_i^K(x^0, y) \leqq 0 \quad \forall i \in I(x^0)\}$$

is called the *linearizing cone* to S at x^0.

2) The set

$$D^K(x^0) = \{y \in \mathbb{R}^n \mid g_i^K(x^0, y) < 0 \quad \forall i \in I(x^0)\}$$

is called the *cone of decreasing directions* to S at x^0.

For $I(x^0) = \emptyset$ by definition we set $C^K(x^0) = D^K(x^0) = \mathbb{R}^n$. Obviously we have the simple inclusion

$$D^K(x^0) \subseteq C^K(x^0) .$$

Regarding also the local cone approximations $\text{int}(K(.,.)$ and $\overline{K(.,.)}$ we can derive directly from the definitions

$$\begin{array}{ccccc} D^{\text{int}(K)}(x^0) & \subseteq & D^K(x^0) & \subseteq & D^{\bar{K}}(x^0) \\ \cap\vert & & \cap\vert & & \cap\vert \\ C^{\text{int}(K)}(x^0) & \subseteq & C^K(x^0) & \subseteq & C^{\bar{K}}(x^0) . \end{array} \tag{3}$$

Moreover, because of the upper semi-continuity of the directional derivatives $g_i^{\text{int}(K)}(x^0, .)$, $i \in I(x^0)$, the cone $D^{\text{int}(K)}(x^0)$ is open. In the same manner because of the lower semi-continuity of the directional derivatives $g_i^{\bar{K}}(x^0, .)$, $i \in I(x^0)$, the cone $C^{\bar{K}}(x^0)$ is closed.

More results can be stated if the local cone approximation $K(.,.)$ is assumed to be convex. Then all cones contained in (3) are convex.

Theorem 4.8.4. Let $K(.,.)$ be a convex local cone approximation. Then it holds:

i) $D^{\text{int}(K)}(x^0) \quad = \text{int}(D^K(x^0)) = \text{int}(D^{\bar{K}}(x^0))$.

ii) $\text{int}(C^{\text{int}(K)}(x^0)) = \text{int}(C^K(x^0)) = \text{int}(C^{\bar{K}}(x^0))$.

iii) If $D^K(x^0) \neq \emptyset$, then

$$\begin{aligned} \operatorname{int}(D^K(x^0)) &= \operatorname{int}(C^K(x^0)), \\ \overline{D^K(x^0)} &= \overline{C^K(x^0)}. \end{aligned}$$

iv) If $D^{\operatorname{int}(K)}(x^0) \neq \emptyset$, then

$$\begin{aligned} D^{\operatorname{int}(K)}(x^0) &= \operatorname{int}(C^{\bar{K}}(x^0)), \\ \overline{D^{\operatorname{int}(K)}(x^0)} &= C^{\bar{K}}(x^0). \end{aligned}$$

(Thus the interior, but also the closure of all cones contained in the table (3), coincide.)

Proof.

1) Because of (3) and since $D^{\operatorname{int}(K)}(x^0)$ is open, it is sufficient to show that

$$\operatorname{int}(D^{\bar{K}}(x^0)) \subseteq D^{\operatorname{int}(K)}(x^0).$$

For this let $y \in \operatorname{int}(D^{\bar{K}}(x^0))$ and $\varepsilon \geqq 0$. Then, because of the convexity, for all $i \in I(x^0)$ we have

$$(y, \varepsilon) \in \operatorname{int}(\overline{K(\operatorname{epi} g_i, (x^0, g_i(x^0)))}) = \operatorname{int}(K(\operatorname{epi} g_i, (x^0, g_i(x^0))))$$

i.e. we have $y \in \operatorname{dom}(g_i^{\operatorname{int}(K)}(x^0, .)$.
Using Theorem 4.6.4 we get

$$g_i^{\operatorname{int}(K)}(x^0, y) = g_i^{\bar{K}}(x^0, y) < 0 \qquad \forall i \in I(x^0)$$

and $y \in D^{\operatorname{int}(K)}(x^0)$.

2) Since it is sufficient to show that

$$\operatorname{int}(C^{\bar{K}}(x^0)) \subseteq C^{\operatorname{int}(K)}(x^0)$$

the proof is analogous to the proof of the first part.

3) Let $y \in C^K(x^0)$. By assumption we can find a vector $y^0 \in D^K(x^0)$ and for all $i \in I(x^0)$ we have

$$g_i^K(x^0, y) \leqq 0 \quad \text{and} \quad g_i^K(x^0, y^0) < 0 \ .$$

Because of the convexity we have for all numbers $\lambda \in (0,1)$

$$g_i^K(x^0, \lambda y + (1-\lambda)\, y^0) < 0$$

which means $y \in \overline{D^K(x^0)}$. Hence we have

$$C^K(x^0) \subseteq \overline{D^K(x^0)} \ .$$

According to our previous remarks and the convexity, this inclusion means

$$\overline{C^K(x^0)} = \overline{D^K(x^0)}$$

and

$$\operatorname{int}(C^K(x^0)) = \operatorname{int}(\overline{C^K(x^0)}) = \operatorname{int}(\overline{D^K(x^0)}) = \operatorname{int}(D^K(x^0)) \ .$$

4) Using the first assertion of the theorem and the third assertion for $\overline{K(.,.)}$ we get

$$D^{\operatorname{int}(K)}(x^0) = \operatorname{int}(D^{\bar{K}}(x^0)) = \operatorname{int}(C^{\bar{K}}(x^0)) \neq \emptyset$$

and

$$\overline{D^{\operatorname{int}(K)}(x^0)} = \overline{\operatorname{int}(C^{\bar{K}}(x^0))} = C^{\bar{K}}(x^0) \ . \qquad \square$$

For the linearizing cone and the cone of decreasing directions the decreasing behavior of the constraint functions is now considered in terms of the K-directional derivatives.

By means of Theorem 4.6.7 it appears that there are close connections between these K-directional derivatives and the K-subdifferentials of a given function. We shall use these connections for the characterization of the last introduced cones.

So as a generalization of the cone of gradients introduced in Section 3.6, we give the following

Definition 4.8.2. Let $K(.,.)$ be a local cone approximation. The set

$$B_K(x^0) = \Big\{ \sum_{i \in I(x^0)} \lambda_i u^i \mid u^i \in \partial_K g_i(x^0),\ \lambda_i \geqq 0 \Big\}$$

is called the *cone of K-subgradients.*

We agree that $B_K(x^0) = \{0\}$ in case of $I(x^0) = \emptyset$. The cone of subgradients is a convex cone (possibly empty) since it is a sum of convex cones accordingly

$$B_K(x^0) = \sum_{i \in I(x^0)} \operatorname{cone}(\partial_K g_i(x^0)) \ .$$

In general, however, this cone is not closed.

To ensure that $B_K(x^0) \neq \emptyset$ in the following we shall assume that $g_i^{\bar{K}}(x^0, 0) = 0$ for all $i \in I(x^0)$. Then by Theorem 4.6.7 all K-subdifferentials are not empty. Comparing the cones of K-subgradients using the local cone approximations $\operatorname{int}(K(.,.))$ and $\overline{K(.,.)}$ we get

$$B_{\bar{K}}(x^0) = B_K(x^0) \ .$$

If we assume that $K(.,.)$ is convex and that $\bigcap_{i \in I(x^0)} \operatorname{dom}(g_i^{\operatorname{int}(K)}(x^0, .)) \neq \emptyset$, then it holds even (see Theorem 4.6.6)

$$B_{\operatorname{int}(K)}(x^0) = B_K(x^0) \ .$$

The last assumption is fulfilled, e.g. in case of $D^{\operatorname{int}(K)}(x^0) \neq \emptyset$.

Now we give a dual connection between the linearizing cone and the cone of K-subgradients.

Theorem 4.8.5. Let $K(.,.)$ be a convex local cone approximation. If $g_i^K(x^0, 0) = 0$ for all $i \in I(x^0)$, then the following relations hold

i) $(C^{\bar{K}}(x^0))^* = \overline{B_K(x^0)}$,

ii) $(B_K(x^0))^* = C^{\bar{K}}(x^0)$.

Proof. By Theorem 4.6.7 for $i \in I(x^0)$ we have

$$\{y \in \mathbb{R}^n \mid g_i^{\bar{K}}(x^0, y) \leqq 0\}^* = \{y \in \mathbb{R}^n \mid uy \leqq 0 \ \ \forall u \in \partial_K g_i(x^0)\}^* = \\ = (\partial_K g_i(x^0))^{**} = \overline{\operatorname{cone}}(\partial_K g_i(x^0)) \ .$$

Since all cones are convex and closed we can use the bipolar theorem (Theorem 2.3.3) to get the first assertion by

$$(C^{\bar{K}}(x^0))^* = \Big(\bigcap_{i \in I(x^0)} \{y \in \mathbb{R}^n \mid g_i^{\bar{K}}(x^0, y) \leq 0\}\Big)^* = \\ = \overline{\sum_{i \in I(x^0)} \{y \in \mathbb{R}^n \mid g_i^{\bar{K}}(x^0, y) \leq 0\}^*} = \\ = \overline{\sum_{i \in I(x^0)} \overline{\operatorname{cone}}(\partial_K g_i(x^0))} = \overline{B_K(x^0)}$$

and the second assertion respectively by

$$C^{\bar{K}}(x^0) = (C^{\bar{K}}(x^0))^{**} = (\overline{B_K(x^0)})^* = (B_K(x^0))^* \ . \qquad \square$$

Moreover, we have even

Theorem 4.8.6. Let $K(.,.)$ be a convex local cone approximation. If $g_i^{\operatorname{int}(K)}(x^0, 0) = 0$ for all $i \in I(x^0)$ and $D^{\operatorname{int}(K)}(x^0) \neq \emptyset$, then it holds

i) $(C^K(x^0))^* = B_K(x^0)$,

ii) $(B_K(x^0))^* = C^K(x^0)$.

Proof. First we should state that with the above assumption we have

$$g_i^{\operatorname{int}(K)}(x^0, .) = g_i^K(x^0, .) = g_i^{\bar{K}}(x^0, .) \qquad \forall i \in I(x^0)$$

(see Theorem 4.6.5) such that

$$C^{\operatorname{int}(K)}(x^0) = C^K(x^0) = C^{\bar{K}}(x^0)$$

and

$$B_{\operatorname{int}(K)}(x^0) = B_K(x^0) = B_{\bar{K}}(x^0) \ .$$

Moreover, by the assumption again, for all $i \in I(x^0)$ the K-subdifferentials $\partial_K g_i(x^0)$ are compact sets (see Theorem 4.6.7) and $0 \notin \partial_K g_i(x^0)$. Otherwise we would get for some $i \in I(x^0)$

$$g_i^K(x^0,.) = g_i^{\mathrm{int}(K)}(x^0,.) \geqq 0$$

which is a contradiction since $D^{\mathrm{int}(K)}(x^0) \neq \emptyset$. Thus, the conical hulls of the K-subdifferentials, i.e. the cones $\mathrm{cone}(\partial_K g_i(x^0))$, $i \in I(x^0)$, are closed.

Finally, by a conclusion of the bipolar theorem, we get (the proof is similar to that of Theorem 4.8.5)

$$\begin{aligned}(C^K(x^0))^* &= \Big(\bigcap_{i \in I(x^0)} \{y \in \mathbb{R}^n \mid g_i^K(x^0, y) \leqq 0\} \Big)^* = \\ &= \sum_{i \in I(x^0)} \{y \in \mathbb{R}^n \mid g_i^K(x^0, y) \leqq 0\}^* = \\ &= \sum_{i \in I(x^0)} \mathrm{cone}(\partial_K g_i(x^0)) = B_K(x^0) \ .\end{aligned}$$

The second equality is straightforward. □

Now we are able to describe generalized multiplier rules for the optimization problem (P). To avoid confusion, we shall use the following declarations:

$$0 * (\pm\infty) = 0 \ , \qquad 0 * \emptyset = \{0\} \ .$$

In our first approach we use Theorem 4.8.3; by introducing the general constraint qualification

$$(K(S, x^0))^* \subseteq B_K(x^0) \ , \tag{C.Q.1}$$

we can formulate the following Kuhn/Tucker type necessary optimality criterion for the problem (P).

<u>Theorem 4.8.7</u>. Let $K(.,.)$ be a convex local cone approximation with $K(.,.) \subseteq T(.,.)$ and $\mathrm{int}(K(.,.)) \subseteq I(.,.)$. Further let one of the conditions $(B1)$ and $(B2)$ of Theorem 4.8.3 and the constraint qualification

(CQ$_1$) be fulfilled.
If $x^0 \in S$ is a local minimum point of (P), then there exist multipliers $\lambda_i \geqq 0$, $i \in I(x^0)$, such that

i) $f^K(x^0, y) + \sum_{i \in I(x^0)} \lambda_i g_i^K(x^0, y) \geqq 0, \ \forall y \in \mathbb{R}^n$,

ii) $0 \in \partial_K f(x^0) + \sum_{i \in I(x^0)} \lambda_i \partial_K g_i(x^0)$.

Proof. Using the constraint qualification (C.Q.1), the second assertion is a consequence of Theorem 4.8.3. Using Theorem 4.6.9 (first part) we get the first assertion. □

We can give other constraint qualifications which are sufficient for (C.Q.1):

$(K(S, x^0))^* \subseteq (C^{\bar{K}}(x^0))^*, \quad g_i^{\bar{K}}(x^0, 0) = 0 \ \ \forall i \in I(x^0)$
and $B_K(x^0)$ is closed. (C.Q.2)

$C^{\bar{K}}(x^0) \subseteq \overline{K(S, x^0)}, \quad g_i^{\bar{K}}(x^0, 0) = 0 \ \ \forall i \in I(x^0)$
and $B_K(x^0)$ is closed. (C.Q.3)

$D^{\text{int}(K)}(x^0) \neq \emptyset, \quad D^{\text{int}(K)}(x^0) \subseteq K(S, x^0)$
and $g_i^{\text{int}(K)}(x^0, 0) = 0 \ \ \forall i \in I(x^0)$. (C.Q.4)

In this representation, (C.Q.1) and (C.Q.2) can be interpreted as generalized Gould/Tolle constraint qualifications. (C.Q.3) is a generalized Abadie constraint qualification and (C.Q.4) is a generalized constraint qualification in the sense of Mangasarian/Fromovitz.

Analogous to the classical case for smooth optimization problems it holds

Theorem 4.8.8. If $K(.,.)$ is a convex local cone approximation, then

(C.Q.4) $\Rightarrow$ (C.Q.3) $\Leftrightarrow$ (C.Q.2) $\Rightarrow$ (C.Q.1) .

Proof.

1) Let (C.Q.4) be fulfilled. Then with Theorem 4.8.4 iv) we have

$$C^{\bar{K}}(x^0) = \overline{D^{\operatorname{int}(K)}(x^0)} \subseteq \overline{K(S,x^0)} \ .$$

From Theorem 4.8.6 we get

$$(C^K(x^0))^* = B_K(x^0)$$

i.e. $B_K(x^0)$ is closed.
Since $g_i^{\operatorname{int}(K)}(x^0,0) = 0$, $\forall i \in I(x^0)$, according to Theorem 4.6.5 we have

$$g_i^{\bar{K}}(x^0,0) = g_i^{\operatorname{int}(K)}(x^0,0) = 0 \qquad \forall i \in I(x^0)$$

and (C.Q.3) holds.

2) The equivalence of (C.Q.2) and (C.Q.3) follows from the convexity of the cones.

3) The last implication is a consequence of Theorem 4.8.5 according to

$$(K(S,x^0))^* \subseteq (C^{\bar{K}}(x^0))^* = \overline{B_K(x^0)} = B_K(x^0) \ . \qquad \square$$

The constraint qualifications introduced above are too complicated and not very practical. Therefore we shall give another approach for obtaining an assertion of the Kuhn/Tucker type. Using Theorem 4.8.2, we have to replace the cones $K_1(S,x^0)$ and $K_2(S,x^0)$ by the associated cones of decreasing directions. So it would be useful to have any relations in the form

$$D^{\operatorname{int}(K_1)}(x^0) \subseteq K_1(S,x^0)$$

$$D^{\operatorname{int}(K_2)}(x^0) \subseteq K_2(S,x^0)$$

which are not generally true. For the cone of interior directions $I(S,x^0)$, however, we can formulate

Theorem 4.8.9. It holds $D^I(x^0) \subseteq I(S, x^0)$.

Proof. If $y \in D^I(x^0)$, then by Theorem 4.7.3 for all $i \in I(x^0)$ we have

$$g_i^I(x^0, y) = \limsup_{\substack{t \downarrow 0 \\ z \longrightarrow y}} \frac{g_i(x^0 + tz) - g_i(x^0)}{t} < 0 ,$$

i.e. for all sequences $\{y^k\} \in \mathbb{R}^n$ and $\{t_k\} \in \mathbb{R}$ with $y^k \to y$ and $t_k \downarrow 0$ we get $g_i(x^0 + t_k y^k) < 0$ for k sufficiently large. Since the other constraints g_i, $i \notin I(x^0)$, are continuous, we have $x^0 + t_k y^k \in S$ for large k. Thus, by definition we have $y \in I(S, x^0)$. □

Now we can sharpen the condition of Theorem 4.8.2 in the following manner:

Theorem 4.8.10. Let $K(.,.)$ be a local cone approximation with $K(.,.) \subseteq T(.,.)$ and $\operatorname{int}(K(.,.)) \subseteq I(.,.)$. If $x^0 \in S$ is a local minimum point of (P), then

$$f^{\bar{K}}(x^0, y) \geqq 0 \qquad \forall\, y \in D^{\operatorname{int}(K)}(x^0) .$$

Proof. The assertion can be derived from Theorem 4.8.2 setting

$$K_1(.,.) = T(.,.) , \qquad K_2(.,.) = I(.,.)$$

and using the relations

$$D^{\operatorname{int}(K)}(x^0) \subseteq D^I(x^0) \subseteq I(S, x^0) ,$$

$$f^{\bar{K}}(x^0, .) \geqq f^T(x^0, .) .$$

□

The optimality condition given in Theorem 4.8.10 can be expressed by the equivalent condition that the inequality system

$$\begin{cases} f^{\bar{K}}(x^0, y) < 0 \\ g_i^{\operatorname{int}(K)}(x^0, y) < 0 , \qquad i \in I(x^0) \end{cases}$$

has no solution. Thus, if all the above directional derivatives are proper convex then we can use the Fan/Glicksberg/Hoffmann alternative theorem

(Theorem 2.9.1) to get the following generalized multiplier rule of the Fritz John type.

Theorem 4.8.11. Let $K(.,.)$ be a convex local cone approximation with $K(.,.) \subseteq T(.,.)$ and $\operatorname{int}(K(.,.)) \subseteq I(.,.)$. Further let the condition

$$\operatorname{dom}(f^{\bar{K}}(x^0,.)) \cap \bigcap_{i\in I(x^0)} \operatorname{dom}(g_i^{\operatorname{int}(K)}(x^0,.)) \neq \emptyset \tag{B3}$$

be fulfilled.
If $x^0 \in S$ is a local minimum point of (P) then there exist multipliers $\lambda_i \geq 0$, $i \in \{0\} \cup I(x^0)$, not all vanishing, such that

i) $\lambda_o f^K(x^0,y) + \sum_{i\in I(x^0)} \lambda_i g_i^K(x^0,y) \geqq 0 \quad \forall y \in \mathbb{R}^n$,

ii) $0 \in \lambda_0 \partial_K f(x^0) + \sum_{i\in I(x^0)} \lambda_i \partial_K g_i(x^0)$.

Proof. Let x^0 be a local minimum point of (P). With respect to Theorem 4.8.10 and the Fan/Glicksberg/Hoffmann alternative theorem we get the existence of multipliers $\lambda_i \geqq 0$, $i \in \{0\} \cup I(x^0)$, not all vanishing, such that

$$\lambda_0 f^{\bar{K}}(x^0,y) + \sum_{i\in I(x^0)} \lambda_i g_i^{\operatorname{int}(K)}(x^0,y) \geqq 0$$

for all $y \in \operatorname{dom}(f^{\bar{K}}(x^0,.)) \cap \bigcap_{i\in I(x^0)} \operatorname{dom}(g_i^{\operatorname{int}(K)}(x^0,.))$. Naturally, this inequality can be extended for $y \in \mathbb{R}^n$.
By Theorem 4.6.6, Theorem 4.6.9 and the associated remark we can derive

$$0 \in \lambda_0 \partial_{\bar{K}} f(x^0) + \sum_{i\in I(x^0)} \lambda_i \partial_{\operatorname{int}(K)} g_i(x^0) =$$

$$= \lambda_0 \partial_K f(x^0) + \sum_{i\in I(x^0)} \lambda_i \partial_K g_i(x^0)$$

and

$$\lambda_0 f^K(x^0,y) + \sum_{i\in I(x^0)} \lambda_i g_i^K(x^0,y) \geqq 0 \qquad \forall y \in \mathbb{R}^n . \qquad \square$$

If we want to derive a generalized Kuhn-Tucker assertion we have to assume of course a constraint qualification to ensure that $\lambda_0 \neq 0$ in the relations i) and ii) of Theorem 4.8.11. We can state immediately

Theorem 4.8.12. Let $K(.,.)$ be a convex local cone approximation with $K(.,.) \subseteq T(.,.)$ and $\text{int}(K(.,.)) \subseteq I(.,.)$. Further let the condition $(B3)$ of Theorem 4.8.11 and the constraint qualification

$$D^K(x^0) \neq \emptyset \tag{C.Q.5}$$

be fulfilled.
If $x^0 \in S$ is a local minimum point of (P) then there exist multipliers $\lambda_i \geq 0$, $i \in I(x^0)$, such that

i) $f^K(x^0, y) + \sum\limits_{i \in I(x^0)} \lambda_i g_i^K(x^0, y) \geqq 0 \quad \forall\, y \in I\!R^n$.

ii) $0 \in \partial_K f(x^0) + \sum\limits_{i \in I(x^0)} \lambda_i \partial_K g_i(x^0)$.

Proof. We use the assertion of Theorem 4.8.11. Assuming $\lambda_0 = 0$ we would get

$$\sum_{i \in I(x^0)} \lambda_i g_i^K(x^0, y) \geqq 0 \qquad \forall\, y \in I\!R^n$$

which contradicts the condition (C.Q.5). □

We shall now apply the formulated approach for the Lipschitzian case, i.e. for problems of the kind

$$\min_{x \in S} f(x) \tag{P_L}$$

with

$$S = \{x \in I\!R^n \mid g_i(x) \leqq 0, \ i = 1, ..., m\}$$

where all functions are assumed to be Lipschitzian around the point x^0.

If we choose for the local cone approximation $K(.,.)$ the Clarke tangent cone $TC(.,.)$, we have seen in Section 3.4 that in case of $E(S,x^0) \neq \emptyset$ the equality

$$E(S,x^0) = \operatorname{int}(H(S,x^0)) = \operatorname{int}(TC(S,x^0))$$

holds; we recall that all three cones are convex. Moreover, regarding the associated K-directional derivatives we have for a function f which is Lipschitzian around x^0, the relation $f^E(x^0,0) = 0$ and

$$f^{TC}(x^0,y) = f^H(x^0,y) = f^E(x^0,y) = \limsup_{\substack{t \downarrow 0 \\ x \to x^0}} \frac{f(x+ty) - f(x)}{t}$$

for all $y \in I\!R^n$ (see Theorems 4.7.10, 4.7.13 and 4.7.16). We shall denote this directional derivative, which is finite and continuous on $I\!R^n$, by $f^0(x^0)$ (the generalized Clarke derivative) as we did in Section 4.4.

For the K-subdifferentials we have

$$\partial_{TC} f(x^0) = \partial_H f(x^0) = \partial_E f(x^0) =$$
$$= \{u \in I\!R^n \mid uy \leqq f^0(x^0,y) \ \ \forall y \in I\!R^n\} \ .$$

As in Section 4.5 we denote this subdifferential (the Clarke subdifferential) by $\partial_{Cl} f(x^0)$. By Theorem 4.6.7 this set is bounded and it holds

$$f^0(x^0,y) = \max\{uy \mid u \in \partial_{Cl} f(x^0) \ \ \forall y \in I\!R^n\} \ .$$

Obviously, the conditions $(B1)$ of Theorem 4.8.3 but also the condition $(B3)$ of Theorem 4.8.11 are fulfilled since $\operatorname{dom}(f^0(x^0,.)) = I\!R^n$, $\operatorname{dom}(g_i^0(x^0,.)) = I\!R^n$, $i = 1,...,m$, and $0 \in TC(S,x^0)$. Thus for the Lipschitzian problem (P_L) Theorem 4.8.3 can be formulated in the simple manner:

Theorem 4.8.13. Let f and g_i be Lipschitzian around $x^0 \in S$. If $x^0 \in S$ is a local minimum point of (P) then

i) $f^0(x^0,y) \geqq 0 \ \ \forall y \in TC(S,x^0)$,

ii) $0 \in \partial_{Cl} f(x^0) + (TC(S,x^0))^*$.

To formulate also multiplier rules for Lipschitzian problems we shall discuss the linearizing cone and the cone of decreasing directions. Obviously we have

$$C^E(x^0) = C^H(x^0) = C^{TC}(x^0) =$$
$$= \{y \in \mathbb{R}^n \mid g_i^0(x^0, y) \leqq 0 \quad \forall i \in I(x^0)\} ,$$

which is a closed convex cone, and

$$D^E(x^0) = D^H(x^0) = D^{TC}(x^0) =$$
$$= \{y \in \mathbb{R}^n \mid g_i^0(x^0, y) < 0 \quad \forall i \in I(x^0)\} ,$$

which is an open convex cone. Clearly for the cone of subgradients we get

$$B_E(x^0) = B_H(x^0) = B_{TC}(x^0) = \sum_{i \in I(x^0)} \operatorname{cone}(\partial_{Cl} g_i(x^0)) .$$

As in Chapter 3 we denote these cones by $C(x^0)$, $D(x^0)$ and $B(x^0)$ respectively and Theorems 4.8.4, 4.8.5 and 4.8.6 can be simplified considerably. So we get

$$(C(x^0))^* = \overline{B(x^0)} , \qquad (B(x^0))^* = C(x^0)$$

and in case of $D^K(x^0) \neq \emptyset$

$$(D(x^0))^* = (C(x^0))^* = B(x^0) ,$$
$$(B(x^0))^* = \overline{D(x^0)} \quad = C(x^0) .$$

Using the constraint qualification

$$(TC(S, x^0))^* \subseteq B(x^0) \tag{C.Q.1}$$

we can formulate

Theorem 4.8.14. Let f and g_i be Lipschitzian around $x^0 \in S$. If $x^0 \in S$ is a local minimum point of (P) then there exist multipliers $\lambda_i \geq 0$, $i \in I(x^0)$, such that

i) $f^0(x^0, y) + \sum_{i \in I(x^0)} \lambda_i g_i^0(x^0, y) \geqq 0 \quad \forall y \in \mathbb{R}^n$.

ii) $0 \in \partial_{Cl} f(x^0) + \sum_{i \in I(x^0)} \lambda_i \partial_{Cl} g_i(x^0)$.

The other constraint qualifications which are sufficient for (C.Q.1) can be simplified in the following manner:

$$(TC(S,x^0))^* \subseteq (C(x^0))^* \text{ and } B_K(x^0) \text{ is closed.} \tag{C.Q.2}$$

$$C(x^0) \subseteq TC(S,x^0) \text{ and } B_K(x^0) \text{ is closed.} \tag{C.Q.3}$$

Regarding (C.Q.4) and (C.Q.5) we state the following assertion which can be proved analogously to Theorem 4.8.9 (without Lipschitz condition).

Theorem 4.8.15. It holds $D^E(x^0) \subseteq E(S,x^0)$.
Therefore we have

$$D(x^0) = D^E(x^0) \subseteq E(S,x^0) \subseteq TC(S,x^0) ,$$

i.e. (C.Q.4) and (C.Q.5) are equivalent to $D(x^0) \neq \emptyset$.

Finally, we want to make a remark about convex problems. If in (P) all the functions are convex, then of course the feasible set is convex and for any point $x^0 \in S$ we have by Theorem 3.4.18

$$TC(S,x^0) = T(S,x^0) \text{ and } E(S,x^0) = I(S,x^0) ,$$

i.e. the Clarke tangent cone can be replaced by the classical contingent cone. Moreover, because convex functions are locally Lipschitzian on the interior of the domain we get

$$f^0(x^0,.) = f^T(x^0,.) = f^I(x^0,.)$$

and this directional derivative coincides with the classical directional derivative according to

$$f'(x^0,y) = \lim_{t \downarrow 0} \frac{f(x^0+ty) - f(x^0)}{t} .$$

Also the Clarke subdifferential $\partial_{Cl} f(x^0)$ coincides with the classical subdifferential $\partial f(x^0)$ in the sense of convex analysis. Therefore all the assertions can be formulated especially for convex functions and we get the classical results of the Kuhn/Tucker theory for convex problems mentioned in Section 4.3.

References to Chapter IV

V.I. AVERBUCH and O.G. SMOLYANOV (1967), Theory of differentiation in linear topological spaces, Russian Math. Surveys, 22, 201–258.

A. BASTIANI (1964), Applications différentiables et varietés différentiables de dimension infinie, J. Analyse Math., 13, 1–114.

A. BEN-TAL and J. ZOWE (1982), A unified theory of first and second order conditions for extremum problems in topological vector spaces, Math. Progr. Study, 19, 39–76.

A. BEN TAL and J. ZOWE (1982), Necessary and sufficient optimality conditions for a class of nonsmooth optimization problems, Math. Progr., 24, 70–91.

J.M. BORWEIN (1984), Subgradients of convex operators, Math. Operationsforschung und Statistik, Ser. Optimization, 15, 179–191.

J.M. BORWEIN and H. WOLKOWICZ (1982), Characterizations of optimality without constraint qualification for the abstract convex program, Math. Progr. Study, 19, 77–100.

F.H. CLARKE (1975), Generalized gradients and applications, Trans. Amer. Math. Soc., 205, 247–262.

F.H. CLARKE (1976), A new approach to Lagrange multipliers, Math. Oper. Res., 1, 165–174.

F.H. CLARKE (1983), Optimization and Nonsmooth Analysis, J. Wiley & Sons, New York.

V.F. DEM'YANOV and A.M. RUBINOV (1980), On quasidifferentiable functionals, Soviet Mat. Doklady, 21, 14–17.

V.F. DEM'YANOV and A.M. RUBINOV (1983), On quasidifferentiable mappings, Math. Operationsforschung und Statistik, Ser. Optimization, 14, 3–21.

V.F. DEM'YANOV and A.M. RUBINOV (1986), Quasidifferentiable Calculus, Optimization Software Inc., New York.

V.F. DEM'YANOV and VASIL'IEV (1985), Nondifferentiable Optimization, Optimization Software Inc., New York.

S. DOLECKI (1982), Tangency and differentiation: Some applications of convergence theory, Ann. Math. Pura Appl., 130, 223–255.

I. EKELAND and R. TEMAM (1976), Convex Analysis and Variational Problems, North-Holland, Amsterdam.

K.H. ELSTER and J. THIERFELDER (1985a), On cone approximations of sets; in: P. Kenderov (Ed.), Mathematical Methods in Operations Research (Summer School in Operations Research – Primorsko 1984), Sofia, 33–59.

K.H. ELSTER and J. THIERFELDER (1985b), A general concept of cone approximations in nondifferentiable optimization; in: V.F. Dem'yanov and D. Pallaschke (Eds.), Nondifferentiable Optimization – Motivations and Applications, Springer Verlag, Berlin, 170–189.

K.H. ELSTER and J. THIERFELDER (1988a), On cone approximations and generalized directional derivatives; in: F.H. Clarke, V.F. Dem'yanov and F. Giannessi (Eds.), Nonsmooth Optimization and Related Topics, Plenum Press, New York, 133–154.

K.H. ELSTER and J. THIERFELDER (1988b), Generalized notions of directional derivatives, Publ. N. 155, Group of Optimization and Operations Research, Univ. of Pisa (Italy).

K.H. ELSTER and J. THIERFELDER (1988c), Abstract cone approximations and generalized differentiability in nonsmooth optimization, Optimization, 19, 315–341.

A. FRÖHLICHER and W. BUCHER (1966), Calculus in Vector Spaces without Norm, Springer Verlag, Berlin.

S. GÄHLER (1978), Grundstrukturen der Analysis, Akademie-Verlag, Berlin.

F. GIANNESSI (1985), Theorems of the alternatives for multifunctions with applications to optimization – General results, Publ. N. 127, Group of Optimization and Operations Research, Univ. of Pisa (Italy).

F. GIANNESSI (1986), Theorems of the alternatives for multifunctions with applications to optimization – Necessary conditions, Publ. N. 131, Group of Optimization and Operations Research, Univ. of Pisa (Italy).

F. GIANNESSI (1989), Semidifferentiable functions and necessary optimality conditions, J.O.T.A., 60, 191–241.

G. GIORGI and A. GUERRAGGIO (1992), On the notion of tangent cone in mathematical programming, Optimization, 25, 11–23.

J.B. HIRIART-URRUTY (1978), On optimality conditions in nondifferentiable programming, Math. Progr., 14, 73–86.

J.B. HIRIART-URRUTY (1979a), New concepts in nondifferentiable programming, Bull. Soc. Math. France, 60, 57–85.

J.B. HIRIART-URRUTY (1979b), Tangent cones, generalized gradients and mathematical programming in Banach spaces, Math. Oper. Res., 4, 79–97.

J.B. HIRIART-URRUTY (1980), Extension of Lipschitz functions, J. Math. Anal. Appl., 77, 539–554.

J.B. HIRIART-URRUTY (1982), Approximating a second-order directional derivative for nonsmooth convex functions, SIAM J. Control Optim., 20, 783–807.

J.B. HIRIART-URRUTY (1983), The approximate first-order and second-order directional derivatives for a convex function; in: J.P. Cecconi and T. Zolezzi (Eds.), Mathematical Theories of Optimization, Springer Verlag, Berlin, 144–177.

A.D. IOFFE (1979), Necessary and sufficient conditions for a local minimum. 1: A reduction theorem and first order conditions. 2: Conditions of Levitin-Miljutin-Osmolovskij-type. 3: Second order conditions and augmented duality. SIAM J. Control Optim., 17, 245–288.

A.D. IOFFE (1986), On the theory of subdifferential; in: J.B. Hiriart-Urruty (Ed.), Fermat Days 85: Mathematics for Optimization, North-Holland, Amsterdam.

A.D. IOFFE and V.M. TICHOMIROV (1979), Theorie der Extremalaufgaben, Verlag der Wissenschaften, Berlin, 1979.

V. JEYAKUMAR (1987), On optimality conditions in nonsmooth inequality constrained minimization, Numer. Funct. Anal. and Opt., 9, 535–546.

A. JOFRE and J.P. PENOT (1989), Comparing new notions of tangent cones, J. London Math. Soc. 40, 280–290.

S. KOMLOSI and M. PAPPALARDO (1993), A general scheme for first order approximations in optimization, Optimization Meth. & Software, 41, 153–161.

L.A. LJUSTERNIK and V.I. SOBOLEV (1968), Elemente der Funktionalanalysis, Akademie-Verlag, Berlin.

G. MARINESCU (1957), Différentielles de Gâteaux et Fréchet dans les espaces localement convexes, Bull. Math. Soc. Sci. Math. Phys. de la R.P.R., 49, 77–86.

H. MAURER and J. ZOWE (1979), First and second order necessary and sufficient optimality conditions for infinite-dimensional programming problems, math. Progr. Study, 16, 98–110.

P. MICHEL and J.P. PENOT (1984), Calcul sous-différentiel pour des fonctions lipschitziennes et non lipschitziennes, C.R. Acad. Sci. Paris, t. 298, 269–272.

M.Z. NASHED (1969), Higher order differentiability of nonlinear operators on normed spaces, Comm. Math. Univ. Carolinae, 10, 509–557.

D. PALLASCHKE, B. RECHT and R. URBANSKI (1986), On locally Lipschitz quasidifferentiable functions in Banach spaces, Optimization, 17, 287–295.

M. PAPPALARDO (1990), On semidifferentiable functions, Publ. N. 168, Group of Optimization and Operations Research, Univ. of Pisa (Italy).

J.P. PENOT (1978), Calcul sous-différentiel et optimisation, J. of Funct. Anal., 27, 248–276.

J.P. PENOT (1982), On regularity conditions in mathematical programming, Math. Progr. Study, 19, 167–199.

B.N. PSENICHNYI (1972), Notwendige Optimalitätsbedingungen, Teubner, Leipzig.

B.N. PSENISHNYI (1980), Convex Analysis and Extremum Problems, Nauka.

R.T. ROCKAFELLAR (1970), Convex Analysis, Princeton Univ. Press, Princeton.

R.T. ROCKAFELLAR (1979), Clarke's tangent cones and the boundaries of closed sets in $I\!R^n$, Nonlinear Analysis, Th. Meth. Appl., 3, 145–154.

R.T. ROCKAFELLAR (1979), Directionally Lipschitzian functions and subdifferential calculus, Proc. London Math. Soc., 39, 331–355.

R.T. ROCKAFELLAR (1980), Generalized directional derivatives and subgradients of nonconvex functions, Canadian J. of Math., 32, 257–280.

R.T. ROCKAFELLAR (1981), The Theory of Subgradients and Its Applications to Problems of Optimization: Convex and Nonconvex Functions, Heldermann-Verlag, Berlin.

R.T. ROCKAFELLAR (1985), Extensions of subgradient calculus with applications to optimization, Nonlinear Analaysis, Th. Meth. Appl., 9, 665–698.

E. SACHS (1978), Differentiability in optimization theory, Math. Operationsforschung und Statistik, Ser. Optimization, 9, 497–513.

A. SHAPIRO (1984), Quasidifferential calculus and first-order optimality conditions in nonsmooth optimization, Math. Progr. Study, 29, 56–68.

A. SHAPIRO (1990), On concepts of directional differentiability, J.O.T.A., 66, 477–487.

M. STUDNIARSKI (1986), Necessary and sufficient conditions for isolated local minima of nonsmooth functions, SIAM J. Control Optim. 24, 1044–1049.

L. THIBAULT (1980), Subdifferentials of compactly Lipschitzian vector-valued functions, Ann. Math. Pura Appl. 125, 157–192.

L. THIBAULT (1982), Subdifferentials of nonconvex vector-valued functions, J. Math. Anal. Appl., 86, 319–344.

L. THIBAULT (1983), Tangent cones and quasi interiorly tangent cones to multifunctions, Trans. Amer. Math. Soc., 277, 601–621.

C. URSESCU (1982), Tangent sets' calculus and necessary conditions for extremality, SIAM J. Control Optim., 20, 563–574.

M. VALADIER (1972), Sous-différentiabilité de fonctions convexes à valeurs dans espace vectoriel ordonné, Math. Scand., 30, 65–74.

M. VLACH (1981), Approximations operators in optimization theory, Zeitschrift für Operations Research (Theory), 25, 15–23.

D.E. WARD (1987), Convex subcones of the contingent cone in nonsmooth calculus and optimization, Trans. Amer. Math. Soc., 302, 661–682. (Corrigendum in Trans. Amer. Math. Soc. 311, 429–431, 1989.)

D.E. WARD (1988), The quantificational tangent cones, Canadian J. of Math., 40, 666–694.

D.E. WARD (1989), Directional derivative calculus and optimality conditions in nonsmooth mathematical programming, J. of Information & Optimization Sciences, 10, 81–96.

D.E. WARD and J.M. BORWEIN (1987), Nonsmooth calculus in finite dimensions, SIAM J. Control Optim., 25, 1312–1340.

CHAPTER V. DUALITY

5.1. Preliminary remarks

In different areas of mathematics it is customary to associate a given problem (the primal problem) to a further problem (the dual problem). Such dual problems are constructed by means of the structure of the primal problem with the aim to get additional information about the quality and/or quantity of the solution of the original problem. Naturally it is desirable to introduce dual problems with relatively simple structure but in many cases this request is incompatible with the demand for strong duality assertions.

Especially in mathematical optimization it is usual to associate the given primal optimization problem (a minimum problem)

$$\operatorname*{Min}_{x \in S} \ \psi(x) \tag{P}$$

to a dual optimization problem (a maximum problem)

$$\operatorname*{Max}_{u \in T} \ \varphi(u) \ . \tag{D}$$

To make it useful, the dual objective function φ and the dual feasible set T shall be constructed in such a manner that a *weak duality* assertion holds, i.e. that for the optimal values of both problems the following relation holds

$$\sup \{\varphi(u) \mid u \in T\} \leqq \inf \{\psi(x) \mid x \in S\} \ . \tag{1}$$

Naturally these optimal values can be finite or infinite.

In the case of empty feasible sets S or T, we would declare in the following that

$\sup \emptyset = -\infty$ and $\inf \emptyset = +\infty$

such that the inequality (1) holds trivially. Otherwise (i.e. in case of $S \neq \emptyset$ and $T \neq \emptyset$) relation (1) is equivalent to

$$\varphi(u) \leqq \psi(x) \qquad \forall x \in S \ \forall u \in T \ . \tag{2}$$

Thus, for any feasible point $u \in T$ of the dualproblem (D) the value $\varphi(u)$ is a lower bound for the optimal value of the primal problem (P). Conversely, the value $\psi(x)$ of a feasible point $x \in S$ of the primal problem (P) is an upper bound for the optimal value of the problem (D).

An assertion which guarantees equality in relation (1) is called *strong duality* assertion. We want to point out that in this form strong duality does not demand the existence of any solution of both problems. However, if there exist points $x^0 \in S$ and $u^0 \in T$ such that $\psi(x^0) = \varphi(u^0)$, then naturally strong duality is fulfilled and x^0 and u^0 are optimal solutions of (P) and (D) respectively.

There are many approaches to the construction of dual optimization problems. However, each approach is restricted to a special class of problems. In this manner more and more generalized duality concepts are constructed using aspects of nonsmooth analysis and generalized convexity.

In this chapter we shall give a short survey about the most important duality concepts. Starting with the well-known duality assertions of linear and convex optimization, we describe the Lagrange duality in connection with saddle-point assertions for the Lagrangian and the perturbation concept using the Fenchel conjugation. We shall see that with each step the preceeding concept can be identified as a special case of the subsequent one.

5.2. Duality in Linear Optimization

We regard the linear optimization problems

$$\operatorname*{Min}_{x \in S} cx \tag{P_{LIN}}$$

where $S = \{x \in I\!R^n \mid : Ax \geqq b, \ x \geqq 0\}$,

$$\underset{u \in T}{\text{Max}}\ bu \qquad (D_{LIN})$$

where $T = \{u \in \mathbb{R}^m \mid uA \leqq c,\ u \geqq 0\}$.

Here A is a fixed (m,n)-matrix, $x \in \mathbb{R}^n$ and $b \in \mathbb{R}^m$ respectively are fixed vectors. Obviously we have

$$ub \leqq u(Ax) = (uA)\,x \leqq cx \qquad \forall s \in S\ \ \forall u \in T\ , \tag{1}$$

i.e. the weak duality assertion holds.

Moreover, if we transform the dual problem in a minimum problem according to

$$\underset{u \in T}{\text{Min}}\ (-bu)$$

with $T = \{u \in \mathbb{R}^m \mid -uA \geqq -c,\ u \geqq 0\}$,

then the dual problem of this minimum problem – the bidual problem of (P_{LIN}) – has the representation

$$\underset{x \in S}{\text{Max}}\ (-cx)$$

with $S = \{x \in \mathbb{R}^n \mid -Ax \leqq -b,\ x \geqq 0\}$,

which is equivalent to (P_{LIN}). Thus (P_{LIN}) and (D_{LIN}) form a dual pair.

We can formulate the strong duality assertion.

Theorem 5.2.1. If one of the dual problems (P_{LIN}) and (D_{LIN}) has a solution, then the other one has a solution too, and both optimal values are equal.

Proof. Let $x^0 \in S$ be a minimum point of (P_{LIN}). Writing the constraints in the form

$$g_I(x) \;= b - Ax \leqq 0$$

$$g_{II}(x) = -x \leqq 0\ ,$$

by the Kuhn/Tucker theorem (Theorem 3.5.5) we get the existence of (multiplier) vectors $u^0 \in I\!R^m$, $u^0 \geqq 0$ and $v^0 \in I\!R^n$, $v^0 \geqq 0$ such that holds

$$c - u^0 A - v^0 = 0$$
$$u^0(b - Ax^0) = 0$$
$$v^0 x^0 = 0 \ .$$

Eliminating v^0 we get

$$u^0 A \leqq c$$
$$u^0(b - Ax^0) = 0$$
$$(c - u^0 A)\, x^0 = 0 \ .$$

Since $u^0 \geqq 0$, that means that $u^0 \in T$ and

$$cx^0 = (u^0 A)\, x^0 = u^0(Ax^0) = u^0 b \ .$$

With respect to the weak duality, u^0 is an optimal point of (D_{LIN}). Conversely, if $u^0 \in T$ is a solution of (D_{LIN}), then in the same manner we can find a solution x^0 of (P_{LIN}) since both problems are dual to each other.□

It is an interesting remark that the Kuhn/Tucker conditions of (P_{LIN}) and (D_{LIN}) coincide and can be described in the form

$$\begin{aligned} &Ax^0 \geqq b \ , \qquad x^0 \geqq 0 \ , \\ &u^0 A \leqq c \ , \qquad u^0 \geqq 0 \ , \\ &u^0(Ax^0 - b) = 0 \ , \\ &(u^0 A - c)\, x^0 = 0 \ , \end{aligned} \tag{2}$$

which is equivalent to

$$\begin{aligned} &x^0 \in S \ , \qquad u^0 \in T \ , \\ &u^0(Ax^0 - b) = 0 \ , \\ &(u^0 A - c)\, x^0 = 0 \ . \end{aligned} \tag{3}$$

Thus (since the Kuhn/Tucker conditions are also sufficient) $x^0 \in \mathbb{R}^n$ and $u^0 \in \mathbb{R}^m$ are solutions of the optimization problems (P_{LIN}) and (D_{LIN}) respectively if and only if x^0 and u^0 are solutions of the complementarity problem (2) or (3).

Introducing the slackness vectors $y^0 = Ax^0 - b$ and $v^0 = c - u^0 A$ we can formulate an optimality criterion for both dual problems in a simple form by complementary slackness conditions.

Theorem 5.2.2. The points $x^0 \in \mathbb{R}^n$ and $u^0 \in \mathbb{R}^m$ are solutions of (P_{LIN}) and (D_{LIN}) respectively iff

$$x^0 \geqq 0\,, \quad y^0 \geqq 0\,, \quad u^0 \geqq 0\,, \quad v^0 \geqq 0$$

$$x^0 v^0 = 0\,, \quad y^0 u^0 = 0\;.$$

In this case the optimal values are equal.

Because of the special structure of linear optimization problems we can give the following

Theorem 5.2.3. If both problems (P_{LIN}) and (D_{LIN}) have nonempty feasible sets (i.e. $S \neq \emptyset$, $T \neq \emptyset$), then both the problems have optimal solutions and the optimal values are equal.

Proof. It is well known that for linear optimization problems either the feasible set is empty or an optimal solution exists or the objective function is unbounded on the feasible set. Regarding the assumptions of the theorem and the weak duality we get the existence of optimal solutions and by Theorem 5.2.1 the optimal values are equal. □

5.3. Duality in Convex Optimization (Wolfe Duality)

We regard the convex optimization problem

$$\operatorname*{Min}_{x \in S} f(x) \qquad (P_C)$$

where $S = \{x \in \mathbb{R}^n \mid g(x) \leqq 0\}$.

Here $f : \mathbb{R}^n \to \mathbb{R}$ and $g : \mathbb{R}^n \to \mathbb{R}^m$ are assumed to be convex and differentiable. The latter means that in the representation $g = (g_1, ..., g_m)$ all components $g_i : \mathbb{R}^n \to \mathbb{R}$ are differentiable convex functions. Using the Lagrangian

$$L(x,u) = f(x) + ug(x) = f(x) + \sum_{i=1}^{m} u_i g_i(x)$$

we introduce the *Wolfe dual problem* according to

$$\operatorname*{Max}_{(x,u)\in T} \; L(x,u) \tag{D_C}$$

$$\text{where } T = \{(x,u) \in \mathbb{R}^n \times \mathbb{R}^m \mid \nabla_x L(x,u) = 0, \;\; u \geqq 0\} \; .$$

We shall remark that x *and* u are the dual variables.

Let $x \in S$ and $(\bar{x}, \bar{u}) \in T$. Then, because of $g(x) \leqq 0$, $\bar{u} \geqq 0$, $\nabla_x L(\bar{x}, \bar{u}) = 0$ and the convexity of $L(., \bar{u})$ we get

$$\begin{aligned} f(x) \; &\geqq \; f(x) + \bar{u}g(x) = L(x,\bar{u}) \; \geqq \\ &\geqq \; L(\bar{x},\bar{u}) + (x - \bar{x}) \, \nabla_x L(\bar{x},\bar{u}) = L(\bar{x},\bar{u}) \; , \end{aligned} \tag{1}$$

i.e. weak duality holds.

We shall remark that we can weaken the convexity assumption to the functions f and g. For the weak duality it is sufficient that $L(.,u)$ is pseudoconvex for all fixed $u \geqq 0$. In fact, because of $\nabla_x L(\bar{x}, \bar{u}) = 0$ we get in this case

$$L(x,\bar{u}) \geqq L(\bar{x},\bar{u}) \qquad \forall \, x \in \mathbb{R}^n$$

and the inequality (1) remains true.

The weak duality can fail, however, if we assume only pseudoconvexity of f and quasiconvexity of g.

<u>Example 5.3.1</u> (Mangasarian). Let the objective function $f : \mathbb{R} \to \mathbb{R}$ and the constraint function $g : \mathbb{R} \to \mathbb{R}$ be given by

$$f(x) = -e^{-x^2} \; , \qquad g(x) = -x + 1 \; .$$

Then, f is pseudoconvex (even strictly pseudoconvex), g is quasiconvex (even linear) and the point $x^0 = 1$ is the (global) minimum point of the problem

$$\operatorname{Min} f(x) \qquad \text{s.t. } g(x) \leqq 0$$

with the optimal value $f(x^0) = -e^{-1}$. The Wolfe dual problem has the representation

$$\operatorname{Max} \{-e^{-x^2} - ux + u\} \qquad \text{s.t. } 2xe^{-x^2} - u = 0\,, \quad u \geqq 0\,.$$

Eliminating u, we see that this problem is equivalent to

$$\operatorname{Max} \{-e^{-x^2}(2x^2 - 2x + 1)\} \qquad \text{s.t. } x \geqq 0\,.$$

Obviously, the objective function is negative and tends to zero with increasing x. Especially for $x = 10$ (and therefore $u = 20e^{-100}$) we have

$$-181e^{-100} > -e^{-1}$$

which contradicts the weak duality.

The following strong duality assertion is a conclusion of the Kuhn/Tucker theorem formulated in Chapter 3.

Theorem 5.3.1 (Wolfe). Let $x^0 \in S$ be an optimal solution of (P_C) and let a constraint qualification be fulfilled. Then there exists a vector $u^0 \geqq 0$ such that (x^0, u^0) is an optimal solution of (D_C) and the optimal values of (P_C) and (D_C) are equal.

Proof. From the Kuhn/Tucker theorem we get the existence of a multiplier vector $u^0 \geq 0$ such that

$$\nabla_x L(x^0, u^0) = 0 \quad \text{and} \quad u^0 g(x^0) = 0\,.$$

Thus $(x^0, u^0) \in T$ and we have

$$f(x^0) = f(x^0) + u^0 g(x^0) = L(x^0, u^0)\,.$$

Because of the weak duality, the point (x^0, u^0) is a solution of (D_C). □

With respect to our previous remarks concerning generalized convexity, we can weaken also in this theorem the convexity assumption by the demand that the Lagrangian $L(.,u)$ be pseudoconvex for each fixed $u \geqq 0$.

If we regard the linear optimization problem (P_{LIN}) of the former section it is easy to show that the Wolfe dual problem (D_C) is equivalent to (D_{LIN}). In fact, we get by definition

$$\text{Max}\,\{cx - u(Ax - b) - vx\}$$

$$\text{s.t. } c - uA - v = 0\,, \quad u \geqq 0\,, \quad v \geqq 0\,. \qquad (D_C)$$

Eliminating v, this means

$$\text{Max}\, ub \qquad \text{s.t. } c - uA \geqq 0\,, \quad u \geqq 0$$

which is the dual problem of linear optimization. Thus (regarding the fact that for polyhedral sets no constraint qualification is necessary), Theorem 5.2.1 is a special case of Theorem 5.3.1.

To formulate an inverse duality assertion, all functions are assumed to be twice differentiable.

Theorem 5.3.2 (Hanson, Huard). Let (x^0, u^0) be an optimal solution of (D_C) and let the Hessian $\nabla_x^2 L(x^0, u^0)$ be not singular. Then x^0 is an optimal solution of (P_C) and the optimal values of (P_C) and (D_C) are equal.

Proof. Since $\nabla_x^2 L(x^0, u^0)$ is not singular, by the implicit function theorem there exists a neighborhood of (x^0, u^0) in which the nonlinear equation $\nabla_x L(x, u) = 0$ is solvable in terms of $x = x(u)$, i.e. it holds $x(u^0) = x^0$ and $\nabla_x L(x(u), u) \equiv 0$. We have assumed that (x^0, u^0) is a maximum point of the dual problem. Thus u^0 is also a maximum point of the problem

$$\text{Max}\, L(x(.), .) \qquad \text{s.t. } u \geqq 0\,.$$

Applying the Kuhn/Tucker theorem to this special problem, we get by derivation of this composite function with respect to u

$$\nabla_x L(x(u^0), u^0)\, \nabla_u x(u^0) + \nabla_u L(x(u^0), u^0) \leqq 0$$
$$u^0(\nabla_x L(x(u^0), u^0)\, \nabla_u x(u^0) + \nabla_u L(x(u^0), u^0)) = 0 \ ,$$

which means (since $\nabla_x L(x(u^0), u^0) = \nabla_x L(x^0, u^0) = 0$) that

$$\nabla_u L(x^0, u^0) = g(x^0) \leqq 0$$
$$u^0 \nabla_u L(x^0, u^0) = u^0 g(x^0) = 0 \ .$$

Thus, x^0 is a feasible point of the primal problem and by the complementary slackness condition we get the equality

$$f(x^0) = f(x^0) + u^0 g(x^0) = L(x^0, u^0) \ .$$

Regarding the weak duality, the point x^0 is a solution of (P_C). □

We can weaken the convexity assertion if we only assume that the objective function is pseudoconvex and the constraint function is quasiconvex. Even if weak duality fails, the proof is similar. Because of the relations

$$\nabla_u L(x^0, u^0) = g(x^0) \leqq 0$$
$$u^0 \nabla_u L(x^0, u^0) = u^0 g(x^0) = 0$$

and the assumptions

$$u^0 \geqq 0 \ , \qquad \nabla_x L(x^0, u^0) = 0 \ ,$$

the Kuhn/Tucker conditions of the primal problem are fulfilled. Now in Theorem 3.9.2 we have stated that by the mentioned assumptions these conditions are also sufficient optimality conditions. Thus, x^0 is a solution of the primal problem (P_C).

Similarly, the assertion remains true if the Lagrangian $L(., u^0)$ is assumed to be pseudoconvex. This is a consequence of Theorem 3.9.3.

We can state that if the strong duality relation

$$f(x^0) = f(\bar{x}) + \bar{u}^T g(\bar{x})$$

holds, then x^0 and $(\bar{x}, \bar{u})$ are optimal solutions of (P_C) and (D_C) respectively. The question arises if the converse assertion is true, especially if the equality $x^0 = \bar{x}$ holds. For this we have

Theorem 5.3.3 (Mangasarian). Let x^0 be an optimal solution of (P_C) and $(\bar{x}, \bar{u})$ be an optimal solution of (D_C). If a constraint qualification in x^0 is fulfilled and if $L(., \bar{u})$ is strict convex in a neighborhood of $\bar{x}$, then we have $x^0 = \bar{x}$ and the optimal values of (P_C) and (D_C) are equal.

Proof. Analogous to Theorem 5.3.1 we can find a vector $u^0 \geq 0$ such that (x^0, u^0) is also a solution of the dual problem and it holds

$$f(x^0) = L(x^0, u^0) = L(\bar{x}, \bar{u}) \ .$$

We only have to show that $x^0 = \bar{x}$.
If $x^0 \neq \bar{x}$, then by the strict convexity we would get (since $\nabla_x L(\bar{x}, \bar{u}) = 0$)

$$L(x^0, \bar{u}) > L(\bar{x}, \bar{u}) + (x^0 - \bar{x}) \nabla_x L(\bar{x}, \bar{u}) = L(\bar{x}, \bar{u}) = f(x^0)$$

i.e.

$$f(x^0) + \bar{u}g(x^0) > f(x^0)$$

or equivalently

$$\bar{u}g(x^0) > 0$$

which is absurd since $\bar{u} \geq 0$ and $g(x^0) \leqq 0$. □

Also in this assertion the convexity assumptions can be weakened. Indeed, if the Lagrangian $L(., u)$ is pseudoconvex for each fixed $u \geq 0$, then the first part of the theorem remains true. Assuming strict pseudoconvexity for $L(., \bar{u})$, then because of $\nabla_x L(\bar{x}, \bar{u}) = 0$ we get, analogously, for $x^0 \neq \bar{x}$

$$L(x^0, \bar{u}) > L(\bar{x}, \bar{u}) = f(x^0)$$

which would be a contradiction to $\bar{u} \geqq 0$ and $g(x^0) \leqq 0$.

5.4. Lagrange Duality

Let $X \subseteq \mathbb{R}^n$ and $U \subseteq \mathbb{R}^m$ be arbitrary sets and $L : X \times U \to \mathbb{R}$ be an arbitrary function. Then we can define two extended real valued functions $\psi : X \to \bar{\mathbb{R}}$ and $\varphi : U \to \bar{\mathbb{R}}$ according to

$$\varphi(u) = \inf \{L(x, u) \mid x \in X\} , \qquad u \in U ,$$

$$\psi(x) = \sup \{L(x, u) \mid u \in U\} , \qquad x \in X .$$

Obviously we have

$$\varphi(u) \leqq L(x, u) \leqq \psi(x) \qquad \forall u \in U \ \ \forall x \in X , \tag{1}$$

and especially

$$\sup \{\varphi(u) \mid u \in U\} \leqq \inf \{\psi(x) \mid x \in X\} . \tag{2}$$

In terms of the Lagrangian this means

$$\sup_{u \in U} \inf_{x \in X} L(x, u) \leqq \inf_{x \in X} \sup_{u \in U} L(x, u) . \tag{3}$$

Thus, the last inequalities provide an approach for the construction of a pair of dual optimization problems.

First of all we emphasize that the above demand of equality is closely connected with the existence of saddle-points of the function L. We recall that (x^0, u^0) is a saddle point of L (exactly a saddle point of L with respect to $X \times U$) iff

$$L(x^0, u) \leqq L(x^0, u^0) \leqq L(x, u^0) \qquad \forall x \in X \ \ \forall u \in U . \tag{4}$$

Lemma 5.4.1. For the points $x^0 \in X$ and $u^0 \in U$ it holds $\varphi(u^0) = \psi(x^0)$ if and only if (x^0, u^0) is a saddle point of L.

Proof. Let $\varphi(u^0) = \psi(x^0)$. Then because of (1) we get

$$\varphi(u^0) = L(x^0, u^0) = \psi(x^0)$$

and

$$L(x^0, u) \leqq \psi(x^0) = L(x^0, u^0) = \varphi(u^0) \leqq L(x, u^0)$$

for all $x \in X$ and all $u \in U$. Hence, (x^0, u^0) is a saddle point of L. Conversely let (x^0, u^0) be a saddle point of L, i.e. let (4) be fulfilled. Then of course we have

$$\begin{aligned}\psi(x^0) &= \sup\{L(x^0,u) \mid u \in U\} \leqq \\ &\leqq \inf\{L(x,u^0) \mid x \in X\} = \varphi(u^0)\end{aligned}$$

and with (1) we get the equality. □

We should remark that in case of $\varphi(u^0) = \psi(x^0)$, which means

$$\max_{u\in U} \inf_{x\in X} L(x,u) = \min_{x\in X} \sup_{u\in U} L(x,u) \ , \tag{5}$$

the points x^0 and u^0 are minimum and maximum point of ψ and φ respectively. The converse implication, however, is not true. In fact, even if minimum and maximum of ψ and φ respectively exist, then in general (5) does not hold.

Example 5.4.1. Let $X = U = [0, 2\pi] \subseteq I\!R$ and the function L be given by

$$L(x,u) = \sin(x+u) \ .$$

Then

$$\begin{aligned}\varphi(u) &= \inf\{\sin(x+u) \mid x \in [0,2\pi]\} = \\ &= \min\{\sin(x+u) \mid x \in [0,2\pi]\} = -1\end{aligned}$$

$$\begin{aligned}\psi(x) &= \sup\{\sin(x+u) \mid u \in [0,2\pi]\} = \\ &= \max\{\sin(x+u) \mid u \in [0,2\pi]\} = +1\end{aligned}$$

for all $x \in X$ and all $u \in U$. Thus

$$\max_{u\in U} \varphi(u) < \min_{x\in X} \psi(x)$$

or equivalently

$$\max_{u\in U} \min_{x\in X} L(x,u) < \min_{x\in X} \max_{u\in U} L(x,u)$$

i.e. the function L does not admit any saddle point.

To ensure the equality (5) which is equivalent with the existence of a saddle point of L, we have to assume special algebraic and topological properties. We shall give the following assertion regarding the structure of ψ and φ.

Lemma 5.4.2.

i) If $L(.,u)$ is lower semi continuous for all fixed $u \in U$, then ψ is a lower semi continuous function. If $L(x,.)$ is upper semi continuous for all fixed $x \in X$, then φ is an upper semi continuous function.

ii) If X, U are compact sets and $L(.,.)$ is continuous, then φ and ψ are even finite and continuous functions.

iii) Let X and U be convex sets. If $L(.,u)$ is convex for all fixed $u \in U$, then ψ is a convex function. If $L(x,.)$ is concave for all fixed $x \in X$, then φ is a concave function.

Proof. By the semi continuity assumptions of $L(.,.)$ the sets

$$\operatorname{epi}\psi = \bigcap_{u \in U} \operatorname{epi} L(.,u)$$

and

$$\operatorname{hypo}\varphi = \bigcap_{x \in X} \operatorname{hypo} L(x,.)$$

are closed. Hence, ψ and φ are lower semi continuous and upper semi continuous respectively. In the same manner we get the convexity properties of these functions.

Now let X and U be compact sets and $L(.,.)$ be continuous. Then by the Weierstrass theorem the functions φ and ψ are finite and it holds even

$$\psi(x) = \max\{L(x,u) \mid u \in U\}, \qquad \varphi(u) = \min\{L(x,u) \mid x \in X\} \quad (6)$$

for all $x \in X$ and all $u \in U$. We show the continuity of ψ. (The proof that φ is continuous is the same.) Above we mentioned that ψ is lower semi continuous. So we have only to show that it holds

$$\limsup_{x \to x^0} \psi(x) \leqq \psi(x^0)$$

for all $x^0 \in X$. Assuming the opposite we could find a sequence $\{x^k\} \subseteq X$ tending to x^0 such that

$$\lim_{k\to\infty} \psi(x^k) > \psi(x^0) .$$

Now by (6) there exists also a sequence $\{u^k\} \subseteq U$ with

$$\psi(x^k) = \max\{L(x^k, u) \mid u \in U\} = L(x^k, u^k)$$

for all $k \in \mathbb{N}$. Moreover, since U is compact, without loss of generality we can assume that this sequence is convergent with $u^k \to u \in U$. Then, however, because of the continuity of L we would get

$$\begin{aligned} L(x^0, u^0) &= \lim_{k\to\infty} L(x^k, u^k) = \\ &= \lim_{k\to\infty} \psi(x^k) > \psi(x^0) = \sup\{L(x^0, u) \mid u \in U\} \end{aligned}$$

which is absurd. □

Now we are able to formulate the following Minimax theorem.

Theorem 5.4.1 (v. Neumann). Let X and U be convex compact sets, $L(., u)$ be convex and lower semi continuous for all fixed $u \in U$, and $L(x, .)$ be concave and upper semi continuous for all fixed $x \in X$. Then there exist points $x^0 \in X$ and $u^0 \in X$ such that $\varphi(u^0) = \psi(x^0)$.

Proof. First of all we should remark that because of the compactness of the sets and the semi continuity properties, the least upper bounds and the least lower bounds are reached, i.e.

$$\psi(u) = \inf_{x\in X} L(x, u) = \min_{x\in X} L(x, u) ,$$

$$\varphi(x) = \sup_{u\in U} L(x, u) = \max_{u\in U} L(x, u)$$

for all $u \in U$, $x \in X$ and

$$\sup_{u\in U} \inf_{x\in X} L(x, u) = \max_{u\in U} \min_{x\in X} L(x, u) = \max_{u\in U} \psi(u) ,$$

$$\inf_{x\in X} \sup_{u\in U} L(x, u) = \min_{x\in X} \max_{u\in U} L(x, u) = \min_{x\in X} \varphi(x) .$$

Now, because of

$$\max_{u\in U}\ \min_{x\in X}\ L(x,u) \leqq \min_{x\in X}\ \max_{u\in U}\ L(x,u)$$

we only have to show only the converse relation. For this let γ be a real number with

$$\gamma < \min_{x\in X}\ \max_{u\in U}\ L(x,u)\ . \tag{6}$$

Then for each point $x \in X$ there exists a point $u \in U$ such that $L(x,u) > \gamma$. This means that

$$\bigcap_{u\in U} \{x \in X \mid L(x,u) \leqq \gamma\} = \emptyset\ .$$

Since all sets are compact we also can choose a finite number of such sets which are disjoint, i.e. there are points $u^1, ..., u^p \in U$ with

$$\bigcap_{i=1}^{p} \{x \in X \mid L(x,u^i) \leqq \gamma\} = \emptyset\ .$$

Thus the convex inequality system (with respect to x)

$$L(x,u^i) - \gamma < 0\ , \qquad i = 1, ..., p$$

admits no solution. Using the Fan/Glicksberg/Hoffmann theorem (Theorem 2.9.1) we can find nonnegative numbers $\lambda_1, ..., \lambda_p \geq 0$, not all vanishing, such that

$$\sum_{i=1}^{p} \lambda_i (L(x,u^i) - \gamma) \geqq 0 \qquad \forall\, x \in X\ .$$

Without loss of generality we can assume $\sum\limits_{i=1}^{p} \lambda_i = 1$ and regarding the concavity of $F(x,.)$ we get

$$\gamma = \sum_{i=1}^{p} \lambda_i \gamma \leqq \sum_{i=1}^{p} \lambda_i L(x,u^i) \leqq L\Big(x, \sum_{i=1}^{p} \lambda_i u^i\Big) \qquad \forall\, x \in X$$

which means that

$$\gamma \leqq \min_{x\in X}\ L\Big(x, \sum_{i=1}^{p} \lambda_i u^i\Big) \leqq \max_{u\in U}\ \min_{x\in X}\ L(x,u)\ .$$

Since this relation is true for all numbers γ fulfilling (6) it holds even

$$\min_{x \in X} \max_{u \in U} L(x,u) \leqq \max_{u \in U} \min_{x \in X} L(x,u) . \qquad \square$$

We should remark that the assumptions of the theorem can be weakened. So one can show that the assertion even holds if convexity and concavity is replaced by quasiconvexity and quasiconcavity respectively. We shall not extend our considerations beyond this. For further results we refer to e.g. Sion (1957), (1958), Karlin (1959), Berge/Ghouila-Houri (1962) and Dem'yanov/Malozemov (1972).

Now we regard the optimization problem

$$\operatorname*{Min}_{x \in S} f(x) \tag{P}$$

where $S := \{x \in X \mid g_i(x) \leqq 0, \ i = 1, ..., m\}$.

Here $X \subseteq I\!R^n$ is an open set and $f : X \to I\!R$ and $g_i : X \to I\!R$, $i = 1, ..., m$, are arbitrary functions defined on X. Also, here for simplification, we declare the vector-valued function $g : X \to I\!R^m$ by $g(x) = (g_1(x), ..., g_m(x))$. To apply the above-discussed results in connection with the construction of a duality concept, we specify with $L(.,.)$ the Lagrange function of the problem (P) introduced in Chapter 3 according to

$$L(x,u) = f(x) + ug(x) = f(x) + \sum_{i=1}^{m} u_i g_i(x)$$

and which is defined on the set $X \times U$ with $U = I\!R^m_+$.

By definition we get for $x \in X$

$$\psi(x) = \sup_{u \geqq 0} L(x,u) = \sup_{u \geqq 0} \{f(x) + ug(x)\} = \\ = \begin{cases} f(x) & \text{if } g(x) \leqq 0 \\ +\infty & \text{else .} \end{cases}$$

Thus, on the feasible set the objective function f coincides with the function ψ. We get

$$\inf_{x \in X} \psi(x) = \inf_{x \in S} f(x)$$

and the optimization problem (P) can be described equivalently by

$$\underset{x \in X}{\text{Min}}\ \psi(x)\ . \tag{P_L}$$

In the same manner we can formulate a dual optimization problem (the *Lagrange dual problem*) by

$$\underset{u \geqq 0}{\text{Max}}\ \varphi(u) \tag{D_L}$$

with the dual objective function

$$\varphi(u) = \inf \{L(x, u) \mid x \in X\}\ .$$

Obviously, according to our former remarks the weak duality relation holds since

$$\sup_{u \geqq 0} \varphi(u) \leqq \inf_{x \in X} \psi(x) = \inf_{x \in S} f(x)$$

or equivalently (in terms of the Lagrangian)

$$\sup_{u \geqq 0} \inf_{x \in X} L(x, u) \leqq \inf_{x \in X} \sup_{u \geqq 0} L(x, u) = \inf_{x \in S} f(x)\ .$$

Moreover, since $L(x, .)$ is linear, the dual objective function φ is concave and upper semi continuous (see Lemma 5.4.2). Thus the dual optimization problem is a maximization problem of a concave objective function with respect to a convex (even polyhedral) feasible set, i.e. it is equivalent to a convex optimization problem.

On the other side, if (P) is a convex optimization problem (i.e. if all functions f and g_i, $i = 1, ..., m$, are assumed to be convex and continuous), then the Lagrangian $L(., u)$ (with respect to x) is a convex and continuous function and therefore the function ψ is convex and lower semi continuous.

To construct strong duality assertions, by Lemma 5.4.1 we have to ensure the existence of saddle points of the Lagrangian. In this connection, with the Minimax theorem of v. Neumann (Theorem 5.4.1) we have provided an approach for strong duality, but we cannot apply this assertion since compactness assumptions fail. In Section 3.14, however, some global optimality conditions for convex optimization problems are described in the form of saddle point assertions by means of additional constraint qualifications. We can extend these results by discussing the solvability not only of the primal but also of the dual problem.

First, as an extension of Theorem 3.14.1 we get

Theorem 5.4.2. Let (x^0, u^0), $x^0 \in X$, $u^0 \geqq 0$ be a saddle point of L with respect to $X \times \mathbb{R}^m_+$. Then x^0 solves (P), u^0 solves (D_L) and the optimal values are equal.

Proof. The assertion is a consequence of Lemma 5.4.1, the weak duality and the above-mentioned property that (P) and (P_L) are equivalent. □

The proof can also be derived directly from the definition of the Lagrangian. Indeed, let (x^0, u^0) be a saddle point of L. Then the saddle point inequalities in (4) can be described by

$$f(x^0) + ug(x^0) \leqq f(x^0) + u^0 g(x^0) \leqq f(x) + u^0 g(x) \quad \forall x \in X \ \ \forall u \geqq 0 \ .$$

Obviously the first inequality is equivalent to

$$ug(x^0) \leqq u^0 g(x^0) \qquad \forall u \geqq 0$$

which means that $g(x^0) \leqq 0$ and (setting $u = 0$) $u^0 g(x^0) \geqq 0$, i.e. even (since $u^0 \geqq 0$)

$$g(x^0) \leqq 0 \ , \qquad u^0 g(x^0) = 0 \ .$$

The second inequality is equivalent to

$$L(x^0, u^0) = \min_{x \in X} L(x, u^0) \ .$$

Thus, (x^0, u^0) is a saddle point of L if and only if

$$g(x^0) \leqq 0 \ ,$$

$$u^0 g(x^0) = 0 \ ,$$

$$L(x^0, u^0) = \min_{x \in X} L(x, u^0) \ .$$

We have that the global Kuhn/Tucker conditions (see Section 3.14) are fulfilled and because of

$$f(x^0) = f(x^0) + u^0 g(x^0) = L(x^0, u^0) =$$
$$= \min_{x \in X} L(x^0, u) = \varphi(u^0)$$

the points x^0 and u^0 are solutions of (P) and (D_L) respectively and strong duality holds.

With respect to the existence of saddle points we can use Theorem 3.14.5 for formulating the following direct duality assertion for convex optimization problems.

Theorem 5.4.3. Let f and g be convex functions. If $x^0 \in S$ is an optimal solution of (P) and if the Slater condition is fulfilled (i.e. there is a point $\bar{x} \in X$ with $g_i(\bar{x}) < 0$ for all $i = 1, ..., m$), then there exists a vector $u^0 \in \mathbb{R}^m_+$, such that u^0 is an optimal solution of (D_L) and the optimal values of (P) and (D_L) are equal.

Proof. By Theorem 3.14.5 we get the existence of the vector $u^0 \geqq 0$ such that (x^0, u^0) is a saddle point of L with respect to $X \times \mathbb{R}^m_+$. Then of course u^0 is a solution of (D_L) and the optimal values are equal. □

In the convex case, Wolfe duality and Lagrange duality are closely connected. Let us compare the dual problems (D_C) and (D_L). For this we have to assume that all functions f and g_i, $i = 1, ..., m$, are convex and differentiable on $\mathbb{R}^n$. Let $(x, u) \in T$ be a feasible point of the Wolfe dual problem, i.e. let

$$\nabla_x L(x, u) = 0 , \qquad u \geqq 0 .$$

Because of the convexity this means that the Lagrangian $L(., u)$ (as a function of x) admits a minimum at x, such that

$$L(x, u) = \varphi(u) .$$

Thus the Wolfe dual objective function and the Lagrange dual objective function coincide for such points. This means

$$\sup \{L(x, u) \mid (x, u) \in T\} = \sup \{\varphi(u) \mid u \geq 0\} \leqq$$
$$\leqq \inf \{f(x) \mid x \in S\}$$

and therefore, strong Wolfe duality implies strong Lagrange duality. The converse assertion is not correct, even in the differentiable convex case. In fact, for the calculation of the Lagrange dual objective function we have to regard

$$\varphi(u) = \inf_{x \in X} L(x, u)$$

which is not equivalent to $\nabla_x L(x, u) = 0$ for any $x \in I\!R^n$.

<u>Example 5.4.2.</u> Let $f : I\!R \to I\!R$ and $g : I\!R \to I\!R$ be given by $f(x) = e^x$ and $g(x) = x$. Then the primal problem

$$\operatorname{Min} e^x \qquad \text{s.t. } x \leqq 0$$

has no solution. The optimal value, however, is bounded and it holds

$$\inf \{e^x \mid x \leqq 0\} = 0 \ .$$

Regarding the Lagrange dual problem we get

$$\varphi(u) = \inf_{x \in X} L(x, u) = \inf_{x \in X} \{e^x + ux\} = \\ = \begin{cases} 0 & \text{if } u = 0 \\ -\infty & \text{if } u > 0 \end{cases}$$

and $\max\limits_{u \geqq 0} \varphi(u) = 0$.
Hence we have strong duality in the sense of

$$\max_{u \geqq 0} \varphi(u) = \inf_{x \in S} f(x) = 0 \ .$$

For the Wolfe dual problem, however, the feasible set T is empty since $u \geqq 0$ and $\nabla_x L(x, u) = e^x + u = 0$ are incompatible. Hence

$$-\infty = \sup_{(x,u) \in T} L(x, u) < \inf_{x \in S} f(x) = 0 \ .$$

To obtain further duality assertions using special tools of convex analysis, we shall interpret the dual problem geometrically. Therefore we regard the so-called upper image set of the problem (P) according to

$$\mathcal{M}(P) = \{(y, z) \in I\!R^m \times I\!R \mid \exists x \in I\!R^n : g(x) \leqq y, \ f(x) \leqq z\}$$

as a subset in the image space of the vector-valued function $(g, f) : \mathbb{R}^n \to \mathbb{R}^{m+1}$.

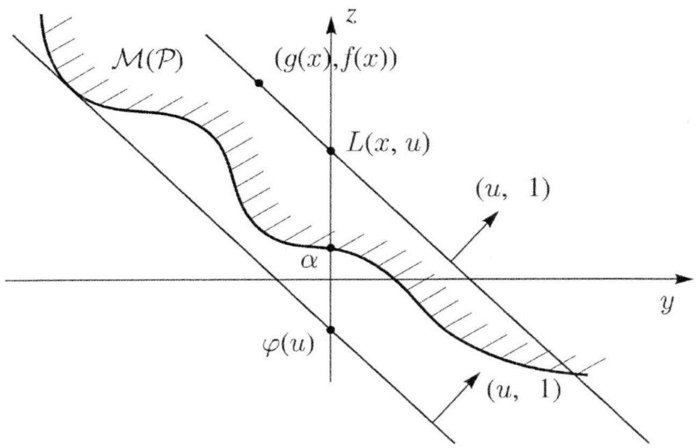

Figure 1.

Obviously the part of $\mathcal{M}(P)$ associated to the feasible set of the primal problem (P) is situated at the left-hand side, i.e. in the left of the z-axis. Denoting the optimal value of (P) by the number α, this value can be represented then by

$$\begin{aligned}\alpha &= \inf\{z \in \mathbb{R} \mid \exists y \leqq 0 : (y, z) \in \mathcal{M}(P)\} = \\ &= \inf\{z \in \mathbb{R} \mid (0, z) \in \mathcal{M}(P)\}\ ,\end{aligned}$$

i.e. it is the smallest z-component of the left part of $\mathcal{M}(P)$.

Now let $u \in \mathbb{R}^m$, $u \geqq 0$ and $x \in X$. We have $(g(x), f(x)) \in \mathcal{M}(P)$ and we can place a nonvertical hyperplane with the normal vector $(u, 1)$ and containing the point $(g(x), f(x))$. This hyperplane is described by the equation

$$z + u^T y = f(x) + ug(x) = L(x, u)\ .$$

Obviously, the value $L(x, u)$ can be recognized as the z-component of the point $(0, z)$ of this hyperplane, i.e. the intersection point of the hyperplane and the z-axis (see Figure 1).

Regarding the dual objective function, in the case of $\varphi(u) > -\infty$ we can construct a parallel hyperplane described by the equation

$$z + u^T y = \varphi(u) = \inf\, L(x, u)$$

which is a support hyperplane to the set $\mathcal{M}(P)$. Also here the value $\varphi(u)$ of the dual objective function can be found by the intersection of this hyperplane with the z-axis (cf. Figure 1). Obviously $\varphi(u) \leqq \alpha$.

If we want to find the optimal value β of the dual problem (D_L) we have to compare all "feasible" supporting hyperplanes to $\mathcal{M}(P)$, i.e. all (nonvertical) support hyperplanes with normal vectors of the form $(u, 1)$, $u \geqq 0$. In Figure 2 this value is illustrated by a special hyperplane with the normal vector $(u^0, 1)$. Thus the vector u^0 is a solution of the dual problem. We see once more that $\beta \leqq \alpha$, which means weak duality.

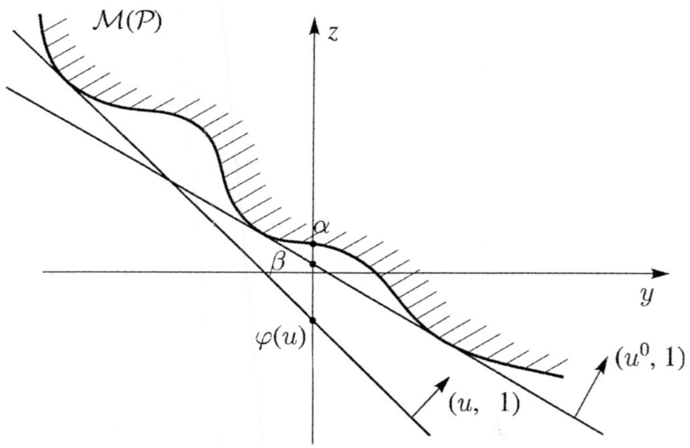

Figure 2.

Moreover, strong duality is equivalent to the existence of a (nonvertical) hyperplane which supports the set $\mathcal{M}(P)$ at the point $(0, \alpha)$ where α is the optimal value of the primal problem. Of course, if (P) is a convex optimization problem, then the set $\mathcal{M}(P)$ is a convex set (the proof is left to the reader) which can be separated by a hyperplane from the point $(0, \alpha)$. Obviously, this hyperplane is a support hyperplane. Assuming a Slater condition, which means that there exists a point $(\bar{y}, \bar{z}) \in \mathcal{M}(P)$

with $\bar{y} < 0$ (i.e. left of the z-axis), we can ensure that this hyperplane is not vertical, i.e. the associated normal vector can be given in the form $(u^0, 1)$, $u^0 \geqq 0$.

In this case, u^0 is a solution of the dual problem (D_L) and it holds strong duality in the sense of

$$\max\{\varphi(u) \mid u \geqq 0\} = \inf\{f(x) \mid x \in S\} .$$

If also the primal problem admits a solution $x^0 \in S$, then we have even

$$\max\{\varphi(u) \mid u \geqq 0\} = \min\{f(x) \mid x \in S\}$$

and (x^0, u^0) is a saddle point of L.

It is useful to introduce the following function.

Definition 5.4.1. The function $q : \mathbb{R}^m \to \bar{\mathbb{R}}$, according to

$$q(y) = \inf\{f(x) \mid x \in X : g(x) \leqq y\} ,$$

is called *marginal function* of the problem (P).

Especially $q(0)$ is the optimal value of the primal problem.

We remember that in case of $\{x \in X \mid g(x) \leqq y\} = \emptyset$ we have $q(y) = +\infty$. Further we can state that the epigraph of this function is closely connected with the set $\mathcal{M}(P)$. Really, it is easy to show that

$$\operatorname{epi} q = \{(y, z) \in \mathbb{R}^m \times \mathbb{R} \mid \forall \varepsilon > 0 \;\; \exists \bar{z} \leqq z + \varepsilon : (y, \bar{z}) \in \mathcal{M}(P)\}$$

i.e. $\operatorname{epi} q$ is the vertical closure of $\mathcal{M}(P)$. Thus, a supporting hyperplane to $\mathcal{M}(P)$ is also a supporting hyperplane to $\operatorname{epi} q$ and we can emphasize that strong duality is closely connected with the subdifferentiability of the marginal function at the point $y = 0$.

First we shall give some remarks about the structure of this function.

Lemma 5.4.3.

i) $q(.)$ is monotonically decreasing, i.e. $y^1 \leqq y^2$ implies $q(y^1) \geqq q(y^2)$.

ii) If (P) is a convex optimization problem, then $q(.)$ is a convex function.

iii) If (P) is a convex optimization problem with finite optimal value and if the Slater condition is fulfilled, then $q(.)$ is a properly convex function with $0 \in \operatorname{int}(\operatorname{dom}(q))$. (Thus q is even continuous at $y = 0$.)

Proof.

i) Let $y^1 \leqq y^2$. Then

$$q(y^1) = \inf \{f(x) \mid x \in X : g(x) \leqq y^1\} \geqq$$
$$\geqq \inf \{f(x) \mid x \in X : g(x) \leqq y^2\} = q(y^2) .$$

ii) Let $y^1, y^2 \in \operatorname{dom}(q)$, i.e. $q(y^1) < \infty$, $q(y^2) < \infty$. Then for each number γ_1, γ_2 with $q(y^1) < \gamma_1$, $q(y^2) < \gamma_2$, by definition of the marginal function we can find points $x_1, x_2 \in X$ such that

$$g(x^1) \leqq y^1 , \qquad f(x^1) \leqq \gamma_1 ,$$
$$g(x^2) \leqq y^2 , \qquad f(x^2) \leqq \gamma_2 .$$

Now because of the convexity of f and g_i, $i = 1, ..., m$, for any $\lambda \in (0, 1)$ we get

$$g(\lambda x^1 + (1 - \lambda) x^2) \leqq \lambda g(x^1) + (1 - \lambda) g(x^2) \leqq \lambda y^1 + (1 - \lambda) y^2$$

and

$$f(\lambda x^1 + (1 - \lambda) x^2) \leqq \lambda f(x^1) + (1 - \lambda) f(x^2) \leqq \lambda \gamma_1 + (1 - \lambda) \gamma_2 ,$$

i.e.

$$q(\lambda y^1 + (1 - \lambda) y^2) = \inf \{f(x) \mid x \in X : g(x) \leqq \lambda y^1 + (1 - \lambda) y^2\} \leqq$$
$$\leqq f(\lambda x^1 + (1 - \lambda) x^2) \leqq \lambda \gamma_1 + (1 - \lambda) \gamma_2 .$$

Since this inequality holds for all $\gamma_1 > q(y^1)$ and $\gamma_2 > q(y^2)$ we get also

$$q(\lambda x^1 + (1 - \lambda) x^2) \leqq \lambda q(x^1) + (1 - \lambda) q(x^2) ,$$

i.e. q is convex.

iii) If the Slater condition is fulfilled, then there exists a feasible point $\bar{x} \in X$ with $g(\bar{x}) < 0$. Then for $\bar{y} = g(\bar{x})$ we get

$$q(\bar{y}) = \inf\{f(x) \mid x \in X : g(x) \leqq \bar{y}\} \leqq f(\bar{x}) < \infty ,$$

i.e. $\bar{y} \in \mathrm{dom}(q)$.
Now by monotonicity, for all points $y \geq \bar{y}$ we have

$$q(y) \leqq q(\bar{y}) < \infty ,$$

i.e. we also have $y \in \mathrm{dom}(q)$, hence $0 \in \mathrm{int}(\mathrm{dom}(q))$.
It remains to show that q is proper. By the finiteness of $q(0)$ and the monotonicity of $q(.)$ again, for all vectors $y \in [\bar{y}, 0] = \{y \in I\!R^m \mid \bar{y} \leqq y \leqq 0\}$ we get

$$-\infty < q(0) \leqq q(y) \leqq q(\bar{y}) < \infty$$

such that q is finite on this set.
Now assuming a point $y^0 \in I\!R^m$ with $q(y^0) = -\infty$, then taking a special point y of the set $\mathrm{int}\,[\bar{y}, 0] = \{y \in I\!R^m \mid \bar{y} < y < 0\}$ we would get

$$q(\lambda y^0 + (1-\lambda)\,y) \leqq \lambda q(y^0) + (1-\lambda)\,q(y) = -\infty$$

for all numbers $\lambda \in (0,1)$. For sufficient small numbers λ, however, we have $\lambda y^0 + (1-\lambda)\,y \in [\bar{y}, 0]$ and we have a contradiction to the finiteness of q on this set. Hence q is proper. □

There is an interesting relation between the dual objective function and the marginal function. In fact, using the Fenchel conjugation we can formulate

Theorem 5.4.4.

$$-q^*(-u) = \begin{cases} \varphi(u) , & \text{if } u \geqq 0 \\ -\infty , & \text{else} . \end{cases}$$

Proof. By definition of the conjugate function we get for $u \geqq 0$

$$\begin{aligned} -q^*(-u) &= -\sup\{-uy - q(y) \mid y \in \mathbb{R}^m\} = \\ &= \inf\{uy + q(y) \mid y \in \mathbb{R}^m\} = \\ &= \inf\{uy + f(x) \mid y \in \mathbb{R}^m,\ x \in X : g(x) \leqq y\} = \\ &= \inf\{ug(x) + f(x) \mid x \in X\} = \\ &= \inf\{L(x,u) \mid x \in X\} = \\ &= \varphi(u)\ . \end{aligned}$$

In the same manner we can see that for $u \ngeqq 0$ we have $-q^*(-u) = -\infty$.□

We have pointed out above that

$$q(0) = \inf\{f(x) \mid x \in S\}$$

which is the optimal value of the primal problem.

Now with Theorem 5.4.4 we see that

$$\begin{aligned} q^{**}(0) &= \sup\{-q^*(u) \mid u \in \mathbb{R}^m\} = \\ &= \sup\{-q^*(-u) \mid u \in \mathbb{R}^m\} = \\ &= \sup\{\varphi(u) \mid u \in \mathbb{R}^m\} \end{aligned}$$

which is the optimal value of the dual problem. So because of

$$q^{**}(0) \leqq q(0)$$

we can recognize once more that weak duality holds.

Let us observe strong duality which means the equality

$$q^{**}(0) = q(0)\ .$$

If $q(0) = -\infty$, then the equality holds trivially. We shall assume therefore that $q(0) > -\infty$. Then it is well known that the existence of a subgradient of $q(.)$ at $y = 0$ is sufficient for this equality. Indeed, according to the definition of the subgradients and the conjugate function the relation

$$-u^0 = \partial q(0)$$

is equivalent to

$$q^*(-u^0) + q(0) = -u^0 0 = 0$$

and

$$\varphi(u^0) = -q(-u^0) = q(0) \ .$$

Hence, u^0 is a solution of the dual problem and it holds

$$\begin{aligned} q^{**}(0) &= \max\{-q^*(-u) \mid u \in \mathbb{R}^m\} = \\ &= -q^*(-u^0) = \varphi(u^0) = \\ &= q(0) = \inf\{f(x) \mid x \in S\} \ . \end{aligned}$$

Conversely, if u^0 is a solution of (D_L), then because of the equivalent conditions above we have $-u^0 \in \partial q(0)$

We remark that for all $u \in -\partial q(0)$ we have $u \geqq 0$, such that it is actually a feasible point of the dual problem. Really, since q is monotonically decreasing, we have

$$-uy \leqq q(y) - q(0) \qquad \forall\, y \in \mathbb{R}^m$$

and especially

$$-uy \leqq q(y) - q(0) \leqq 0 \qquad \forall\, y \geqq 0 \ .$$

Hence $-u \leqq 0$, i.e. $u \geqq 0$.

Summarizing this result we can formulate

Theorem 5.4.5. Let $q(0) \neq -\infty$. Then strong duality in the sense of

$$\max\{\varphi(u) \mid u \geqq 0\} = \inf\{f(x) \mid x \in S\}$$

holds if and only if $\partial q(0) \neq \emptyset$.
In this case $-\partial q(0)$ is the set of optimal solutions of (D_L).

For convex optimization problems, we can give a sufficient condition for the subdifferentiability of q at $y = 0$ and therefore a sufficient condition for strong duality. This assertion is closely connected with Theorem 5.4.3.

Theorem 5.4.6. Let f, g be convex functions and let the Slater condition be fulfilled. If $q(0) \neq -\infty$, then $\partial q(0) \neq \emptyset$, i.e.

$$\max\{\varphi(u) \mid u \geqq 0\} = \inf\{f(x) \mid x \in S\}$$

or equivalently

$$\max_{u \geqq 0} \inf_{x \in X} L(x,u) = \inf_{x \in X} \sup_{u \leqq 0} L(x,u) .$$

If additionally $x^0 \in S$ is a solution of (P), then we even have

$$\max\{\varphi(u) \mid u \geqq 0\} = \min\{f(x) \mid x \in S\}$$

or equivalently

$$\max_{u \geqq 0} \inf_{x \in X} L(x,u) = \min_{x \in X} \sup_{u \leqq 0} L(x,u) .$$

Proof. Because of the assumption we can use Lemma 5.4.3 such that q is a properly convex function with $0 \in \mathrm{int}(\mathrm{dom}(q))$. Now by Theorem 2.6.6 $\partial q(0) \neq \emptyset$. The rest is a consequence of Theorem 5.4.5. □

5.5. Perturbed Optimization Problems

The Lagrange duality concept is used mainly for convex optimization problems. In nonconvex optimization usually strong duality fails and therefore other duality concepts are developed. Most of these concepts are based on the modification of the classical Lagrangian.

In this section we give a general approach for the construction of dual optimization problems by means of a perturbation of the given problem and using the theory of conjugate functions. In this manner generalized Lagrangians can be associated which permit the formulation of strong duality theorems in the form of saddle point assertions. We only give the basic ideas of this approach. For further details we point to the references.

Analogous to the previous section, let the optimization problem (P) be given according to

$$\operatorname*{Min}_{x \in S} f(x) \qquad (P)$$

where $S = \{x \in X \mid g(x) \leqq 0\}$.

Also here, $X \subseteq \mathbb{R}^n$ is an open set, $f : X \to \mathbb{R}$ and $g : X \to \mathbb{R}^m$ with $g(x) = (g_1(x), ..., g_m(x))$ are arbitrary functions defined on X. Now, we assume that the problem is embedded in a family of perturbed optimization problems with the parameter $y \in Y \subseteq \mathbb{R}^m$ according to

$$\operatorname*{Min}_{x \in X} F(x, y) \qquad (P)_{(y)}$$

where $F : X \times Y \to \mathbb{R}$ is a suitable perturbation function with the property that

$$F(x, 0) = f(x) \qquad \forall x \in S \tag{1}$$

and

$$\inf \{F(x, 0) \mid x \in X\} = \inf \{f(x) \mid x \in S\} \ . \tag{2}$$

Thus, by definition the problem $(P)_{(0)}$ is equivalent to (P).

The function $F(.,.)$ is called *perturbation function*. The vector $y \in Y$ is called *perturbation parameter*.

Example 5.5.1. Let $Y = \mathbb{R}^m$ and

$$F_0(x, y) = \begin{cases} f(x) & \text{iff } g(x) \leqq y \\ +\infty & \text{else} \end{cases}$$

$$F_r(x, y) = \begin{cases} f(x) + r \cdot \|y\|^2 & \text{iff } g(x) \leqq y \\ +\infty & \text{else} \end{cases} \qquad (r > 0) \ .$$

Then in both cases relations (1) and (2) are fulfilled. Frequently $F_0(.,.)$ is called *standard perturbation function*.

Analogous to the Lagrange duality we shall introduce the marginal function by the following

Definition 5.5.1. The function $q_F : \mathbb{R}^m \to \bar{\mathbb{R}}$, according to

$$q_F(y) = \inf \{F(x, y) \mid x \in X\} \ ,$$

is called *generalized marginal function* of the problem (P).

Also here $q_F(0)$ is the optimal value of the primal problem. Using the special perturbations discussed in Example 5.5.1 above, we can state close connections with the marginal function for Lagrange duality discussed in the previous section.

Example 5.5.2. For the standard perturbation function $F_0(.,.)$ of Example 5.5.1 we have

$$q_{F_0}(y) = \inf\{F_0(x,y) \mid x \in X\} =$$
$$= \inf\{f(x) \mid x \in X : g(x) \leqq y\}$$

which is the marginal function of Lagrange duality. Using $F_r(.,.)$ we get

$$q_{F_r}(y) = \inf\{F_r(x,y) \mid x \in X\} =$$
$$= \inf\{f(x) + r \cdot \|y\|^2 \mid x \in X : g(x) \leqq y\} =$$
$$= q_{F_0}(y) + r \cdot \|y\|^2 .$$

Hence, if (P) is a convex optimization problem, then q_{F_0} but also q_{F_r}, $r > 0$, are convex functions (see Lemma 5.4.3). Otherwise, if convexity fails, one can try to generate a convex marginal function by adding the term $r \cdot \|y\|^2$ with r sufficiently large.

For the general case we can formulate

Lemma 5.5.1. If $F(.,.)$ is a convex function, then also $q_F(.)$ is convex.

Proof. The proof is similar to that of Lemma 5.4.3. Let $y^1, y^2 \in \operatorname{dom}(q_F)$, i.e. $q_F(y^1) < \infty$, $q_F(y^2) < \infty$. Then for each number γ_1, γ_2 with $q_F(y^1) < \gamma_1$, $q_F(y^2) < \gamma_2$, by definition of the marginal function we can find points $x_1, x_2 \in X$ such that

$$q_F(y^1) \leqq F(x^1, y^1) \leqq \gamma_1 ,$$
$$q_F(y^2) \leqq F(x^2, y^2) \leqq \gamma_2 .$$

Now because of the convexity, for any $\lambda \in (0,1)$ we get

$$q_F(\lambda y^1 + (1-\lambda)\, y^2) = \inf\{F(x, \lambda y^1 + (1-\lambda)\, y^2 \mid x \in X\} \leqq$$

$$\begin{aligned} &\leqq F(\lambda x^1 + (1-\lambda)\, x^2, \lambda y^1 + (1-\lambda)\, y^2) \leqq \\ &\leqq \lambda F(x^1, y^1) + (1-\lambda)\, F(x^2, y^2) \leqq \\ &\leqq \lambda \gamma_1 + (1-\lambda)\, \gamma_2 \,. \end{aligned}$$

Since this inequality holds for all $\gamma_1 > q_F(y^1)$ and $\gamma_2 > q_F(y^2)$ we get even

$$q_F(\lambda y^1 + (1-\lambda)\, y^2) \leqq \lambda q_F(y^1) + (1-\lambda)\, q_F(y^2) \,,$$

i.e. q_F is convex. □

Now we construct a dual optimization problem using results of conjugate functions. By definition we have

$$F^*(v, u) = \sup \{vx + uy - F(x, y) \mid x \in X,\ y \in Y\}$$

and especially

$$\begin{aligned} -F^*(0, u) &= -\sup \{uy - F(x, y) \mid x \in X,\ y \in Y\} = \\ &= \inf \{F(x, y) - uy \mid x \in X,\ y \in Y\} = \\ &= \inf \{q_F(y) - uy \mid y \in Y\} = \\ &= -\sup \{uy - q_F(y) \mid y \in Y\} = \\ &= -q_F^*(u) \,. \end{aligned} \tag{3}$$

Regarding the inequality

$$q_F(0) + q_F^*(u) \geqq -u0 = 0 \qquad \forall\, u \in \mathbb{R}^m$$

and taking into consideration that $q_F(0)$ is the optimal value of the primal problem (P), the relation

$$-q_F^*(u) \leqq q_F(0) \qquad \forall\, u \in \mathbb{R}^m$$

can be regarded as weakly duality assertion by using the following dual problem

$$\operatorname*{Max}_{u \in \mathbb{R}^m} \ -q_F^*(u) = -F^*(0, u) \,. \tag{D_F}$$

To allow a better comparison of this duality concept with the Lagrange duality concept discussed in the previous section, we shall replace the argument u by the term $-u$. Naturally, the negative sign has no influence on the optimal value and therefore on the duality conditions. Thus we write

$$\operatorname*{Max}_{u \in \mathbb{R}^m} \; -q_F^*(-u) = -F^*(0,-u)\;. \qquad (D_F)$$

In this manner, taking the standard perturbation function $F_0(.,.)$ the dual problem (D_{F_0}) coincides with the Lagrange dual problem (see Theorem 5.4.4). Also here the dual objective function is concave and upper semi continuous and it holds

$$\begin{aligned} q_F^{**} &= \sup\{0u - q_F^*(u) \mid u \in \mathbb{R}^m\} = \\ &= \sup\{-q_F^*(-u) \mid u \in \mathbb{R}^m\} \end{aligned}$$

i.e. $q_F^{**}(0)$ is the optimal value of (D_F).

Summarizing these results we have the pair of dual optimization problems

$$\operatorname*{Min}_{x \in \mathbb{R}^n} \; F(x,0) \qquad (P)$$

$$\operatorname*{Max}_{u \in \mathbb{R}^m} \; -F^*(0,-u) \qquad (D_F)$$

with the optimal values $q_F(0)$ and $q_F^{**}(0)$ respectively and the weak duality relation

$$q_F^{**}(0) \leqq q_F(0)$$

holds.

Before we give some assertions regarding strong duality, we shall remark that also the dual problem (D_F) is imbedded in the family of perturbed problems according to

$$\operatorname*{Max}_{u \in \mathbb{R}^m} \; -F^*(-v,-u) \qquad (D_F)_{(v)}$$

which are equivalent to

$$\operatorname*{Min}_{u \in \mathbb{R}^m} \; F^*(-v, -u) \; .$$

Thus, in the same manner we can construct the bidual problem (DD_F) by

$$\operatorname*{Max}_{x \in \mathbb{R}^n} \; -F^{**}(x, 0) \qquad (DD_F)$$

or equivalently

$$\operatorname*{Min}_{x \in \mathbb{R}^n} \; F^{**}(x, 0) \; .$$

If the perturbation function F is properly convex and lower semi continuous on X, then it holds $F^{**} = F$ and the bidual problem is equivalent to the primal problem (P).

In the following we shall discuss strong duality, i.e. we shall give necessary and sufficient conditions for the equality of the optimality values. Since this is equivalent to

$$q_F^{**}(0) = q_F(0)$$

we have to discuss the increasing behaviour of the marginal function q in a neighborhood of the point $y = 0$. We introduce the following notations.

Definition 5.5.2. The problem (P) (precisely: the family of disturbed problems) is called

i) *normal*, if $q_F(0)$ is finite and q_F is lower semi continuous at $y = 0$,

ii) *stable*, if $q_F(0)$ is finite and $\partial q_F(0) \neq \emptyset$.

Obviously, (P) is normal if it is stable. In fact, in case of $\partial q_F(0) \neq \emptyset$ we have $q_F(0) = q_F^{**}(0)$. Hence, since q_F^{**} is lower semi continuous and $q_F \geqq q_F^{**}$, also q_F is lower semi continuous at $y = 0$. Moreover, analogous to Theorem 5.4.5, we can formulate the following assertion.

Theorem 5.5.1. It holds strong duality in the sense of

$$\max\{-q_F^*(-u) \mid u \in I\!R^m\} = \inf\{f(x) \mid x \in S\}$$

if and only if (P) is stable. In this case $-\partial q_F(0)$ is the set of optimal solutions of (D_F).

Proof. The proof is similar to that of Theorem 5.4.5. Also here the condition

$$-u^0 \in \partial q_F(0)$$

is equivalent to

$$q_F^*(-u^0) + q_F(0) = -u0 = 0$$

and

$$-q_F^*(-u^0) = q_F(0) \ .$$

Hence, this means that u^0 is a solution of the dual problem and it holds

$$\begin{aligned} q_F^{**}(0) &= \max\{-q_F^*(-u) \mid u \in I\!R^m\} = \\ &= -q_F^*(-u^0) = \\ &= q_F(0) = \inf\{f(x) \mid x \in S\} \ . \end{aligned}$$

□

The assertions can be sharpened if the perturbation function $F(.,.)$ is assumed to be convex. Then it holds

Theorem 5.5.2. Let $F(.,.)$ be convex and $q_F(0)$ be finite. Then it holds strong duality in the sense of

$$\sup\{-q_F^*(-u) \mid u \in I\!R^m\} = \inf\{f(x) \mid x \in S\}$$

if and only if (P) is normal.

Proof. Using Lemma 5.5.1, the marginal function q_F is convex by assumption. Thus it holds $q_F^{**}(0) = q_F(0)$, i.e.

$$\sup\{-q_F^*(-u) \mid u \in I\!R^m\} = \inf\{f(x) \mid x \in S\} \ ,$$

if and only if q_F is lower semi continuous at $y = 0$. □

Combining Theorems 5.5.1 and 5.5.2 we get

Theorem 5.5.3. Let $F(.,.)$ be convex and $q_F(0)$ be finite. Then it holds strong duality in the sense of

$$\max\{-q_F^*(-u) \mid u \in I\!R^m\} = \inf\{f(x) \mid x \in S\}$$

if and only if (P) is stable
if and only if (P) is normal and (D_F) admits any solution.
In all cases $-\partial q_F(0)$ is the set of optimal solutions of (D_F).

A sufficient criterion which guarantees that (P) is stable shall be given in the following assertion. For this we have to ensure the subdifferentiability of $q_F(.)$ at the point $y = 0$. In Lagrange duality, i.e. by means of the standard perturbation function $F_0(.,.)$, this could be guaranteed by the Slater condition (see Theorem 5.4.6). Now we use a more general condition. It holds

Theorem 5.5.4. Let $F(.,.)$ be convex and $q_F(0)$ be finite. Further we assume that there exists a point $\bar{x} \in X$ such that $F(\bar{x},.)$ is continuous at $y = 0$. Then (P) is stable.

Proof. By assumption, $F(\bar{x},0)$ is continuous at $y = 0$ and therefore it is bounded on a neighborhood $N(0)$ of $y = 0$. Thus there exists a number $K > 0$ such that

$$q_F(y) = \inf\{F(x,y) \mid x \in X\} \leqq F(\bar{x},y) \leqq K$$

for all $y \in N(0)$.
Since q_F is convex and because $q_F(0) = 0$, we can state that q_F is finite on this neighborhood. Hence it is continuous at $y = 0$ and by Theorem 2.6.6 we have $\partial q_F(0) \neq \emptyset$. □

Finally we introduce a generalization of the classical Lagrangian. Doing so we shall also be able to formulate saddle point assertions for this duality concept by means of the perturbation function $F(.,.)$.

Definition 5.5.3. The function $L_F : X \times I\!R^m \to \bar{I\!R}$, according to

$$L_F(x,u) = \inf\{uy + F(x,y) \mid y \in Y\}$$

is called *generalized Lagrangian* of (P).

Let us calculate the Lagrangians associated to the special perturbation functions discussed in Example 5.5.1.

Example 5.5.3. Using the standard perturbation function $F_0(.,.)$ we get

$$\begin{aligned} L_{F_0}(x,u) &= \inf\{uy + F_o(x,y) \mid y \in I\!R^m\} = \\ &= \inf\{uy + f(x) \mid y \in I\!R^m : y \geqq g(x)\} = \\ &= \begin{cases} f(x) + ug(x) & \text{if } u \geqq 0 \\ -\infty & \text{else} \end{cases} \end{aligned}$$

which is the classical Lagrangian extended by $-\infty$ for $u \ngeqq 0$. For $F_r(.,.)$ $(r > 0)$ we get

$$\begin{aligned} L_{F_r}(x,u) &= \inf\{uy + F_r(x,y) \mid y \in I\!R^m\} = \\ &= \inf\{uy + f(x) + r \cdot \|y\|^2 \mid y \in I\!R^m : y \geqq g(x)\} = \\ &= f(x) + \inf\{uy + r \cdot \|y\|^2 \mid y \in I\!R^m : y \geqq g(x)\} = \\ &= f(x) + \sum_{i=1}^m \inf\{u_i y_i + r \cdot y_i^2 \mid y_i \in I\!R : y_i \geqq g_i(x)\} = \\ &= f(x) + \sum_{i=1}^m u_i \cdot \max\left\{g_i(x), -\frac{u_i}{2r}\right\} + \\ &\quad + \sum_{i=1}^m r \cdot \left(\max\left\{g_i(x), -\frac{u_i}{2r}\right\}\right)^2 = \\ &= f(x) + \sum_{i=1}^m \sigma_i(x,u) \end{aligned}$$

where

$$\sigma_i(x,u) = \begin{cases} -\dfrac{1}{4r}\,u_i^2 & \text{if } g_i(x) \leqq -\dfrac{u_i}{2r} \\ u_i g_i(x) + r\cdot(g_i(x))^2 & \text{else .} \end{cases}$$

Frequently this function is called *augmented Lagrangian.*

Special convexity properties of the generalized Lagrangian can be stated.

Lemma 5.5.2.

i) $L_F(x,.)$ is concave and upper semi continuous for fixed $x \in X$.

ii) If $F(.,.)$ is a convex function, then also $L_F(.,u)$ is convex for fixed $u \in U$.

Proof.

i) It holds

$$-L_F(x,u) = \sup\{-uy - F(x,y) \mid y \in Y\} = [F(x,.)]^*(-u) \ .$$

Thus, for fixed $x \in X$ the function $-L_F(x,.)$ is a special conjugate function. Therefore it is convex and lower semi continuous as a function of u. Hence $L_F(x,.)$ is concave and upper semi continuous.

ii) The proof is similar to those of Lemmas 5.4.3 and 5.5.1. Let $u \in I\!R^m$ and $x^1, x^2 \in \mathrm{dom}(L_F(.,u))$, i.e. $L_F(x^1,u) < \infty$, $L_F(x^2,u) < \infty$. Then for each number γ_1, γ_2 with $L_F(x^1,u) < \gamma_1$, $L_F(x^2,u) < \gamma_2$, we can, by definition of the generalized Lagrangian, find points $y_1, y_2 \in Y$ such that

$$L_F(x^1,u) \leqq uy^1 + F(x^1,y^1) \leqq \gamma_1 \ ,$$
$$L_F(x^2,u) \leqq uy^2 + F(x^2,y^2) \leqq \gamma_2 \ .$$

Now because of the convexity assumption, for any $\lambda \in (0,1)$ we get

$$L_F(\lambda x^1 + (1-\lambda)\,x^2, u) =$$
$$= \inf\{uy + F(\lambda x^1 + (1-\lambda)\,x^2, y) \mid y \in Y\} \ \leqq$$

$$\begin{aligned}
&\leqq u(\lambda y^1 + (1-\lambda)\, y^2) + F(\lambda x^1 + (1-\lambda)\, x^2, \lambda y^1 + (1-\lambda)\, y^2) \leqq \\
&\leqq u(\lambda y^1 + (1-\lambda)\, y^2) + \lambda F(x^1, y^1) + (1-\lambda)\, F(x^2, y^2) = \\
&= \lambda(uy^1 + F(x^1, y^1)) + (1-\lambda)(uy^2 + F(x^2, y^2)) \leqq \\
&\leqq \lambda\gamma_1 + (1-\lambda)\, \gamma_2 \ .
\end{aligned}$$

Since this inequality holds for all $\gamma_1 > L_F(x^1, u)$ and $\gamma_2 > L_F(x^2, u)$ we get also

$$L_F(\lambda x^1 + (1-\lambda)\, x^2, u) \leqq \lambda L_F(x^1, u) + (1-\lambda)\, L_F(x^2, u) \ ,$$

i.e. $L_F(.,u)$ is convex. □

Now we are able to represent the dual problem also by means of the generalized Lagrangian. In fact, taking the conjugate of the perturbation function $F(.,.)$ we get

$$\begin{aligned}
-F^*(0, -u) &= -\sup\{-uy - F(x,y) \mid x \in X,\ y \in Y\} = \\
&= \inf\{F(x,y) + uy \mid x \in X,\ y \in Y\} = \\
&= \inf\{L_F(x,u) \mid x \in X\} \ .
\end{aligned}$$

Thus, the dual objective function has the representation

$$-q_F(-u) = -F^*(0,-u) = \inf\{L_F(x,u) \mid x \in X\}$$

and for the dual problem we can write

$$\operatorname*{Max}_{u \in I\!R^m} \ \inf\{L_F(x,u) \mid x \in X\} \ .$$

On the other side we can calculate the biconjugate function of $F(x,.)$ according to

$$\begin{aligned}
[F(x,.)]^{**}(y) &= \sup\{uy - [F(x,.)]^*(u) \mid u \in I\!R^m\} = \\
&= \sup\{-uy - [F(x,.)]^*(-u) \mid u \in I\!R^m\} \ .
\end{aligned}$$

Taking into consideration the equality

$$-L_F(x,u) = \sup\{-uy - F(x,y) \mid y \in Y\} = [F(x,.)]^*(-u)$$

we get even

$$[F(x,.)]^{**}(y) = \sup\{-uy + L_F(x,u) \mid u \in \mathbb{R}^m\} ,$$

that means especially

$$[F(x,.)]^{**}(0) = \sup\{L_F(x,u) \mid u \in \mathbb{R}^m\} .$$

If F is properly convex and lower semi continuous, then of course

$$F(x,0) = [F(x,.)]^{**}(0) = \sup\{L_F(x,u) \mid u \in \mathbb{R}^m\}$$

and also the primal problem can be represented in the form

$$\operatorname*{Min}_{x\in X} \ \sup\{L_F(x,u) \mid u \in \mathbb{R}^m\} .$$

Thus, in this case strong duality is closely connected with the existence of saddle points of the generalized Lagrangian. We shall summarize this result in

Theorem 5.5.5. Let $F(.,.)$ be properly convex and lower semi continuous. Then it holds strong duality in the form

$$\max\{-q_F^*(-u) \mid u \in \mathbb{R}^m\} = \min\{f(x) \mid x \in S\}$$

if and only if (x^0, u^0) is a saddle point of L_F.

Proof. By Lemma 5.4.1 the point (x^0, u^0) is a saddle point of L_F if and only if

$$\max_{u\in\mathbb{R}^m} \ \inf_{x\in X} \ L_F(x,u) = \min_{x\in X} \ \sup_{u\in\mathbb{R}^m} \ L_F(x,u) .$$

By our above remarks, however, this is equivalent to

$$\begin{aligned}\max\{-q_F(-u) \mid u \in \mathbb{R}^m\} &= \min\{F(x,0) \mid x \in X\} = \\ &= \min\{f(x) \mid x \in S\} .\end{aligned}$$

□

Finally we can formulate the following direct duality assertion.

Theorem 5.5.6. Let $F(.,.)$ be properly convex and lower semi continuous. Further we assume that the optimization problem (P) is stable. If x^0 is a solution of (P), then there also exists a point $u^0 \in I\!R^m$ which is a solution of (D_F) and the optimal values are equal.

Proof. Let x^0 be a solution of (P). Because of the assumption we can use Theorem 5.5.1 and we get

$$\max\{-q_F^*(-u) \mid u \in I\!R^m\} = \inf\{f(x) \mid x \in S\} = \\ = \min\{f(x) \mid x \in S\} .$$

By Theorem 5.5.5, (x^0, u^0) is a saddle point of L_F. Hence, u^0 is a solution of the dual problem and strong duality holds. □

References to Chapter V

K.J. ARROW, F.J. GOULD and S.M. HOWE (1973), A general saddle-point result for constrained optimization, Math. Progr., 5, 225–234.

J.P. AUBIN (1980), Further properties of Lagrange multipliers in non-smooth optimization, Appl. Math. and Opt., 6, 79–90.

J.P. AUBIN (1993), Optima and Equilibria: An Introduction to Nonlinear Analysis, Springer, Berlin, Heidelberg.

M.S. BAZARAA and C.M. SHETTY (1979), Nonlinear Programming – Theory and Algorithms, John Wiley, New York.

C. BERGE and A. GHOUILA-HOURI (1962), Programmes, Jeux et Réseaux de Transport, Dunod, Paris.

V.F. DEM'YANOV and V.N. MALOZEMOV (1974), Introduction in Minimax, John Wiley.

I. EKELAND and R. TEMAM (1976), Convex Analysis and Variational Problems, North-Holland, Amsterdam.

A.V. FIACCO and G.P. McCORMICK (1968), Nonlinear Programming: Sequential Unconstrained Minimization Techniques, John Wiley, New York.

J. GAUVIN (1979), The generalized gradients of a marginal function in mathematical programming, Math. of Op. Res., 4, 458–463.

A.M. GEOFFRION (1971), Duality in nonlinear programming: A simplified applications-oriented development, SIAM-Review, 13, 1–37.

J. GLICKSBERG (1952), A further generalization of the Kakutani fixed point theorem with applications to Nash equilibrium points, Proc. A.M.S., 3, 170–174.

E.G. GOL'STEIN (1971), The Theory of Duality in Mathematical Theory and its Applications, Nauka.

E.G. GOL'STEIN (1972), Theory of Convex Programming, A.M.S. Translation Series.

F.J. GOULD (1969), Extensions of Lagrange multipliers in nonlinear programming, SIAM J. Appl. Math., 17, 1280–1297.

M.R. HESTENES (1975), Optimization Theory – The Finite Dimensional Case, John Wiley, New York.

A.D. IOFFE (1979), Necessary and sufficient conditions for a local minimum. 1: A reduction theorem and first order conditions. 2: Conditions of Levitin-Miljutin-Osmolovskij-type. 3: Second order conditions and augmented duality, SIAM J. Control Optim., 17, 245–288.

S. KAKUTANI (1941), A generalization of Browers fixed point theorem, Duke Math. J. 8, 457–459.

S. KARLIN (1960), Mathematical Methods and Theory in Games, Programming and Economics, McGraw-Hill, New York.

P.J. LAURENT (1972), Approximation et Optimisation, Hermann, Paris.

D.G. MAHAJAN and M.N. VARTAK (1977), Generalization of some duality theorems in nonlinear programming, Math. Progr., 12, 293–317.

O.L. MANGASARIAN (1969), Nonlinear Programming, McGraw-Hill, New York.

G.P. McCORMICK (1976), Optimality criteria in nonlinear programming, SIAM-A.M.S. Proc. 9, 27–38.

J. v. NEUMANN (1928), Zur Theorie der Gesellschaftsspiele, Math. Ann., 100, 295–320.

B.N. PSHENICHNYI (1980), Convex Analysis and Extremum Problems, Nauka.

R.T. ROCKAFELLAR (1970), Convex Analysis, Princeton Univ. Press, Princeton.

R.T. ROCKAFELLAR (1973), A dual approach to solving nonlinear programming problems by unconstrained optimization, Math. Progr., 5, 354–373.

R.T. ROCKAFELLAR (1974), Conjugate duality and optimization, SIAM-Publications.

R.T. ROCKAFELLAR (1974), Augmented Lagrange multiplier functions and duality in nonconvex programming, SIAM J. Control, 12, 268–285.

R.T. ROCKAFELLAR (1976), Augmented Lagrangians and Applications of the proximal point algorithm in convex programming, Math. of Op. Res., 1, 97–116.

R.T. ROCKAFELLAR (1976), Lagrange multipliers on Optimization, SIAM-A.M.S. Proc. 9, 145–168.

R.T. ROCKAFELLAR (1982), Lagrange multipliers and subderivatives of optimal value functions in nonlinear programming, Math. Progr. Study, 17, 28–66.

R.T. ROCKAFELLAR (1985), Extensions of subgradient calculus with applications to optimization, Nonlinear Analysis, Th. Meth. Appl., 9, 665–698.

M. SION (1957), Sur une generalisation du theoreme minimax, C.R. Academic des Sci., 244, 2120–2123.

M. SION (1958), On general minimax theorems, Pac. J. Math., 1958.

J. TIND and L.A. WOLSEY (1981), An elementary survey on general duality theory in mathematical programming, Math. Progr. 21, 241–261.

CHAPTER VI. VECTOR OPTIMIZATION

6.1 Vector Optimization Problems

We say that an optimization problem is a *vector* or *multiobjective* or *multicriteria* problem when the objective function is a vector function $f = (f_1, f_2, ..., f_p)$. The variable x ranges over a set $X \subset \mathbb{R}^n$ and may be required to satisfy some further constraints. For the sake of brevity, these further conditions will be considered only through functional constraints and only through inequalities:

$$\operatorname*{Min}_{x \in S} f(x) \tag{V.P.}$$

with $f : X \subset \mathbb{R}^n \to \mathbb{R}^p$ and $S = \{x \in X : g_j(x) \leqq 0\,;\, j = 1, ..., m\}$. In this chapter, X will always denote an open set. The set S will be named as the *feasible decision set.*

A vector optimization problem is implicit in any decision-making process, when the final choice depends on conflicting goals, namely different criteria which one wants to minimize (or to maximize) and which are not reducible to a single one. These criteria are conflictive and none of them has higher priority over others. Politics, business and in general group decision making are always concerned with satisfying many different view points. It is not surprising to suppose that almost any real-world application of mathematics is able to offer conflictive multiple criteria. Mathematical economics game theory, welfare theory, production theory, the theory of equilibria and many other frameworks deal with applications of vector optimization.

A point $x \in S$ could be defined as an optimal solution for $(V.P.)$ when it simultaneously minimizes all the objective functions $f_1, f_2, ..., f_p$ in the feasible decision set S; by this definition, a vector optimization problem would have no further mathematical interest with respect to a scalar one. But this approach would mean that criteria have no conflicting nature and are no more than one single index in different forms. So this definition is too narrow to be of practical use. Unlike mathematical programming problems (with a single objective function) an optimal solution in the previous sense does not necessarily exist in multiobjective optimization problems. This fact heavily depends on the kind of $(V.P.)$ problems and not on the hypotheses that we can introduce on the function f and/or on the feasible set S.

We will also consider the points that simultaneously minimize all p objectives; we will call them ideal or utopia points. Yet we need a broader notion than this extreme and fortunate case. The crucial point is to define an optimality notion that takes into account the different objectives $f_1, ..., f_p$.

The difficulty of a vector optimization problem arises from the incomparability for some pairs of alternatives. From a mathematical point of view, a vector optimization problem consists of the research of those values which will be defined as optimal in the *partially* ordered set $Z = f(S) \subset \mathbb{R}^p$. The set Z is called the *image space* under the mapping f or the *outcome space*.

Historically the first satisfactory definition of an optimal solution for $(V.P.)$ was the *Pareto optimum*. This definition was formulated by the Italian mathematical economist Vilfredo Pareto in his *Manuale di Economia Politica* (1906) and in some previous papers by the same author at the beginning of the century. A point $x^0 \in S$ is said to be a Pareto optimum (here a Pareto minimum) or a Pareto efficient point when it is nondominated or there is no $x \in S$ with $f_i(x) \leqq f_i(x^0)$, $\forall\, i = 1, ..., p$, and $f_j(x) < f_j(x^0)$ at least for one index j. Hence for every $x \in S$ we cannot have $\Delta f = f(x) - f(x^0) \in \mathbb{R}^p_- \backslash \{0\}$.

The theory of vector optimization is at the crossroads of many subjects. The terminology feels the effects of such a position. The terms "minimum"/"maximum" and "optimum" are in line with a mathematical tradition while words such as "efficient" or "nondominated" find a larger use in business-related topics. In any case we will use all these words synonymously.

The notions of Paretian optimality is based on the *componentwise order.* So we can extend this definition of efficiency by replacing the componentwise order with a general ordering relation or by replacing $\mathbb{R}^p_-$ with some set C (a cone, as we will see) when we are requiring that there be no $x \in S$ with $\Delta f \in \mathbb{R}^p_- \backslash \{0\}$.

In consideration of these remarks, the next section will be devoted to some definitions and to some general properties of binary relations in its connection with (convex, closed and pointed) cones.

6.2. Conical Preference Orders

Definition 6.2.1. Let R be a binary relation on an arbitrary set $Z \subset \mathbb{R}^p$. We say that R is:

a) *reflexive* when xRx, $\forall\, x \in Z$;

b) *irreflexive* when $x \not{R} x$, $\forall\, x \in Z$ (the notation $x \not{R} y$ denotes the negation of xRy);

c) *symmetric* when xRy implies yRx, $\forall\, x, y \in Z$;

d) *asymmetric* when xRy implies $y \not{R} x$, $\forall\, x, y \in Z$;

e) *antisymmetric* when xRy and yRx imply $x = y$, $\forall\, x, y \in Z$;

f) *transitive* when xRy and yRz imply xRz, $\forall\, x, y, z \in Z$;

g) *negatively transitive* when $x \not{R} y$ and $y \not{R} z$ imply $x \not{R} z$, $\forall\, x, y, z \in Z$;

h) *complete* (or *connected*) when $\forall\, x, y \in Z$ we have xRy or yRx (possibly both);

i) *weakly complete* (or *weakly connected*) when $\forall\, x, y \in Z$, $x \neq y$, we have xRy or yRx;

j) *linear* when xRy implies $(cx+z)\, R(cy+z)$, $\forall\, c \in \mathbb{R}_+$ and $\forall\, x, y, z \in Z$ such that $cx + z, cy + z \in Z$.

Remark 6.2.1. The linearity is a compatibility condition between the order and the algebraic structures. It requires that Z be a subset of a linear space (e.g. $Z \subset \mathbb{R}^p$). All the other previous definitions can be given for arbitrary spaces Z without linear structure.

Definition 6.2.2. A binary relation R on a set Z is said to be:

a) a *preorder* when it is reflexive and transitive;

b) a *partial order* when it is reflexive, antisymmetric and transitive;

c) a *total order* when it is reflexive, antisymmetric, transitive and complete.

Definition 6.2.3. A binary relation R on a set Z is said to be a *weak order* when it is asymmetric and negatively transitive.

Remark 6.2.2. A partial order becomes a *strict* partial order by substituting the hypothesis of reflexivity in b) by the irreflexivity one. Then, of course, it is asymmetric and hence antisymmetric since xRy and yRx imply xRx, but R is assumed to be irreflexive.
Analogously a binary relation R is said to be a *strict* total order when it is irreflexive, transitive and weakly complete. So a strict total order is a strict partial order which is weakly complete. The relations between strict total order and strict partial order are the same as between total order and partial order.

Theorem 6.2.1.
i) If R is a strict total order, then it is a weak order.

ii) If R is a weak order, then it is a strict partial order.

Proof.
i) R is asymmetric due to the existence of two elements $x, y \in Z$, such that xRy and yRx would lead to xRx, which is in contrast with the hypothesis of irreflexivity. R is negatively transitive too as $x \not{R} y$ and $y \not{R} z$ mean that zRy and yRx. Thus we have zRx or $x \not{R} z$ as a consequence of the asymmetry of R.

ii) If relation R is a weak order, R must be irreflexive. If there would be a $x \in Z$ such that xRx, R could not be asymmetric. R is antisymmetric since it is asymmetric. R is also transitive; indeed, $\forall x, y, z \in Z$, xRy and yRz imply (owing to the asymmetric hypothesis) $z \not R y$ and $y \not R x$. As R is negatively transitive, we have $z \not R x$ and xRz; in fact, assuming $x \not R z$, by the negative transitivity and $z \not R y$ we would get $x \not R y$ which is a contradiction to the assumption. □

We will denote an order relation (partial or total) by $\geqq$; a strict order relation will be indicated by $\geq$. It is easy to verify that an arbitrary partial order $\geqq$ induces a strict partial order by setting $x \geq y$ when it is $x \geqq y$ while it is not $y \geqq x$ or when it is $x \geqq y$ with $x \neq y$. We will write $x \leqq y$ or $x \leq y$ for $y \geqq x$ or $y \geq x$.

Example 6.2.1.

a) In $\mathbb{R}^p$ we obtain the *lexicographic order* when we put $x \geq y$ if and only if $x_i = y_i$, $\forall i < k$ $(1 \leqq k \leqq p)$, and $x_k > y_k$.

b) In $\mathbb{R}^p$ we obtain the *Pareto order* or the *componentwise* order by setting $x \geqq y$ if and only if $x_i \geqq y_i$, $\forall i = 1, ..., p$. In particular, for the Paretian ordering we will write $x \geq y$ when $x \geqq y$ and $\exists i : x_i > y_i$ and $x > y$ when it is $x_i > y_i$, $\forall i = 1, ..., p$.

c) A partial (componentwise) order in the set of all the real sequences is obtained by setting $x = \{x^k\} \geqq y = \{y^k\}$ if and only if $x^k \geqq y^k$, $\forall k \in \mathbb{N}$.

In the previous examples of componentwise ordering we have $x \geqq y$ if and only if the vector (or the sequence) $x - y$ has all nonnegative components or equivalently if and only if $x - y$ belongs to the convex and pointed cone $\mathbb{R}^p_+$ (respectively: to the convex and pointed cone of the real nonnegative sequences). This connection between a partial order $x \geqq y$ and the belonging of $x - y$ to a convex pointed cone is a general statement. We formulate it explicitly for euclidean spaces, even if it holds for any linear space.

Theorem 6.2.2.

i) If $\mathbb{R}^p$ is partially ordered by a linear relation $\geqq$, then the set $C = \{y \in \mathbb{R}^p : y \geqq 0\}$ is a convex pointed cone (which includes the origin).

ii) If C is a convex and pointed cone of $\mathbb{R}^p$ such that $0 \in C$, then the binary relation so defined:

$$x \geqq y \text{ when } x - y \in C$$

is a partial linear order in $\mathbb{R}^p$.

The proof of the theorem is immediate and it shows a strong connection between the antisymmetry of the relation $\geqq$ and the pointness of the ordering cone C. The same bi-implication exists between the transitive and linear properties and the convexity of C, and between the reflexive property and the assumption that $0 \in C$.

From now on we will consider only *partial orders generated by nontrivial convex and pointed cones* (*including the origin*) and we will write:

$$x \geqq_C y \qquad \text{for } x - y \in C \; ;$$

$$x \geq_C y \qquad \text{for } x - y \in C \backslash \{0\} \; .$$

We will drop the subscript C and more simply write $x \geqq y$ and $x \geq y$ where it is clear that C is the ordering cone.

It is also convenient to assume that $\operatorname{int} C \neq \emptyset$ *and that* C *is a closed cone.* These requirements have not to be merely understood as a technical condition. Indeed, if $\operatorname{int} C = \emptyset$, it may be that a pair of vectors does not admit a minorant, as one can see by considering $C \subset \mathbb{R}^2$, $C = \{(x,y) : x = 0,\, y \geqq 0\}$. The closedness of C is a sufficient condition which assures that the order relation satisfies the *Archimedean property* ($a \geqq nb$ for any $n \in \mathbb{N}$ implies $b \leqq 0$).

Also we will generalize the distinction that we already introduced for the Paretian order by using the notation:

$$x > y$$

which means $x - y \in K = \operatorname{int} C$. It is easy to verify that the relation $x > y$ is also irreflexive and transitive.

6.3. Optimality (or Efficiency) Notions

In the previous section we introduced the general notion of ordering relations, which we will express explicitly through a general closed convex pointed cone C (with $\operatorname{int} C \neq \emptyset$). Now the componentwise order and the set $\mathbb{R}^p_+$ are only a particular case.

In particular, in this section we will give the definitions of *ideal efficiency*, *efficiency* and *weakly efficiency* (or minimality) for a set $Z \subset \mathbb{R}^p$ when $\mathbb{R}^p$ is partially ordered by a closed convex pointed cone C (with $\operatorname{int} C \neq \emptyset$). In Section 4 *proper efficiency* will be introduced. Naturally, maximality notions can be obtained analogously by exchanging the order relations $\geqq$ and $\leqq$.

Definition 6.3.1.

a) An element $z^0 \in Z$ is said to be an *ideal* (*or utopia*) *minimal or efficient value* of the set Z when it is $z \geqq z^0$, $\forall\, z \in Z$; we will write $z^0 \in IE(Z)$;

b) $z^0 \in Z$ is said to be a *minimal* or *efficient value* of Z ($z^0 \in E(Z)$) when $z^0 \geqq z$ for some $z \in Z$ implies $z \geqq z^0$ (i.e. $z = z^0$) or equivalently when $\nexists\, z \in Z$ with $z^0 \geq z$;

c) $z^0 \in Z$ is said to be a *weakly minimal* or *weakly efficient value* of Z ($z^0 \in WE(Z)$) when z^0 is a minimal value of the set Z ordered by the cone $K = \{0\} \cup \operatorname{int} C$, i.e. when $\nexists\, z \in Z$ with $z^0 > z$.

Respectively, we will write $z^0 \in IE_C(Z)$, $z^0 \in E_C(Z)$, $z^0 \in WE_C(Z)$ when we wish to emphasize the ordering cone of the set Z.

For the case where $Z = f(S)$, we have the definitions of ideal (or utopia) *points*, of efficient *points* (or solutions of ($V.P.$)) and weakly efficient *points*. For a point $x^0 \in S$ with $z^0 = f(x^0)$, we will write $x^0 \in IE(S)$, $x^0 \in E(S)$, $x^0 \in WE(S)$ respectively if and only if $z^0 \in IE(Z)$, $z^0 \in E(Z)$, $z^0 \in WE(Z)$.

When it is nonempty the set $IE(Z)$ is singleton for the antisymmetry of the relation $\geqq$. An ideal minimal value dominates any other value of

Z; for $Z \subset I\!R^p$ endowed for example with the componentwise order, z^0 is the value with the minimal components z_i^0 for any $i = 1, ..., p$. The notion of minimal value weakens the previous section: if $z^0 \in E(Z)$, we can have some values of Z which are not comparable with z^0 but z^0 is the best point with respect to those which admit comparability. In other words z^0 is *nondominated* by any value of Z. Of course $IE(Z) \subset E(Z)$; Theorem 6.3.2 will make this relation more exact. The definition of weak minimality can be easily explained with the aid of the particular cone $C = I\!R^p_+$; then $z^0 \in WE(Z)$ when there does not exist any $z \in Z$ with $z_i < z_i^0$ $(\forall\, i = 1, ..., p)$. Theorem 6.3.1 will show the general inclusion of $E(Z)$ in $WE(Z)$.

Remark 6.3.1. Definitions 6.3.1 can be given in *local* form by considering z as ranging only over a neighborhood of $z^0 \in Z$. In particular z^0 is called a local minimal value of Z when there exists $\delta > 0$ such that $z^0 \in E(Z \cap N_\delta(z^0))$. If Z is convex, then $E(Z) = E(Z \cap N_\delta(z^0))$. It suffices to show that $E(Z \cap N_\delta(z^0)) \subset E(Z)$. Suppose that $z^0 \in E(Z \cap N_\delta(z^0)) \backslash E(Z)$. Then there exists $z^1 \in Z$ such that $z^0 - z^1 = c \in C \backslash \{0\}$. Since Z is convex, $\forall\, \delta > 0$ there exists $\lambda \in (0,1)$ for which $z^2 = z^0 - \lambda c \in Z \cap N_\delta(z^0)$. It follows that $z^0 - z^2 \in C \backslash \{0\}$. Thus $z^0 \notin E(Z \cap N_\delta(z^0))$.

Remark 6.3.2. As a minimal value of Z is also a maximal with respect to the partial order induced by $-C$, it is sufficient to limit our research to minimality notions.

The different efficiency definitions have been formulated through the ordering relation $\geqq$. The same definitions can be given emphasizing the geometrical features of the image set Z. So it will often be easier to verify Definitions 6.3.1 since they are reduced to an inclusion or to an intersection between two sets and one of them is a prefixed cone.

Definition 6.3.1 a) can be rewritten in the form $z - z^0 \in C$, $\forall\, z \in Z$, or $Z - z^0 \subset C$.

Equivalently $z^0 \in E(Z)$ if and only if there is no element $z \in Z$ such that $z^0 - z \in C \backslash \{0\}$ or $z - z^0 \in -C \backslash \{0\}$. This inclusion can be replaced by the set equation:

$$(Z - z^0) \cap (-C) = \{0\} \qquad (\text{or } Z \cap (z^0 - C) = \{z^0\})$$

which offers a useful geometric characterization of minimal values. They can be determined by sliding the cone $-C$ to contact the boundary of Z; only those points that can be contacted by the cone sole vertex are candidates.

Analogously $z^0 \in WE(Z) = E_K(Z)$ with $K = \{0\} \cup \operatorname{int} C$ if and only if $Z \cap (z^0 - K) = \{z^0\}$ or equivalently $Z \cap (z^0 - \operatorname{int} C) = \emptyset$.

Example 6.3.1. In $\mathbb{R}^2$ let us consider the ordering cone $C = \mathbb{R}^2_+$. For the set $Z = \{(z_1, z_2) : -1 \leqq z_2 \leqq 0,\ z_2 \geqq -z_1 - 1\}$ it is easy to verify that

$$IE(Z) = \emptyset \ ;$$
$$E(Z) = \{(z_1, z_2) : z_2 = -z_1 - 1,\ -1 \leqq z_2 \leqq 0\} \ ;$$
$$WE(Z) = E(Z) \cup \{(z_1, z_2) : z_2 = -1,\ z_1 \geqq 0\} \ .$$

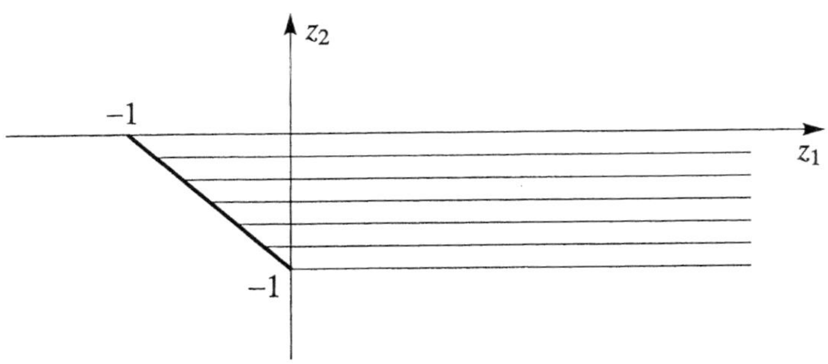

Figure 1.

Theorem 6.3.1. The following inclusions hold:

$$IE(Z) \subset E(Z) \subset WE(Z) \ .$$

Proof. We have already remarked that $IE(Z)$ is contained in $E(Z)$. So we only have to prove the second inclusion. Let $z^0 \in E(Z)$ or $Z \cap (z^0 - C) = \{z^0\}$. Then for $K = \{0\} \cup \operatorname{int} C$, we have $Z \cap (z^0 - K) \subset Z \cap (z^0 - C) = \{z^0\}$ which assures that $z^0 \in WE(Z)$. □

Theorem 6.3.2. If the set $IE(Z)$ is not empty, we have $IE(Z) = E(Z)$ and this set is reduced to a singleton.

Proof. We have to prove that $E(Z) \subset IE(Z)$. If $z \in E(Z)$ and $z^0 \in IE(Z)$, from $z \geqq z^0$ we deduce $z^0 \geqq z$ and $z = z^0$ for the anti-symmetric property. □

6.4. Proper Efficiency

The more restrictive notion of *proper efficiency* was first introduced by H.W. Kuhn and A.W. Tucker in their classical paper of 1951. In order to understand their aim we have to go back to our vector optimization problem:

$$\operatorname*{Min}_{x \in S} f(x) \qquad (V.P.)$$

with $f : X \subset \mathbb{R}^n \to \mathbb{R}^p$, X open and $S = \{x \in X \; : \; g_j(x) \leqq 0; \; j = 1, ..., m\}$. Kuhn and Tucker intended to get rid of those solutions that cannot be satisfactorily characterized by a scalar minimization problem (see the following Section 7) and generally to avoid some undesirable situations.

In Kuhn-Tucker's definition of proper efficiency the Paretian ordering is considered. $I(x^0)$ will denote the set of binding or active constraints at $x^0 \in S : I(x^0) = \{j : g_j(x^0) = 0\}$.

Definition 6.4.1. Let $f, g = (g_1, ..., g_m)$ be differentiable functions. A point $x^0 \in S$ is said to be *Kuhn-Tucker properly efficient* ($x^0 \in PE(S)_{KT}$) when it is efficient and there is no vector $y \in \mathbb{R}^n$ such that:

$$\begin{cases} Jf(x^0)\, y \leq 0 \\ Jg(x^0)\, y \leqq 0 \end{cases}$$

where J is the symbol of the Jacobian matrix and in particular Jg only concerns the functions g_k with $k \in I(x^0)$. We will also write that $z^0 = f(x^0) \in PE(Z)_{KT}$.

Kuhn-Tucker's definition leaves out those efficient points for which in a neighborhood of the same, along the directions y such that $Jg(x^0)\,y \leqq 0$, there are some points $x = x^0 + ty$ such that $\Delta f_i = f_i(x) - f_i(x^0) < 0$ for some i (being $\nabla f_i(x^0)\,y < 0$), even if they make positive those quantities Δf_j for which $\nabla f_j(x^0)\,y = 0$. According to Definition 6.4.1, x^0 is an *improper* efficient point when we are able to move along some "feasible" directions and to minorize some criteria with a marginal gain of a lower order and therefore incomparably higher than the marginal loss of the remaining criteria. The following Kuhn-Tucker example (with $p = 2$) shows a point x^0 which is efficient but not properly efficient (in the sense of Kuhn-Tucker); we can move from x^0 and make $\Delta f_1 < 0$, $\Delta f_2 > 0$, but the latter variation is a higher order infinitesimal quantity.

Example 6.4.1. Let $f(x) = (-x, x^2 - 2x)$, $f : I\!R \to I\!R^2$, and $g(x) = -x$. Any point $x \geqq 1$ is efficient for this $(V.P.)$ but $x^0 = 1$ does not satisfy Definition 6.4.1; indeed any $y > 0$ is a solution of the previous system.

Kuhn-Tucker's pioneer definition was followed by a number of other definitions of proper efficiency. In this section we wish to produce a rather large picture of this limitation to optimality notion, but with some guiding principle which will avoid a pure list of definitions. So we have gathered these definitions into two groups. In the first we have inserted those approaches which follow Kuhn-Tucker's definition or that in any case produce like features. The first slight change of Kuhn-Tucker's definition is due to [Klinger, 1967]. His definition does not require the differentiability of the function g.

Definition 6.4.2 Let f be a differentiable function. A point $x^0 \in S$ is said to be *Klinger properly efficient* ($x^0 \in PE(S)_K$ or $z^0 = f(x^0) \in PE(Z)_K$) when it is efficient and for no vector $y \in I\!R^n$ such that $Jf(x^0)\,y \leq 0$ there exists a sequence $\{t_k\} \subset I\!R_+$ converging to 0 with $g(x^0 + t_k y) \leqq 0$.

Remark 6.4.1. If $F(S, x^0)$ indicates the radial tangent cone to S at x^0, the above property can be expressed by saying that there is no vector $y \in F(S, x^0)$ such that $Jf(x^0)\, y \leq 0$.

Also the following Geoffrion's notion of proper efficiency is given when $C = I\!R^p_+$. This definition has a global character; the differentiability hypothesis is not even required for the objective function f.

Definition 6.4.3. A point $x^0 \in S$ is a *Geoffrion properly efficient point* ($x^0 \in PE(S)_G$ or $z^0 = f(x^0) \in PE(Z)_G$) when it is efficient and there is some real $M > 0$ such that for any i and for any $x \in S$ satisfying $f_i(x) < f_i(x^0)$ there exists at least one $j = j(i, x)$ with $f_i(x^0) - f_i(x) \leqq M \cdot [f_j(x) - f_j(x^0)]$.

Then for Geoffrion a point x^0 is properly efficient when it minimizes all the objective functions or there exists another point x that minorizes some criterion f_i but with a gain $f_i(x^0) - f_i(x)$ which is controlled as it is above bounded from $-M\Delta f_j$ (for some j and some $M > 0$). So Definition 6.4.3 eliminates unbounded trade-offs between the objectives. In other words, x^0 is a properly Geoffrion efficient point when it is an ideal point or there exists at least a pair of criteria, say f_i and f_j, for which a finite improvement of the first objective is possible only at the expense of some reasonable degradation of the second.

Remark 6.4.2. Geoffrion's definition was generalized in [Singh-Hanson, 1991]. A point x^0 is said to be *conditionally* properly efficient when it satisfies a modified Definition 6.4.3 in which $M = M(x)$ depends on x. The concept of conditional proper efficiency has been used by [Mishra-Mukherjee, 1995] to develop duality results.

Example 6.4.2. Let us consider the following $(V.P.)$: $f(x) = (x^3, -x^2)$, $f : I\!R \to I\!R^2$, and $g(x) = -x$. The origin is not Geoffrion properly efficient because there is no $M > 0$ such that $x^2 \leqq Mx^3$, $\forall x \geqq 0$. (On the contrary we can notice that $x^0 = 0$ satisfies Definitions 6.4.1 and 6.4.2.)

Geoffrion's definition was generalized by R. Hartley to any ordering cone C. In both definitions the explicit request that x^0 be an efficient point is indeed superfluous. With Hartley's definition we renounce the explicit reference to $(V.P.)$. Let us again be concerned with efficient points of any set $Z \subset \mathbb{R}^p$. We recall that $C_{\geqq}$ is the *nonnegative polar cone* of C: $C_{\geqq} = \{u : u{\cdot}y \geqq 0,\ \forall y \in C\}$; later we will use the symbol $C_{>}$ in order to denote the *strict positive polar* of C: $C_{>} = \{u : u \cdot y > 0,\ \forall y \in C \backslash \{0\}\}$.

Definition 6.4.4. A value $z^0 \in Z$ is said to be a *Hartley properly efficient* ($z^0 \in PE(Z)_{Ha}$) when it is efficient and there exists $M > 0$ such that for any $z \in Z$ and $\lambda \in C_{\geqq}$, with $\lambda(z - z^0) < 0$, there exists $\mu \in C_{\geqq}$ such that $\dfrac{\lambda}{\|\lambda\|}(z^0 - z) \leqq M \dfrac{\mu}{\|\mu\|}(z - z^0)$.

The second group of proper efficiency definitions follows a geometrical approach. It goes on focusing the image set Z without any more reference to the objective function f and to its components. The ordering cone is here a general cone. The first definition of this group is due to Hurwicz.

Definition 6.4.5. A value $z^0 \in Z$ is said to be a *Hurwicz properly efficient point* ($z^0 \in PE(Z)_{Hu}$) when it is efficient and:

$$\operatorname{cl}\operatorname{conv}\operatorname{cone}\left[(Z - z^0) \cup C\right] \cap (-C) = \{0\} .$$

Example 6.4.3. If we refer Hurwicz's definition to the particular componentwise order, we obtain that the minimal closed convex cone containing $(Z - z^0) \cup \mathbb{R}^p_+$ (i.e. its closed convex conical hull) cannot interesect $\mathbb{R}^p_- \backslash \{0\}$. This condition is not satisfied e.g. in the following figure (when the cone generated by $(Z - z^0) \cup \mathbb{R}^2_+$ is not convex) since $\operatorname{cl}\operatorname{conv}\operatorname{cone}\left[(Z - z^0) \cup \mathbb{R}^2_+\right] = \mathbb{R}^2$. Thus $z^0 \notin PE(Z)_{Hu}$ but $z^0 \in E(Z)$.

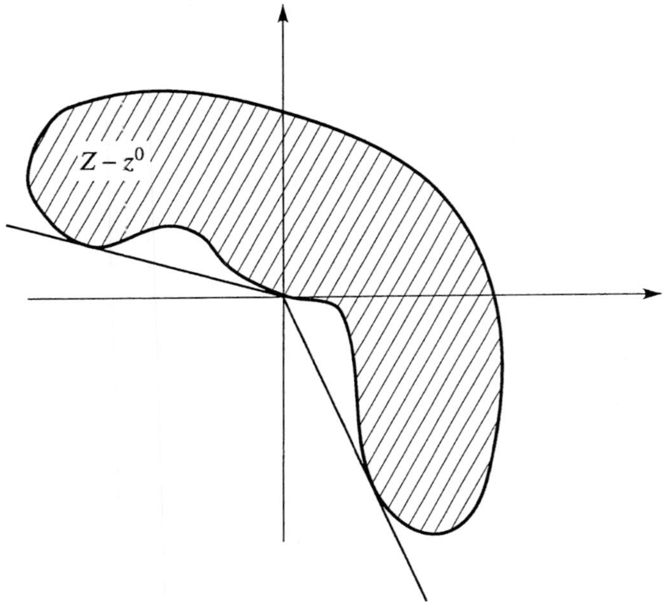

Figure 2.

The geometrical approach to Hurwicz's definition was developed by [Benson, 1979], [Borwein, 1977] and [Henig, 1982]. Roughly speaking, all these geometrical definitions call z^0 a properly efficient value when the sets $Z - z^0$ or $Z + C - z^0$ or $(Z - z^0) \cup C$ are "well" separated from $-C\backslash\{0\}$ (and not only generically disjoint, as happened for the efficient values). Namely, a cone containing the previous sets or a local conical approximation of these sets cannot intersect $-C\backslash\{0\}$.

Definition 6.4.6. A value $z^0 \in Z$ is said to be a *Benson properly efficient point* ($z^0 \in PE(Z)_{Be}$) when it is efficient and:

$$\text{cl cone}\,[Z + C - z^0] \cap (-C) = \{0\}\ .$$

Definition 6.4.7.

i) A value $z^0 \in Z$ is said to be a *Borwein properly efficient* ($z^0 \in PE(Z)_{Bo}$) when it is efficient and

$$T(Z+C,z^0)\cap(-C)=\{0\}$$

where T denotes the Bouligand tangent cone;

ii) $z^0 \in Z$ is said to be a *global Borwein properly efficient* value ($z^0 \in PE(Z)_{GBo}$) when it is efficient and:

$$\mathrm{cl\,cone}\,[Z-z^0]\cap(-C)=\{0\}\ ;$$

iii) $z^0 \in Z$ is said to be a *local Borwein properly efficient* value ($z^0 \in PE(Z)_{LBo}$) when it is efficient and:

$$T\,[Z,z^0]\cap(-C)=\{0\}\ .$$

Definition 6.4.8. A value $z^0 \in Z$ is said to be a *Henig properly efficient* ($z^0 \in PE(Z)_{He}$) when $z^0 \in E_{C'}(Z)$ where C' is a (closed, pointed and convex) cone with $C\backslash\{0\} \subset \mathrm{int}(C')$.

Remark 6.4.3. In Definitions 6.4.5, 6.4.6 and 6.4.7 ii), the requirement that $z^0 \in E(Z)$ is superfluous as we have $Z-z^0 \subset \mathrm{cl\,conv\,cone}\,[(Z-z^0)\cup C]$, $Z-z^0 \subset \mathrm{cl\,cone}\,[Z+C-z^0]$ and $Z-z^0 \subset \mathrm{cl\,cone}\,[Z-z^0]$.
The same requ rement is superfluous in Definition 6.4.7 i). Indeed let us suppose that there exists a point $z \in Z$ such that $z-z^0 \in -C\backslash\{0\}$. Let $z^k = z-\lambda_k(z-z^0) \in Z+C$ with $0<\lambda_k<1$ and $\lim\limits_{k\to+\infty}\lambda_k = 1$. Then $\lim\limits_{k\to+\infty} z^k = z^0$. But $\lim\limits_{k\to+\infty}\dfrac{1}{1-\lambda_k}(z^k-z^0) = \lim\limits_{k\to+\infty}\dfrac{1}{1-\lambda_k}[(1-\lambda_k)\,z-(1-\lambda_k)\,z^0] = z-z^0 \in T(Z+C,z^0)\cap(-C\backslash\{0\})$, while we supposed $z^0 \in PE(Z)_{B_C}$.
Again the requirement that z^0 be efficient is superfluous in Definition 6.4.7 iii) when Z is convex. Indeed, in this case, we have $z-z^0 \in T(Z,z^0)$. Then by the proper efficiency we get $z-z^0 \notin -C\backslash\{0\}$ or $z^0 \in E(Z)$.

Remark 6.4.4. A value z^0 satisfying Definition 6.4.8 is called *global* by [Henig, 1982a] in order to distinguish it from a *local* version of the same definition.

By means of Definition 6.4.7 iii) we can give an interpretation of weakly minimal values in terms of proper efficiency.

Theorem 6.4.1. Let $K = \operatorname{int} C \cup \{0\}$. Then $WE_C(Z) = PE_K(Z)_{LBo}$.

Proof. From $WE_C(Z) = E_K(Z)$ it suffices to show that $E_K(Z) \subset PE_K(Z)_{LBo}$. Let $z^0 \in E_K(Z) \setminus PE_K(Z)_{LBo}$. Then $(-\operatorname{int} C) \cap T(Z, z^0) \neq \emptyset$ or there exists $y \in \operatorname{int} C$ such that $y = \lim_{k \to +\infty} \lambda_k (z^0 - z^k)$ with $\lambda_k > 0$, $z^k \in Z$ and $\lim_{k \to +\infty} z^k = z^0$. It follows that there exists an integer N such that, $\forall\, k \geq N$, $\lambda_k (z^0 - z^k) \in K \setminus \{0\}$ or $z^0 - z^k \in K \setminus \{0\}$. By definition $z^0 \notin E_K(Z)$, which is a contradiction. □

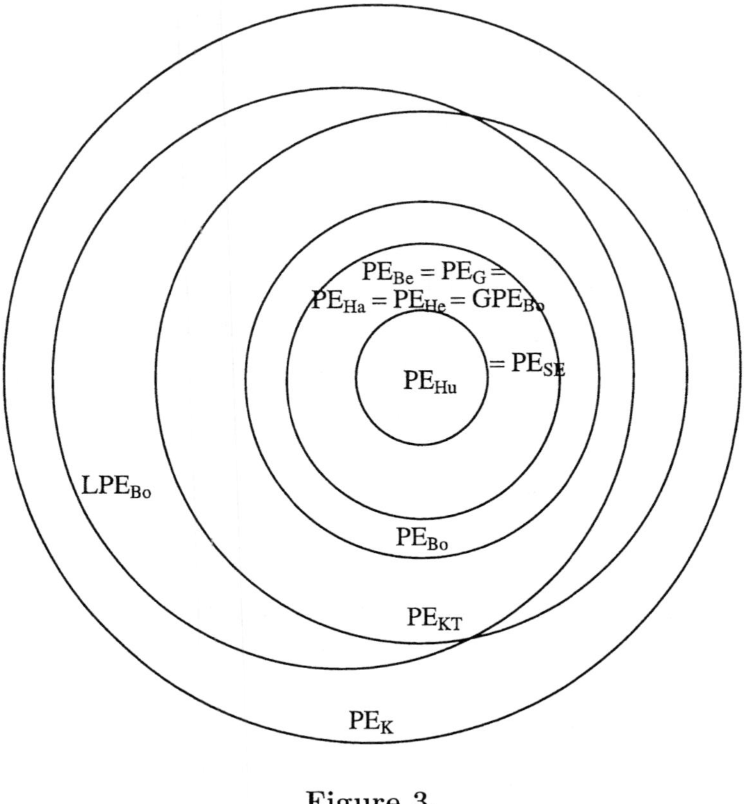

Figure 3.

The last definition of proper efficiency that we wish to quote was more recently introduced by J.M. Borwein and D. Zhuang.

Definition 6.4.9. A value $z^0 \in Z$ is said to be *superefficient* ($z^0 \in PE(Z)_{SE}$) when there exists $m > 0$ such that $\mathrm{cl}\ \mathrm{cone}(Z - z^0) \cap (B - C) \subset mB$ where B is the closed unit ball of $\mathbb{R}^p$.

The above figure draws the inclusion relations among the previous definitions proven by Theorem 6.4.2. The shortened symbols used in this picture are obvious. A few cautionary remarks need to be made. Quoted definitions were stated as they appeared in the reference; slight variations were made only on the symbols. But now in order to compare the different notions of proper efficiency, the ordering cone in Theorem 6.4.2 will always be given by $\mathbb{R}^p_+$ when we will involve the Definitions of Kuhn-Tucker, Klinger and Geoffrion. The inclusion relations do not need any additional hypothesis. Only two exceptions are necessary. When we treat Kuhn-Tucker and Klinger properly efficient points, these definitions require implicitly some differentiability conditions. Moreover, the statement that $PE(Z)_{LBo} \subset PE(Z)_{KT}$ holds under a suitable constraint qualification hypothesis; thanks to Abadie's constraint qualification we may claim a coherence between a geometrical and an analytical description of the proper efficiency. Kuhn-Tucker's definition, later improved by Klinger, turns out the most general.

The proof b$_5$) uses the important notion of a base for a cone.

Definition 6.4.10. A set Λ is said to be a *base* for the cone C when any $y \in C \backslash \{0\}$ can be written as $y = t\lambda$ (with unique $\lambda \in \Lambda$, $t > 0$) and $0 \notin \Lambda$.

In finite-dimensional spaces, a convex cone C has a convex compact base if and only if it is closed and pointed. For $C = \mathbb{R}^p_+$ a trivial base is given by $\Lambda = \mathrm{conv}\,\{e^1, ..., e^p\}$, where $\{e^k\}$ is the canonical basis of $\mathbb{R}^p$.

<u>Theorem 6.4.2.</u>

a) $PE(Z)_{Hu} \subset PE(Z)_{GBo}$;

b) $PE(Z)_{Ha} = PE(Z)_G = PE(Z)_{Be} = PE(Z)_{GBo} = PE(Z)_{He} = PE(Z)_{SE}$;

c) $PE(Z)_{Be} \subset PE(Z)_{Bo}$;

d) $PE(Z)_{Bo} \subset PE(Z)_{LBo}$;

e) $PE(Z)_{LBo} \subset PZ(Z)_{KT}$ if f and g_i $(i \in I(x^0))$ are differentiable and the Abadie constraint qualification holds: $T(S, x^0) = C(x^0) = \{y : y\nabla g_i(x^0) \leqq 0,\ i \in I(x^0)\}$;

f) $PE(Z)_{KT} \subset PE(Z)_K$ if f and g_i $(i \in I(x^0))$ are supposed differentiable.

<u>Proof.</u>

a) The statement immediately follows from these inclusions: $\mathrm{cl\,cone}(Z - z^0) \cap (-C) \subset \mathrm{cl\,conv\,cone}\,[(Z - z^0) \cup C] \cap (-C) = \{0\}$.

b$_1$) $PE(Z)_{Ha} \subset PE(Z)_G$. Let $z^0 \in PE(Z)_{Ha}$ and $z \in Z$ such that $z_i = f_i(x) < f_i(x^0) = z_i^0$ or $e^i(z - z^0) < 0$ where $\{e^k\}$ is the canonical basis of $I\!R^p$. Hartley's definition assures the existence of $\mu \geqq 0$ ($\|\mu\| = 1$) such that:

$$\begin{aligned} e^i(z^0 - z) &\leqq M\mu(z - z^0) = M \sum_{j=1}^{p} \mu_j(z_j - z_j^0) \leqq \\ &\leqq M \sum_{j \in J^+} \mu_j(z_j - z_j^0) \leqq M \sum_{j \in J^+} (z_j - z_j^0) \leqq \\ &\leqq M\,p(z_{jo} - z_{jo}^0) \end{aligned}$$

where J^+ is the (nonempty) index set for which $z_j - z_j^0 > 0$ and $z_{jo} - z_{jo}^0 = \max_{j \in J^+} (z_j - z_j^0)$. We so have proven that $z^0 \in PE(Z)_G$.

$PE(Z)_G \subset PE(Z)_{Ha}$. Let $f(x^0) = z^0 \in PE(Z)_G$, $z \in Z$ and $\lambda \geqq 0$ with $\|\lambda\| = 1$ and $\lambda(z - z^0) < 0$. If, for the i-th coordinate, we have $z_i < z_i^0$, we can find an index $j = j(i, z^0)$ such that $(z_i^0 - z_i) \leqq M(z_j - z_j^0)$. In case of $z_i \geqq z_i^0$, we consider $j = i$. Multiplying by $\lambda_i \geqq 0$ and summing up, we get:

$$\begin{aligned}\lambda(z^0 - z) = \sum_{i=1}^{p} \lambda_i(z_i^0 - z_i) &\leqq \\ \leqq \sum_{i=1}^{p} \lambda_i\, M(z_{j(i,z^0)} - z^0_{j(i,z^0)}) &= M\mu(z - z^0)\end{aligned}$$

where $\mu \geqq 0$ and we can always consider $\|\mu\| = 1$. Hence $z^0 \in PE(Z)_{Ha}$.

b$_2$) $PE(Z)_G \subset PE(Z)_{Be}$. By contradiction suppose $z^0 \in PE(Z)_G$ and $z^0 \notin PE(Z)_{Be}$. Then $\mathrm{cl\,cone}(Z + I\!R^p_+ - z^0) \cap I\!R^p_- \neq \{0\}$ or there exists $y \in I\!R^p_-$ with $y = \lim_{k \to +\infty} \lambda_k(z^k + c^k - z^0) \neq 0$ ($z^k \in Z$, $c^k \in I\!R^p_+$, $\lambda_k \geqq 0$).
Without loss of generality we can always suppose $y = (y_1, ..., y_p)$ with $y_1 = -1$. So from $z^k - z^0 \leqq \dfrac{y + o(1)}{\lambda_k}$, for any k large enough we can deduce $z_1^k - z_1^0 < -\dfrac{1}{2\lambda_k}$ and $z_j^k - z_j^0 \leqq \dfrac{1}{2M\lambda_k}$ (any $M > 0$; $j = 2, ..., p$). We have $J_k = \{j : z_j^k > z_j^0\} \neq 0$ because z^0 is an efficient value. Then, $\forall j \in J_k$, we obtain $0 \leqq z_j^k - z_j^0 \leqq \dfrac{1}{2M\lambda_k}$ and $\dfrac{z_1^0 - z_1^k}{z_j^k - z_j^0} > \dfrac{1/2\lambda_k}{1/2M\lambda_k} = M$. This conclusion contradicts $z^0 \in PE(Z)_G$ since this is true for each $M > 0$.

b$_3$) $PE(Z)_{Be} \subset PE(Z)_{GBo}$. It easily follows:

$$\mathrm{cl\,cone}(Z - z^0) \cap (-C) \subset \mathrm{cl\,cone}(Z + C - z^0) \cap (-C) = \{0\}\ .$$

b$_4$) $PE(Z)_{GBo} \subset PE(Z)_{He}$. If $z^0 \in PE(Z)_{GBo}$ we have $\mathrm{cl\,cone}(Z - z^0) \cap (-C) = \{0\}$. From a well-known theorem of separation between cones (see Lemma 6.5.5) it follows that there exists a (closed) convex and pointed cone C' such that $(-C)\backslash\{0\} \subset \mathrm{int}(-C')$ and

$\text{cl cone}(Z - z^0) \cap (-C') = \{0\}$. All the more reason for having $(Z - z^0) \cap (-C') = \{0\}$ or $z^0 \in PE(Z)_{He}$.

b$_5$) $PE(Z)_{He} \subset PE(Z)_{SE}$. If $z^0 \in PE(Z)_{He}$ and C' is the cone of Definition 6.4.8, there exists $\varepsilon > 0$ such that $(Z - z^0) \cap \text{cl cone}(-\Lambda + \varepsilon B) = \{0\}$ where B is the closed unit ball and Λ is a base for C. So we are able to deduce $\text{cone}\,(Z - z^0) \cap (-\Lambda + \varepsilon B) = \emptyset$.
Now for any fixed $z \in Z$, let us choose any $\sigma \in C + (z - z^0)$. From $z - z^0 = \sigma - t\lambda$ for some $t \geqq 0$ and some $\lambda \in \Lambda$, we deduce $\|z - z^0\| = \|\sigma\|$ when $t = 0$. For $t > 0$ we have $\dfrac{\|\sigma\|}{t} \geqq \varepsilon$; indeed $\dfrac{\|\sigma\|}{t} < \varepsilon$, with $\dfrac{z - z^0}{t} = \dfrac{\sigma}{t} - \lambda$, would imply $\text{cone}(Z - z^0) \cap (\varepsilon B - \Lambda) \neq \emptyset$.
Let us set $M = \sup_{\lambda \in \Lambda} \|\lambda\| < +\infty$:

$$\|z - z^0\| = \|\sigma - t\lambda\| \leqq \|\sigma\| + tM \leqq \|\sigma\| \left(1 + \frac{M}{\varepsilon}\right) \leqq m \|\sigma\| .$$

From this, for any $z \in Z$ and for any $\sigma \in C + (z - z^0)$, we obtain $\text{cone}(Z - z^0) \cap (B - C) \subset mB$. Indeed for any $\omega = t(z - z^0) = b - y$ $(t > 0,\ z \in Z,\ b \in B,\ y \in C)$ one has $\dfrac{1}{t} b \in C + (z - z^0)$ and hence $\|z - z^0\| \leqq m \left\| \dfrac{1}{t} b \right\|$ or $\|t(z - z^0)\| \leqq m$.

Lastly, $\text{cl cone}(Z - z^0) \cap (B - C) \subset mB$. By contradiction let us suppose $w \in \text{cl cone}(Z - z^0) \cap (B - C)$ with $\|w\| > m$. From $w = \omega_k + o(1) = b - y$ $(\omega_k \in \text{cone}(Z - z^0),\ b \in B,\ y \in C)$ it follows that $\omega_k \in B - C + \dfrac{1}{k} B = \dfrac{k+1}{k} B - C$ or $\dfrac{k}{k+1} \omega_k \in B - C$. So we would obtain $\dfrac{k}{k+1} \omega_k \in (B - C) \cap \text{cone}(Z - z^0)$ or $\left\| \dfrac{k}{k+1} \omega_k \right\| \leqq m$ but $\lim_{k \to +\infty} \dfrac{k}{k+1} \omega_k = w$ with $\|w\| > m$.

b$_6$) $PE(Z)_{SE} \subset PE(Z)_{Ha}$. By contradiction let us suppose that $z^0 \in PE(Z)_{SE} \backslash PE(Z)_{Ha}$. Thus $\forall \varepsilon > 0$ there exists $\lambda \in C^*_{\geqq}$ with $\|\lambda\| = 1$ and $\lambda(z - z^0) < 0$ for some $z \in Z$ such that $\lambda(z - z^0) < -(m + \varepsilon)\, \mu(z - z^0)$ for any $\mu \in C^*_{\geqq}$, $\|\mu\| = 1$ (where m is the positive number of Definition 6.4.9). From this inequality we can deduce the following one:

$$\frac{\mu(z-z^0)}{\|z-z^0\|} < \frac{1}{m+\varepsilon} \ .$$

Then we have:

$$z - z^0 \in \frac{\|z-z^0\|}{m+\varepsilon}\,(B-C) \subset \frac{\|z-z^0\|}{m+\frac{\varepsilon}{2}}\,(B-C) \ .$$

Indeed, if $\dfrac{z-z^0}{\|z-z^0\|} \notin \dfrac{1}{m+\varepsilon}\,(B-C)$, there exists μ with $\|\mu\| = 1$ such that $\mu \cdot \dfrac{z-z^0}{\|z-z^0\|} > \dfrac{1}{m+\varepsilon}\,\mu\cdot(b-y)$ for any $b \in B$ and for any $y \in C$. It is easy to verify $\mu \in C_{\geqq}$. As the previous inequality holds $\forall\, b \in B$ and $\forall\, y \in C$ we get the contradiction $\mu \cdot \dfrac{z-z^0}{\|z-z^0\|} > \dfrac{1}{m+\varepsilon}$. Therefore we obtain that:

$$\frac{m+\varepsilon/2}{\|z^0-z\|}\,(z-z^0) \in (B-C) \cap \operatorname{cone}(Z-z^0) \ .$$

Then, as $z^0 \in PE(Z)_{SE}$, we have

$$\left\| \frac{m+\varepsilon/2}{\|z-z^0\|}\,(z-z^0) \right\| \leqq m$$

and hence the absurdum:

$$\|z-z^0\| \leqq m \cdot \frac{\|z-z^0\|}{m+\frac{\varepsilon}{2}} < \|z-z^0\| \ .$$

c) $PE(Z)_{Be} \subset PE(Z)_{Bo}$. This statement is true as:

$$T(Z+C, z^0) \cap (-C) \subset \operatorname{cl}\operatorname{cone}(Z+C-z^0) \cap (-C) = \{0\} \ .$$

d) $PE(Z)_{Eo} \subset PE(Z)_{LBo}$. Also this inclusion follows readily from the inclusion:

$$T(Z, z^0) \cap (-C) \subset T(Z + C, z^0) \cap (-C) .$$

e) $PE(Z)_{LBo} \subset PE(Z)_{KT}$. Since:

$$[Jf(x^0) \cdot C(x^0)] \cap \mathbb{R}^p_- =$$
$$= [Jf(x^0) \cdot T(S, x^0)] \cap \mathbb{R}^p_- \subset T(Z, z^0) \cap \mathbb{R}^p_- = \{0\}$$

it follows immediately that there is no $y \in \mathbb{R}^n$ with $\nabla g_j(x^0)\, y \leqq 0$ ($j \in I(x^0)$) and $Jf(x^0)\, y \leq 0$. So we have that $z^0 \in PE(Z)_{KT}$.

f) $PE(Z)_{KT} \subset PE(Z)_K$. By contradiction suppose $z^0 \in PE(Z)_{KT} \setminus PE(Z)_K$. Then there exists a vector $y \in \mathbb{R}^n$ such that $Jf(x^0)\, y \leq 0$ and $g(x^0 + t_k y) \leqq 0$ for some sequence $\{t_k\} \subset \mathbb{R}_+$ converging to 0. Then we would have $g_j(x^0 + t_k y) - g_j(x^0) \leq 0$ for any $j \in I(x^0)$ and hence $\nabla g_j(x^0)\, y \leqq 0$. This conclusion contradicts the hypothesis $z^0 \in PE(Z)_{KT}$. □

Remark 6.4.6. The comparison between the different definitions of properly efficient points in infinite dimension spaces can be found in [Guerraggio-Molho-Zaffaroni, 1993].

Remark 6.4.7. Without requiring Abadie constraint qualification, we are still able to prove the inclusion $PE(Z)_{LBo} \subset PE(Z)_K$. Indeed from $T(Z, z^0) \cap \mathbb{R}^p_- = \{0\}$ we obtain

$$Jf(x^0) \cdot T(S, x^0) \cap \mathbb{R}^p_- = \{0\}$$

and

$$Jf(x^0)\, F(S, x^0) \cap \mathbb{R}^p_- = \{0\}$$

as the radial tangent cone to S at x^0 is contained in Bouligand tangent cone. Then $z^0 \in PE(Z)_K$ (see Remark 6.4.1).

Now we provide some examples that testify that the inclusions of Theorem 6.4.2 are strict. Afterwards we will prove some sufficient conditions that assure the equivalence between some of the previous definitions of proper efficiency.

Example 6.4.4. In Example 6.4.3 Hurwicz's definition was not satisfied. The same figure shows that $z^0 \in PE(Z)_{Be}$.

Example 6.4.5. In the following figure, Definition 6.4.7 i) is satisfied by $z^0 = 0$ while $z^0 = 0 \notin PE(Z)_{Be}$ as $\mathrm{cl\,cone}(Z + I\!R^2_+) \cap I\!R^2_- = \{(z_1, z_2) : z_1 \leqq 0,\ z_2 = 0\} \cup \{(z_1, z_2) : z_1 = 0,\ z_2 \leqq 0\}$.

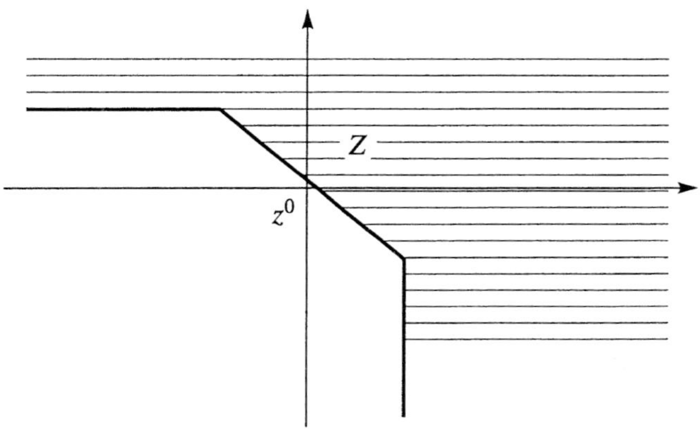

Figure 4.

Example 6.4.6. Let $Z = \{(z_1, z_2) : z_2 \geqq -z_1 e^{z_1}\}$. For $(z_1, z_2) = (0, 0)$ it is easy to check that Definition 6.4.7 iii) is satisfied but $(0, 0) \notin PE(Z)_{Bo}$ since $T(Z + I\!R^2_+, 0) \cap I\!R^2_- = \{(z_1, z_2) : z_1 \leqq 0,\ z_2 = 0\}$.

Example 6.4.7. Now let us consider the following $(V.P.)$:

$$f(x) = (f_1(x), f_2(x)) = ((1 - x^2)^3, (1 - x^2)^6)$$

with $g(x) = 1 - x \leqq 0$. The point $x^0 = 1$ is Kuhn-Tucker properly efficient as $f_1'(1) = f_2'(1) = 0$ but $z^0 = 0 \notin PE(Z)_{LBo}$ since $Z = \{(z_1, z_2) : z_1 \leqq 0,\ z_2 = z_1^2\}$ and $T(Z, 0) = \{(z_1, z_2) : z_1 \leqq 0,\ z_2 = 0\}$.

Example 6.4.8. The point $x^0 = (1, 0)$ is Klinger properly efficient for $(V.P.)$ with

$$f(x) = (f_1(x_1, x_2), f_2(x_1, x_2)) = (-x_1, -x_2)$$

and

$$g(x) = (g_1(x_1, x_2), g_2(x_1, x_2), g_3(x_1, x_2)) = (x_1^2 + x_2^2 - 1, -x_1, -x_2) .$$

But $x^0 \notin PE(S)_{KT}$.

Generally sufficient conditions for ensuring the equality between some of the properly efficient solution sets require some convexity conditions. So we recall a few convexity definitions for vector functions that generalize well-known notions in the scalar case. This generalization is quite natural when one recalls that $x - y \in C$ means $x \geqq y$.

<u>Definition 6.4.11</u>.

a) A function $f : X \subset I\!R^n \to I\!R^p$, X being a convex set, is said to be *C-convex* in X when

$$tf(x^1) + (1-t)\, f(x^2) \in C + f(tx^1 + (1-t)\, x^2)$$

$\forall t \in [0,1]$ and $\forall x^1, x^2 \in X$.

b) A function $f : X \subset I\!R^n \to I\!R^p$ is said to be *C-pre-invex* when there exists a function $\eta : X \times X \to X$ such that $\forall x^1, x^2 \in X$ and $\forall t \in [0,1]$, $x^2 + t\eta(x^1, x^2) \in X$:

$$tf(x^1) + (1-t)\, f(x^2) \in C + f(x^2 + t\eta(x^1, x^2)) .$$

c) A differentiable function $f : X \subset I\!R^n \to I\!R^p$ is said to be *C-invex* when there exists a function $\eta : X \times X \to X$ such that, $\forall x^1, x^2 \in X$:

$$f(x^2) - f(x^1) \in C + Jf(x^1)\, \eta(x^1, x^2) ,$$

Jf being the Jacobian matrix of f.

d) A differentiable function $f : X \subset I\!R^n \to I\!R^p$ is said to be *C-pseudoconvex* when, $\forall x^1, x^2 \in X$, $f(x^2) - f(x^1) \in -C\backslash\{0\}$ implies $Jf(x^1)(x^2 - x^1) \in -\text{int}\, C$.

e) A differentiable function $f : X \subset \mathbb{R}^n \to \mathbb{R}^p$ is said to be *C-quasiconvex* when, $\forall x^1, x^2 \in X$, $f(x^2) - f(x^1) \in -C$ implies $Jf(x^1) \cdot (x^2 - x^1) \in -C$.

The concavity notions can be easily obtained. From the previous definitions, we have the following implications: a) $\Rightarrow$ b) $\Rightarrow$ c) when f is a differentiable function. When $\mathbb{R}^p$ is componentwise ordered, a vector function works out to be $\mathbb{R}^p_+$-convex or $\mathbb{R}^p_+$-pre-invex or $\mathbb{R}^p_+$-invex if and only if all its components are convex or pre-invex or invex. This claim is not true for $\mathbb{R}^p_+$-pseudoconvex (and $\mathbb{R}^p_+$-quasiconvex) functions: the given definitions 6.4.11 d) and e) are more general than the ones obtained by requiring that all components of f are pseudoconvex (or quasiconvex). In the following we will especially take into account the C-pseudoconvex (or C-quasiconvex) functions at $x^1 \in X$ (x^1 being a fixed point). It is easy to show that the function $f : \mathbb{R}^2 \to \mathbb{R}^3$, $f = (x_1^2 + x_2^2, -x_1^2 - x_2^2, -x_1 + x_2)$ is $\mathbb{R}^3_+$-pseudoconvex at $(0, 0)$ but its second component is not pseudoconvex at the same point.

Definition 6.4.12. A set $Z \subset \mathbb{R}^p$ is said to be *C-convex* when $Z + C$ is a convex set.

It is easy to check that C-convexity is a more general notion than convexity and is preserved under addition and multiplication by a positive scalar. When $Z = f(S)$ is the image set of S under the function f, one can verify that Z is C-convex in X if f is a C-convex function. But the class of functions with this property is larger.

Definition 6.4.13. A function $f : X \subset \mathbb{R}^n \to \mathbb{R}^p$ is said to be *C-convexlike* when, $\forall x^1, x^2 \in X$ and $\forall t \in [0, 1]$, $\exists \bar{x} \in X$ such that $tf(x^1) + (1 - t) f(x^2) \in C + f(\bar{x})$.

Theorem 6.4.3. A function $f : X \subset \mathbb{R}^n \to \mathbb{R}^p$ is C-convexlike if and only if $f(X)$ is a C-convex set.

Proof. It is immediate. □

Example 6.4.9. Every function $f : \mathbb{R} \to \mathbb{R}$, $C = \mathbb{R}_+$ is C-convexlike even if it is not C-convex.

Definition 6.4.13 was later generalized by [Li-Wang, 1994] in order to give a sufficient condition for the equivalence of the Benson and Borwein proper efficiency.

Definition 6.4.14. A function $f : X \subset \mathbb{R}^n \to \mathbb{R}^p$ is said to be *C-subconvexlike* when $\forall c \in \operatorname{int} C$, $\forall x^1, x^2 \in X$ and $\forall t \in [0,1]$, $\exists \bar{x} \in X$ such that $c + tf(x^1) + (1-t)\, f(x^2) \in \operatorname{int} C + f(\bar{x})$.

In [Jeyakumar, 1986] one can find an example of a C-subconvexlike function which is not C-convexlike.

Theorem 6.4.4. If the function f is C-convexlike or the set Z is C-convex, we have $PE(Z)_{Bo} = PE(Z)_{Hu}$.

Proof. Let $z^0 \in PE(Z)_{Bo}$. Then $z^0 \in PE(Z)_{Be}$, too. Since $Z+C$ is convex, $\operatorname{cl}\operatorname{cone}(Z+C-z^0) = T(Z+C, z^0)$. We have $(Z-z^0) \cup C \subset Z+C-z^0$ and hence $\operatorname{cone}[(Z-z^0) \cup C] \subset \operatorname{cone}(Z+C-z^0)$. Since the sets $Z+C$ and $Z+C-z^0$ are convex, also $\operatorname{cone}(Z+C-z^0)$ is convex. So we obtain that $\operatorname{cl}\operatorname{conv}\operatorname{cone}[(Z-z^0) \cup C] \subset \operatorname{cl}\operatorname{cone}(Z+C-z^0)$ which guarantees that $z^0 \in PE(Z)_{Hu}$. □

Theorem 6.4.5. If the function f is C-subconvexlike, we have $PE(Z)_{Be} = PE(Z)_{Bo}$.

Proof. We only have to prove that $\operatorname{cl}\operatorname{cone}[Z+C-z^0] \subset T(Z+C, z^0)$ or $\operatorname{cone}[Z+C-z^0] \subset T(Z+C, z^0)$ as Bouligand tangent cone is closed. Let $z \in \operatorname{cone}[Z+C-z^0]$. There exists $t \geqq 0$, $x \in S$ and $c \in C$ such that $z = t(f(x) + c - f(x^0))$. By the C-subconvexlikeness of f, for any $\bar{c}_k = \dfrac{\bar{c}}{k^2}$ $(\bar{c} \in \operatorname{int} C,\ k = 1, 2, \ldots)$ there exists $x_k \in X$ satisfying

$$\bar{c}_k + \frac{1}{k} f(x) + \left(1 - \frac{1}{k}\right) f(x^0) \in \operatorname{int} C + f(x^k)\ .$$

Also, for k large enough,

$$c_k = \frac{1}{k}\,c + \bar{c}_k + \frac{1}{k}\,f(x) + \left(1 - \frac{1}{k}\right) f(x^0) - f(x^k) \in \operatorname{int} C \ .$$

Then $z \in T(Z + C, z^0)$ as $z = \lim_{k \to +\infty} \tau_k(f(x^k) + c_k - f(x^0))$ with $\tau_k = tk \geqq 0$. □

We can improve Theorem 6.4.4 and arrive at the equality between $PE(Z)_{KT}$ (and all the intermediate sets) and $PE(Z)_{Hu}$ under convexity assumptions which, however, are not comparable with the hypothesis of the previous theorem. Of course involving Definition 6.4.1 necessarily implies the choice of the particular ordering cone $\mathbb{R}^p_+$. Moreover the equality $PE(Z)_{KT} = PE(Z)_{Hu}$ will suppose the inclusion $PE(Z)_{Hu} \subset PE(Z)_{KT}$ which was proved under some assumptions, namely the differentiability of the functions f and g_j $(j \in I(x^0))$ and the Abadie constraint qualification.

<u>Theorem 6.4.6</u>. If the functions f and g are $\mathbb{R}^p_+$-invex with respect to a same function and the Abadie constraint qualification holds, we have $PE(Z)_{KT} = PE(Z)_{Hu}$.

<u>Proof</u>. If $z^0 = f(x^0) \in PE(Z)_{KT}$, then none of the p systems

$$\begin{cases} \nabla f_i(x^0)\, y < 0 \\ \nabla f_j(x^0)\, y \leqq 0 \qquad (j \neq i) \\ Jg(x^0)\, y \leqq 0 \end{cases}$$

(where Jg only concerns the functions g_j with $j \in I(x^0)$) has a solution $y \in \mathbb{R}^n$. By applying the theorem of alternative of Motzkin to each, we get some numbers $u^i_j \geqq 0$ and $v^i_j \geqq 0$ such that

$$\nabla f_i(x^0) + \sum_{j \neq i} u^i_j \nabla f_j(x^0) + \sum_{j \in I(x^0)} v^i_j \nabla g_j(x^0) = 0 \ .$$

Summing over i yields

$$\sum_{j=1}^{p} \Big(1 + \sum_{i \neq j} u^i_j\Big) \nabla f_j(x^0) + \sum_{j \in I(x^0)} \Big(\sum_{i=1}^{p} v^i_j\Big) \nabla g_j(x^0) = 0$$

or

$$\nabla(\lambda f(x^0) + \mu g(x^0)) = 0$$

where $\lambda_i = 1 + \sum_{i \neq j} u_j^i$ and $\mu_j = \sum_{i=1}^{p} v_j^i$ for $j \in I(x^0)$ and $\mu_j = 0$ for $j \notin I(x^0)$. Since the function $L(x) = \lambda f(x) + \mu g(x)$ is invex with respect to η, it achieves an unconstrained minimum at x^0; moreover, it is $\lambda > 0$, $\mu \geqq 0$ and $\mu g(x^0) = 0$. From $L(x) \geqq L(x^0)$ we obtain $\lambda f(x) \geqq \lambda f(x^0)$. Then the point x^0 is a solution of the scalar problem $(S.P_1)$ which we will consider in Section 7. Theorem 6.7.1 will ensure that $z^0 \in PE(Z)_{Hu}$. □

If we strengthen convexity assumptions by also involving the function g, then $PE(S)_K$ coincides with $PE(S)_{KT}$. In this case, all the sets previously introduced for describing the proper efficiency are the same and we may claim that convexity assumptions give a coherence between a local and a global description of this notion.

Theorem 6.4.7. If the functions g_j $(j \in I(x^0))$ are pseudo-concave at x^0 and the functions g_j $(j \notin I(x^0))$ are continuous at x^0, then $PE(S)_{KT} = PE(S)_K$.

Proof. By contradiction, suppose that there exists a point $x^0 \in PE(S)_K \setminus PE(S)_{KT}$. Then there exists a vector $y \in \mathbb{R}^n$ such that $Jf(x^0)\, y \leq 0$ and $\nabla g_j(x^0)\, y \leqq 0$ $(j \in I(x^0))$. From the hypotheses on g we obtain $Jf(x^0)\, y \leq 0$ and $g(x^0 + ty) \leqq 0$ for any $t \in \mathbb{R}_+$ small enough. □

Remark 6.4.8. In the previous theorem it is not possible to weaken the pseudo-concavity assumption by requiring e.g. the invexity of $-g_j$ or the quasi-concavity of g_j at x^0. Consider the following $(V.P.)$:

$$\min f(x) = \min(f_1(x_1, x_2), f_2(x_1, x_2)) = \min(x_1, x_2)$$

and $-x_1^3 - x_2 \leqq 0$. The point $(0,0) \in PE(S)_K$ but $(0,0) \notin PE(S)_{KT}$. It is easy to verify that $-g = x_1^3 + x_2$ is not pseudo-convex but is invex with respect e.g. to $\eta = (\eta_1, \eta_2) = (x_1, +x_1^3 + x_2)$. The same conclusion can be obtained for the point $x^0 = 0$ and for the problem

$\min f(x) = \min(f_1(x), f_2(x)) = \min(-x^3)$ with the quasi-concave function $g(x) = -x^3$.

Any properly efficient solution set introduced in this section is a subset of $E(Z)$. The aim of these different notions of proper efficiency is to retain most of the efficient points and to eliminate some anomalous situations. Thus theorems which assure that the closure of the set of proper efficient points contains the set $E(Z)$ are very desirable. Later on, we will prove (see Theorem 6.5.13) that generally the properly efficiency is a good approximation of solutions of $(V.P.)$. When the hypotheses of Theorem 6.4.6 and of the following Theorem 6.4.8 are fulfilled, we can state that $PE(Z)_{Hu} = E(Z)$ or that all the sets of proper efficiency are the same and coincide with $E(Z)$. It is the case when the problem is linear or $S = \mathbb{R}^n$ and all the functions f_i and g_j are linear or affine. *Any optimal solution of a linear vector problem is proper* with respect to all the definitions of this section.

Theorem 6.4.8. If the functions f_i and g_j $(j \in I(x^0))$ are pseudo-concave at x^0 and the function g_j $(j \notin I(x^0))$ are continuous at x^0, we have $PE(Z)_{KT} = E(Z)$.

Proof. It is quite similar to Theorem 6.4.7. □

The following theorem can be particularly compared with Theorem 6.4.4.

Theorem 6.4.9. If Z is a convex set, then we have $PE(Z)_{LBo} = PE(Z)_{Hu}$.

Proof. By the convexity of Z we have $T(Z, z^0) = \mathrm{cl\ cone}(Z - z^0)$. Hence $PE(Z)_{LBo} = PE(Z)_{GBo} = PE(Z)_{Hu}$, where the last equality follows from Theorem 6.4.4. □

The analysis of the relation between $PE(Z)_{LBo}$ and $PE(Z)_{Be}$ needs the notion of asymptotic cone. We introduce this definition with a few properties which will be used later on.

<u>Definition 6.4.15</u>. The *asymptotic cone* of the set $Z \subset \mathbb{R}^p$ is the cone

$$As(Z) = \{y \in \mathbb{R}^p : \exists \{\lambda_k\} \subset \mathbb{R}_+, \ \lim_{k \to +\infty} \lambda_k = 0 ; \\ \exists \{z_k\} \subset Z : y = \lim_{k \to +\infty} \lambda_k z^k\} .$$

It is immediate to verify that $As(Z)$ is a closed cone and that $As(Z) = As(Z + \alpha)$, $\forall \alpha \in \mathbb{R}^p$. From $Z_1 \subset Z_2$ it works out $As(Z_1) \subset As(Z_2)$. When Z is a closed convex set, we have (see [Rockafellar, 1970], Theorem 8.2) the equality between the asymptotic and the recession cone (see Definition 3.4.9):

$$As(Z) = 0^+(Z) = \{y : z + \alpha y \in Z ; \forall z \in Z, \ \forall \alpha \geqq 0\} .$$

<u>Lemma 6.4.1</u>. A set $Z \subset \mathbb{R}^p$ is bounded if and only if $As(Z) = \{0\}$.

<u>Proof</u>. Let Z be a bounded set with $\|z\| \leq K$, $\forall z \in Z$, and let $\bar{z} \in As(Z) : \bar{z} = \lim\limits_{k \to +\infty} t_k z^k$ with $\{z^k\} \subset Z$ and $\{t_k\} \subset R_+$ with $\lim\limits_{k \to +\infty} t_k = 0$. From $\|t_k z^k\| \leq t_k K$ we obtain $\bar{z} = 0$. Conversely, let $As(Z) = \{0\}$. If Z would be unbounded there would exist a sequence $\{z^k\} \subset Z$ such that $\lim\limits_{k \to +\infty} z^k = \infty$. Hence we would have (perhaps by choosing a subsequence) $0 \neq z = \lim\limits_{k \to +\infty} \dfrac{z^k}{\|z^k\|} \in As(Z)$. □

<u>Lemma 6.4.2</u>. For $Z_1, Z_2 \subset \mathbb{R}^p$, it is $As(Z_1 \cap Z_2) \subset As(Z_1) \cap As(Z_2)$.

<u>Proof</u>. It is a straightforward application of the monotonic property of the asymptotic cone. □

<u>Remark 6.4.9</u>. In general the equality $As(Z_1 \cap Z_2) = As(Z_1) \cap As(Z_2)$ does not hold. A sufficient condition is obtained by requiring that Z_1 and Z_2 be closed convex sets with $Z_1 \cap Z_2$ being nonempty. Indeed in this case we have $As(Z_1 \cap Z_2) = 0^+(Z_1 \cap Z_2)$ and the same claim is true for Z_1 and Z_2.

<u>Theorem 6.4.10</u>. If Z is a closed set and $As(Z) \cap (-C) = \{0\}$, then $PE(Z)_{LBo} = PE(Z)_{Be}$.

Proof. By contradiction suppose $z^0 \in PE(Z)_{LBo} \backslash PE(Z)_{Be}$ or $T(Z, z^0) \cap (-C) = \{0\}$ with $\text{cl cone}(Z + C - z^0) \cap (-C) \neq \{0\}$. Then there exist $\{\lambda_k\} \subset I\!R_+$ and $\{z^k\} \subset Z$ such that $\lim_{k \to +\infty} \lambda_k(z^k - z^0) = z \in -C \backslash \{0\}$. The sequence $\{z^k\}$ cannot converge to z^0 since otherwise we would get $z \in T(Z, z^0)$ which contradicts $z^0 \in PE(Z)_{LBo}$. Also the sequence $\{\lambda_k\}$ cannot contain a subsequence converging to zero since we would get $z \in As(Z - z^0) \cap (-C) = As(Z) \cap (-C)$, which is a contradiction to our assumption. Hence $\{z^k\}$ is a bounded sequence. In any case by choosing a subsequence, we can suppose that $\{z^k\}$ converges to a value $z^* \neq z^0$. Thus $\{\lambda_k\}$ converges to $\lambda_0 > 0$. But we would have $\lim_{k \to +\infty} z^k = z^0 + z/\lambda_0 \in Z$ as Z is a closed set and this conclusion contradicts the hypothesis that z^0 be efficient. $\square$

6.5. Theorems of Existence

In ordinary scalar optimization problems, for existence theorems the classical reference is given by the Weierstrass theorem which guarantees the existence of extremal points for a function $f : S \subset I\!R^n \to I\!R$ if it is continuous and S is a compact set. A well-known generalization assures the existence of a minimal point under the hypotheses that f is a lower semicontinuous and S is a compact set. If we turn our attention to the image space $Z = f(S)$, then we can state that Z has a minimal value when $Z + I\!R_+$ is closed and bounded below.

The similar situation happens in vector optimization problems. Since the compactness assumption can be too strict for applications, one seeks to weaken this hypothesis possibly in favor of that kind of semi-compactness that we meet in generalized Weierstrass theorem. The goal of avoiding this restriction is generally achieved by introducing the so-called C-semicompact, C-compact, C-closed, C-bounded and C-quasibounded sets, by using Zorn's lemma and by imposing adequate (topological) conditions on the ordering cone. In our case these conditions are always satisfied. All the previous definitions are based differently upon the easy equality $E(Z) = E(Z + C)$ and upon the generalization of classical notions by considering only a cone-dependent part of Z.

This section begins with the various generalizations of compactness, that we have just mentioned, and with their relationships. Later we will give the more general definition of C-complete set introduced by [Luc, 1989]; so the main existence theorem will be Theorem 6.5.6. Proofs and examples of this section have a definite reference in [Corley, 1980], who introduced the cone semicompactness condition as a light generalization of a previous definition of [Wagner, 1977], in [Hartley, 1978] who defined the C-compactness, in [Sawaragi-Nakayama-Tanino, 1985] and in [Cambini-Martein, 1994] where one can find the more general notion of C-quasibounded sets. For readers interested in more recent developments in the infinite-dimensional case we quote some existence results in particular spaces ordered by supernormal cones (see [Isac, 1983] and [Postolica, 1993]) and by cones introduced by [Ha, 1994] with the property that any bounded set which is contained in a complete subset of Z has a limit.

Definition 6.5.1. A set $Z \subset \mathbb{R}^p$ is said to be:

a) *C-compact* when the set $(z - C) \cap Z$ is compact, $\forall\, z \in Z$;

b) *C-semicompact* when any cover of Z of the form $\{(z^\alpha - C)^c;\ \alpha \in I,\ z^\alpha \in Z\}$, when I is an index set, admits a finite subcover.

Theorem 6.5.1. If Z is C-compact, then Z is also C-semicompact.

Proof. Let us consider any open cover of Z of the form $\{(z^\alpha - C)^c;\ \alpha \in I,\ z^\alpha \in Z\}$ and any $z^0 \in Z$. The subfamily $\{(z^\alpha - C)^c;\ \alpha \in I,\ z^\alpha \in Z,\ z^\alpha \neq z^0\}$ is an open cover of $(z^0 - C) \cap Z$. As Z is C-compact, the set $(z^0 - C) \cap Z$ is compact and hence its previous cover has a finite subcover. If we add $(z^0 - C)^c$ to this subcover, we have a finite subcover of Z. □

Of course any compact set Z satisfies Definition 6.5.1 a). Example 6.5.1 will show a set Z which is C-compact but not compact. The C-semicompactness condition is still weaker; it only concerns the particular covers of the form $\{(z^\alpha - C)^c\}$. See Example 6.5.2 for a C-semicompact set which is not C-compact. Theorem 6.5.4 will give a sufficient condition for a C-semicompact set to be also C-compact.

Example 6.5.1. The statement that a compact set is also C-compact (and therefore C-semicompact) cannot be reversed. For example, take $C = \mathbb{R}^2_+$ and $Z = \{(z_1, z_2) : z_1 + z_2 \geqq 0\}$.

Example 6.5.2. The inclusion relation of Theorem 6.5.2 is proper. The set $Z = \{(z_1, z_2) : z_1^2 + z_2^2 \leq 1, \ z_1 > 0, \ z_2 > 0\} \cup \{(0,0)\}$ is C-semicompact but not C-compact with respect to $C = \mathbb{R}^2_+$.

Remark 6.5.1. [Luc, 1989] gives another definition of C-compactness. Here a set $Z \subset \mathbb{R}^p$ is said to be C-compact when any cover of Z of the form $\{U_\alpha + C \, ; \, \alpha \in I, \ U_\alpha \text{ are open}\}$ admits a finite subcover. The two definitions are not comparable. The set $Z = \{(z_1, z_2) : z_1 = -z_2\}$ satisfies Definition 6.5.1 a) with respect to $C = \mathbb{R}^2_+$ but not $\mathbb{R}^2_+$-compact for Luc. An example in the opposite direction is given by the set of Example 6.5.2. However, Luc's definition of C-compactness also extends that of compactness. Also we have that a C-compact set (in the meaning of Luc) is C-semicompact; indeed let us consider any cover of Z of the form $\{(z^\alpha - C)^c \, ; \, \alpha \in I, \ z^\alpha \in Z\}$. Since the sets $(z^\alpha - C)^c + C$ form a cover of Z which admits a finite subcover, say $(\alpha_1, ..., \alpha_s)$, we also have that $Z \subset \bigcup_{i=1}^{s} (z^{\alpha_i} - C)^c$.

In existence theorems the hypothesis of C-semicompactness of Z can be replaced by the condition that Z is a C-closed and C-bounded set. So we arrive at a second generalization of the notion of compactness.

Definition 6.5.2. A set $Z \subset \mathbb{R}^p$ is said to be:

a) *C-closed* when $Z + C$ is closed;

b) *C-bounded* when $As(Z) \cap (-C) = \{0\}$.

Remark 6.5.2. It is of interest to note that there is no implication between the closure of Z and its C-closure. The set $Z = \{(z_1, z_2) : z_1 z_2 = -1, \ z_1 > 0\}$ is closed but not $\mathbb{R}^2_+$-closed; the set $Z = \{(z_1, z_2) : z_1^2 + z_2^2 <$

$1\} \cup \{(z_1, z_2) : z_1^2 + z_2^2 = 1,\ z_1 \leqq 0,\ z_2 \leqq 0\}$ is not closed while it satisfies Definition 6.5.2 a) for $C = \mathbb{R}^2_+$. A sufficient condition for a closed set to be also a C-closed set is supplied by the following Remark 6.5.3; for $Z_2 = C$, in particular, it states that a closed and C-bounded set is also C-closed.

Remark 6.5.3. A bounded set is also C-bounded. Definition 6.5.2 b) can be motivated by the property that, for nonempty closed sets Z_1 and Z_2, the equality $As(Z_1) \cap As(-Z_2) = \{0\}$ implies that $Z_1 + Z_2$ is a closed set. In fact, let $z = \lim_{k \to +\infty} (z_1^k + z_2^k)$ with $z_1^k \in Z_1$, $z_2^k \in Z_2$. If $\{z_1^k\}$ has no convergent subsequence, $\|z_1^k\| \to +\infty$ and $\dfrac{z_1^k + z_2^k}{\|z_1^k\|} \to 0$ as $k \to +\infty$. In any case, by taking a subsequence we can assume that $\dfrac{z_1^k}{\|z_1^k\|} \to z_1 \in As(Z_1)$. Moreover, $\dfrac{z_2^k}{\|z_1^k\|} \to -z_1 \in As(Z_2)$. This conclusion cannot be true since $z_1 \neq 0$. Therefore we have to suppose that $\{z_1^k\}$ converges to some $\bar{z}_1$. Then $\{z_2^k\}$ converges to $z - \bar{z}_1$. Since Z_1 and Z_2 are closed, $z = \bar{z}_1 + (z - \bar{z}_1) \in Z_1 + Z_2$.

Remark 6.5.4. [Luc, 1989] gives another definition of C-boundedness: a set $Z \subset \mathbb{R}^p$ is C-bounded when, for any neighborhood U of the origin, there exists $t > 0$ such that $Z \subset tU + C$. Also in this case the two definitions are not generally comparable. The set $Z = C = \{(z_1, z_2) : z_1 = z_2\}$ satisfies Luc's definition but there exist some points of $As(Z)$ that belong to $(-C) \backslash \{0\}$: an example of such a point is given by $(1, 1)$. The set $Z = \{(z_1, z_2) : z_1 \geqq 0,\ z_2 \geqq 0\}$ does not satisfy Luc's definition with $C = \{(z_1, z_2) : z_1 = z_2 \geqq 0\}$ while $As(Z)$ can intersect $-C$ only at zero.

Also we wish to point out the relations between a C-closed and C-bounded set and the previous definitions of compact, C-compact, C-semicompact sets.

We observe that Definition 6.5.2 gives a further generalization of the compactness notion. Indeed for a compact (and hence bounded) set Z we have $As(Z) = \{0\} = As(Z) \cap (-C)$ and Remark 6.5.3 assures that Z is a C-closed set.

Theorem 6.5.2 will show that a C-bounded and C-closed set is also a C-semicompact set.

On the contrary there is no implication with the C-compactness notion. The already quoted set $Z = \{(z_1, z_2) : z_1^2 + z_2^2 < 1\} \cup \{(z_1, z_2) : z_1^2 + z_2^2 = 1,\ z_1 \leqq 0,\ z_2 \leqq 0\}$ is $\mathbb{R}^2_+$-closed and $\mathbb{R}^2_+$-bounded but not $\mathbb{R}^2_+$-compact. The set $Z = \{(z_1, z_2) : z_1^2 + z_2^2 \leqq 1,\ z_1 < 0,\ z_2 < 0\}$ is $\mathbb{R}^2_+$-compact and $\mathbb{R}^2_+$-bounded but not $\mathbb{R}^2_+$-closed. Finally the set $Z = \{(z_1, z_2) : z_1 z_2 = 1,\ z_1 < 0\}$ is $\mathbb{R}^2_+$-compact and $\mathbb{R}^2_+$-closed but not a $\mathbb{R}^2_+$-bounded set. (An example of a point of $-C\backslash\{0\}$ which belongs to $As(Z)$ is given by $(-1, 0)$).

The following figure draws the proven inclusions:

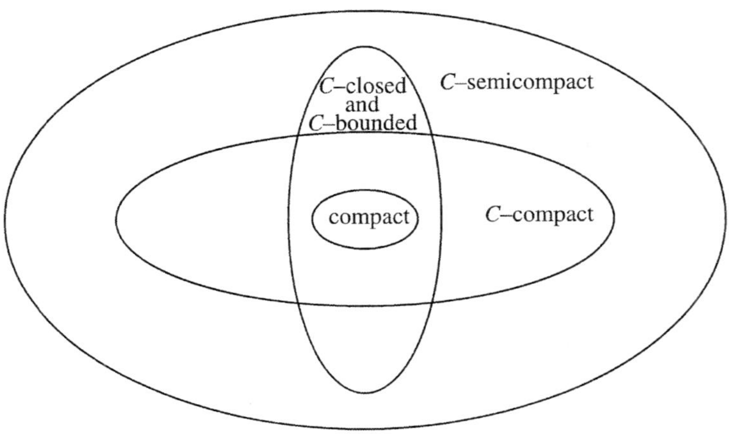

Figure 5.

Lemma 6.5.1. The set Z is C-bounded if and only if also $Z + C$ is C-bounded.

Proof. From $As(Z) \subset As(Z + C)$ one gets immediately that Z is C-bounded if $As(Z + C) \cap (-C) = \{0\}$. Conversely, contrary to what we have to prove, let us suppose that $As(Z + C) \cap (-C) \neq \{0\}$. Then there exist sequences $\{t_k\} \subset \mathbb{R}_+$ with $\lim_{k \to +\infty} t_k = 0$ and $\{z^k + c^k\} \subset Z + C$

such that $\lim_{k \to +\infty} t_k(z^k + c^k) = -c \in (-C)\backslash\{0\}$.
First let us suppose that $\{t_k c^k\}$ has a convergent subsequence. We may always assume that $\lim_{k \to +\infty} t_k c^k = c' \in C$. Therefore $\{t_k z^k\}$ converges to $-c - c'$ which is a nonzero vector in $As(Z) \cap (-C)$ since C is a pointed cone. So we have that $As(Z) \cap (-C) \neq \{0\}$.
If we suppose that $\{t_k c^k\}$ has no convergent subsequence, $\{t_k c^k\}$ is unbounded. From Lemma 6.4.1 we have $As(\{t_k c^k\}) \neq \{0\}$. Namely, we may assume by taking a subsequence of $\{t_k c^k\}$ that there exists another sequence $\{\tau_k\} \subset \mathbb{R}_+$ with $\lim_{k \to +\infty} \tau_k = 0$ and $\lim_{k \to +\infty} \tau_k(t_k c^k) = \bar{c} \neq 0$. Naturally $\bar{c} \in C$. From

$$\|\tau_k t_k z^k + \bar{c}\| =$$

$$= \|\tau_k[t_k(z^k + c^k) + c] - (\tau_k t_k c^k - \bar{c}) - \tau_k c\|$$

$$\leqq \tau_k \|t_k(z^k + c^k) + c\| + \|\tau_k t_k c^k - \bar{c}\| + \tau_k \|c\|$$

it follows that $\lim_{k \to +\infty} \tau_k t_k z^k = -\bar{c}$ or $As(Z) \cap (-C) \neq \{0\}$. □

Lemma 6.5.2. If $Z + C$ is C-semicompact, then Z is also C-semicompact.

Proof. Let $\{(z^\alpha - C)^c ;\ \alpha \in I,\ z^\alpha \in Z\}$ be an open cover of Z. For any $z \in Z$, let $z \in (z^{\alpha_0} - C)^c$ with $z^{\alpha_0} \in Z$. Since C is a convex cone we have $z + C \subset (z^{\alpha_0} - C)^c$. In fact, if $z + c \in z^{\alpha_0} - C$ for some $c \in C$, then $z \in z^{\alpha_0} - C$, which is a contradiction. Hence $\{(z^\alpha - C)^c ;\ \alpha \in I,\ z^\alpha \in Z\}$ is also an open cover of $Z + C$. Since this last set is C-semicompact, this cover has a finite subcover which is, of course, a subcover of Z. □

Theorem 6.5.2. If Z is a C-closed and C-bounded set, then Z is also C-semicompact.

Proof. For any $z \in Z + C$, let us consider the set $(z - C) \cap (Z + C)$. By virtue of Lemmas 6.4.2 and 6.5.1 we have $As\,[(z - C) \cap (Z + C)] \subset As(z - C) \cap As(Z + C) = (-C) \cap As(Z + C) = \{0\}$ since Z is C-bounded. Therefore $(z - C) \cap (Z + C)$ is a closed and bounded set, i.e. a compact set $\forall\, z \in Z + C$. By Definition 6.5.1 a), $Z + C$ is C-compact and, by Theorem 6.5.1, also C-semicompact. Hence by Lemma 6.5.2, Z is C-semicompact. □

The C-boundedness notion has been generalized by [Cambini-Martein, 1994] by means of the definition of C-quasibounded sets.

Definition 6.5.3. A set $Z \subset \mathbb{R}^p$ is said to be a *C-quasibounded set* when the section $(z - C) \cap Z$ is a bounded set for any $z \in Z$.

Theorem 6.5.3. If $Z \subset \mathbb{R}^p$ is C-bounded, then Z is C-quasibounded.

Proof. In contradiction to what we wish to prove, we suppose that $(z^0 - C) \cap Z$ is nonbounded for $z^0 \in Z$. Then there exists a sequence $\{z^k\} = \{z^0 - c^k\}$, with $c^k \in C$ and $\lim\limits_{k \to +\infty} \|z^k\| = +\infty$. Without any loss of generality we can assume that $\lim\limits_{k \to +\infty} \dfrac{z^k}{\|z^k\|} = z^*$. Then also $\left\{-\dfrac{c^k}{\|z^k\|}\right\}$ converges to the same element. Contrary to the hypothesis we have found a point $z^* \in -C \cap As(Z)$, $z^* \neq 0$. □

The following example will show that the class of C-bounded sets is strictly contained in the class of sets which satisfy Definition 6.5.3.

Example 6.5.3. We have already noticed that the set $Z = (z_1, z_2) : z_1 z_2 = 1,\ z_1 < 0\}$ is not $\mathbb{R}^2_+$-bounded. But this set is C-quasibounded.

Under some additional algebraic and topological assumptions, all the previous concepts (C-compactness, C-semicompactness, C-boundedness and C-closure, C-quasiboundedness) coincide.

Theorem 6.5.4. Let $Z \subset \mathbb{R}^p$ be a closed and a convex set. Then the following statements are *equivalent*:

a) Z is C-quasibounded;

b) Z is C-closed and C-bounded;

c) Z is C-compact;

d) Z is C-semicompact.

Proof. a) → b). Assume that there exists $c \in As(Z) \cap (-C \backslash \{0\})$. The hypotheses about the set Z assure $As(Z) = O^+(Z)$ and therefore $z + tc \in Z$, $\forall\, z \in Z$ and $\forall\, t \geqq 0$. As $z + tc \in z - C$ too, then the set $Z \cap (z - C)$ is unbounded, contradicting the C-quasiboundedness of Z. By Remark 6.5.2 the set Z is also C-closed.
b) → c). In Theorem 6.5.3 we have already proved that the set $(z - C) \cap Z$ is bounded $\forall\, z \in Z$. Therefore it is compact and Z is C-compact.
c) → d). See Theorem 6.5.1.
d) → b). Suppose, to the contrary, that some $-c \neq 0$ belongs to the set $As(Z) \cap (-C)$. From $As(Z) = O^+(Z)$ we have $z - c \in Z$, for each $z \in Z$, and so $E(Z) = \emptyset$, in contradiction to the following Theorem 6.5.8 b). As Z is a closed set, the C-boundedness implies its C-closure. □

The following notion of C-completeness is rather far from the previous generalizations of compactness conditions, but in a more general way, it still guarantees the existence of efficient points.

Definition 6.5.4. A set $Z \subset I\!R^p$ is said to be *C-complete* whence it has no covers of the form $\{(z^\alpha - C)^c;\ \alpha \in I\}$ when I is an index set and $\{z^\alpha\} \subset Z$ is a decreasing net (such that $\alpha, \beta \in I$ and $\beta > \alpha$ imply $z^\alpha - z^\beta \in C \backslash \{0\}$).

Theorem 6.5.5. If Z is a C-semicompact set, then it is also a C-complete set.

Proof. Suppose to the contrary that Z has a cover of the form $\{(z^\alpha - C)^c\}$, where $\{z^\alpha\}$ is a decreasing net. As Z is C-semicompact, we have $Z \subset \bigcup_1^n (z^\alpha - C)^c$ or $Z \subset (z^n - C)^c$. So we arrive at the contradiction that $z^n \notin (z^n - C)$. □

Example 6.5.4. There are examples of sets which are C-complete but not C-semicompact. One is provided by the set $Z = \{(z_1, z_2) : z_2 < 0,\ z_1 z_2 = 1\} \cup \{(z_1, z_2) : z_2 \leqq 0,\ z_2 = -z_1\}$. The set Z is $I\!R^2_+$-complete, but it does not satisfy Definition 6.5.1 b) since its cover $\{(z^n - I\!R^+_2)^c\}$, with $z^n = (n, -n)$, has no finite subcovers.

Theorem 6.5.6. We have $E(Z) \neq \emptyset$ if Z is a nonempty C-complete set.

Proof. Let us consider the set P consisting of decreasing nets in Z. For $a, b \in P$, if we write aRb when the net b is contained in a, we have that the binary relation R satisfies Definition 6.2.2 b). So P is a partially ordered set. Since any subset of P has an upper bound, P is inductively ordered. By Zorn's lemma, we get a maximal element $\{\bar{z}^\alpha\}$.

Now we are able to prove that $E(Z) \neq \emptyset$. Suppose to the contrary that the set of efficient points is empty and consider the net $\{(\bar{z}^\alpha - C)^c\}$. It forms a cover of Z. Indeed, if this is not the case, there exists $z \in Z$ such that $z \in \bar{z}^\alpha - C$ for any α, or $\bar{z}^\alpha \geqq z$. As $z \notin E(Z)$, there is some $z_1 \in Z$ with $z \geq z_1$. We conclude that $\bar{z}^\alpha \geq z_1$, for any α, and we obtain a decreasing net $\{\bar{z}^\alpha\} \cup \{z_1\}$ which contradicts the maximality of $\{\bar{z}^\alpha\}$. Through $Z \subset \bigcup_\alpha \{(\bar{z}^\alpha - C)^c\}$, we arrive at the contradiction that Z is not C-complete. □

Theorem 6.5.7. We have $E(Z) \neq \emptyset$ if and only if there is some $z \in Z$ such that its section $Z \cap (z - C)$ is C-complete.

Proof. From Theorem 6.5.6, if $Z \cap (z - C)$ is C-complete for some $z \in Z$, we get that $E(Z \cap (z - C))$ is nonempty. It suffices to prove that $E(Z \cap (z - C)) \subset E(Z)$. Let $z_0 \in E(Z \cap (z - C))$. If there is some $\bar{z} \in Z$ with $z_0 \geq \bar{z}$, then $\bar{z} \in Z \cap (z - C)$ and hence $\bar{z} \geq z_0$. Conversely, if $z \in E(Z)$, the section $Z \cap (z - C)$ is obviously a C-complete set because there are no decreasing nets in $Z \cap (z - C)$. □

Theorem 6.5.6 and the inclusions proven through Theorems 6.5.1, 6.5.2 and 6.5.5 furnish other sufficient conditions for $E(Z) \neq \emptyset$.

Theorem 6.5.8. We have $E(Z) \neq \emptyset$ if one of the following conditions is satisfied:

a) Z is a compact or a C-compact set;

b) Z is a C-semicompact set;

c) Z is a C-closed and C-bounded set.

Theorem 6.5.9. Also we have $E(Z) \neq \emptyset$ if either of the following conditions holds:

i) Z is a closed and C-quasibounded set;

ii) Z is a C-closed and C-quasibounded set.

Proof.

i) Z is C-quasibounded if and only if it is C-compact. The thesis follows immediately from Theorem 6.5.8 a).

ii) Consider the set $(z - C) \cap (Z + C)$ for any $z \in Z$. First we will prove that this set is bounded. In fact if it is unbounded, then there exists a sequence $\{z + c^k\} = \{z^k - \gamma^k\}$ with $c^k, \gamma^k \in -C$, $z^k \in Z$ and $\lim_{k \to +\infty} \|c^k\| = +\infty$. The hypothesis about the cone C assures that $\exists\, \alpha \in \mathbb{R}^p$ such that $\alpha c < 0$, $\forall\, c \in -C \backslash \{0\}$. Let B be the closed unit ball and $\bar{c}^k = \dfrac{c^k}{\|c^k\|}$, $\bar{\gamma}^k = \dfrac{\gamma^k}{\|\gamma^k\|}$. As there exists $c^0 \in B \cap (-C)$ such that $\alpha c^0 \geqq \alpha c$, $\forall\, c \in B \cap (-C)$, it works out

$$\begin{aligned} \alpha z^k &= \alpha(z + c^k + \gamma^k) = \alpha z + \alpha \bar{c}^k \, \|c^k\| + \alpha \bar{\gamma}^k \, \|\gamma^k\| \ \leqq \\ &\leqq \ \alpha z + \alpha c^0 (\|c^k\| + \|\gamma^k\|) \end{aligned}$$

which runs to $-\infty$. Consequently we have $\lim_{k \to +\infty} \|z^k\| = +\infty$. This is absurd because $z^k \in Z \cap (z - C)$ and Z is a C-quasibounded set. Now the set $(z - C) \cap (Z + C)$ is closed and bounded and it is easy to verify that $E(Z + C) \neq \emptyset$. The thesis follows from the inclusion $E(Z + C) \subset Z$. □

In a particular case the conditions a), b), c), d) of Theorem 6.5.4 become also necessary for $E(Z) \neq \emptyset$.

Theorem 6.5.10. Let $Z \subset \mathbb{R}^p$ be a closed convex set. Then $E(Z) \neq \emptyset$ *if and only if* any one of the conditions in Theorem 6.5.4 is satisfied.

Proof. It is enough to prove that $E(Z) \neq \emptyset$ implies that Z is C-closed and C-bounded. To this purpose we can repeat the proof of the quoted theorem (in the implication d) $\rightarrow$ b)). $\square$

Theorems 6.5.6, 6.5.8 and 6.5.9 give some sufficient conditions for $E(Z) \neq \emptyset$ under some hypotheses on the image space $Z = f(S)$. A further theorem of existence involves, instead of the image space, the set S and the function f.

Definition 6.5.5. A function $f : \mathbb{R}^n \longrightarrow \mathbb{R}^p$ is said to be *C-semicontinuous* when the set $f^{-1}(z - C)$ is closed $\forall\, z \in \mathbb{R}^p$.

Remark 6.5.5. For $p = 1$ and $C = \mathbb{R}^p_+$ (resp. $C = \mathbb{R}^p_-$) the C-semicontinuity definition collapses to the usual definition of lower (respectively upper) semicontinuity.

Theorem 6.5.11. If $S \subset \mathbb{R}^n$ is a compact set and f is a C-semicontinuous function, then $E(Z) \neq \emptyset$.

Proof. If $\bigcup\limits_{\alpha} \{(z^\alpha - C)^c\, ;\ \alpha \in I,\ z^\alpha \in Z\}$ is an open cover of Z, then Definition 6.5.5 assures that $\bigcup\limits_{\alpha} \{f^{-1}[(z^\alpha - C)^c]\, ;\ \alpha \in I,\ z^\alpha \in Z\}$ is an open cover of S. But S is compact; hence this cover admits a finite subcover whose image forms a finite subcover of Z. Then this set is C-semicompact. The thesis follows from Theorem 6.5.8 b). $\square$

Up until now we have been concerned with existence theorems for the set $E(Z)$. A necessary condition for $WE(Z) \neq \emptyset$ and $PE(Z)_{He} \neq \emptyset$ can be given in terms of asymptotic cone.

Lemma 6.5.3. A nonzero vector $z \in As(Z)$ if and only if $\mathrm{cone}(z + U) \cap Z \cap V \neq \emptyset$ for every neighborhood U of zero and for every neighborhood V of ∞.

Proof. As a simple consequence of the definition, we have $z \in As(Z) \backslash \{0\}$ if and only if $z \in \mathrm{cl\,cone}(Z \cap V)$ for every neighborhood V of ∞, or $(z + U) \cap \mathrm{cone}(Z \cap V) \neq \emptyset$ for every $U(0)$ and for every V. $\square$

Theorem 6.5.12.

i) $WE(Z) \neq \emptyset$ implies $As(Z) \cap (-\operatorname{int} C) = \emptyset$;

ii) $PE(Z)_{He} \neq \emptyset$ implies $As(Z) \cap (-C) = \{0\}$.

Proof.

i) Suppose to the contrary that there is some $z \in As(Z) \cap (-\operatorname{int} C)$. Of course $z \in As(Z - z^0)$ for each $z^0 \in Z$ and then $\operatorname{cone}(z + U) \cap (Z - z^0) \cap V \neq \emptyset$ for any neighborhood V of ∞, by virtue of Lemma 6.5.3. Since $z \in (-\operatorname{int} C)$, we may choose a neighborhood $U(0)$ small enough such that $\operatorname{cone}(z + U) \backslash \{0\} \subset (-\operatorname{int} C)$. As V does not contain zero, the above relations imply $(Z - z^0) \cap (-\operatorname{int} C) \neq \emptyset$. This shows that no $z^0 \in Z$ can be weakly efficient.

ii) If $z^0 \in PE(Z)_{He}$ there exists a cone C' with $C \backslash \{0\} \subset \operatorname{int} C'$ such that $z^0 \in E_{C'}(Z)$. By i) we have $As(Z) \cap (-\operatorname{int} C') = \emptyset$ and consequently $As(Z) \cap (-C) = \{0\}$. □

We conclude this section by utilizing Definition 6.5.2 in order to prove the remark at the end of Section 6.4. In particular Theorem 6.5.13 will hold for compact sets.

Lemma 6.5.4. If $Z \subset I\!R^p$ is C-bounded and C-closed, we have $Z \subset E(Z) + C$.

Proof. The set $Z' = (z - C) \cap (Z + C)$ for any $z \in Z$ is compact (see the proof of Theorem 6.5.9 ii)). In view of Theorem 6.5.8 a), $E(Z') \neq \emptyset$; moreover, $E(Z') \subset E(Z)$. Indeed, if there would exist an element $\tilde{z} \in E(Z') \backslash E(Z)$, we would have a value $z' \in Z$ such that $\tilde{z} - z' \in C$. We can also obtain $z' \in Z'$ which leads to a contradiction with respect to $\tilde{z} \in E(Z')$.
The conclusion $\emptyset \neq E(Z') \subset E(Z)$ is sufficient for proving $Z' \cap E(Z) \neq \emptyset$. Then, for any $z \in Z$, we have found a value $\bar{z} \in E(Z)$ such that $\bar{z} \in z - C$ or $Z \subset E(Z) + C$. □

Lemma 6.5.5. Let A be a closed cone such that $A \cap C = \{0\}$. Then there exists a sequence $\{C_k\}$ of closed convex pointed cones, $C_k \neq C$, such that:

i) $C_{k+1} \subset C_k$, $\forall k$;

ii) $\bigcap_{k=1}^{+\infty} C_k = C$;

iii) $A \cap C_k = \{0\}$, $\forall k$;

iv) $C \backslash \{0\} \subset \operatorname{int} C_k$, $\forall k$.

Proof. Let $D = \left\{d : d = \dfrac{c}{\|c\|}\,;\ c \in C,\ c \neq 0\right\}$. We have $D \cap A = \emptyset$ with $C = \operatorname{cone} D \cup \{0\}$ and D a compact set. Then the distance ε of C from the closed set A is positive. Set $C_k = \operatorname{cl}\operatorname{cone}(D + U_{\varepsilon/k}(0))$. Since A is a closed cone, for k large enough, from $A \cap (D + U_{\varepsilon/k}(0)) = \emptyset$ it also works out $A \cap C_k = \{0\}$. It is easy to verify that the cones C_k also fulfill the other requirements of our thesis. □

Theorem 6.5.13. Let $Z \in I\!R^p$ be a C-bounded and C-closed set. Then $E(Z) \subset \operatorname{cl} PE(Z)_{He}$.

Proof. From Definition 6.5.2 b) and from Lemma 6.5.1 we get $As(Z + C) \cap (-C) = \{0\}$. Then, in view of Lemma 6.5.5, there exists a sequence of pointed closed cones $\{C_k\}$ such that $\cap C_k = C$, $C_{k+1} \subset C_k$, $As(Z+C) \cap (-C_k) = \{0\}$ and $C \backslash \{0\} \subset \operatorname{int} C_k$. Since the set $Z + C + C_k$ is closed, we have that the set $Z + C$ is C_k-closed and C_k-bounded; then Lemma 6.5.4 gives $Z \subset E_{C_k}(Z) + C_k$ for any k.
Now let $\tilde{z} \in E(Z)$. From the previous inclusion we obtain a sequence $\{z^k\}$ such that $z^k \in E_{C_k}(Z) \cap (\tilde{z} - C_k)$. Then $z^k \in PE(Z)_{He}$ (see Definition 6.4.8). Now it suffices to prove that a subsequence of $\{z^k\}$ converges to $\tilde{z}$.
Since $C_{k+1} \subset C_k$, $z^k \in (Z + C) \cap (\tilde{z} - C_1)$ for every k. This section is closed and bounded since $As[(Z + C) \cap (\tilde{z} - C_1)] \subset As(Z + C) \cap (-C_1)$ and we can assume without loss of generality that $\{z^k\}$ converges to some $\bar{z} \in (Z + C) \cap (\tilde{z} - C_1)$. Since $\tilde{z} - z^k \in C_k$, we have $\tilde{z} - \bar{z} \in C$ and this implies $\tilde{z} = \bar{z}$. □

6.6. Optimality Conditions

In the same spirit of scalar optimization we are going to produce a few necessary or/and sufficient conditions related to several optimality notions. All these conditions will require the differentiability of the functions f and g; the nondifferentiable case will be analyzed in Section 8 by taking into account weaker assumptions. All the conditions of this section are first-order conditions. In the literature also some second-order conditions are known, both in the smooth and nonsmooth case. In general, they use the notions of Hessian matrix, of second-order tangent cone or of some other particular cones. The generalization to the nonsmooth case is often fulfilled with the reference to the functional class $C^{1,1}$ or to functions whose derivatives are locally Lipschitz. With reference to these derivatives one can apply the approach we will introduce in Section 8.

First consider the problem $(V.P.)$ when *no functional (inequality) constraint* is considered. The reference for the scalar case is given here by the well-known necessary condition: if x^0 is a local minimum point for $f : S \subset \mathbb{R}^n \to \mathbb{R}$, then $\nabla f(x^0)\, y \geqq 0$, $\forall\, y \in T(S, x^0)$.

<u>Theorem 6.6.1</u>. If x^0 is a weakly local efficient point for $(V.P.)$ with f differentiable at x^0, then $Jf(x^0)\, y \notin -\operatorname{int} C$, $\forall\, y \in T(S, x^0)$.

<u>Proof</u>. For $y \in T(S, x^0)$ we may find a sequence $\{x^k\} \subset S$, $\lim_{k \to +\infty} x^k = x^0$ with $\lim_{k \to +\infty} \dfrac{x^k - x^0}{\|x^k - x^0\|} = y$. By Taylor's expansion $f(x^k) - f(x^0) = Jf(x^0)(x^k - x^0) + o(\|x^k - x^0\|)$, we get

$$\frac{f(x^k) - f(x^0)}{\|x^k - x^0\|} = Jf(x^0)\, \frac{x^k - x^0}{\|x^k - x^0\|} + o(1)$$

and hence:

$$\lim_{k \to +\infty} \frac{f(x^k) - f(x^0)}{\|x^k - x^0\|} = Jf(x^0)\, y \; .$$

So we have proved the theorem since $f(x^k) - f(x^0) \notin -\operatorname{int} C$. $\square$

<u>Theorem 6.6.2</u>. The condition $Jf(x^0)\,y \notin -C$, $\forall y \in T(S,x^0)\backslash\{0\}$ is sufficient for x^0 to be a local efficient point for $(V.P.)$.

<u>Proof</u>. Ab absurdo suppose that x^0 is not a local efficient point. Then there exists a sequence $\{x^k\} \subset S$ converging to x^0 with $f(x^k) - f(x^0) \in -C\backslash\{0\}$. From:

$$\frac{f(x^k) - f(x^0)}{\|x^k - x^0\|} = Jf(x^0)\,\frac{x^k - x^0}{\|x^k - x^0\|} + o(1)\ ,$$

by taking the limits for $k \to +\infty$ and possibly by choosing a subsequence of $\{x^k\}$, we get $Jf(x^0)\,y \in -C$ for an element $y \in T(S,x^0)\backslash\{0\}$ which leads to a contradiction. $\square$

We can note some differences with the guideline of the scalar case. For $p > 1$ and S open set, Theorem 6.6.1 does not lead to a result like Fermat's proposition, a generalization which will be given by Theorem 6.6.3. Still, in the case that S is an open set, the condition expressed in Theorem 6.6.2 is never satisfied for $p = 1$; this is so because, if one has $\nabla f(x^0)\,y > 0$ for some $y \in I\!R^n$, then $\nabla f(x^0)(-y) < 0$.

Before proceeding to the particular case of S open set, we give another sufficient condition for the same problem handled by Theorems 6.6.1 and 6.6.2. The proof can be obtained as in Theorem 6.6.2.

<u>Theorem 6.6.3</u>. The point x^0 is a local efficient solution of $(V.P.)$ if $\sup\limits_{\vartheta \in C_{\geqq}} \vartheta \cdot Jf(x^0)\,y > 0$ for every $y \in T(S,x^0)\backslash\{0\}$.

<u>Theorem 6.6.4</u>. If x^0 is a weakly local efficient point for $(V.P.)$ with f differentiable at x^0 and S an open set, then $\exists\,\vartheta^0 \in C_{\geqq}\backslash\{0\}$ such that $\vartheta^0 \cdot Jf(x^0) = 0$.

<u>Proof</u>. By Theorem 6.6.1 we immediately get $Jf(x^0)\,y \notin -\operatorname{int} C$, $\forall y \in I\!R^n$. Since the sets $\{Jf(x^0)\,y\,;\ y \in I\!R^n\}$ and $-\operatorname{int} C$ can be separated by a hyperplane, there exists an element $\vartheta^0 \in C_{\geqq}\backslash\{0\}$ such that $\vartheta^0 Jf(x^0)\,y \geqq 0$, $\forall y \in I\!R^n$, or $\vartheta^0 Jf(x^0) = 0$. $\square$

The following example will show that generally Theorem 6.6.4 is not a sufficient condition for x^0 to be a local minimal point. This goal will be achieved under a suitable generalized convexity assumption.

Example 6.6.1. Consider the problem $(V.P.)$ with $S = \mathbb{R}^2$, $C = \mathbb{R}^2_+$ and $f(x) = f(x_1, x_2) = (x_1 - x_2, 1 - x_1 - e^{-x_2})$. We have, for $\vartheta^0 = (1,1) \in C_{\geqq} \backslash \{0\}$, $\vartheta^0 Jf(0,0) = 0$ but $(0,0)$ is not a local efficient point as $f(x_1, x_1) \in f(0,0) - (C \backslash \{0\})$, $\forall x_1 \neq 0$.

Theorem 6.6.5. The conditions $\vartheta^0 Jf(x^0) = 0$, for some $\vartheta^0 \in C_{\geqq} \backslash \{0\}$, and f C-pseudoconvex at x^0 are sufficient for x^0 to be an efficient point for $(V.P.)$.

Proof. Ab absurdo there exists a point $x \in S$ such that $f(x) - f(x^0) \in -C \backslash \{0\}$. The C-pseudoconvexity of f at x^0 implies that $Jf(x^0)(x - x^0) \in -\operatorname{int} C$. So we get the contradiction $\vartheta Jf(x^0)(x^k - x^0) < 0$, $\forall \vartheta \in C_{\geqq} \backslash \{0\}$. □

In the remainder of the section we turn our attention to vector problems $(V.P.)$, with the inequality constraints $g_j(x) \leqq 0$. We will provide a few necessary conditions à la F. John or à la Kuhn-Tucker, satisfied, respectively, by an efficient, a proper efficient and a weakly efficient point. The necessary conditions will become sufficient under a suitable generalized convexity hypothesis. As in Theorem 6.6.5, these hypotheses are also sufficient for assuring the global feature of efficient points. The necessary (and sufficient) conditions for properly efficient points will take into account the large definition by Kuhn-Tucker. So in this case the domination structure is given by $C = \mathbb{R}^p_+$.

Theorem 6.6.6. If x^0 is a weakly local efficient point for $(V.P.)$ where f and g are differentiable at x^0, then there exist $\vartheta^0 \in C_{\geqq}$ and $\lambda^0 \in \mathbb{R}^m_+$ with $(\vartheta^0, \lambda^0) \neq 0$ such that $\vartheta^0 Jf(x^0) + \lambda^0 Jg(x^0) = 0$ and $\lambda^0 g(x^0) = 0$.

Proof. The following system has no solution in $\mathbb{R}^n$:

$$\begin{cases} Jf(x^0)(x-x^0) \in -\operatorname{int} C \\ g(x^0) + Jg(x^0)(x-x^0) \in \operatorname{int} \mathbb{R}^m_- \end{cases}$$

since, from $f(x^0+t(x-x^0)) - f(x^0) \in -\operatorname{int} C$ and $g(x^0+t(x-x^0)) < 0$

for some $x \in \mathbb{R}^n$ and t positive, small enough, we would be able to deduce the existence of an element $x^0+t(x-x^0) \in S$ which contradicts the hypothesis of weakly local efficiency of x^0. Then there exists a hyperplane that separates the sets $\{Jf(x^0)(x-x^0);\ x \in \mathbb{R}^n\} \times \{g(x^0)+Jg(x^0)(x-x^0)\ ;\ x \in \mathbb{R}^n\}$ and $-\operatorname{int} C \times \operatorname{int} \mathbb{R}^m_-$ or there exists a vector $(\vartheta^0, \lambda^0) \neq 0$ such that $\vartheta^0 Jf(x^0)(x-x^0) + \lambda^0 [Jg(x^0)(x-x^0) + g(x^0)] \geqq 0$, $\forall\, x \in \mathbb{R}^n$, and $-\vartheta^0 c + \lambda^0 u \leqq 0$, $\forall\,(c,u) \in (C \times \mathbb{R}^m_-)$. By putting in the last inequality, respectively, $c = 0$ and $u = 0$, we get $\lambda^0 \in \mathbb{R}^m_+$ and $\vartheta^0 \in C_{\geqq}$. The first inequality, for $x = x^0$, gives $\lambda^0 g(x^0) = 0$. The same inequality, which is true for any $x \in \mathbb{R}^n$, now gives $\vartheta^0 Jf(x^0) + \lambda^0 Jg(x^0) = 0$. □

Theorem 6.6.7. If there exists a vector $(\vartheta^0, \lambda^0) \neq 0$ with $\vartheta^0 \in C_{\geqq}$, $\lambda^0 \in \mathbb{R}^m_+$, $\vartheta^0 Jf(x^0) + \lambda^0 Jg(x^0) = 0$, $\lambda^0 g(x^0) = 0$, f a C-pseudoconvex function at x^0 and $\lambda^0 g$ quasiconvex at x^0, then x^0 is an efficient point.

Proof. Ab absurdo there exists a point $x \in S$ such that $f(x) - f(x^0) \in -C\backslash\{0\}$. Since f is C-pseudoconvex at x^0, we have $Jf(x^0)(x-x^0) \in -\operatorname{int} C$ or $\vartheta^0 Jf(x^0)(x-x^0) < 0$. In virtue of quasiconvexity of $\lambda^0 g$, from $g(x) \leqq 0$ or $\lambda^0 g(x) = \lambda^0 g(x) - \lambda^0 g(x^0) \leqq 0$ we obtain $\lambda^0 Jg(x^0)(x-x^0) \leqq 0$. Then $[\vartheta^0 Jf(x^0) + \lambda^0 Jg(x^0)]\,(x-x^0) < 0$ which contradicts the hypothesis $\vartheta^0 Jf(x^0) + \lambda^0 Jg(x^0) = 0$. □

Remark 6.6.1. The proof of theorems like Theorem 6.6.7 remains substantially unchanged under slight variations concerning both the hypotheses of generalized convexity and the assumptions on the multipliers. In particular the proof is the same if the assumption of C-pseudoconvexity of f is replaced with the hypothesis of pseudoconvexity about the scalar function $\vartheta^0 f$ with $\vartheta^0 \in C_>$.

Remark 6.6.2. We still obtain the same result if the assumption of C-pseudoconvexity of f at x^0 and of quasiconvexity of $\lambda^0 g$ at x^0 are replaced with the condition that f and g are C-invex at x^0 with respect to a same η and $\vartheta^0 \in C_>$. Indeed, from $f(x) - f(x^0) - Jf(x^0)\,\eta(x, x^0) \in C$, $\forall x \in S$, it follows:

$$\begin{aligned}\vartheta^0[(f(x) - f(x^0)] &\geqq \vartheta^0 Jf(x^0)\,\eta(x, x^0) = -\lambda^0 Jg(x^0)\,\eta(x, x^0) \geqq \\ &\geqq -\lambda^0 g(x) + \lambda^0 g(x^0) = -\lambda^0 g(x) \geqq 0\ .\end{aligned}$$

Then the point x^0 is minimal for the scalar function $\vartheta^0 f$. From Theorem 6.7.1 it will follow that $x^0 \in PE(S)_{Hu} \subset E(S)$.

Theorem 6.6.8. If $x^0 \in S$ is a properly efficient solution to $(V.P.)$ in the sense of Kuhn-Tucker, then there exist $\vartheta^0 \in \operatorname{int} I\!R^p_+$ and $\lambda^0 \in I\!R^m_+$ such that:

$$\begin{cases} \vartheta^0 \cdot Jf(x^0) + \lambda^0 \cdot Jg(x^0) = 0 \\ \lambda^0 \cdot g(x^0) = 0 \end{cases}$$

Proof. From Definition 6.4.1 there is no $y \in I\!R^n$ such that $Jf(x^0)\,y \leq 0$ and $\nabla g_j(x^0)\,y \leqq 0$ $(j \in I(x^0))$. Then, from the theorem of the alternative of Tucker (see point 20) in Section 2.4), there exist $\vartheta^0 > 0$ and $\lambda^0 \geqq 0$ such that

$$\sum_1^p \vartheta_i^0 \nabla f_i(x^0) + \sum_{j \in I(x^0)} \lambda_j^0 \nabla g_j(x^0) = 0\ .$$

Letting $\lambda_j^0 = 0$ for $j \notin I(x^0)$, we can immediately establish the theorem. □

Remark 6.6.3. From Theorem 6.4.2 it follows that in a Paretian framework the conditions of the above theorem are also necessary (without any other assumption) for all the other definitions of proper efficiency with the only exceptions being local proper efficiency in the sense of Borwein and proper efficiency in the sense of Klinger.

Theorem 6.6.9. If f and g are differentiable at x^0 and the functions $\vartheta^0 f$ and $\lambda^0 g$ are pseudoconvex and quasiconvex at x^0, respectively, the conditions in Theorem 6.6.8 are also sufficient for $x^0 \in S$ to be a Kuhn-Tucker properly efficient solution to $(V.P.)$.

Proof. In virtue of the quasiconvexity of the function $\lambda^0 g$, from $\lambda^0 g(x) \leqq 0 = \lambda^0 g(x^0)$ we obtain $\lambda^0 Jg(x^0)(x-x^0) \leqq 0$, $\forall x \in S$. Then $\vartheta^0 Jf(x^0)(x-x^0) \geqq 0$ and hence $\vartheta^0 f(x) \geqq \vartheta^0 f(x^0)$ in view of the pseudoconvexity of $\vartheta^0 f$. The conclusion is as in Remark 6.6.3. □

Remark 6.6.4. We obtain the same conclusion of Theorem 6.6.9 if the request of invexity of f_i and g_j $(i = 1,...,p;\ j = 1,...,m)$ with respect to a same η is made (instead of the pseudoconvexity of $\vartheta^0 f$ and of the quasiconvexity of $\lambda^0 g$). See Remark 6.6.2.

Example 6.6.2. Let $f : I\!R^3 \to I\!R^2$ with $f_1(x,y,z) = y-z$ and $f_2(x,y,z) = x+y+y^2+z^2$; let $g_1(x,y,z) = -x-y \leqq 0$ and $g_2(x,y,z) = -y+z^4 \leqq 0$. The point $(0,0,0)$ satisfies the conditions expressed by Theorem 6.6.8 for $\vartheta^0 = \lambda^0 = (1,1)$. Hence the origin can be a properly efficient point (according to Kuhn-Tucker definition). This conjecture is confirmed by Theorem 6.6.9 since $\vartheta^0 f = f_1 + f_2 = x + 2y - z + y^2 + z^2$ and $\lambda^0 g = g_1 + g_2 = -x - 2y + z^4$ are convex functions.

If we also relate Theorem 6.6.6 to the componentwise order of $I\!R^p$ we may compare the signs of the components of the multiplier ϑ^0. When $z^0 \in E(Z)$ we have $\vartheta^0 \geqq 0$, but the stronger hypothesis $z^0 \in PE(Z)_{KT}$ guarantees $\vartheta^0 > 0$. The same approach to the following theorem says for $z^0 \in WE(Z)$ that we have to add a *constraint qualification* in order to have $\vartheta^0 \neq 0$.

Theorem 6.6.10. Suppose that $(V.P.)$ satisfies the Kuhn-Tucker constraint qualification at $x^0 \in S$. Then a necessary condition for x^0 to be weakly local minimal point for $(V.P.)$ is that there exist $\vartheta^0 \in I\!R^p$ and $\lambda^0 \in I\!R^m$ such that

$$\begin{cases} \vartheta^0 Jf(x^0) + \lambda^0 Jg(x^0) = 0 \\ \lambda^0 g(x^0) = 0 \\ \vartheta^0 \geq 0 \\ \lambda^0 \geqq 0 \,. \end{cases}$$

Proof. Let x^0 be a weakly local minimum point of $(V.P.)$. Now we will prove that there is no $y \in I\!R^n$ such that $\nabla f_i(x^0)\, y < 0$ $(i = 1, ..., p)$ and $\nabla g_j(x^0)\, y \leqq 0$ $(j \in I(x^0))$. Then the theorem will follow from Motzkin's theorem of alternative (see point 19) in Section 2.4).
Suppose, to the contrary, that there exists a vector y satisfying the above inequalities. From the Kuhn-Tucker constraint qualification there exists a function $\omega \in C^1[0, \bar{t}]$ such that $\omega(0) = x^0$, $g(\omega(t)) \leqq 0$ for any $t \in [0, \bar{t}]$ and $\omega'(0) = \alpha y$ with $\alpha > 0$. From $f_i(\omega(t)) = f_i(x^0) + t\nabla f_i(x^0)\, \alpha y + o(t)$ and $\nabla f_i(x^0)\, y < 0$, $\forall i = 1, ..., p$, it follows $f_i(\omega(t)) < f_i(x^0)$ for t sufficiently small. This contradicts the locally weak optimality of x^0. □

Theorem 6.6.11. If the functions f and g are $I\!R^p_+$-invex at x^0, with respect to a same η, the conditions in Theorem 6.6.10 are also sufficient for the weak minimality of the point x^0.

Proof. Absurdly suppose that there is a feasible point x such that $f(x) - f(x^0) \in -\mathrm{int}\, I\!R^p_+$, i.e. $\vartheta^0[f(x) - f(x^0)] < 0$. The proof goes on as in Remark 6.6.2. □

The optimality conditions we considered for $(V.P.)$ with the explicit consideration of the inequality constraints $g_j(x) \leqq 0$, have a rather common structure. In general a solution of our problem satisfies the equality $\vartheta Jf(x) + \lambda Jg(x) = 0$ for some ϑ and λ or $JL(x) = 0$ for $L = \vartheta f + \lambda g$ (and some other conditions). Some further convexity requirements turn these necessary conditions to be also sufficient. The fact that the solutions of $(V.P.)$ are contained in the solution set of the equality $JL(x) = 0$, can be useful to introduce some duality theorems, as a conclusion of this section.

Also in the vector case duality theory is widely developed. Various approaches have been introduced and many duality problems have been analyzed in order to solve the original problem or to obtain some further informations about its solutions. Generally these approaches have a common structure composed of a *weak* assertion, which states an order relation for the optimal values of both problems, and a *strong* duality theorem which assures the equality between these optimal values.

In this section we will not treat the general duality theory in multiobjective programming. We limit ourselves to deal with the so-called *Mond-Weir vector duality*, which has the advantage of the same objective function as the primal problem. Besides, this duality naturally resorts to some aspects of the generalized convexity. So in this section we associate a *dual* problem to the *primal* problem $(V.P.)$:

$$\underset{(u,\vartheta,\lambda)\in S'}{\text{Max}} \quad f(u) \tag{D.V.P.}$$

with

$$S' = \{(u,\vartheta,\lambda) : \vartheta Jf(u) + \lambda Jg(u) = 0\,;\ \vartheta \in C_{\geqq} \backslash \{0\},$$
$$\lambda \in \mathbb{R}^m_+\,;\ \lambda g(u) \geqq 0\}\ .$$

Of course the functions f and g are assumed to be differentiable.

Theorem 6.6.11 (*weak duality*). Let x be any feasible point for $(V.P.)$ and (u,ϑ,λ) be any feasible point for $(D.V.P.)$. If the function ϑf is pseudoconvex at x and if λg is quasiconvex at u, then $f(x) - f(u) \notin -\operatorname{int} C$.

Proof. Since x and (u,ϑ,λ) are feasible points, respectively, for $(V.P.)$ and $(D.V.P.)$, we have $\lambda g(x) - \lambda g(u) \leqq 0$. From the quasiconvexity of λg, we get:

$$\nabla \lambda g(u)(x-u) = \lambda Jg(u)(x-u) \leqq 0$$

or:

$$\vartheta Jf(u)(x-u) = \nabla \vartheta f(u)(x-u) \geqq 0\ .$$

The pseudoconvexity of the function ϑf guarantees $\vartheta(f(x) - f(u)) \geqq 0$. Since $\vartheta \in C_{\geqq}$, $\vartheta \neq 0$, $f(x) - f(u)$ cannot belong to $-\operatorname{int} C$. □

Remark 6.6.4. The generalized convexity assumptions of the previous theorem are satisfied if f is C-convex at u and g_j is convex at u, $\forall j = 1, ..., m$. Also appropriate invex type hypotheses can be made. We have the same result as Theorem 6.6.11 if f is C-invex at u and g_j is invex at u, $\forall j = 1, ..., m$ (for the same function η), or if the Lagrangian function $\vartheta f + \lambda g$ is invex (or pseudoconvex) at u.

The so-called strong duality theorems state that, if x^0 is a feasible point for $(V.P.)$ which satisfies some optimality notions, then there exist ϑ^0 and λ^0 such that $(x^0, \vartheta^0, \lambda^0)$ is a feasible point for $(D.V.P.)$. An additional generalized convexity assumption implies that also $(x^0, \vartheta^0, \lambda^0)$ satisfies some optimality notions for $(D.V.P.)$.

Theorem 6.6.12 (*strong duality*). If $x^0 \in WE(S)$, then there exist $\vartheta^0 \in C_{\geqq}$ and $\lambda^0 \in I\!R^m_+$ with $(\vartheta^0, \lambda^0) \neq 0$ and $(x^0, \vartheta^0, \lambda^0)$ is feasible for $(D.V.P.)$. If also $\vartheta^0 f$ and $\lambda^0 g$ are respectively pseudoconvex and quasiconvex functions, then $(x^0, \vartheta^0, \lambda^0)$ is a weakly efficient point for $(D.V.P.)$ and the objective values of the primal and the dual problem are equal.

Proof. From Theorem 6.6.6 we obtain the first part of the theorem. Now suppose that $(x^0, \vartheta^0, \lambda^0)$ is not weakly efficient for $(D.V.P.)$; then there exists $(u, \vartheta, \lambda) \in S'$ such that $f(x^0) - f(u) \in -\operatorname{int} C$. This conclusion would contradict the weak duality. Of course the objective values of the two problems are the same. □

6.7. Scalarization

In Section 6.1 we underlined the main differences between scalar and vector optimization problems. In the scalar case where a complete order is given, we can always decide, for each pair of alternatives, which is preferred. This important feature is no longer valid in the vector case because the preference order is only partial and a value function, that can order the alternatives, lacks. To overcome this difficulty, techniques which convert vector problems into appropriate scalar problems can be applied. Since the scalar optimization is widely developed, the importance of such a step is clear. Scalarization just means the replacement of a vector optimization problem by a suitable scalar problem with a real objective function which aggregates the criteria. This scalar function is called *scalarizing function*; in Economics it is called *utility* or *welfare function*. In this section different scalarization schemes will be presented and relations between the solutions of scalar problems and the different optimality notions that we introduced for the original $(V.P.)$ will be investigated.

The *first* scalarization technique goes back to Kuhn and Tucker. To our $(V.P.)$ we now associate the following scalar problem:

$$\min_{x \in S} \lambda f(x) \tag{SP_1}$$

with $\lambda \in C_>$ and $\|\lambda\| = 1$. In this manner, instead of the vector function f, we are led to minimize the scalar function $\sum_{i=1}^{p} \lambda_i f_i$ with a normalized weight vector $\lambda = (\lambda_1, ..., \lambda_p)$ of the strict positive polar $C_>$. If $C = \mathbb{R}^p_+$, this means then that we have $\lambda_i > 0$, $\forall\, i = 1, ..., p$. The vector λ can be considered as an index of the relative importance of the weights of the criteria. We will denote the points $x^0 \in S$ that solve (SP_1) by $Pos^>(S)$; their values will compose the set $Pos^>(Z)$. Then a value $z^0 = f(x^0) \in Pos^>(Z)$ when there exists some $\lambda_0 \in C_>$ ($\|\lambda_0\| = 1$) such that $\lambda_0 z^0 \leqq \lambda_0 z$, $\forall\, z \in Z$, or $\lambda_0 f(x^0) \leqq \lambda_0 f(x)$, $\forall\, x \in S$.

Theorem 6.7.1. $Pos^>(Z) = PE(Z)_{Hu}$.

Proof. If $z^0 \in Pos^>(Z)$, then there exists $\lambda_0 \in C_>$ such that $\lambda_0(z^0 - z) \leqq 0$, $\forall\, z \in Z$. Then we have $\lambda_0 z \geqq 0$, $\forall\, z \in Z - z^0$ and $\lambda_0 c > 0$, $\forall\, c \in C \backslash \{0\}$. The set $\{z : \lambda_0 z \geqq 0\}$ is a closed convex cone which contains $Z - z^0$ and C. Then the same set contains $\mathrm{cl\,conv\,cone}\,[(Z - z^0) \cup C]$. Consequently $\lambda_0 z \geqq 0$, $\forall\, z \in \mathrm{cl\,conv\,cone}\,[(Z - z^0) \cup C]$. The existence of an element $\bar{z} \in \mathrm{cl\,conv\,cone}\,[(Z - z^0) \cup C] \cap [-C \backslash \{0\}]$ is absurd because we have $\lambda_0 \bar{z} < 0$.

On the other hand let $z^0 \in PE(Z)_{Hu}$ and λ_0 an element of the nonnegative polar cone of $\mathrm{cl\,conv\,cone}\,[(Z - z^0) \cup C]$. Then we have $\lambda_0 z \geqq 0$, $\forall\, z \in Z - z^0$, and $\lambda_0 c \geqq 0$, $\forall\, c \in C$. In particular, if $c \in C \backslash \{0\}$, then

$c \in \mathrm{cl\,conv\,cone}\,[(Z-z^0)\cup C]$ and the hypothesis $z^0 \in PE(Z)_{Hu}$ assures that $c \notin -\mathrm{cl\,conv\,cone}\,[(Z-z^0)\cup C]$. Then $\lambda_0 c > 0$. In this manner we have proved that $z^0 \in Pos^{>}(Z)$. □

Remark 6.7.1. The hypotheses of Theorems 6.4.4, 6.4.6 and 6.4.9, when they are fulfilled, assure that $Pos^{>}(Z)$ can also be identified, respectively, with $PE(Z)_{Be} = PE(Z)_{G} = PE(Z)_{Ha} = PE(Z)_{He} = PE(Z)_{GBo} = PE(Z)_{SE}$ and with $PE(Z)_{Bo}$; with $PE(Z)_{KT}$; with $PE(Z)_{LBo}$.

If we can identify the solutions of (SP_1) with the set of Hurwicz's properly efficient values, it follows that $Pos^{>}(Z)$ is contained in $E(Z)$ and in $WE(Z)$. Theorem 6.7.2 is a density theorem of $Pos^{>}(Z)$ in $E(Z)$. The inclusion of $Pos^{>}(Z)$ in $E(Z)$ cannot be reversed; for the set $Z = \{(x,y) : y \geqq x^2\}$ and the cone $C = \mathbb{R}^2_+$, the point $(0,0) \in E(Z)$ but it does not minimize any linear function λz (with $\lambda \in C_{>}$, $\|\lambda\| = 1$ and $z \in Z$).

Theorem 6.7.2. If Z is a C-closed and C-convex set, then $E(Z) \subset \mathrm{cl}\,Pos^{>}(Z)$.

Proof. From Theorem 6.4.4 it follows that $PE(Z)_{He} = PE(Z)_{Hu}$ and hence $\mathrm{cl}\,PE(Z)_{He} = \mathrm{cl}\,PE(Z)_{Hu} = \mathrm{cl}\,Pos^{>}(Z)$. We can suppose $E(Z) = E(Z+C) \neq \emptyset$. Then from Theorem 6.5.10 we obtain that $Z+C$ is C-closed and C-bounded. Theorem 6.5.13 guarantees the thesis $E(Z) = E(Z+C) \subset \mathrm{cl}\,PE(Z+C)_{He} = \mathrm{cl}\,PE(Z)_{He} = \mathrm{cl}\,Pos^{>}(Z)$. □

Roughly speaking, the scalarization (SP_1) is too narrow. We have that any solution of (SP_1) is also a solution of $(V.P.)$ and is characterized as a proper value (in Hurwicz's sense). But with this scalarization we may lose some solutions of $(V.P.)$; under the hypothesis of Theorem 6.7.2 the loss is to some extent compensated as every $z^0 \in E(Z)$ can be obtained as $\lim_{k\to+\infty} z^k$, with $z^k \in Pos^{>}(Z)$.

We can associate another scalar problem to $(V.P.)$ by allowing that the multiplier λ be in the larger set $C_{\geqq}$ or that some of the weights of

f be zero (in the particular case of $Z \subset \mathbb{R}^p$ ordered in a componentwise sense):

$$\min_{x \in S} \lambda f(x) \tag{SP_2}$$

with $\lambda \in C_{\geqq}$, $\|\lambda\| = 1$. We will denote the point and the optimal value sets that solve (SP_2) by $Pos^{\geqq}(S)$ and by $Pos^{\geqq}(Z)$. It is obvious that $Pos^{>}(Z) \subset Pos^{\geqq}(Z)$.

<u>Theorem 6.7.3</u>. If Z is a closed set, then $\mathrm{cl}\, Pos^{>}(Z) \subset Pos^{\geqq}(Z)$.

<u>Proof</u>. If $\bar{z} \in \mathrm{cl}\, Pos^{>}(Z)$, there are two sequences $\{z^k\}$ and $\{\lambda_k\}$ with $z^k \in Pos^{>}(Z)$, $\lim_{k \to +\infty} z^k = \bar{z}$ and $\lambda_k \in C_{>}$, $\|\lambda_k\| = 1$, $\lambda_k z^k \leqq \lambda_k z$ for any $z \in Z$. Without loss of generality, we may assume that $\{\lambda_k\}$ converges to some $\bar{\lambda} \in C_{\geqq}$ with $\|\bar{\lambda}\| = 1$. We have that $\bar{z} \in Pos^{\geqq}(Z)$ since $\lambda_k(z^k - z) \leqq 0$, $\forall k$ and $\forall z \in Z$; taking the limit as $k \to +\infty$, we have $\bar{\lambda}(\bar{z} - z) \leqq 0$ for any $z \in Z$. □

<u>Theorem 6.7.4</u>. $Pos^{\geqq}(Z) \subset WE(Z)$.

<u>Proof</u>. If $z^0 \in Pos^{\geqq}(Z)$, then there exists a vector $\lambda_0 \in C_{\geqq}$ $(\|\lambda\| = 1)$ such that $\lambda_0(z - z^0) \geqq 0$, $\forall z \in Z$. Since we cannot have $\lambda_0 c = 0$ for $c \in -\mathrm{int}\, C$, we obtain $\lambda_0 c < 0$, $\forall c \in -\mathrm{int}\, C$. Hence it follows that $(Z - z^0) \cap (-\mathrm{int}\, C) = \emptyset$, i.e. $z^0 \in WE(Z)$. □

Without requiring further assumptions, the set $Pos^{\geqq}(Z)$ is placed between $Pos^{>}(Z) = PE(Z)_{Hu}$ and $WE(Z)$. This new result is more adequate if the ordering cone would be open. Some further hypotheses allow us to specify the placement of $Pos^{\geqq}(Z)$ with regard to $WE(Z)$. When convexity assumptions will be assumed, then for any given weakly efficient point x^0 we always will be able to find a weight vector $\lambda_0 \in C_{\geqq}$ such that z^0 minimizes the linear function $\lambda^0 z$. This weight vector is not necessarily unique; there may be more than one weight producing a solution of $(V.P.)$.

<u>Theorem 6.7.5</u>. Let Z be a C-convex set. Then $Pos^{\geqq}(Z) = WE(Z)$.

<u>Proof</u>. It is enough to prove that any $x^0 \in WE(S)$ is also a solu-

tion of (SP_2) for some $\lambda \in C_{\geqq}$, $\|\lambda\| = 1$. If $x^0 \in WE(S)$, then $Z \cap (z^0 - \operatorname{int} C) = \emptyset$. Then also $(Z + C) \cap (z^0 - \operatorname{int} C) = \emptyset$; in fact, if there exist $z \in Z$ and $c \in C$ with $z + c \in z^0 - \operatorname{int} C$, we would have $z \in z^0 - \operatorname{int} C$ contradicting the hypothesis $x^0 \in WE(S)$. As the set $(Z + C)$ is convex, we can separate the two sets $z^0 - \operatorname{int} C$ and $Z + C$ by a nonzero vector $\bar{\lambda} \in I\!R^p$, $\|\bar{\lambda}\| = 1$. So we obtain $\bar{\lambda}(z^0 - c') \leqq \bar{\lambda}(z + c'')$ for any $c' \in \operatorname{int} C$, $c'' \in C$, $z \in Z$. In particular this shows that $\bar{\lambda} \in C_{\geqq}$. From $\bar{\lambda}(z^0 - z - c'' - c') \leqq 0$, we obtain $\bar{\lambda} z^0 \leqq \bar{\lambda} z$, $\forall z \in Z$ or that $x^0 \in Pos^{\geqq}(S)$. □

Remark 6.7.2. From Theorem 6.7.5 it works out that $E(Z) = Pos^{\geqq}(Z)$ if C is an open cone and Z is a C-convex set.

Theorem 6.7.6. If there exists $\lambda_0 \in C_{\geqq}$ such that z^0 *uniquely* (or strongly) minimizes $\lambda_0 z$ over Z, then $z^0 \in E(Z)$.

Proof. If $z^0 \notin E(Z)$, then there would exist $\bar{z} \in Z$ such that $z^0 - \bar{z} \in C \backslash \{0\}$. Then $\lambda_0(z^0 - z) \geqq 0$ which contradicts the hypothesis about z^0. □

The linear scalarization can be generalized by taking into account a general $C_1 - C_2$-*monotone transformation* (not necessarily linear) of the objective function f or of the image space Z. These definitions for functions $u : I\!R^p \to I\!R^s$ depend on the ordering of $I\!R^p$ and of $I\!R^s$. For $z^1, z^2 \in I\!R^p$ we will write $z^1 \geqq z^2$ or $u(z^1) \geqq u(z^2)$ when $z^1 - z^2 \in C_1$ or $u(z^1) - u(z^2) \in C_2$ where C_1 and C_2 are the ordering cones of $I\!R^p$ and $I\!R^s$, respectively. For $s = 1$ and $C_2 = I\!R_+$ we get the particular definitions of *scalar* C-*monotone* transformations ($C = C_1$).

Definition 6.7.1. We say that a function $u : Z \subset I\!R^p \to I\!R^s$ is:

a) $C_1 - C_2$-*increasing* on Z when, for any $z^1, z^2 \in Z$, $z^1 \geq z^2$ implies $u(z^1) \geq u(z^2)$;

b) *strictly* $C_1 - C_2$-*increasing* on Z when $z^1 \geq z^2$ implies $u(z^1) \geqq u(z^2)$ and $z^1 > z^2$ implies $u(z^1) \geq u(z^2)$;

c) *weakly* $C_1 - C_2$*-increasing* on Z when $z^1 > z^2$ implies $u(z^1) \geq u(z^2)$;

d) *properly* $C_1 - C_2$*-increasing* on Z when $C_1' - C_2$-increasing with respect to some (closed pointed and convex) cone $C_1' \neq I\!R^p$ such that $C \backslash \{0\} \subset \text{int}\, C_1'$.

The definitions of vector *decreasing* functions are analogous. The proof of the following theorem is immediate.

Theorem 6.7.7. For the Definitions 6.7.1, we get: d) $\Rightarrow$ a) $\Rightarrow$ b) $\Rightarrow$ c).

Now let u be a *scalar* function which satisfies some Definitions 6.7.1. We introduce the set $Pos(u(Z))$ for the values of the solutions of the scalar problem:

$$\min_{x \in S} u\,[f(x)]\ . \qquad (SP_3)$$

Theorem 6.7.8. The set $Pos(u(Z))$ is a subset of:

a) $PE(Z)_{He}$ if u is a properly C-increasing function on Z;

b) $E(Z)$ if u is C-increasing on Z;

c) $WE(Z)$ if u is weakly C-increasing on Z.

Proof. It follows immediately from the definitions involved in the proposition. □

We are able to compare these results with the linear scalarization. Indeed, the function $u(z) = \lambda z$ with $\lambda \in C_>$ satisfies Definitions 6.7.1 a), b), c). The generalization of the statement $Pos^>(Z) = PE(Z)_{Hu}$ is only partial and we can assume the weaker inclusion $Pos^>(u(Z)) \subset E(Z)$ for all the C-increasing transformations. As for the scalar problem (SP_2) the function $u(z) = \lambda z$ with $\lambda \in C_{\geqq}$ satisfies Definitions 6.7.1 b) and c). Then Theorem 6.7.8 c) generalizes Theorem 6.7.4 to all the weak C-increasing transformations.

Now let us go back to the particular linear scalarization. The most expressive results were stated by Theorems 6.7.2 and 6.7.5, even if the solutions of the scalar problem only led to approximate solutions of $(V.P.)$ or to weakly efficient solutions. However, both these theorems required a convexity assumption on the set Z. When this hypothesis is not satisfied one can imagine convexifying the set of outcomes with a C-increasing function $H : Z \subset \mathbb{R}^p \to \mathbb{R}^p$ $(C_1 = C_2 = C)$. We will follow [Henig, 1988] by considering, for $\lambda \in C_>$ or $\lambda \in C_{\geqq}$:

$$\min_{x \in S} \lambda H\,[f(x)] \,. \qquad (SP_4)\text{-}(SP_5)$$

Theorem 6.7.9. If the following hypotheses are satisfied:

i) H is any C-increasing function on Z;

ii) there exists the inverse function H^{-1} and also H^{-1} is C-increasing on $H(Z)$;

iii) the set $H(Z)$ is C-convex and C-closed;

then the set of points that minimize $\lambda H(Z)$ for some $\lambda \in C_>$ is dense in $E(Z)$.

Proof. The hypotheses i) and ii) assure the following relations, $\forall\, z \in Z$: $H(Z) \cap (H(z) - C) = H(Z) \cap H(z - C) = H(Z \cap (z - C))$. Then $z^0 \in E(Z)$ if and only if $H(z^0) \in E(H(Z))$. Now the conclusion follows from Theorem 6.7.2. □

Theorem 6.7.9 generalizes the previous theorem 6.7.2. In any case it is not necessarily true that properly efficient points (in Hurwicz's sense) can be obtained by minimizing the functions $\lambda H(z)$ with $\lambda \in C_>$.

Example 6.7.1. For $Z = \{(z_1, z_2) : z_1 \leqq 0,\ z_2 \leqq 0,\ z_1 + z_2 \geqq -1\}$ we have $E(Z) = PE(Z)_{Hu} = \{(z_1, z_2) : z_1 \leqq 0,\ z_2 \leqq 0,\ z_1 + z_2 = -1\}$. Let us now consider the $\mathbb{R}^2_+$-increasing function $H = (H_1, H_2)$ with $H_1(z) = -\sqrt{-z_1}$ and $H_2(z) = -\sqrt{-z_2}$. If we minimize $\sum_{i=1}^{2} \lambda_i H_i(z)$

with $\lambda_i > 0$ $(i = 1, 2)$, we are not able to find out the proper solutions $(0, -1)$ and $(-1, 0)$.

Theorem 6.7.10. Let H be a C-increasing function on Z with also the inverse function H^{-1} C-increasing on $H(Z)$ and the set $H(Z)$ C-convex. If $z^0 \in E(Z)$, then there exists $\lambda_0 \in C_{\geqq}$ such that z^0 minimizes $\lambda_0 H(Z)$.

Proof. We have already noticed (see proof of Theorem 6.7.9) that, if $z^0 \in E(Z)$, then $H(z^0) \in E(H(Z))$. The claim is verified by Theorem 6.7.5. □

The second scalarization scheme (see [Lin, 1976c] and [Lin, 1977]) that we are going to introduce is also based upon the remark that linear scalarization is actually successful only under a certain directional convexity condition. This remark leads us to work out a different approach where no convexity condition is required at all in order to obtain the entire desired set of the *quasi efficient solutions*. As we will see, this notion is different but practically as good as the notion of efficient solutions. The main idea is the conversion of all but one of the multiple objectives to equality constraints and the determination of the optimal solutions of the resultant single-objective problem. This scalarization technique was called by [Lin, 1976c] the *method of proper equality constraints*. Its features make quite natural the reference to the componentwise ordering.

Definition 6.7.2. Let $\lambda \in \mathbb{R}^p_+$, $\|\lambda\| = 1$. We call a vector z^0 *λ-lineally minimum* for the set $Z \subset \mathbb{R}^p$ when it minimizes the λ-projection of the lineal subset containing z^0, given by the intersection of $\operatorname{cl} Z$ with the straight line $\{z^0 + a\lambda\,;\ a \in \mathbb{R}\}$ or $\lambda z^0 = \min\{\lambda z\,;\ z \in \operatorname{cl} Z,\ z = z^0 + a\lambda,\ a \in \mathbb{R}\}$. We will write $z^0 \in \mathcal{L}(\lambda)$.

Remark 6.7.3. It is easy to verify that z^0 satisfies the previous definition if and only if there is no negative number a such that $z^0 + a\lambda \in \operatorname{cl} Z$. Geometrically $\mathcal{L}(\lambda)$ represents those points of $\operatorname{cl} Z$ that are *exposed* in the direction λ.

Definition 6.7.3. A vector z^0 is said to be *quasi-minimum* when it is λ-lineally minimum $\forall \lambda \in I\!R^p_+$, $\|\lambda\| = 1$. We will write

$$z^0 \in Q(Z) = \bigcap_{\substack{\lambda \in I\!R^p_+ \\ \|\lambda\|=1}} \mathcal{L}(\lambda) .$$

Theorem 6.7.11. $z^0 \in Q(Z)$ if and only if $z^0 \in E(\mathrm{cl}\, Z)$.

Proof. Assume that $z^0 \in Q(Z)$, but $z^0 \notin E(\mathrm{cl}\, Z)$. Then there exists $z' \in \mathrm{cl}\, Z$ such that $z' \leq z^0$ or $z^0 - z' \in I\!R^p_+ \backslash \{0\}$ or $z^0 = z' + \lambda$, with $\lambda \in I\!R^p_+ \backslash \{0\}$. But from $\lambda z^0 = \lambda(z' + \lambda) > \lambda z'$, being always able to suppose $\|\lambda\| = 1$, we deduce that z^0 is not an element of $Q(Z)$.
On the other hand, $z^0 \notin Q(Z)$ implies that z^0 is not a λ-lineally minimum vector for some $\lambda \in I\!R^p_+$, $\|\lambda\| = 1$. Then, for some $a \in I\!R$, there exists some $z' = z^0 + a\lambda \in \mathrm{cl}\, Z$ such that $\lambda z' < \lambda z^0$. Then a must be negative. Consequently $z^0 \geq z'$ and z^0 cannot belong to $E(\mathrm{cl}\, Z)$. □

This theorem makes clear our interest in the set $Q(Z)$. If Z is a closed set, we have $Q(Z) = E(Z)$. In any case, a quasiminimum vector is a solution of $(V.P.)$ if it is inside Z; generally, it is only infinitesimally distant from Z and it possesses all the characteristics of minimal values.

So we can focus our attention on the set $Q(Z)$. To obtain it is by definition to find the interesection of the sets $\mathcal{L}(\lambda)$ with λ ranging over an infinite number of vectors of $I\!R^p_+$. It is a nearly impossible task. Then we can consider the fundamental vectors e^i $(i = 1, ..., p)$ and the sets $\mathcal{L}(e^i)$. But again the determination of only a finite number of these sets can present some difficulties (and the inclusion $E(\mathrm{cl}\, Z) = Q(Z) \subset \bigcap_{i=1}^{p} \mathcal{L}(e^i)$ is not satisfactory). Then we will consider *a particular set*, say $\mathcal{L}(e^k)$, by supposing that just the set of e^k-lineally minimum vectors is of specific interest, and we will try *to extract* the set $Q \subset \mathcal{L}(e^k)$ with some conditions that assure the equality $Q(Z) = \mathcal{L}(e^k)$.

Remark 6.7.4. Generally the inclusion $Q(Z) \subset \mathcal{L}(e^k)$ cannot be reversed. For the set $Z = \{z \in I\!R^2 : z_1 = z_2,\ 0 \leqq z_1 \leqq 1\}$ we have $E(\mathrm{cl}\, Z) =$

$Q(Z) = \{(0,0)\}$ but $\mathcal{L}(e^1) = Z$.

This process is included in scalarization techniques since it is deeply linked with the following scalar optimization problem:

$$\begin{cases} \min z_k \\ z \in Z_k(\alpha) = \{z \in \operatorname{cl} Z : z_i = \alpha_i \ (i = 1, ..., p;\ i \neq k),\ \alpha_i \in I\!R\} \end{cases} \qquad (SP_6)$$

If Q_k denotes the set of all vectors $\alpha \in I\!R^{p-1}$ for which, unless an obvious reordering of its components, there is some $z \in \operatorname{cl} Z$ such that $z_i = \alpha_i$ $(i \neq k)$ and $\varphi_k(\alpha)$ denotes $\min\{z_k\,;\ z_i = \alpha_i \ (i = 1, ..., p,\ i \neq k),\ z \in \operatorname{cl} Z\}$, we have for instance that $\mathcal{L}(e^1) = \{(\varphi_1(\alpha), \alpha)\,;\ \alpha \in Q_1\}$. So the determination of the set $\mathcal{L}(e^k)$ has been achieved by means of a scalar problem. Now we have to strengthen the relationships between e^k-lineally minimum vectors and quasi-minimum vectors.

<u>Theorem 6.7.12</u>. An element $(\varphi_k(\alpha^0), \alpha^0) \in \mathcal{L}(e^k)$ becomes a quasi-minimum vector for Z if and only if the scalar function φ_k is $I\!R^{p-1}_+$-left-decreasing at α^0 or $\varphi_k(\alpha) > \varphi_k(\alpha^0)$, $\forall\, \alpha \leq \alpha^0$ $(\alpha \in Q_k)$.

<u>Proof</u>. Suppose $\varphi_k(\alpha') \leqq \varphi_k(\alpha^0)$ for some $\alpha' \in Q_k$, $\alpha' \leq \alpha^0$. Then $\bar{\lambda} = (\varphi_k(\alpha^0), \alpha^0) - (\varphi_k(\alpha'), \alpha') \in I\!R^p_+ \backslash \{0\}$ and the vector $(\varphi_k(\alpha^0), \alpha^0)$ cannot be a $\bar{\lambda}$-lineally minimum for Theorem 6.7.11.
On the other hand, suppose that there exists $\bar{z} \in \operatorname{cl} Z$ such that $\bar{z} \leq (\varphi_k(\alpha^0), \alpha^0)$. Let $\bar{\alpha} = (\bar{z}_1, ..., \bar{z}_{k-1}, \bar{z}_{k+1}, \bar{z}_p)$. The case $\bar{\alpha} = \alpha^0$ and $\bar{z}_k < \varphi_k(\alpha^0)$ is impossible by definition of φ_k. Also the case $\bar{\alpha} \leq \alpha^0$ and $\bar{z}_k \leqq \varphi_k(\alpha^0)$ is impossible because of the hypothesis and of the definition of φ_k. Then the vector $(\varphi_k(\alpha^0), \alpha^0)$ is a quasi-minimum vector. □

<u>Remark 6.7.5</u>. [Lin, 1976c] devises some tests for a vector to be quasi-minimum by developing necessary and sufficient conditions for φ_k to be pointwise $I\!R^{p-1}_+$-left-decreasing. Conditions on derivatives are natural candidates for this kind of property; so, in these contexts, φ_k is supposed a.e. differentiable.

The previous approach can be generalized by considering the following parametric *inequality* constrained scalar problem:

$$\begin{cases} \min z_k \\ z \in W_k(\alpha) = \{z \in Z \, : \, z_i \leqq \alpha_i \ (i = 1, ..., p; \ i \neq k), \ \alpha_i \in I\!R\} \end{cases} \qquad (SP_7)$$

This variant has the disadvantage that optimization problems with inequality constraints are usually more difficult to solve than problems with equality constraints. But also in this case we can avoid any convexity and in spite of this we succeed in obtaining the entire set of the (weakly) efficient points of the set Z (and not of its closure).

Theorem 6.7.13. $z^0 \in E(Z)$ if and only if z^0 is a solution of (SP_7) for every $k = 1, ..., p$ with $\alpha = (\alpha_1, ..., \alpha_p) = z^0$.

Proof. Assume $z^0 \in E(Z)$. Suppose that z^0 does not solve (SP_7) for some k. Thus there exists $z \in Z$ such that $z_k < z_k^0$ and $z_i \leqq z_i^0$ $(i \neq k)$. This conclusion implies that $z^0 \notin E(Z)$, a contradiction.
Now assume that z^0 solves (SP_7), $\forall\, k = 1, ..., p$, with $\alpha = z^0$. If $z^0 \notin E(Z)$, then there exists $z \in Z$ such that $z \leq z^0$ or $z_k < z_k^0$ for some k, contradicting the hypothesis. □

Theorem 6.7.14. If z^0 solves (SP_7) for some k with $\alpha = z^0$ and this optimal solution is unique, then $z^0 \in E(Z)$.

Proof. If $z^0 \notin E(Z)$, then there exists $z \in Z$ with $z \leq z^0$. So we would get $z_k \leqq z_k^0$ for every $k = 1, ..., p$, a contradiction. □

In order to give a characterization of the weakly efficient values we can involve the objective function f and refer to the notion of (scalar) semistrictly quasi-convex functions. Theorem 6.7.15 can be directly compared with the previous Theorems 6.7.13 and 6.7.14.

Theorem 6.7.15. Let f_i $(i = 1, ..., p)$ be lower semicontinuous and semistrictly quasi-convex functions. Then $z^0 = f(x^0) \in WE(Z)$ if and only if z^0 solves (SP_7), with $\alpha = z^0$, for some k.

Proof. Assume $z^0 \in WE(Z)$. If z^0 does not solve (SP_7), with $\alpha = z^0$, for any k, there exist $z^1, ..., z^p \in Z$ such that we have, $\forall\, k = 1, ..., p : z_k^k < z_k^0$, $z_i^k \leqq z_i^0 \; (i \neq k)$.
The inequalities $f_k(x^k) < f_k(x^0)$ and $f_i(x^k) \leqq f_i(x^0)$, $i \neq k$, imply $f_k\Big(\frac{1}{p}\sum_{i=1}^{p} x^i\Big) < f_k(x^0)$, $\forall\, k = 1, ..., p$, contradicting $z^0 = f(x^0) \in WE(Z)$. Indeed, if $f_k(x^k) \geqq \underset{i \neq k}{\text{Max}}\; f_k(x^i)$, we have $f_k\Big(\frac{1}{p}\sum_{i=1}^{p} x^i\Big) \leqq \underset{i}{\text{Max}}\; f_k(x^i) = f_k(x^k) < f_k(x^0)$. Also, in the other case, $f_k(x^k) < \underset{i \neq k}{\text{Max}}\; f_k(x^i)$, we obtain $f_k\Big(\frac{1}{p}\sum_{i=1}^{p} x^i\Big) < \underset{i}{\text{Max}}\; f_k(x^i) \leqq f_k(x^0)$.
If z^0 is a solution of (SP_7), with $\alpha = z^0$, for some k and $z^0 \notin WE(Z)$, there exists $z \in Z$ such that $z < z^0$, contradicting the hypothesis. □

One can already find the idea of minimizing one single objective and of incorporating the others into some parametric (equality or inequality) constraints in some pioneer writings of the Forties of the last century. More recently [Wendell and Lee, 1977] and [Benson, 1978] use the method of parametric inequality and equality constraints with a scalar function which is the sum of the single objectives. [Soland, 1979] considers as objective function the scalar function $u[f]$ with some assumption of monotonicity about u. [Corley, 1980] incorporates features of both schemes of scalarization which we exposed by studying the scalar function λf $(\lambda \in C_>)$ with some inequality parametric constraints. Lin's papers were also investigated by [White, 1985]. The relations established by [Chankong and Haimes, 1983], about $(V.P.)$ and (SP_7), were extended to weakly efficient solutions by [Ruiz-Canales and Rufián-Lizana, 1995] and [Luc-Schaible, 1997].

In a *third* scheme of scalarization techniques, we group those approaches which do not resort either to weighting coefficients or to some value functions but do use a reference objective, that is, any desirable point in the objective space (no matter whether it is attainable or not). In this case the scalar function is given by a kind of distance, in the objective space, of the set Z from a value which is indicative of some aspiration levels. Then the former $(V.P.)$ is changed to minimize this kind of distance, induced by a norm. Particularly, one does not use just one norm but a family of indexed norms or a parametric norm so to obtain the (properly, weakly) efficient values as approximations to the fixed value with respect to some norm of the family.

Usually these characterizations of the solutions of $(V.P.)$, which use a distance measure concept, are quoted with the term of *compromise programming*. See [Hiriart-Urruty, 1979], [Wierzbicki, 1980], [Gearhart, 1983], [Ciligot-Travain, 1984]. We will proceed according to [Jahn, 1987].

Definition 6.7.4. For any $z^1, z^2 \in Z$ with $z^1 \leqq z^2$, the set:

$$[z^1, z^2] = (z^1 + C) \cap (z^2 - C)$$

is called the *order interval* between z^1 and z^2.

Definition 6.7.5. For any $a \in \operatorname{int} C$ the *Minkowski functional* $\|z\|_a$ on Z of the order interval $[-a, a]$ is defined by:

$$\|z\|_a = \inf \{t > 0 \; : \; t^{-1} z \in [-a, a]\} \; .$$

It is easy to check that the Minkowski functional is well defined for every $z \in Z$ (it is indeed a norm for each $a \in \operatorname{int} C$) and this parametric norm is chosen in such a way that its unit ball equals the order interval: $[-a, a] = \{z \in Z \; : \; \|z\|_a \leqq 1\}$. We will use this remark in the proof of the following theorem.

Theorem 6.7.16. Let the set Z have a strictly lower bound $\bar{z}$ or $Z \subset \bar{z} + \operatorname{int} C$. Then $z^0 \in E(Z)$ if and only if $\|z^0 - \bar{z}\|_a < \|z - \bar{z}\|_a$ for some $a \in \operatorname{int} C$ and for every $z \in Z$. More generally $z^0 \in WE(Z)$ if and only if $\|z^0 - \bar{z}\|_a \leqq \|z - \bar{z}\|_a$ for some $a \in \operatorname{int} C$ and for every $z \in Z$.

Proof. If $z^0 \in E(Z)$ one has $(z^0 - C) \cap Z = \{z^0\}$ and then $(z^0 - \bar{z} - C) \cap (Z - \bar{z}) = \{z^0 - \bar{z}\}$. As $\bar{z} - z^0 \in -\operatorname{int} C$, we obtain $Z - \bar{z} \subset \operatorname{int} C \subset$

$(\bar{z}-z^0)+\operatorname{int} C \subset (\bar{z}-z^0)+C$. So we conclude that $(\bar{z}-z^0+C)\cap(z^0-\bar{z}-C)\cap(Z-\bar{z})=\{z^0-\bar{z}\}$ or $[-z^0+\bar{z}, z^0-\bar{z}]\cap(Z-\bar{z})=\{z^0-\bar{z}\}$. The value z^0 strictly minimizes $\|z-\bar{z}\|_a$ for $a=z^0-\bar{z}$.
Conversely, let us assume that $z^0 \notin E(Z)$. Then there exists some $\tilde{z} \neq z^0$ with $\tilde{z} \in Z \cap (z^0-C)$. Consequently we have $\tilde{z}-\bar{z} \in z^0-\bar{z}-C$ which implies (by the definition of the Minkowski functional) the contradiction $\|\tilde{z}-\bar{z}\|_a \leqq \|z^0-\bar{z}\|_a$ for any $a \subset \operatorname{int} C$.
Now assume that $z^0 \in WE(Z)$. In a similar way to efficient points we obtain $(\bar{z}-z^0+\operatorname{int} C)\cap(z^0-\bar{z}-\operatorname{int} C)\cap(Z-\bar{z})=\emptyset$ or $\operatorname{int}[\bar{z}-z^0, z^0-\bar{z}]\cap(Z-\bar{z})=\emptyset$. This equality implies $\{z \in Z : \|z-\bar{z}\|_a < 1\}\cap(Z-\bar{z})=\emptyset$, for $a=z^0-\bar{z}$, or z^0 minimizes the a-distance from $\bar{z}$.
For the converse implication, let us suppose that $z^0 \in Z$ minimizes $\|z-\bar{z}\|_a$ for some $a \in \operatorname{int} C$ but $z^0 \notin WE(Z)$, Then there exists some $\tilde{z} \in Z \cap (z^0-\operatorname{int} C)$. We obtain $\tilde{z}-\bar{z} \in z^0-\bar{z}-\operatorname{int} C$ and the contradiction $\|\tilde{z}-\bar{z}\|_a < \|z^0-\bar{z}\|_a$ for any $a \in \operatorname{int} C$. □

Theorem 6.7.16 allows us to characterize the efficient and the weakly efficient values for the problem $(V.P.)$ as minimal solutions of a scalar problem with a parametric norm. No convexity assumption is required but only that the set Z has a strictly lower bound. The Minkowski functional was more recently used by [Zheng, 2000] in order to characterize the Henig properly efficient values and the superefficient values. Jahn's approach was also generalized by other authors; for instance [Luc, 1987a] does not consider a norm but a general function which satisfies a monotonicity assumption.

Remark 6.7.6. When $\mathbb{R}^p$ is componentwise ordered, it is easy to verify that (for every vector $a \in \mathbb{R}^p$ with positive components) the Minkowski functional is defined by $\|z\|_a = \max_i |z_i|/a_i$.

6.8. The Nondifferentiable Case

In Chapter Four, nonsmooth calculus was introduced for scalar functions $f : X \subset \mathbb{R}^n \to \mathbb{R}$ with its applications to optimization. In particular we recall the definition of the *Clarke generalized derivative* for a Lipschitzian function around $x^0 \in \operatorname{int} X$

$$f^0(x^0, y) = \limsup_{\substack{t \downarrow 0 \\ x \to x^0}} \frac{f(x+ty) - f(x)}{t} =$$

$$= \inf_{\substack{\varepsilon > 0 \\ \delta > 0}} \sup_{\substack{0 < t < \varepsilon \\ \|x - x^0\| < \delta}} \frac{f(x+ty) - f(x)}{t}, \tag{$*$}$$

and the definition of the *Clarke subdifferential* of f at x^0

$$\partial^0 f(x^0) = \{u \in I\!R^n \,:\, uy \leqq f^0(x^0, y),\ \forall\, y \in I\!R^n\}\ .$$

Rademacher's theorem claims that a locally Lipschitzian function is a.e. differentiable. So we were able to define the notion of the Clarke subdifferential directly:

$$\begin{aligned} J^0 f(x^0) = \operatorname{conv} \{u \in I\!R^n \mid \exists\, \{x^k\} \subset I\!R^n :\ & \lim_{k \to +\infty} x^k = x^0, \\ & f \text{ is differentiable at } x^k,\ \forall\, k \in I\!N, \\ & \lim_{k \to +\infty} \nabla f(x^k) = u\} = \operatorname{conv} \bar{J} f(x^0) \end{aligned}$$

and to prove the equality $J^0 f(x^0) = \partial^0 f(x^0)$. Finally the same notions of Clarke generalized derivatives and of Clarke subdifferential were characterized as a particular case of the definitions of K-directional derivative and of K-subdifferential

$$f^K(x^0, y) = \inf \{\xi \in I\!R \,:\, (y, \xi) \in K(\operatorname{epi} f, (x^0, f(x^0))),\ y \in I\!R^n\}$$

$$\partial^K f(x^0) = \{u \in I\!R^n \,:\, uy \leqq f^K(x^0, y),\ \forall\, y \in I\!R^n\}\ .$$

It was proved that $f^H(x^0, y) = f^0(x^0, y)$ and that $\partial^K f(x^0) = \partial^0 f(x^0) = J^0 f(x^0)$ where H is the Clarke tangent cone.

Now we wish to extend the nonsmooth calculus to vector functions by generalizing Clarke's approach that we have recalled for single-valued functions. The starting point is the definition $(*)$. Its generalization is not immediate since it involves the conical order structure of the (extended) p-dimensional space. Then we have to define the notions of the infimum and the supremum of a subset of $I\!R^p$. The easy example of the set $X = \{(x, y) : x^2 + y^2 = 1\}$ of $I\!R^2$, componentwise ordered, hints that these elements can now be "far" from the set. For the definitions of infimum and of supremum we will assume that the image space is provided with two elements that are, respectively, greater and lower than any other point of $I\!R^p$. We will call these elements $+\infty$ and $-\infty$ and we will write $\mathrm{cl}\, I\!R^p = I\!R^p \cup \{\pm\infty\}$. After this generalization, in any case, we will not yet be able to assure that the equality which appears in $(*)$ remains true.

Definition 6.8.1. An *upper* (*lower*) *bound* of the set $Z \subset I\!R^p$ is a value $\bar{z} \in \mathrm{cl}\, I\!R^p$ such that $\bar{z} \in z + C$ ($\bar{z} \in z - C$), $\forall z \in Z$. The *supremum* (*infimum*) of the set $Z \subset I\!R^p$, when it exists, is the lowest upper (greatest lower) bound of Z. We will denote it by $\sup Z$ ($\inf Z$).

Remark 6.8.1. It is obvious that the infimum $\bar{z}$ of the set Z, when it exists, is an ideal or utopia minimal value of Z (see Definition 6.3.1 a)) if $\bar{z} \in Z$.

A set Z may not possess the supremum (infimum). So we will assume a *completeness axiom*: the conical order structure of $I\!R^p$ has to assure that the supremum (infimum) always exists in $I\!R^p$ for every upper (lower) bounded set in $I\!R^p$. Note that not every conical order is complete. Consider for instance $I\!R^3$ ordered by the cone $C = \{(x, y, z) : z \geqq \max(|x|, |y|)\}$ and the set $Z = \{(x, y, z) : x = 0,\ -2 \leqq y \leqq 2,\ z = 2\sqrt{2}\}$; both the points $z_1 = (1, 0, 0)$ and $z_2 = (-1, 0, 0)$ are a lower bound for Z but any other point, which is greater than z_1 and than z_2, is not a lower bound for Z. Then the infimum of Z does not exist in $I\!R^3$.

The *set of the limit points* (or *limit class*) of a function $f : X \subset I\!R^n \to I\!R^p$ at a point $x^0 \in \mathrm{cl}\, X$ is defined as

$$L_f(x^0) = \bigcap_{N(x^0) \in \mathcal{N}} \mathrm{cl}\, \{f(x),\ x \in N(x^0)\}$$

where $\mathcal{N}$ denotes the family of the neighborhoods $N(x^0)$ of x^0. Then, if $L_f(x^0) \neq \emptyset$ the following inequalities hold:

$$\sup_{N(x^0)\in\mathcal{N}} \inf_{x\in N(x^0)} f(x) \leqq \inf L_f(x^0) \leqq \sup L_f(x^0) \leqq$$

$$\leqq \inf_{N(x^0)\in\mathcal{N}} \sup_{x\in N(x^0)} f(x) .$$

We will confine ourselves to proving the last inequality. Fix $N \in \mathcal{N}$ and let $s_N = \sup f(N)$. It follows that

$$\sup f(N) = s_N = \sup \operatorname{cl} f(N) \geqq \sup\Big\{ \bigcap_{N(x^0)\in\mathcal{N}} \operatorname{cl} f(N)\Big\} .$$

By taking the infimum with respect to $N(x^0)$, we obtain

$$\inf_{N(x^0)\in\mathcal{N}} \sup f(N) \geqq \inf_{N(x^0)\in\mathcal{N}} \sup\Big\{ \bigcap_{N(x^0)\in\mathcal{N}} \operatorname{cl} f(N)\Big\} = \sup L_f(x^0) .$$

Example 6.8.1. The now proved inequality can be strict. Consider the function $f : (-1,1) \subset \mathbb{R} \to \mathbb{R}^2$, with $\mathbb{R}^2$ componentwise ordered and the images $f\Big(-\frac{1}{n}, \frac{1}{n}\Big) = \{0\times(-\infty,-n)\}\cup\{(-\infty,-n)\times 0\}\cup\{(-1,-1)\}$, $n \in \mathbb{N}$. It works out $\sup L_f(0) = (-1,-1)$ and $\sup_{x\in(-\frac{1}{n},\frac{1}{n})} f(x) = (0,0)$, $\forall n \in \mathbb{N}$.

By the previous inequality we deduce that generally the notion of maximum limit can be defined differently, in the vector case. In order to extend the definition of the Clarke generalized derivative $(*)$ one needs to specify the approach. From now on we will intend maximum limit as *the supremum of the set of the limit points of f at x^0*. This choice has the quality of also involving the topological structure of the image space Z. Moreover, the previous inequality hints at a more careful analysis if it is worked out by means of $\sup L_f(x^0)$ rather than $\inf_{N(x^0)\in\mathcal{N}} \sup_{x\in N(x^0)} f(x)$.

Let $q(x,t,y) = t^{-1}[f(x+ty) - f(x)]$. Them the set

$$Df(x^0,y) = \bigcap_{\substack{N(x^0)\in\mathcal{N}\\ \varepsilon>0}} \operatorname{cl}\{q(x,t,y)\,;\, x \in N(x^0),\ 0 < t < \varepsilon\}$$

represents the limit class of q or the limits of sequences such as $t_k^{-1}[f(x^k + t_k y) - f(x^k)]$ with $\lim_{k \to +\infty} x^k = x^0$ and $\lim_{k \to +\infty} t_k = 0^+$.

Lemma 6.8.1. If $f : X \subset I\!R^n \to I\!R^p$ is a locally Lipschitz function at $x^0 \in \operatorname{int} X$ (with Lipschitz constant L) then there exist a neighborhood $N(x^0)$ and $\varepsilon > 0$ such that $q(x, t, y) \in L \|y\| B$, $\forall x \in N(x^0)$ and $\forall t \in (0, \varepsilon)$, where B denotes the unit ball.

Proof. Let $\bar{N}(x^0)$ be a neighborhood where f is a Lipschitzian function. It is always possible to find a neighborhood $N(x^0)$ and $\varepsilon > 0$ such that $N(x^0) + \varepsilon V(0) \subset \bar{N}(x^0)$ for any neighborhood V of the origin. From the definition of Lipschitzian function, $\forall x \in N(x^0)$ and $\forall t \in (0, \varepsilon)$, we get $\|q(x, t, y)\| \leqq L \|y\|$. □

Theorem 6.8.1. Let $f : X \subset I\!R^n \to I\!R^p$ be a locally Lipschitz function at $x^0 \in \operatorname{int} X$. Then, $\forall y \in I\!R^n$:

a) $Df(x^0, y) \neq \emptyset$;

b) $Df(x^0, y)$ is a compact set;

c) $Df(x^0, \alpha y) = \alpha Df(x^0, y)$, $\forall \alpha > 0$;

d) $D(\alpha f(x^0, y)) = \alpha Df(x^0, y)$, $\forall \alpha > 0$;

e) $D(-f(x^0, y)) = Df(x^0, -y)$;

f) $Df(x^0, y^1 + y^2) \subset Df(x^0, y^1) + Df(x^0, y^2)$, $\forall y^1, y^2 \in I\!R^n$;

g) $D(f_1 + f_2)(x^0, y) \subset Df_1(x^0, y) + Df_2(x^0, y)$ with f_1 and f_2 locally Lipschitzian at x^0.

Proof.

a) It follows immediately from Lemma 6.8.1.

b) Also this statement follows immediately from the definition of $Df(x^0, y)$ and from the previous lemma.

c) and d) See the definition of $q(x, t, y)$.

e) Let $L \in D(-f(x^0,y))$. Then there exist $\{x^k\}$ and $\{t_k\}$ with $\lim_{k\to+\infty} x^k = x^0$ and $\lim_{k\to+\infty} t_k = 0^+$ such that $\lim_{k\to+\infty} t_k^{-1}[-f(x^k+t_ky)+f(x^k)] = L$. It follows that $\lim_{k\to+\infty} t_k^{-1}[f(x^k+t_ky-t_ky)-f(x^k+t_ky)] = L \in Df(x^0,-y)$. In the same manner one can prove $Df(x^0,-y) \subset D(-f(x^0,y))$.

f) If $L \in Df(x^0,y^1+y^2)$, there exist $\{x^k\}$ and $\{t_k\}$ with $\lim_{k\to+\infty} x^k = x^0$, $\lim_{k\to+\infty} t_k = 0^+$ and $\lim_{k\to+\infty} q(x^k,t_k,y^1+y^2) = L$. On the other hand we have

$$\begin{aligned} q(x^k,t_k,y^1+y^2) &= t_k^{-1}[f(x^k+t_ky^1+t_ky^2)-f(x^k)] = \\ &= t_k^{-1}[f(x^k+t_ky^1+t_ky^2)-f(x^k+t_ky^1)] + \\ &\quad + t_k^{-1}[f(x^k+t_ky^1)-f(x^k)] = \\ &= q(x^k+t_ky^1,t_k,y^2)+q(x^k,t_k,y^1)\ . \end{aligned}$$

Thanks to Lemma 6.8.1 this last sequence (or eventually a subsequence) converges to an element of $Df(x^0,y^2)+Df(x^0,y^1)$.

g) The proof is the same as in f). □

Definition 6.8.2. Let $f : I\!R^n \to I\!R^p$ be locally Lipschitz at $x^0 \in \operatorname{int} X$. The function $f^0 : I\!R^n \to I\!R^p$

$$f^0(x^0,y) = \sup Df(x^0,y)$$

is said to be the *generalized directional derivative* of f at x^0.

Remark 6.8.1. The Lipschitz assumption implies (see also Theorem 6.8.1 a)) $f^0(x^0,y) > -\infty$. The usual hypothesis $\operatorname{int} C \neq \emptyset$ allows us to refuse the case $f^0(x^0,y) = +\infty$. Indeed from Theorem 6.8.1 b) we have the boundedness of $Df(x^0,y)$ or the existence of a neighborhood $U_n(0)$ such that $Df(x^0,y) \subset U_n(0)$. If $c \in \operatorname{int} C$, there exists $U_r(0)$ with $c - U_r(0) \subset C$ or $Df(x^0,y) \subset \dfrac{n}{r}c - C$. The set $Df(x^0,y)$ is above bounded and the completeness axiom assures $f^0(x^0, y) < +\infty$.

Theorem 6.8.2. Let $f, f_1, f_2 : X \subset \mathbb{R}^n \to \mathbb{R}^p$ be locally Lipschitz functions at $x^0 \in \operatorname{int} X$. The following properties hold (for any $y, y^1, y^2 \in \mathbb{R}^n$):

a) $f^0(x^0, \alpha y) = \alpha f^0(x^0, y)\,, \quad \forall\, \alpha > 0\,;$

b) $(\alpha f)^0(x^0, y) = \alpha f^0(x^0, y)\,, \quad \forall\, \alpha > 0\,;$

c) $(-f)^0(x^0, y) = f^0(x^0, -y)\,;$

d) $f^0(x^0, y^1 + y^2) \leqq f^0(x^0, y^1) + f^0(x^0, y^2)\,;$

e) $(f_1 + f_2)^0(x^0, y) \leqq f_1^0(x^0, y) + f_2^0(x^0, y)\,.$

Proof. One can proceed as in Theorem 6.8.1. □

In the same spirit of the scalar case, we now introduce the notion of generalized subdifferential. It comes down to the definition in Chapter Four when f is a scalar function. In Definition 6.8.3, $L(\mathbb{R}^n, \mathbb{R}^p)$ will denote the space of the linear maps from $\mathbb{R}^n$ to $\mathbb{R}^p$.

Definition 6.8.3. The set

$$\partial^0 f(x^0) = \{A \in L(\mathbb{R}^n, \mathbb{R}^p) : Ay \leqq f^0(x^0, y),\ \forall\, y \in \mathbb{R}^n\}$$

is said to be the *generalized subdifferential* at $x^0 \in \operatorname{int} X$ of the function $f : X \subset \mathbb{R}^n \to \mathbb{R}^p$, which is locally Lipschitzian at x^0.

Theorem 6.8.3. The generalized subdifferential $\partial^0 f(x^0)$ is a nonempty convex set with

$$f^0(x^0, y) = \max_{A \in \partial^0 f(x^0)} Ay\ .$$

Proof. Fix any $y \neq 0$. We can define a map $T(\alpha y)$, linear with respect to α and such that $T(0) = 0$ and $T(y) = f^0(x^0, y)$. For its linearity we have $T(\alpha y) = f^0(x^0, \alpha y)$, $\forall\, \alpha \geqq 0$. From $0 \leqq f^0(x^0, y) + f^0(x^0, -y)$ it follows, for any $\alpha \geqq 0$, $T(-\alpha y) = -T(\alpha y) = -f^0(x^0, \alpha y) \leqq f^0(x^0, -\alpha y)$. The inequality $T(\alpha y) \leqq f^0(x^0, \alpha y)$, $\forall\, \alpha \in \mathbb{R}$, and the Hahn-Banach theorem (extended to functions with values in a complete ordered linear space) produce a linear map A such that $Ay \leqq f^0(x^0, y)$, $\forall\, y \in \mathbb{R}^n$. We have $\partial^0 f(x^0) \neq \emptyset$, as $A \in \partial^0 f(x^0)$, and $f^0(x^0, y) = \max\{Ay\,;\ A \in \partial^0 f(x^0)\}$. The convexity of the set $\partial^0 f(x^0)$ is immediate. □

Remark 6.8.2. [Thibault, 1980] has deepened the analogy with the scalar case by proving, under some additional slight hypotheses, that $\partial^0 f(x^0)$ is a compact set also in infinite dimensional spaces.

Another useful property of $\partial^0 f(x^0)$ is given by its operator convexity.

Definition 6.8.4. A set $U \subset L(\mathbb{R}^n, \mathbb{R}^p)$ is said to be *operator convex* when, $\forall\, A_1, A_2 \in U$ and $\forall\, \Lambda \in L_{\geqq}(\mathbb{R}^p, \mathbb{R}^p)$ with $(I_p - \Lambda) \in L_{\geqq}(\mathbb{R}^p, \mathbb{R}^p)$, we have

$$\Lambda A_1 + (I_p - \Lambda)\, A_2 \in U$$

where I_p is the identity map in $\mathbb{R}^p$ and $L_{\geqq}(\mathbb{R}^p, \mathbb{R}^p)$ denotes the set of the linear maps $A : \mathbb{R}^p \to \mathbb{R}^p$ such that $Ac \in C$, $\forall\, c \in C$.

Definition 6.8.5. The set

$$\operatorname{coo} U = \{\sum_{i=1}^{n} \Lambda_i A_i\,;\ A_i \in U,\ \Lambda_i \in L_{\geqq}(\mathbb{R}^p, \mathbb{R}^p),\ \sum \Lambda_i = I_p\,;\ n \in \mathbb{N}\}$$

is called the *operator convex hull* of the set $U \subset L(\mathbb{R}^p, \mathbb{R}^p)$.

In the convex hull of a set $U \subset L(\mathbb{R}^n, \mathbb{R}^p)$ the weighting multipliers can always be considered as a diagonal matrix whose elements are $\lambda_j^{(i)} = \lambda^{(i)}$, $j = 1, \ldots, p$, $0 \leqq \lambda^{(i)} \leqq 1$. Then an easy comparison leads to the inclusion $\operatorname{conv} U \subset \operatorname{coo} U$. From Definition 6.8.4 we arrive at this proposition immediately.

Theorem 6.8.4. The generalized subdifferential $\partial^0 f(x^0)$ is an operator convex set.

Remark 6.8.3. When $C = I\!R^p_+$, the generalized subdifferential of the vector function f is given by the product of the subdifferentials $\partial^0 f_i(x^0)$, defined in Chapter Four for scalar functions. To this end, one has to prove that $\sup\, Df(x^0, y) = \prod_{i=1}^{p} \sup\, Df_i(x^0, y)$ for any fixed y. Then $Ay \leqq f^0(x^0, y)$ if and only if $A_i y \leqq f_i^0(x^0, y)$ or $A \in \partial^0 f(x^0)$ if and only if $A_i \in \partial^0 f_i(x^0)$. Let $l = (l_1, ..., l_p) \in Df(x^0, y)$. Then $l_i \in Df_i(x^0, y)$. Hence $Df(x^0, y) \subset \prod_{i=1}^{p} Df_i(x^0, y)$ and consequently we get

$$\sup\, Df(x^0, y) \leqq \sup \prod_{i=1}^{p} Df_i(x^0, y) = \prod_{i=1}^{p} \sup\, Df_i(x^0, y)\ .$$

Consider the case $\sup\, Df(x^0, y) \leq \sup \prod_{i=1}^{p} Df_i(x^0, y)$. There exist $\omega \in I\!R^p$ with $\sup\, Df(x^0, y) \leq \omega \leq \sup \prod_{i=1}^{p} Df_i(x^0, y)$, an index i and some sequences $\{x^k\}$ and $\{t_k\}$ converging to x^0 and 0^+ such that $\omega_i < s_i = \lim_{k \to +\infty} t_k^{-1}[f_i(x^k + t_k y) - f_i(x^k)]$. Eventually by taking some subsequences one can get an element $s = (s_1, ..., s_p) \in Df(x^0, y)$. It works out $s \leqq \sup\, Df(x^0, y) \leq \omega$ or $s_j \leqq \omega_j$ $(\forall j = 1, ..., p)$ but $\omega_i < s_i$.

For scalar functions, $f^0(x^0, y)$ was also defined as a particular case of k-directional derivative. We wish to show that this equivalence is still valid for the vector case. In the following definition, H denotes the Clarke tangent cone; the epigraph of $f : X \subset I\!R^n \to I\!R^p$ (denoted by epi f) is the set $\{(x, z) : x \in X\, ;\ z - f(x) \in C\}$.

Definition 6.8.6. The function $f^H(x^0, y) : I\!R^n \to I\!R^p$

$$f^H(x^0, y) = \inf \{\xi \in I\!R^p\ :\ (y, \xi) \in H(\text{epi}\, f, (x^0, f(x^0))),\ y \in I\!R^n\}$$

is called *directional derivative* of $f : X \subset \mathbb{R}^n \to \mathbb{R}^p$ at $x^0 \in \operatorname{int} X$.

Theorem 6.8.5. Let $f : X \subset \mathbb{R}^n \to \mathbb{R}^p$ be a locally Lipschitz function at $x^0 \in \operatorname{int} X$. Then $f^0(x^0, y) = f^H(x^0, y)$, $\forall y \in \mathbb{R}^n$.

Proof. Fix any $y \in \mathbb{R}^n$. Let ξ such that $(y, \xi) \in H(\operatorname{epi} f, (x^0, f(x^0)))$ and $l \in Df(x^0, y)$. Then there exist $\{x^k\}$ and $\{t_k\}$ with $\lim_{k \to +\infty} x^k = x^0$ and $\lim_{k \to +\infty} t_k = 0^+$ such that $l = \lim_{k \to +\infty} t_k^{-1}[f(x^k + t_k y) - f(x^k)]$. By the continuity of f, $\{f(x^k)\}$ converges to $f(x^0)$. By the definition of Clarke's tangent cone there exists a sequence $\{(y^k, \xi^k)\}$ converging to (y, ξ) with $(x^k, f(x^k)) + t_k(y^k, \xi^k) \in \operatorname{epi} f$. Then $\xi^k - t_k^{-1}[f(x^k + t_k y^k) - f(x^k)] \in C$ or $\xi - l \in C$. So we have obtained $\inf \xi \geqq \sup l$ or $f^H(x^0, y) \geqq f^0(x^0, y)$. Now we have to prove that any upper bound ξ of $f^0(x^0, y)$ is also an upper bound of $f^H(x^0, y)$. Consider a sequence $\{t_k, x^k, z^k\}$ converging to $(0, x^0, f(x^0))$ with $(x^k, z^k) \in \operatorname{epi} f$. Eventually by taking some subsequences, $t_k^{-1}[f(x^k + t_k y) - f(x^k)]$ converges to some $l \in Df(x^0, y)$. Then

$$z^k + t_k(\xi + o(1)) \in f(x^k) + C + t_k(l + C + o(1)) \subset f(x^k + t_k y) + C \ .$$

It works out that $[(x^k, z^k) + t_k(y, \xi + o(1))] \in \operatorname{epi} f$ or $(y, \xi) \in H(\operatorname{epi} f, (x^0, f(x^0)))$ or $\xi \geqq f^H(x^0, y)$. □

As a consequence of this theorem we have that the geometrical approach of Definition 6.8.6 leads to the same definition and to the same properties for the generalized subdifferential $\partial^0 f(x^0)$. The equality of Theorem 6.8.5 was extended by [Thibault, 1982] to non-Lipschitzian functions.

In the scalar case the generalized subdifferential can also be characterized through Rademacher's theorem. This theorem is still valid with vector functions. Then we are able to introduce the definition of the so-called *generalized Jacobian* and to show the relationships with Definition 6.8.3. The definition of generalized Jacobian already appears in some papers of the early Seventies and even today this definition is the most widespread and quoted.

Lemma 6.8.2 (*Rademacher's theorem*). A vector Lipschitzian function on an open set $A \subset \mathbb{R}^n$ is a.e. differentiable or it is differentiable on a subset $\Omega \subset A$ with $m(A \backslash \Omega) = 0$.

Definition 6.8.7. Let $f : X \subset \mathbb{R}^n \to \mathbb{R}^p$ be a locally Lipschitz function at $x^0 \in \operatorname{int} X$. The set:

$$J^0 f(x^0) = \operatorname{conv} \bar{J} f(x^0) = \operatorname{conv} \{A \in L(\mathbb{R}^n, \mathbb{R}^p) : A = \lim_{\substack{x^k \to x^0 \\ x^k \in \Omega}} Jf(x^k)\}$$

is said to be the *generalized Jacobian* of f at x^0.

Theorem 6.8.6. The following properties hold:

i) $J^0 f(x^0) \neq \emptyset$;

ii) $J^0 f(x^0)$ is a convex set;

iii) $J^0 f(x^0)$ is a compact subset in $L(\mathbb{R}^n, \mathbb{R}^p)$;

iv) $J^0 f(x^0) \subset \prod_{i=1}^{p} J^0 f_i(x^0)$.

Proof.

i) It is enough to remark that $\lim_{x^k \to x^0} Jf(x^k)$, $x^k \in \Omega$, exists since f is locally Lipschitzian at x^0 and hence Jf is norm-bounded.

ii) It is an immediate consequence of Definition 6.8.7.

iii) The convex hull of a compact set is compact. Thus it is sufficient to show that $\bar{J} f(x^0)$ is a compact set or a closed set. Let $\{A_i\} \subset \bar{J} f(x^0)$, with $\lim_{i \to +\infty} A_i = A$. Then $A_i = \lim_{k \to +\infty} Jf(x_i^k)$ with $x_i^k \in \Omega$, $\lim_{k \to +\infty} x_i^k = x^0$. We obtain $A \in \bar{J} f(x^0)$.

iv) We can prove the inclusion for $\bar{J}$ but the inclusion $\bar{J} f(x^0) \subset \prod \bar{J} f_i(x^0)$ easily follows from the definition of the sets $\bar{J}$. □

Theorem 6.8.7.

i) $J^0(f+g)(x^0) \subset J^0 f(x^0) + J^0 g(x^0)$;

ii) $J^0(\alpha f)(x^0) = \alpha J^0 f(x^0)$, $\forall\, \alpha \in I\!R$.

Proof.

i) It is sufficient to prove the inclusion for the sets $\bar{J}$. Let $A \in \bar{J}(f+g)(x^0)$, $A = \lim_{k \to +\infty} J(f+g)(x^k) = \lim_{k \to +\infty} [Jf(x^k) + Jg(x^k)]$, with x^k converging to x^0 and $x^k \in \Omega$. By taking some possible subsequences, $Jf(x^k)$ and $Jg(x^k)$ converge to some elements of $\bar{J}f(x^0)$ and of $\bar{J}g(x^0)$, respectively. Then $A \in \bar{J}f(x^0) + \bar{J}g(x^0)$.

ii) This can be proved in an analogous manner. □

But now the generalized Jacobian definition is not equivalent to the definition of the generalized subdifferential. The inclusion relation and its consequences about optimization problems will be deduced from the following theorem.

Theorem 6.8.8. $\partial^0 f(x^0) = \mathrm{cl\;coo}\, J^0 f(x^0)$.

Proof. First we will prove that $f^0(x^0, y) = \sup_{A \in \bar{J}f(x^0)} Ay$, $\forall\, y \in I\!R^n$. Let $A \in \bar{J}f(x^0)$, $A = \lim_{k \to +\infty} Jf(x^k)$, with x^k converging to x^0 and $x^k \in \Omega$. For any $y \in I\!R^n$ we get $Ay = \lim_{k \to +\infty} Jf(x^k)\, y = \lim_{k \to +\infty} \lim_{i \to +\infty} q(x^k, t_i, y)$ $= \lim_{k \to +\infty} q(x^k, t_{i_k}, y)$. It works out $Ay \in Df(x^0, y)$ and hence $\sup_{A \in \bar{J}f(x^0)} Ay \leqq f^0(x^0, y)$. For the opposite inequality consider any $\lim_{k \to +\infty} q(x^k, t_k, y) \in Df(x^0, y)$. From $\lim_{k \to +\infty} q(x^k, t_k, y) \leqq \sup_{x^k \in \Omega} \lim_{x^k \to x^0} \lim_{t_k \to 0+} \dfrac{f(x^k + t_k y) - f(x^k)}{t_k}$, we obtain $\lim_{k \to +\infty} q(x^k, t_k, y) \leqq \sup_{A \in \bar{J}f(x^0)} Ay$ and hence $f^0(x^0, y) \leqq \sup_{A \in \bar{J}f(x^0)} Ay$.

Now $J^0 f(x^0) = \mathrm{conv}\, \bar{J}f(x^0) \subset \mathrm{coo}\, \bar{J}f(x^0)$ implies $\mathrm{coo}\, J^0 f(x^0) \subset \mathrm{coo}\, \bar{J}f(x^0)$. As $\bar{J}f(x^0) \subset J^0 f(x^0)$ implies the opposite inclusion, we have $\mathrm{coo}\, J^0 f(x^0) = \mathrm{coo}\, \bar{J}f(x^0)$ or $\mathrm{cl\;coo}\, J^0 f(x^0) = \mathrm{cl\;coo}\, \bar{J}f(x^0) = \partial^0 f(x^0)$.

The last equality is a direct consequence of the theorem of Zaslavskii (see [Rubinov, 1977, p. 124]). □

Theorem 6.8.8 states that generally $J^0 f(x^0)$ is a subset of $\delta^0 f(x^0)$. This inclusion justifies the choice that necessary conditions for $(V.P.)$ will be given in sharper terms of generalized Jacobian. Generally the last part of this section will propose a non-smooth version of some theorems which were demonstrated in Section 6 in the differentiable case.

Theorem 6.8.9. If f is a locally Lipschitz function at $x^0 \in \operatorname{int} X$ and x^0 is an unconstrained weakly local efficient point of f, then $J^0 f(x^0)y \cap (-\operatorname{int} C)^c \neq \emptyset$ for every $y \in I\!R^n$.

Proof. We suppose that we can find some $\bar{y} \in I\!R^n$ such that $J^0 f(x^0)\bar{y} \subset -\operatorname{int} C$. Since the set $J^0 f(x^0)\bar{y}$ is convex and compact, there exists a closed convex neighborhood V of it contained in $-\operatorname{int} C$. By the upper semi-continuity of the generalized Jacobian, there exists $\delta > 0$ such that $J^0 f(x)\bar{y} \subset V$ for every $x \in \operatorname{conv}\{x^0, x^0 + \delta\bar{y}\}$. As V is convex and closed, we have $\operatorname{cl}\operatorname{conv}\{J^0 f(x)\bar{y}\,;\, x \in \operatorname{conv}\{x^0, x^0 + \delta\bar{y}\}\} \subset V$. By applying the mean value theorem, we obtain:

$$f(x^0 + t\bar{y}) - f(x^0) \in$$

$$\in \operatorname{cl}\operatorname{conv}\{J^0 f(x)\bar{y}\,;\, x \in \operatorname{conv}\{x^0, x^0 + \delta\bar{y}\}\} \subset tV \subset -\operatorname{int} C$$

$\forall t \in (0, \delta)$, which contradicts the assumption of the theorem. □

Remark 6.8.4. The condition $J^0 f(x^0)y \cap (-\operatorname{int} C)^c \neq \emptyset$, $\forall y \in I\!R^n$, is equivalent to $0 \in \vartheta^0 J^0 f(x^0)$ for some $\vartheta^0 \in C_{\geqq} \backslash \{0\}$.

Indeed, if there exists some $\bar{y} \in I\!R^n$ with $J^0 f(x^0)\bar{y} \subset -\operatorname{int} C$, then one has $\vartheta A\bar{y} < 0$, $\forall \vartheta \in C_{\geqq} \backslash \{0\}$ and $\forall A \in J^0 f(x^0)$. Conversely, if $0 \neq \vartheta J^0 f(x^0)$ for every $\vartheta \in C_{\geqq} \backslash \{0\}$, one can separate the origin and the convex compact set $\{\vartheta J^0 f(x^0)\,;\, \vartheta \in C_{\geqq} \backslash \{0\}\}$. Thus, there is some $\bar{y} \in I\!R^n$ such that $\vartheta A\bar{y} < 0$, $\forall \vartheta \in C_{\geqq} \backslash \{0\}$ and $\forall A \in J^0 f(x^0)$. But this inequality implies $J^0 f(x^0)\bar{y} \subset -\operatorname{int} C$.

Theorem 6.8.10. Let f be a locally Lipschitz function at $x^0 \in \operatorname{int} X$ and $J^0 f(x^0)y \in (-C)^c$, $\forall y \in \mathbb{R}^n \backslash \{0\}$. Then x^0 is a local efficient solution of $(V.P.)$.

Proof. If x^0 would not be a local efficient solution of $(V.P.)$, then there would be a sequence $\{x^k\} \subset S$ converging to x^0 such that $f(x^k) - f(x^0) \in -C \backslash \{0\}$. Without loss of generality we may assume that the sequence $\{y^k\}$, with $y^k = (x^k - x^0)/\|x^k - x^0\|$, converges to some $\bar{y} \in \mathbb{R}^n$. Now we can consider the convex compact set $J^0 f(x^0)\bar{y}$ and take a closed convex neighborhood V of it contained in $(-C)^c$. Then there exists $\varepsilon > 0$ such that $J^0 f(x)y \subset V$ whenever $\|x - x^0\| < \varepsilon$ and $\|y - \bar{y}\| < \varepsilon$. For k large enough, we have $\|x^k - x^0\| < \varepsilon$ and $\|y^k - \bar{y}\| < \varepsilon$. So, for such k, by using the mean value theorem, we obtain:

$$
\begin{aligned}
& f(x^k) - f(x^0) \in \\
& \in \operatorname{cl} \operatorname{conv} \{J^0 f(x)(x^k - x^0)\,;\, x \in \operatorname{conv}\{x^0, x^k\}\} \subset \|x^k - x^0\|\, V \subset (-C)^c
\end{aligned}
$$

which contradicts the assumption. □

Theorem 6.8.11. If x^0 is a weakly local efficient point for $(V.P.)$, where f and g are locally Lipschitz functions at x^0, then there exist $\vartheta^0 \in C_{\geqq}$ and $\lambda^0 \geqq 0$, $(\vartheta^0, \lambda^0) \neq 0$, such that $0 \in J^0(\vartheta^0 f + \lambda^0 g)(x^0)$, $\lambda^0 g(x^0) = 0$.

Proof. See the proof of the previous necessary condition: the constrained case can be easily reduced to the unconstrained case. □

Theorem 6.8.12. Let us consider $(V.P.)$ where f is a C-convex function and every g_j is convex. If $x^0 \in S$ satisfies the conditions $0 \in J^0(\vartheta^0 f + \lambda^0 g)(x^0)$, for some $\vartheta^0 \in C_{\geqq}$ and $\lambda^0 \geqq 0$, and $\lambda^0 g(x^0) = 0$, then $x^0 \in WE(S)$.

Proof. Since the scalar function $\vartheta^0 f + \lambda^0 g$ is convex, the condition $0 \in J^0(\vartheta^0 f + \lambda^0 g)(x^0)$ assures that x^0 is a minimum point of the Lagrangian function:

$$
0 \leqq \sum_1^p \vartheta_i^0 (f_i(x) - f_i(x^0)) + \sum_1^m \lambda_j^0 (g_j(x) - g_j(x^0)) =
$$

$$= \sum_1^p \vartheta_i^0 (f_i(x) - f_i(x^0)) + \sum_1^m \lambda_j^0 g_j(x) \leqq$$

$$\leqq \sum_1^p \vartheta_i^0 (f_i(x) - f_i(x^0)) .$$

From this inquality we deduce that no $x \in S$ can exist such that $f(x) < f(x^0)$. □

References to Chapter VI

B. AGHEZZAF and M. HACHIMI (1999), Second order optimality conditions in multiobjective optimization problems, J.O.T.A., 37–50.

A. BAKUNDA (1992), Connexion entre les points efficaces et les point relativement faiblement efficaces in programmation multiobjective par rapport a un cone, Bull. Soc. Roy. Sci. Liege, 377–395.

D. BATHIA and S. AGGARWAL (1992), Optimality and duality for multiobjective nonsmooth programming, Eur. J. Oper. Res., 360–367.

H.P. BENSON (1978), Existence of efficient solutions for vector maximization problems, J.O.T.A., 569–580.

H.P. BENSON (1979), An improved definition of proper efficiency for vector minimization with respect to cones, J. Math. Anal. Appl., 232–241.

H.P. BENSON (1983), Efficiency and proper efficiency in vector optimization with respect to cones, J. Math. Analysis and Appl., 173–189.

E.M. BEDNARCZUK (1995), Berge – type theorems for vector optimization problems, Optimization, 373–384.

G. BITRAN (1978), Duality in nonlinear multiple criteria optimization problems, J.O.T.A., 569–580.

G. BITRAN and T.L. MAGNANTI (1979), The structure of admissible point with respect to cone dominance, J.O.T.A., 573–614.

N. BOISSARD (1995), Characterizations of efficient sets in vector optimization, Bull. Austr. Math. Soc., 43–54.

S. BOLINTINEANU and M. EL MAGHRI (1998), Second-order efficiency conditions and sensitivity of efficient points, J.O.T.A., 569–592.

J.M. BORWEIN (1977), Proper efficiency for minimization with respect to cones, SIAM J. Control Optim., 57–63.

J.M. BORWEIN (1980), The geometry of Pareto efficiency over cones, Math. Operationsforsch. Stat. Ser. Optim., 235–248.

J.M. BORWEIN (1983), On the existence of Pareto efficient points, Math. Op. Res., 64–73.

J.M. BORWEIN and D. ZHUANG (1993), Superefficiency in vector optimization, Trans. Amer. Math. Soc., 105–122.

K. BREKZIS and M. PIRLOT (1991), Proper efficiency in non convex vector-maximization-problems, European J. Op. Res., 74–81.

A. CAMBINI and L. MARTEIN (1992), Generalized concavity and optimality conditions in vector and scalar optimization; in: Proceedings of the IVth International Workshop on Generalized Convexity, Pecs.

A. CAMBINI and L. MARTEIN (1994), On the existence of efficient points, Optimization, 283–290.

E. CASTAGNOLI and P. MAZZOLENI (1989), Scalar and vector generalized convexity; in: F.H. Clarke, V.F. Dem'yanov and F. Giannessi (eds.), Nonsmooth Optimization and Related Topics, Plenum Press, New York.

L. CESARI and M.B. SURYANARAYANA (1978), An existence theorem for Pareto problems, Non Linear Analysis, Theory, Methods and Applications, 225–233.

V. CHANKONG and Y.Y. HAIMES (1983), Multiobjective decision making: theory and methodology, North-Holland.

V. CHANKONG, Y.Y. HAIMES, J. THADATHIL and S. ZIONTS (1985), Multiple criteria optimization: a state of the art review, Springer, Berlin.

M. CILIGOT-TRAVAIN (1995), On Lagrange-Kuhn-Tucker multipliers for Pareto optimization problems, Numer. Funct. Anal. and Optimiz.,

689–693.

H.W. CORLEY (1980a), An existence result for maximization with respect to cones, J.O.T.A., 277–281.

H.W. CORLEY (1980b), A new scalar equivalence for Pareto optimization, IEEE Trans. Automat. Contr., 829–830.

H.W. CORLEY (1985), On optimality conditions for maximizations with respect to cones, J.O.T.A., 67–78.

B.D. CRAVEN (1989a), Vector-valued optimization; in: S. Schaible and W.T. Ziemba (eds.), Generalized Concavity in Optimization and Economics, Academic Press, New York.

B.D. CRAVEN (1989b), Nonsmooth multiobjective programming, Numer. Funct. Anal. and Optimiz., 49–64.

B.D. CRAVEN and X.Q. YANG (1991), A nonsmooth version of alternative theorem and nonsmooth multiobjective programming, Utilitas Mathematica, 117–128.

L.N. DAS and S. NANDA (1995), Proper efficiency conditions and duality for multiobjective programming problems involving semilocally invex functions, Optimization, 43–51.

J.P. DAUER and W. STADLER (1986), A survey of vector optimization in infinite-dimensional spaces, II, J.O.T.A., 265–241.

J.P. DAUER and R.J. GALLAGHER (1990), Positive proper efficient points and related cone results in optimization theory, SIAM J. Control and Opt., 158–172.

R. DURIER (1988), Weighting factors results in vector optimization, J.O.T.A., 411–430.

R.R. EGUDO (1986), Efficiency and generalized convex duality for multiobjective programs, J. Math. Analysis and Appl., 84–94.

R.R. EGUDO (1987), Proper efficiency and multiobjective duality in nonlinear programming, J. Inf. and Opt. Sc., 155–166.

R.R. EGUDO and M.A. HANSON (1987), Multiobjective duality with invexity, J.O.T.A., 469–477.

R.R. EGUDO, T. WEIR and B. MOND (1992), Duality without constraint qualification for multiobjective Programming, J. Austral. Math. Soc., 531–544.

M. EHRGOTT (2000), Multicriteria optimization, Springer Verlag.

W.T. FU (1996), On the density of proper efficient points, Proc. Amer. Math. Soc., 1213–1217.

T. GAL (1983), On efficient sets in vector maximum problems. A brief survey, Essays and survey on multiple criteria decision making, Springer, Berlin.

W.B. GEARHART (1979), Compromise solutions and estimation of the noninferior set, J.O.T.A., 29–47.

W.B. GEARHART (1983), Characterization of properly efficient solution by generalized scalarization methods, J.O.T.A., 491–502.

W.B. GEARHART (1989), Families of differentiable scalarization function, J.O.T.A., 321–332.

A.M. GEOFFRION (1968), Proper efficiency and the theory of vector maximization, J. Math. Anal. Appl., 618–630.

G. GIORGI and A. GUERRAGGIO (1996), The notion of invexity in vector optimization: smooth and nonsmooth case, 5th Int. Symposium on Generalized Convexity, C.I.R.M.-Luminy, Marseille, June 17–21.

A. GUERRAGGIO, and D.T. LUC (2001), Optimality conditions for $C^{1,1}$ vector optimization problems, J.O.T.A., 515–529.

A. GUERRAGGIO, D.T. LUC and N.B. MINH (2001), Second-order optimality conditions for C^1 multiobjective programming problems, Acta Math. Vietnamica, 257–268.

A. GUERRAGGIO, E. MOLHO and A. ZAFFARONI (1990), La nozione di punto propriamente Pareto-efficiente, Ist. Matematica Finanziaria, Univ. Torino, n. 56.

A. GUERRAGGIO, A. MOLHO and A. ZAFFARONI (1994), On the notion of proper efficiency in vector optimization, J.O.T.A., 1–21.

T.R. GULATI and M.A. ISLAM (1990), Efficiency and proper efficiency in nonlinear vector maximum problems, European J. Op. Res., 373–382.

T.R. GULATI and N. TALAAT (1991), Duality in non convex vector minimum problems, Bull. Austr. Math. Soc., 501–509.

T.X.D. HA (1994), On the existence of efficient points in locally convex spaces, J. Global Opt., 265–278.

T.X.D. HA (1994), A note on the class of cones ensuring the existence of efficient points in bounded complete sets, Optimization, 141–152.

R. HARTLEY (1978), On cone efficiency, cone-convexity and cone-compactness, SIAM J. Appl. Math., 211–222.

S. HELBIG (1991), Approximation of the efficient point set by perturbation of the ordering cone, Z. Oper. Res., 197–220.

M.I. HENIG (1982a) Proper efficiency with respect to cones, J.O.T.A., 387–407.

M.I. HENIG (1982b), A cone Separation Theorem, J.O.T.A., 451–455.

M.I. HENIG (1988), Characterizing the nondominated set by separable functions, J.O.T.A., 423–444.

M.I. HENIG (1990), Value functions, domination cones and proper efficiency in multicriteria optimization, Math. Programming, 205–217.

J.B. HIRIART-URRUTY (1979), Tangent cones, generalized gradients and mathematical programming in Banach spaces, Math. Oper. Res., 79–97.

L. HURWICZ (1958), Programming in linear spaces; in: K.J. Arrow , L. Hurwicz and H. Uzawa (eds.), Studies in Linear and Nonlinear Programming.

G. ISAC (1983), Sur l'existence de l'optimum de Pareto, Riv. Mat. Univ. Parma, 303–325.

Y. ISHIZUKA and K. SHIMIZU (1984), Necessary and sufficient conditions for the efficient solutions of nondifferentiable multiobjective problems, IEEE, 625–629.

J. JAHN (1984), Scalarization in vector optimization, Math. Programming, 203–218.

J. JAHN (1985a), Scalarization in multiobjective optimization; in: P. Serafini (ed.), Mathematics of multiobjective optimization, 45–88.

J. JAHN (1985b), A characterization of properly minimal elements of a set, SIAM J. Control and Opt., 649–656.

J. JAHN (1986), Mathematical vector optimization in partially ordered linear spaces, Lang.

J. JAHN (1987), Parametric approximation problems arising in vector optimization, J.O.T.A., 503–516.

J. JAHN and E. SACHS (1986), Generalized quasi-convex mappings and vector optimization, SIAM J. Control and Opt., 306–322.

V. JEYAKUMAR (1986), A generalization of a minimax theorem of Fan via theorem of alternative, J.O.T.A., 525–533.

V. JEYAKUMAR and D.T. LUC (1998), Approximate Jacobian matrices for nonsmooth continuous maps and C^1-optimization, SIAM J. Control and Opt., 1815–1832.

V. JEYAKUMAR and X.Q. YANG (1993), Convex composite multiobjective nonsmooth programming, Math. Programming, 325–343.

V. JEYAKUMAR and B. MOND (1992), On generalized convex mathematical programming, J. Austral. Math. Soc., 43–53.

P.Q. KHANH (1995), Sufficient optimality conditions in vector optimization with invex-convexlike functions, J.O.T.A., 359–378.

A. KLINGER (1967), Improper solution of the vector maximum problems, Oper. Res., 570–572.

H.W. KUHN and A.W. TUCKER (1951), Non linear programming; in: Proceedings of the Second Berkeley Symposium on Math. Statistics and Probability, Berkeley.

J.G. LIN (1976a), Objective problems: Pareto-optimal solution by method of proper equality constraints, IEFE Trans. Automatic Control, 641–650.

J. LIN (1976b), Maximal vectors and multi-objective optimization, J.O.T.A., 41–64.

J. LIN (1976c), Proper equality in constraints and maximization of index vectors, J.O.T.A., 215–244.

J. LIN (1976d), Three methods for determining Pareto-optimal solution of multiple-objective problems; in: Y.C. Ho and S.K. Mitter (eds.), Direction in large scale systems: many persons optimization and decentralized control, Plenum, 117–138.

J. LIN (1977), Proper inequality constraints and maximization of index vector, J.O.T.A., 505–521.

Z.F. LI and S.Y. WANG (1994), Lagrange multipliers and saddle points in multiobjective programming, J.O.T.A., 63–81.

D.T. LUC (1984), On duality theory in multiobjective programming, J.O.T.A., 557–582.

D.T. LUC (1987a), Scalarization of vector optimization problems, J.O.T.A., 85–102.

D.T. LUC (1987b), Convexity and closedness of sets with respect to cones, Optimization, 785–789.

D.T. LUC (1989a), An existence theorem in vector optimization, Math. Oper. Res., 693–699.

D.T. LUC (1989b), Theory of vector optimization, Springer Verlag.

D.T. LUC (1990), Recession cones and the domination property in vector optimization, Math. Prog., 113–122.

D.T. LUC and C. MALIVERT (1992), Invex optimization problems, Bull. Austral. Math. Soc., 47–66.

D.T. LUC and S. SCHAIBLE (1997), Efficiency and generalized concavity, J.O.T.A., 147–153.

T. MAEDA (1994), Constraint qualifications in multiobjective optimization problems: differentiable case, J.O.T.A., 483–500.

C. MAUVERT (1990), First and second order conditions in vector optimization, Ann. Sc. Math. Quebec, 65–79.

I. MARUSIAC (1982), On Fritz John type optimality criterion in multiobjective optimization, L'An. Num. et la Theorie de l'Approx., 109–114.

L. MARTEIN (1994), Soluzioni efficienti e condizioni di ottimalità nell' ottimizzazione vettoriale; in: G. Di Pillo (ed.), Atti della Scuola CIRO, Masson ed.

L. MARTEIN (1994), On the existence of efficient points, Optimization, 283–290.

K. MISHRA and R.N. MUKHERJEE (1995), Generalized convex composite multi-objective nonsmooth programming and conditional proper efficiency, Optimization, 53–66.

A.B. NEMETH (1992), Convex operators: some subdifferentiability results, Optimization, 275–301.

J.W. NIEUWENHUIS (1981), Properly efficient and efficient solutions for vector maximization problems in Euclidean space, J. Math. Anal. and Appl., 311–317.

D. PALLASCHKE and S. ROLEWICZ (1997), Foundations of mathematical optimization, Kluwer Acad. Publ.

A. PASCOLETTI and P. SERAFINI (1984), Scalarizing vector optimization problems, J.O.T.A., 499–524.

V. PREDA (1993), On proper efficiency and duality for multiobjective programming problems, Rev. Roum. Math. Pures Appl., 545–554.

V. POSTOLICA (1993), New existence results for efficient points in locally convex spaces ordered by supernormal cones, J. Global Opti., 233–242.

V. POSTOLICA (1995), Properties of Pareto sets in locally convex spaces, Optimization, 223–229.

T.W. REILAND (1989a), Generalized differentiability for a class of nondifferentiable operators with applications to nonsmooth optimization, J. Austral. Math. Soc., 114–132.

T.W. REILAND (1989b), Generalized invexity for nonsmooth vector-valued mapping, Num. Funct. Anal. and Optimiz., 1191–1202.

T.W. REILAND (1990), Nonsmooth invexity, Bull. Austr. Math. Soc., 437–446.

S. ROLEWICZ (1975), On a norm of scalarization in infinite dimensional Banach spaces, Control and Cyb., 85–89.

S. ROLEWICZ (1984), On sufficient conditions of vector optimization, Oper. Res. Verf., 151–157.

A.M. RUBINOV (1977), Sublinear operators and their applications, Russian Math. Surveys, 115–175.

C.P. RUIZ-CANALES and A. RUFIAN-LIZANA (1995), A characterization of weakly efficient points, Math. Programming, 205–212.

Y. SAWARAGI, H. NAKAYAMA and T. TANINO (1985), Theory of multiobjective optimization, Academic Press.

C. SINGH (1987), Optimality conditions in multiobjective differentiable programming, J.O.T.A., 115–123.

C. SINGH and M.A. HANSON (1991), Generalized proper efficiency in multiobjective programming, J. Inf. and Opt. Sc., 139–144.

R.M. SOLAND (1979), Multicriteria optimization: a general characterization of efficient solutions, Decision Sciences, 26–38.

W. STADLER (1979), A survey of multicriteria optimization or the vector maximum problem, J.O.T.A., 1–51.

T. STAIB (1991), On necessary and sufficient optimality conditions for multicriterial optimization problems, Z. Oper. Res., 231–248.

A. STERNA KARWAT (1989), Approximating families of cones and proper efficiency in vector optimization, Optimization, 809–817.

C. SWARTZ (1987), Pshenichyi's theorem for vector optimization, J.O.T.A., 309–317.

K. TAMURA and S. ARAI (1982), On proper and improper efficient solutions of optimal problems with multicriteria, J.O.T.A., 191–205.

L. THIBAULT (1980), Subdifferentials of compactly Lipschitzian vector valued functions, Ann. Mat. Pure Appl., 157–192.

L. THIBAULT (1982), Subdifferentials of non-convex vector valued functions, J. Math. Anal. Appl., 319–354.

D.H. WAGNER (1977), Semicompactness with respect to a Euclidean cone, Canadian J. Math., 29–36.

S. WANG (1991), Second order necessary and sufficient conditions in multiobjective programming, Numer. Functional Anal. and Optimiz., 237–252.

A.R. WARBURTON (1983), Quasiconcave vector optimizaton: connectedness of the sets of Pareto-optimal and weak Pareto-optimal alternatives, J.O.T.A., 537–557.

P. WEIDNER (1992), The influence of proper efficiency on optimal solutions of scalarizing problems in multicriteria optimization, Fehrenbach.

T. WEIR (1987), Proper efficiency and duality for vector valued optimization problems, J. Austral. Math. Soc., 21–34.

T. WEIR (1988), A note on invex functions and duality in multiple objective optimization, Op. Research, 98–104.

T. WEIR and B. MOND (1988), Pre-invex functions in multiobjective optimization, J. Math. Anal. Appl., 39–53.

T. WEIR, B. MOND and B.D. CRAVEN (1986), On duality for weakly minimized vector valued optimization problems, Optimization, 711–721.

R.E. WENDELL and D.N. LEE (1977), Efficiency in multiple objective optimization, Math. Programming, 406–414.

D.J. WHITE (1983), Concepts of proper efficiency, European J. Oper. Res., 180–188.

D.J. WHITE (1985), Characterizations of efficient sets by constrained objectives, J.O.T.A., 603–629.

D.J. WHITE (1985), Vector maximization and Lagrange multipliers, Math. Programming, 192–205.

A. WIERZBICKI (1980), The use of reference objectives in multiobjective optimization; in: G. Fandel and T. Gal (eds.), Multiple criteria decision making theory and application, Springer, Berlin.

A. WIERZBICKI (1981), A mathematical basis for satisficing decision making; in: J. Morse (ed.), Organizations: multiple agents with multiple criteria, Springer Verlag, New York.

P.L. YU (1974), Cone convexity, cone estreme points and nondominated solutions in decision problems with multiobjectives, J.O.T.A., 319–377.

P.L. YU (1985), Multiple criteria decision making , Plenum Press.

X.Y. ZHENG (2000), Scalarization of Henig proper efficient points in a normed space, J.O.T.A., 233–247.

D.M. ZHUANG (1991), Bases of convex cones and Borwein's proper efficiency, J.O.T.A., 613–320.

D.M. ZHUANG (1994), Density results for proper efficiences, SIAM J. Control Optim., 51–58.

SUBJECT INDEX